SCHAUM'S OUTLINE OF

THEORY AND PROBLEMS

OF

NUMERICAL ANALYSIS

Second Edition

•

BY

FRANCIS SCHEID, Ph.D.
Professor Emeritus of Mathematics
Boston University

•

SCHAUM'S OUTLINE SERIES
McGRAW-HILL BOOK COMPANY

New York St. Louis San Francisco Auckland Bogotá Caracas Colorado Springs Hamburg Lisbon London Madrid Mexico Milan Montreal New Delhi Oklahoma City Panama Paris San Juan São Paulo Singapore Sydney Tokyo Toronto

FRANCIS SCHEID is Emeritus Professor of Mathematics at Boston University where he has been a faculty member since receiving his doctorate from MIT in 1948, serving as department chairman from 1956 to 1968. In 1961–1962 he was Fulbright Professor at Rangoon University in Burma. Professor Scheid has lectured extensively on educational television and the videotapes are used by the U.S. Navy. His research now centers on computer simulations in golf. Among his publications are the Schaum's Outlines of *Numerical Analysis, Computer Science,* and *Computers and Programming*.

Schaum's Outline of Theory and Problems of
NUMERICAL ANALYSIS

2 3 4 5 6 7 8 9 10 11 12 13 14 15 16 17 18 19 20 SHP SHP 8 9 2 1 0 9 8

ISBN 0-07-055221-5

Sponsoring Editor, John Aliano
Production Supervisor, Leroy Young
Editing Supervisor, Marthe Grice
Project Supervision, The Total Book

Library of Congress Cataloging-in-Publication Data

Scheid, Francis J.
Schaum's outline of theory and problems of numerical analysis/by Francis Scheid.—2nd ed.
p. cm.—(Schaum's outline series)
ISBN 0-07-055221-5
1. Numerical analysis—Problems, exercises, etc. I. Title.
II. Title: Theory and problems of numerical analysis.
QA297.S37 1989
519.4—dc 19 87-35348
CIP

Preface

The main goal of numerical analysis remains what it has always been, to find approximate solutions to complex problems using only the simplest operations of arithmetic. In short, it is a business of solving hard problems by doing lots of easy steps. Rather clearly, this means finding procedures by which computers can do the solving for us. The problems come from a variety of mathematical origins, particularly algebra and analysis, the boundaries being at times somewhat indistinct. Much background theory is borrowed from such fields by the numerical analyst, and some must be included in an introductory text for clarity. It is also true that our subject returns more than raw numbers across the boundaries. Numerical method has made important contributions to algebraic and analytic theory.

Many new topics have been added for this second edition. Included are backward error analysis, splines, adaptive integration, fast Fourier transforms, finite elements, stiff differential equations, and the QR method. The chapter on linear systems has been completely rewritten. A number of older topics have been shortened or eliminated, but a representative portion of classical numerical analysis has been retained partly for historical reasons. Some of the cuts have brought a tear to the author's eye, especially that of the constructive proof for the existence of solutions to differential equations. On the whole the new edition is a bit more demanding, but the same can be said of the subject itself.

The presentation and purposes remain the same. There is adequate material for a year course at beginning graduate level. With suitable omissions an introductory semester course can easily be arranged. The problem format allows convenient use as a supplement to other texts and facilitates independent study. Each chapter still begins with a summary of what is to come and should be taken as a table of contents for the chapter. It is not intended to be self-explanatory, and supporting detail is provided among the solved problems.

To repeat the closing paragraph of my original preface, there is no doubt that, in spite of strenuous effort, errors do remain within the text. Numerical analysts are among the world's most error conscious people, probably because they make so many of them. I will be grateful to hear from readers who discover errors. (The response to this request in the first edition was humbling.) There is still no reward except the exhilaration that accompanies the search for the all-too-elusive "truth."

FRANCIS SCHEID

Contents

Chapter 1

What Is Numerical Analysis?

ALGORITHMS

The objective of numerical analysis is to solve complex numerical problems using only the simple operations of arithmetic, to develop and evaluate methods for computing numerical results from given data. The methods of computation are called algorithms.

Our efforts will be focused on the search for algorithms. For some problems no satisfactory algorithm has yet been found, while for others there are several and we must choose among them. There are various reasons for choosing one algorithm over another, two obvious criteria being speed and accuracy. Speed is clearly an advantage, though for problems of modest size this advantage is almost eliminated by the power of the computer. For larger scale problems speed is still a major factor, and a slow algorithm may have to be rejected as impractical. However, other things being equal, the faster method surely gets the nod.

EXAMPLE 1.1. Find the square root of 2 to four decimal places.

More than one algorithm, using only the four basic operations of arithmetic, exists. The favorite is without much doubt

$$x_1 = 1 \qquad x_{n+1} = \frac{1}{2}\left(x_n + \frac{2}{x_n}\right)$$

from which a few mental calculations quickly manage

$$x_2 = \frac{3}{2} \qquad x_3 = \frac{17}{12} \qquad x_4 = \frac{1}{2}\left(\frac{17}{12} + \frac{24}{17}\right)$$

or, rounded to four decimal places,

$$x_2 = 1.5000 \qquad x_3 = 1.4167 \qquad x_4 = 1.4142$$

the last being correct to all four places. This numerical algorithm has a long history, and it will be encountered again in Chapter 25 as a special case of the problem of finding roots of equations.

ERROR

The numerical optimist asks how accurate are the computed results; the numerical pessimist asks how much error has been introduced. The two questions are, of course, one and the same. Only rarely will the given data be exact, since it often originates in measurement processes. So there is probably error in the input information. And usually the algorithm itself introduces error, perhaps unavoidable roundoffs. The output information will then contain error from both of these sources.

EXAMPLE 1.2. Suppose the number .1492 is correct to the four decimal places given. In other words, it is an approximation to a true value that lies somewhere in the interval between .14915 and .14925. The error is then at most five units in the fifth place, or half a unit in the fourth. In such a case the approximation is said to have four significant digits. Similarly, 14.92 has two correct decimal places and four significant digits provided its error does not exceed .005.

EXAMPLE 1.3. The number .10664 is said to be rounded to four decimal places when abbreviated to .1066, while .10666 would be rounded to .1067. In both cases the error made by rounding is no more than .00005, assuming the given figures are correct. The first is an example of rounding down, the second of rounding up. A borderline case such as .10665 is usually rounded to the nearest even digit, here to .1066. This is to avoid long-range prejudice between the ups and downs.

EXAMPLE 1.4. When 1.492 is multiplied by 1.066, the product is 1.590472. Computers work to a fixed "word length," all numbers being tailored to that length. Assuming a fictitious four-digit machine, the above product would be rounded to 1.590. Such roundoff errors are algorithm errors and are made by the unavoidable millions in modern computing.

SUPPORTING THEORY

Though our view of numerical analysis will be applications oriented, we will naturally be concerned with supporting theory, which is used both to discover algorithms and to establish their validity. Often the theory to which we are led has intrinsic interest; it is attractive mathematics. We then have the best of both worlds, but must not forget that our interests are more functional than aesthetic.

EXAMPLE 1.5. Computing values of the trigonometric, exponential, and other nonelementary functions clearly depends upon supporting theory. To get the cosine of x for small x, the classic series is still a good choice.

$$\cos x = 1 - \frac{x^2}{2!} + \frac{x^4}{4!} - \frac{x^6}{6!} + \cdots$$

With $x = .5$ this becomes

$$\begin{aligned}\cos .5 &= 1 - .125 + .0026041 - .0000217 + \cdots \\ &= .877582\end{aligned}$$

which is correct as far as it goes. The error bound in this case is guaranteed by further supporting theory which states that for series such as this the error is no greater than the first omitted term. (See Problem 1.9.) Here the first omitted term is $x^8/8!$, which for $x = .5$ amounts to just less than .0000001.

NUMBER REPRESENTATIONS

Since our ultimate objectives are numerical, a word or two about the representation of numbers will not be out of place. Numerical input will usually be in decimal form, since this is the form with which we are most familiar. As almost everyone knows, however, computers generally find binary representations more convenient, their 0 and 1 digits matching the off and on or high and low states of electrical components. For positive integers the binary form is

$$d_n 2^n + d_{n-1} 2^{n-1} + \cdots + d_1 2^1 + d_0 2^0$$

while for positive numbers less than one it is

$$d_{-1} 2^{-1} + d_{-2} 2^{-2} + d_{-3} 2^{-3} + \cdots$$

with all binary digits d_i either 0 or 1. Such representations are unique.

Floating-point representations have an additional convenience. In this form, numbers are described by three parts, a sign, a mantissa, and an exponent (which itself has a sign). Turning to decimals for the first illustrations, the number .1492 might appear as

$$+.1492\ 10^0$$

the sign being +, the mantissa .1492, and the exponent 0. The alternative $+1.492\ 10^{-1}$ is available, among other possibilities, but standard practice calls for the leading (nonzero) digit to come just after the point. The exponent then takes care of the order of magnitude. Such representations are called normalized. Thus 1492 would be expressed as $+.1492\ 10^4$.

EXAMPLE 1.6. Convert the decimal 13.75 into a binary floating-point form.

More formal conversion methods exist, but even without them the binary equivalent of 13.75 is easily seen to be 1101.11, with $8+4+1$ on the left of the point and $\frac{1}{2}+\frac{1}{4}$ on the right. Now rewrite this as

$$+.110111(+100)$$

where the +100 in parentheses serves as exponent 4. A final conversion to

$$01101110100$$

in which nothing but zeros and ones appear is attractive for electrical purposes, provided that certain conventions are understood. The leading zero is interpreted as a plus sign. (1 would mean minus.) Six binary digits, or bits, then make the mantissa, a binary point being assumed at its head. The zero that follows is another plus sign, this time for the exponent, which then ends the representation. The final form does not look very much like 13.75 but is understandable. In practice both the mantissa and exponent would involve more digits, and the forms of sign and exponent will vary, but floating-point representations are a basic tool of modern computing.

NORMS OF VECTORS AND MATRICES

The Euclidean length of a vector, that is,

$$(v_1^2 + v_2^2 + \cdots + v_n^2)^{1/2}$$

for the vector V with components v_i, is also called a **norm** of V and given the symbol $\|V\|$. Three basic properties of this norm are

1. $\|V\| \geqq 0$, and equals 0 if and only if $V = 0$
2. $\|cV\| = c \cdot \|V\|$ for any number c
3. $\|V + W\| \leqq \|V\| + \|W\|$

The last is known as the **triangle inequality.**

Several other real functions also have these properties and are also called norms. Of particular interest are the L_p norms

$$\|V\|_p = \left(\sum_{i=1}^{n} |v_i|^p\right)^{1/p}$$

for $p \geqq 1$. With $p = 1$, it is the L_1 norm, the sum of the component magnitudes. With $p = 2$, it is the familiar vector length, or Euclidean norm. As p tends to infinity, the dominant v_i takes over and we have the maximum norm

$$\|V\|_\infty = \max_i |v_i|$$

On more than one occasion, we will find use for these norms, particularly in studying the error behavior of algorithms.

EXAMPLE 1.7. Using the L_1 norm, the vectors $(1, 0)$ $(\frac{1}{2}, \frac{1}{2})$ $(0, 1)$ among others have norm one. A plot of such unit vectors is given as Fig. 1-1a, all emanating from the origin. Their terminal points form a square. Figure 1-1b shows the more familiar unit vectors of the Euclidean norm. Using the L_∞ norm, the vectors $(1, 0)$ $(1, 1)$ $(0, 1)$ among others have norm one. Their plot appears as Fig. 1-1c, terminal points again forming a square.

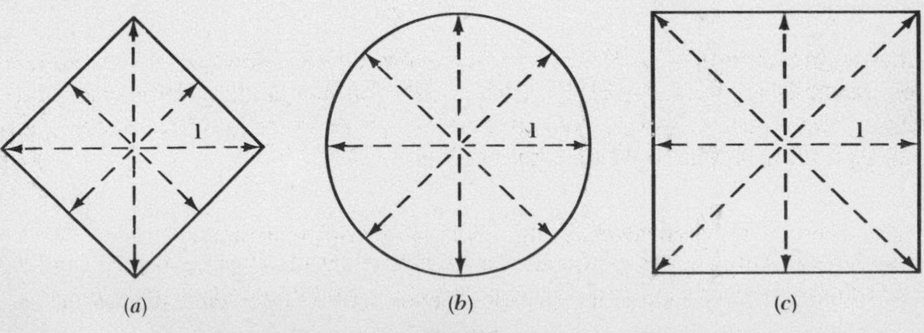

Fig. 1-1

Turning to matrices, we define

$$||A|| = \max ||AV||$$

the maximum being taken over all unit vectors V. The meaning of unit here depends upon the type of vector norm being used. Such matrix norms have parallel properties to those listed above for vectors.

1. $||A|| \geqq 0$, and equals zero if and only if $A = 0$
2. $||cA|| = c \cdot ||A||$ for any number c
3. $||A + B|| \leqq ||A|| + ||B||$

In addition, for matrices A and B and vector V, the properties

4. $||AV|| \leqq ||A|| \cdot ||V||$
5. $||AB|| \leqq ||A|| \cdot ||B||$

will be useful. The L_1 and L_∞ norms have the advantage of being easy to compute, the first being the maximum absolute column sum

$$||A||_1 = \max_j \sum_{i=1}^{n} |a_{ij}|$$

and the second the absolute row sum of A

$$||A||_\infty = \max_i \sum_{j=1}^{n} |a_{ij}|$$

Many of these features will be proved in the solved problems.

EXAMPLE 1.8. Find the L_1, L_2, and L_∞ norms of this matrix:

$$A = \begin{bmatrix} 1 & 1 \\ 1 & 0 \end{bmatrix}$$

The maximum column sum and row sum are found instantly, and we are off to a fast start with

$$L_1 = L_\infty = 2$$

Unfortunately there is no corresponding supporting theory to help with L_2 and this very innocent-appearing matrix does not yield this value without some resistance. By definition, the L_2 norm of A is the maximum L_2 norm of the vector

$$\begin{bmatrix} 1 & 1 \\ 1 & 0 \end{bmatrix} \begin{bmatrix} x \\ y \end{bmatrix} = \begin{bmatrix} x + y \\ x \end{bmatrix}$$

for $x^2 + y^2 = 1$, that is, for (x, y) on the unit circle of Fig. 1-1*b*. The square of this norm is

$$(x + y)^2 + x^2 = 1 + 2xy + x^2 = 1 + 2x\sqrt{1 - x^2} + x^2$$

which can be maximized by elementary calculus. The assumption that y is positive is not restrictive here since the norm takes the same value for (x, y) and $(-x, -y)$. Eventually one finds that a maximum occurs for $x^2 = \frac{1}{2} + \sqrt{5}/10$ and that

$$||A||_2^2 = \frac{3 + \sqrt{5}}{2}$$

Solved Problems

1.1. Calculate the value of the polynomial

$$p(x) = 2x^3 - 3x^2 + 5x - 4$$

for the argument $x = 3$.

Following the natural course, we find $x^2 = 9$, $x^3 = 27$, and putting the pieces together,

$$p(3) = 54 - 27 + 15 - 4 = 38$$

A count finds that five multiplications, an addition, and two subtractions have been performed.

Now rearrange the polynomial as

$$p(x) = [(2x - 3)x + 5]x - 4$$

and try again. From $x = 3$ we have successively 6, 3, 9, 14, 42, and 38. This time only three multiplications have been made, instead of five. The reduction is not dramatic, but it is suggestive. For a general polynomial of degree n, the first algorithm requires $2n - 1$ multiplications, the second just n. In a larger operation, involving many polynomial evaluations, the saving in time and algorithm (roundoff) errors may be significant.

1.2. Define the error of an approximation.

The traditional definition is

$$\text{True value} = \text{approximation} + \text{error}$$

so that, for example,

$$\sqrt{2} = 1.414214 + \text{error}$$

$$\pi = 3.1415926536 + \text{error}$$

1.3. What is relative error?

This is error measured relative to the true value.

$$\text{Relative error} = \frac{\text{error}}{\text{true value}}$$

In the common event that the true value is unknown, or unwieldy, the approximation is substituted for it and the result still is called, somewhat loosely, the relative error. Thus the familiar approximation 1.414 for $\sqrt{2}$ has a relative error of about

$$\frac{.0002}{1.414} = .00014$$

while the cruder approximation 1.41 has a relative error near .003.

1.4. Suppose the numbers $x_1, x_2, \ldots, x_n$ are approximations to $X_1, X_2, \ldots, X_n$ and that in each case the maximum possible error is E. Prove that the maximum possible error in the sum of the x_i is nE.

Since $x_i - E \leqq X_i \leqq x_i + E$

it follows by addition that

$$\sum x_i - nE \leqq \sum X_i \leqq \sum x_i + nE$$

so that

$$-nE \leqq \sum X_i - \sum x_i \leqq nE$$

which is what was to be proved.

1.5. Compute the sum $\sqrt{1} + \sqrt{2} + \cdots + \sqrt{100}$ with all the roots evaluated to two decimal places. By the preceding problem, what is the maximum possible error?

Whether by a few well-chosen lines of programming or by a more old-fashioned appeal to tables, the roots in question can be found and summed. The result is 671.38. Since each root has a maximum error of $E = .005$, the maximum possible error in the sum is $nE = 100(.005) = .5$, suggesting that the sum as found may not be correct to *even one* decimal place.

1.6. What is meant by the probable error of a computed result?

This is an error estimate such that the actual error will exceed the estimate with probability one-half. In other words, the actual error is as likely to be greater than the estimate as less. Since this depends upon the error distribution, it is not an easy target, and a rough substitute is often used, $\sqrt{n}E$, with E the maximum possible error.

1.7. What is the actual error of the result in Problem 1.5, and how does it compare with the maximum and probable errors?

A new computation, with square roots found to five decimal places, resulted in the sum 671.46288. This time the maximum error is 100(.000005) which is .0005, so we have the sum correct to three places as 671.463. The actual error of the earlier result is thus about .08, compared with the maximum .50 and the probable .05. One of our estimates was too pessimistic and the other slightly optimistic.

1.8. Suppose a thousand square roots are to be summed, instead of a mere hundred. If three-place accuracy is wanted, how accurately should the individual roots be computed?

For a solid guarantee it is best to assume the worst, that the maximum possible error might be attained. The formula nE of Problem 1.4 becomes $1000E$, showing that three decimal places may be lost in a summation of this length. Since three are wanted in the output, it may be wise to have six places correct in the input. The point is, in very long computations there is time for very small errors to make a substantial collective contribution.

1.9. Calculate the series

$$1-\frac{1}{2}+\frac{1}{3}-\frac{1}{4}+\cdots$$

correct to three digits.

This series illustrates an often-used theorem of analysis. Because its terms alternate in sign and steadily decrease, the partial sums dodge back and forth across the limit, the series value. This implies that the error at any point will be less than the first omitted term. To get the specified accuracy, we therefore need $1/n \leqq .0005$, or $n \geqq 2000$. Two thousand terms will have to be added. Working to eight decimal places, the 2000 roundoffs may accumulate to

$$nE = 2000(.000000005) = .00001$$

which seems negligible, so we permit the computation to proceed, round the result to three places, and have .693.

Note that in this problem we have no input error, only algorithm errors. First, we take just a partial sum instead of the series, and then we make numerous roundoff errors in trying to evaluate this sum. The first is called a truncation error and it seems to be the larger of the two error sources in this problem. In summary

$$\begin{aligned}\text{Actual error} &= \text{truncation error} + \text{roundoff error}\\ &= .0005 + .00001\end{aligned}$$

more or less. In fact the series value is the natural logarithm of 2, and to three places it is our .693.

1.10. Prove that if the series

$$a_1 - a_2 + a_3 - a_4 + \cdots$$

is convergent, all the a_i being positive, then

$$\frac{1}{2}a_1 + \frac{1}{2}(a_1 - a_2) - \frac{1}{2}(a_2 - a_3) + \frac{1}{2}(a_3 - a_4) + \cdots$$

is also convergent and represents the same number.

With A_n and B_n representing the nth partial sums of the two series, it is easily seen that $A_n - B_n = \pm \frac{1}{2}a_n$. Since the first series is convergent, $\lim a_n$ is zero and the result follows.

1.11. Apply the theorem of the preceding problem to evaluate the series of Problem 1.9, again to three decimal places.

A little algebra finds $B_1 = \frac{1}{2}$, and for $n > 1$

$$B_n = \frac{1}{2} + \sum_{k=2}^{n} (-1)^k \frac{1}{2k(k-1)}$$

This is again an alternating series with monotone terms, so the theorem of Problem 1.9 is again available. For three-digit accuracy we need

$$\frac{1}{2n(n+1)} \leqq .0005$$

or $n \geqq 32$. This is far fewer terms than were needed earlier and roundoff will hardly be an issue on an eight-digit machine. The new algorithm is much faster than the other and manages the same .693 with less effort.

1.12. Given that the numbers .1492 and .1498 are correct as far as they go, that is, the errors are no larger than five units in the fifth place, illustrate the development of relative error by considering the quotient $1/(.1498 - .1492)$.

For the given numbers the relative errors are about 5/15,000 which is near .03 percent. For their sum and difference a maximum error of one unit in the fourth place is possible. In the case of the sum, this again leads to a relative error of about .03 percent, but with the .0006 difference we find an error of one part in six, which is 17 percent. Turning to the required quotient, it may be just as well to take the pessimistic view. As given, a quotient of 1667 would be calculated, to the nearest integer. But conceivably it is $1/(.14985 - .14915)$ which ought to have been found instead, and this would have brought us 1429. At the other extreme is $1/(.14975 - .14925) = 2000$. This very simple example makes it clear that a large relative error generated at some interior stage of a continuing calculation can lead to large absolute errors down the line.

1.13. What is meant by the condition of a numerical problem?

A problem is well-conditioned if small changes in the input information cause small changes in the output. Otherwise it is ill-conditioned. For instance, the system

$$x + y = 1$$

$$1.1x + y = 2$$

presents an obvious difficulty. It represents the intersection of nearly parallel lines and has the solution $x = 10$, $y = -9$.

Now change the 1.1 to 1.05 and solve again. This time $x = 20$ and $y = -19$. A 5 percent change in one coefficient has caused a 100 percent change in the solution.

1.14. What is a *stable* algorithm?

In extended calculations it is likely that many roundoffs will be made. Each of these plays the role of an input error for the remainder of the computation, and each has an impact on the eventual output. Algorithms for which the cumulative effect of all such errors is limited, so that a useful result is generated, are called **stable algorithms.** Unfortunately, there are times when the accumulation is devastating and the solution is overwhelmed by error. Needless to say, such algorithms are called **unstable**.

1.15. Interpret the floating-point decimal $+.1066 * 10^4$.

Clearly the exponent shifts the decimal point four places to the right, to make 1066. Similarly, $+.1066 * 10^{-2}$ is .001066.

1.16. Interpret the floating-point binary symbol $+.10111010 * 2^4$.

The exponent shifts the binary point four places to the right, to make 1011.1010, equivalent to decimal $11 + \frac{5}{8}$ or 11.625. Similarly, $+.10111010 * 2^{-1}$ is .01011101. This is, of course, $\frac{1}{32}$ times the number given originally.

1.17. Interpret the floating-point binary symbol 0101110100100, given that the mantissa uses eight places and the exponent three, apart from their signs.

The zeros in positions one and ten are to be taken as plus signs.

$$\underbrace{0}_{\text{Sign}}\underbrace{10111010}_{\text{Mantissa}}\underbrace{0}_{\text{Sign}}\underbrace{100}_{\text{Exponent}}$$

The binary point is assumed at the head of the mantissa. With these understandings we have once again $+.10111010 * 2^4$. Similarly, and with the same conventions, $+.10111010 * 2^{-1}$ becomes 0101110101001, the last four digits meaning an exponent of -1.

1.18. Add these floating-point numbers, using the conventions of the preceding problem.

$$0101101110010$$

$$0100011001100$$

One way or another, the binary points will have to be "lined up." Interpreting the symbols leads to the following sum:

$$\begin{array}{r} 10.110111 \\ +\quad .000010001100 \\ \hline = 10.111001001100 \end{array}$$

In the form used for the inputs this becomes

$$0101110010010$$

with the mantissa again taking eight places and the exponent three, apart from signs. A roundoff error is made when the last six binary digits are eliminated to conform with machine capabilities.

1.19. What is overflow?

Again using the conventions of our fictitious machine, the largest number that can be expressed is 0111111110111, both the mantissa and the exponent being maximal. Seven shifts of the binary point make this the equivalent of 1111111.1 which comes to decimal $127 + \frac{1}{2}$, or $2^7 - 2^{-1}$. Any number larger than this cannot be represented under the given conventions and is called an overflow.

1.20. What is underflow?

The smallest number that can be represented in the form being used, apart from zero and negatives, is 0000000011111. However, for various reasons it is convenient to insist that the leading digit of a mantissa be a 1. This is known as the **normalized** form, and fixes the exponent. Again an exception must be made for the number zero. If normalization is required, the smallest positive number becomes 0100000001111. In decimal this is $2^{-1} * 2^{-7}$ or 2^{-8}. Any positive number smaller than this cannot be represented and is called an **underflow**. Any floating-point system of number representation will have such limitations and the concepts of overflow and underflow will apply.

1.21. Imagine an even simpler floating-point system, in which mantissas have only three binary digits and exponents are -1, 0, or 1. How are these numbers distributed on the real line?

Assuming normalization, these numbers have the form $.1xx$ apart from exponent. The entire set, therefore, consists of three subsets of four numbers each, as follows:

.0100	.0101	.0110	.0111	(for exponent -1)
.100	.101	.110	.111	(for exponent 0)
1.00	1.01	1.10	1.11	(for exponent 1)

These are plotted in Fig. 1-2. Notice the denser packing of the smaller numbers, the separation increasing from $\frac{1}{16}$ to $\frac{1}{4}$ as we pass from group to group. This is due, of course, to the fact that we have only three significant digits (the leader fixed at 1) with the exponent supplying progressive magnification as it increases. For example, 1.005 is not available here. The set is not that dense in this part of its range. A fourth significant digit would be needed. Realistic floating-point systems have this same feature, in a more complex way, and the ideas of significant digits and relative error are relevant.

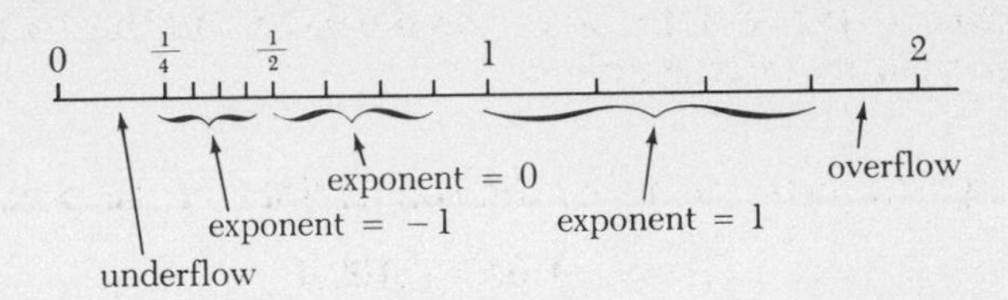

Fig. 1-2

1.22. Assume a number x represented by a floating-point binary symbol, rounded to a mantissa of n bits. Also assume normalization. What are the bounds for the absolute and relative errors caused by the rounding?

Rounding will cause an error of at most a unit in the $(n+1)$th binary place, or half a unit in the nth place. So

$$\text{Absolute error} \leqq 2^{-n-1}$$

while for the relative error we must take into account the true value x. Normalization means a mantissa no smaller than $\frac{1}{2}$ and this leads to the following bound:

$$|\text{Relative error}| \leqq \frac{2^{-n-1}}{2^{-1}} = 2^{-n}$$

It is useful to rewrite this letting $\text{fl}(x)$ represent the floating-point symbol for x. Then

$$\text{Relative error} = \frac{\text{fl}(x) - x}{x} = E$$

or

$$\text{fl}(x) = x(1+E) = x + xE$$

with $|E| \leqq 2^{-n}$. The operation of rounding off can thus be viewed as the replacement of x by a perturbed value $x + xE$, the perturbation being relatively small.

1.23. Find a bound for the relative error made by the addition of two floating-point numbers.

Let the numbers be $x = m_1 * 2^e$ and $y = m_2 * 2^f$ with y the smaller. Then m_2 must be shifted $e-f$ places to the right (lining up the binary points). The mantissas are then added, the result normalized and rounded. There are two possibilities. Either overflow occurs to the left of the binary point (not overflow in the sense of Problem 1.19), or it does not. The first possibility is characterized by

$$1 \leqq |m_1 + m_2 * 2^{f-e}| < 2$$

and the second by

$$\frac{1}{2} \leqq |m_1 + m_2 * 2^{f-e}| < 1$$

If overflow does occur, a right shift of one place will be required, and we have

$$\text{fl}(x+y) = [(m_1 + m_2 * 2^{f-e})2^{-1} + \epsilon] * 2^{e+1}$$

where ϵ is the roundoff error. This can be rewritten

$$\text{fl}(x+y) = (x+y)\left(1 + \frac{2\epsilon}{m_1 + m_2 * 2^{f-e}}\right)$$
$$= (x+y)(1+E)$$

with $|E| \leqq 2\epsilon \leqq 2^{-n}$.

If there is no overflow, then

$$\text{fl}(x+y) = [(m_1 + m_2 * 2^{f-e}) + \epsilon] * 2^e$$
$$= (x+y)\left(1 + \frac{\epsilon}{m_1 + m_2 * 2^{f-e}}\right)$$
$$= (x+y)(1+E)$$

with E bounded as before.

A corresponding result for floating-point subtraction will be found in Problem 1.45.

1.24. Find a bound for the relative error made by multiplying two floating-point numbers.

Again let the two numbers be $x = m_1 * 2^e$ and $y = m_2 * 2^f$. Then $xy = m_1 m_2 * 2^{e+f}$ with $\frac{1}{4} \leqq |m_1 m_2| < 1$ because of normalization. This means that to normalize the product there will be a left shift of at most one place. Rounding will, therefore, produce either $m_1 m_2 + \epsilon$ or $2m_1 m_2 + \epsilon$, with $|\epsilon| \leqq 2^{-n-1}$. This can be summarized as follows:

$$\text{fl}(xy) = \begin{cases} (m_1 m_2 + \epsilon) * 2^{e+f} & \text{if } |m_1 m_2| \geqq \frac{1}{2} \\ (2m_1 m_2 + \epsilon) * 2^{e+f-1} & \text{if } \frac{1}{2} > |m_1 m_2| \geqq \frac{1}{4} \end{cases}$$

$$= m_1 m_2 * 2^{e+f} \begin{cases} 1 + \dfrac{\epsilon}{m_1 m_2} & \text{if } |m_1 m_2| \geqq \frac{1}{2} \\ 1 + \dfrac{\epsilon}{2m_1 m_2} & \text{if } \frac{1}{2} > |m_1 m_2| \geqq \frac{1}{4} \end{cases}$$

$$= xy(1+E)$$

with $|E| \leqq 2|\epsilon| \leqq 2^{-n}$.

A similar result is sketched for the operation of division in Problem 1.46. This means that in all four arithmetic operations, using floating-point numbers, the relative error introduced does not exceed 1 in the least significant place of the mantissa.

1.25. Estimate the error generated in computing the sum

$$x_1 + x_2 + \cdots + x_k$$

using floating-point operations.

We consider the partial sums s_i. Let $s_1 = x_1$. Then

$$s_2 = \text{fl}(s_1 + x_2) = (s_1 + x_2)(1 + E_1)$$

with E_1 bounded by 2^{-n} as shown in Problem 1.23. Rewriting,

$$s_2 = x_1(1 + E_1) + x_2(1 + E_1)$$

Continuing

$$s_3 = \mathrm{fl}(s_2 + x_3) = (s_2 + x_3)(1 + E_2)$$
$$= x_1(1 + E_1)(1 + E_2) + x_2(1 + E_1)(1 + E_2) + x_3(1 + E_2)$$

and eventually

$$s_k = \mathrm{fl}(s_{k-1} + x_k) = (s_{k-1} + x_k)(1 + E_{k-1})$$
$$= x_1(1 + c_1) + x_2(1 + c_2) + \cdots + x_k(1 + c_k)$$

where, for $i = 2, \ldots, k$,

$$1 + c_i = (1 + E_{i-1})(1 + E_i) \cdots (1 + E_{k-1})$$

and $1 + c_1 = 1 + c_2$. In view of the uniform bound on the E_j, we now have this estimate for the $1 + c_i$:

$$(1 - 2^{-n})^{k-i+1} \leqq 1 + c_i \leqq (1 + 2^{-n})^{k-i+1}$$

Summarizing

$$\mathrm{fl}\left(\sum_{j=1}^{k} x_j\right) = \left(\sum_{j=1}^{k} x_j\right)(1 + E)$$

where

$$E = \sum_{j=1}^{k} x_j c_j \Big/ \sum_{j=1}^{k} x_j$$

Note that if the true sum $\sum x_j$ is small compared with the x_j, then the relative error E can be large. This is the cancellation effect caused by subtractions, observed earlier in Problem 1.12.

1.26. Illustrate a forward error analysis.

Suppose the value of $A(B + C)$ is to be computed, using approximations a, b, c which are in error by amounts e_1, e_2, e_3. Then the true value is

$$A(B + C) = (a + e_1)(b + e_2 + c + e_3) = ab + ac + \text{error}$$

where

$$\text{Error} = a(e_2 + e_3) + be_1 + ce_1 + e_1e_2 + e_1e_3$$

Assuming the uniform bound $|e_i| \leqq e$ and that error products can be neglected, we find

$$|\text{Error}| \leqq (2\,|a| + |b| + |c|)e$$

This type of procedure is called forward error analysis. In principle it could be carried out for any algorithm. Usually, however, the analysis is tedious if not overwhelming. Besides, the resulting bounds are usually very conservative, suitable if what is needed is an idea of the worst that might happen. In the present example one point of minor interest does surface. The value of a seems to be twice as sensitive as the values of b and c.

1.27. What is backward error analysis?

The underlying idea of backward error analysis is to take the result of a computation and try to determine the range of input data that could have produced it. It is important not to misunderstand the motive here. There is no intention of modifying the data to accommodate the answer. If a backward error analysis is completed and shows that the result found is consistent with the input data, within the range of observational or roundoff error, then one may have some confidence in the result. If this does not happen, then a major source of error exists elsewhere, presumably within the algorithm itself.

1.28. Show that the error analysis in Problem 1.23 was a backward error analysis.

The result obtained was

$$\text{fl}(x+y)=(x+y)(1+E)$$

with $|E| \leqq 2^{-n}$, where n is the number of binary places in the mantissas. Rewriting this as

$$\text{fl}(x+y)=x(1+E)+y(1+E)$$

and recalling Problem 1.22, we see that the sum as computed, namely $\text{fl}(x+y)$, is also the true sum of numbers differing from the original x and y by no more than the roundoff error bound E. That is, the output can be explained by input data well within the recognized error limit.

1.29. Show that the analysis done in Problem 1.24 was a backward error analysis.

We found

$$\text{fl}(xy)=xy(1+E)$$

which we can think of as the product of x by $y(1+E)$. This means that the computed $\text{fl}(xy)$ is also the true product of numbers differing from the original x and y by no more than the roundoff error. It is consistent with input data well within our recognized error limit.

1.30. What does the backward error analysis made in Problem 1.25 indicate?

First, the equation

$$\text{fl}\left(\sum_{j=1}^{k} x_j\right)=x_1(1+c_1)+\cdots+x_k(1+c_k)$$

shows that the floating-point sum of k numbers x_1 to x_k is also the true sum of k numbers which differ from the x_j by relative errors of size c_j. Unfortunately the estimates then obtained in Problem 1.25 also show that these errors can be much greater than simple roundoffs.

1.31. Prove the triangle property of vector length, the L_2 norm, by first proving the Cauchy–Schwarz inequality.

$$\left(\sum a_i b_i\right)^2 \leqq \left(\sum a_i^2\right)\left(\sum b_i^2\right)$$

One interesting proof begins by noting that $\sum (a_i - b_i x)^2$ is nonnegative, so that the quadratic equation

$$\left(\sum b_i^2\right)x^2 - 2\left(\sum a_i b_i\right)x + \sum a_i^2 = 0$$

cannot have distinct real roots. This requires

$$4\left(\sum a_i b_i\right)^2 - 4\sum a_i^2 \sum b_i^2 \leqq 0$$

and cancelling the 4's we have the Cauchy–Schwarz inequality.

The triangle inequality now follows quite directly, but with a bit of algebra. Written in component form, it states

$$[(v_1+w_1)^2+\cdots+(v_n+w_n)^2]^{1/2} \leqq (v_1^2+\cdots+v_n^2)^{1/2}+(w_1^2+\cdots+w_n^2)^{1/2}$$

Squaring, removing common terms, squaring again, and using Cauchy–Schwarz will bring the desired result (see Problem 1.50).

1.32. Show that the vector L_p norm approaches $\max |v_i|$ for p tending to infinity.

Suppose v_m is the absolutely largest component and rewrite the sum as

$$|v_m|\left(1+\sum_{i\neq m}\left|\frac{v_i}{v_m}\right|^p\right)^{1/p}$$

Within the parentheses all terms but the first approach zero as limit, and the required result follows.

1.33. Show that the definition $\|A\| = \max \|AV\|$ for unit V satisfies properties 1 to 3 as given in the introduction.

These follow rather easily from the corresponding properties of the companion vector norm. Since AV is a vector, $\|AV\| \geqq 0$ and so $\|A\| \geqq 0$. If $\|A\| = 0$ and even one element of A were not zero, then V could be chosen to make a component of AV positive, a contradiction of $\max \|AV\| = 0$. This proves the first.

Next we find

$$\|cA\| = \max \|cAV\| = \max |c| \cdot \|AV\| = |c| \cdot \|A\|$$

proving the second. The third is handled similarly.

1.34. What are the L_1, L_2, and L_∞ norms of the identity matrix?

They are all 1. We have

$$\|I\| = \max \|IV\| = \max \|V\| = 1$$

since V is a unit vector.

1.35. What are the L_1, L_2, and L_∞ norms of the matrix $\begin{bmatrix} 1 & 1 \\ 1 & 1 \end{bmatrix}$?

We have

$$AV = \begin{bmatrix} v_1 + v_2 \\ v_1 + v_2 \end{bmatrix}$$

Assume for simplicity that v_1, v_2 are nonnegative. Then for L_1 we add, and find $\|AV\| = 2(v_1 + v_2) = 2$, since V is a unit vector in L_1 norm. Thus $\|A\|_1 = 2$. For the L_2 norm we must square and add the two components, obtaining $2(v_1^2 + 2v_1v_2 + v_2^2)$. In this norm $v_1^2 + v_2^2 = 1$ so we maximize v_1v_2. Elementary calculus then produces $v_1 = v_2 = 1/\sqrt{2}$ leading quickly to $\|A\|_2 = 2$. Finally $\|AV\|_\infty = v_1 + v_2$, since with this norm we seek the maximum component. But here again the maximum is 2, because with this norm neither v_i can exceed 1. The L_1 and L_∞ norms could have been read instantly using the result of the following problem or its companion.

1.36. Show that

$$\|A\|_\infty = \max_i \sum_{j=1}^{n} |a_{ij}|$$

Choose a vector V with all components of size 1 and signs matching the a_{rj} such that $\sum |a_{rj}|$ is maximal. Then $\sum a_{rj}v_j$ is an element of AV equaling this maximal value and clearly cannot be exceeded. Since this V has norm 1, the norm of A also takes this value. The similar result for the L_1 norm is left as Problem 1.52.

1.37. Prove that $\|AV\| \leqq \|A\| \cdot \|V\|$.

For a unit vector U we have, by definition of $\|A\|$,

$$\|AU\| \leqq \max_U \|AU\| = \|A\|$$

so choosing $U = V/\|V\|$ and applying property 2,

$$\left\| A\left(\frac{V}{\|V\|}\right) \right\| \leqq \|A\| \qquad \|AV\| \leqq \|A\| \cdot \|V\|$$

1.38. Prove $\|AB\| \leqq \|A\| \cdot \|B\|$.

We make repeated use of the result of Problem 1.37:

$$\|AB\| = \max \|ABU\| \leqq \max \|A\| \cdot \|BU\| \leqq \max \|A\| \cdot \|B\| \cdot \|U\| = \|A\| \cdot \|B\|$$

Supplementary Problems

1.39. Compute $1/.982$ using the *supporting theory*

$$\frac{1}{1-x} = 1 + x + x^2 + \cdots$$

with $x = .018$.

1.40. Numbers are accurate to two places when their error does not exceed .005. The following square roots are taken from a table. Round each to two places and note the amount of the roundoff. How do these roundoff errors compare with the maximum of .005?

n	11	12	13	14	15	16	17	18	19	20
$\sqrt{n}$ to three places	3.317	3.464	3.606	3.742	3.873	4.000	4.123	4.243	4.359	4.472
$\sqrt{n}$ to two places	3.32	3.46								
approx. roundoff	+.003	−.004								

The total roundoff error could theoretically be anywhere from $10(-.005)$ to $10(.005)$. Actually what is the total? How does it compare with the "probable error" of $\sqrt{10}(.005)$?

1.41. Suppose N numbers, all correct to a given number of places, are to be summed. For about what size N will the last digit of the computed sum probably be meaningless? The last two digits? Use the probable error formula.

1.42. A sequence $J_0, J_1, J_2, \ldots$ is defined by

$$J_{n+1} = 2nJ_n - J_{n-1}$$

with $J_0 = .765198$ and $J_1 = .440051$ correct to six places. Compute $J_2, \ldots, J_7$ and compare with the correct values which follow. (These correct values were obtained by an altogether different process. See the next problem for explanation of errors.)

n	2	3	4	5	6	7
Correct J_n	.114903	.019563	.002477	.000250	.000021	.000002

1.43. Show that for the sequence of the preceding problem,

$$J_7 = 36{,}767J_1 - 21{,}144J_0$$

exactly. Compute this from the given values of J_0 and J_1. The same erroneous value will be obtained. The large coefficients multiply the roundoff errors in the given J_0 and J_1 values and the combined results then contain a large error.

1.44. To six places the number J_8 should be all zeros. What does the formula of Problem 1.42 actually produce?

1.45. Show that the error introduced by floating-point subtraction is bounded by 2^{-n}. Let $x = m_1 * 2^e$ and $y = m_2 * 2^f$ as in Problem 1.23. Then $x - y = (m_1 - m_2 * 2^{f-e})2^e$ and unless this is zero

$$2^{-n} \leqq |m_1 - m_2 * 2^{f-e}| < 2$$

Normalizing the new mantissa may require up to $n-1$ left shifts, the actual number s being determined by

$$2^{-s-1} \leqq |m_1 - m_2 * 2^{f-e}| < 2^{-s}$$

Now show that

$$\text{fl}(x-y) = [(m_1 - m_2 * 2^{f-e}) * 2^s + \epsilon] * 2^{e-s}$$

and eventually

$$\text{fl}(x-y) = (x-y)(1+E)$$

with $|E| \leqq 2^{-n}$.

1.46. Show that the error introduced during floating-point division is bounded by 2^{-n}. With the conventions of Problem 1.24, let one-half of the numerator mantissa be divided by the denominator mantissa (to avoid quoticnts greater than one) and the exponents subtracted. This gives

$$\frac{x}{y} = \left(\frac{m_1}{2m_2}\right) * 2^{e-f+1}$$

with $\frac{1}{4} \leqq |m_1/2m_2| < 1$. Now follow the remainder of the analysis made for the multiplication operation to show that once again the relative error is bounded as stated.

1.47. Analyze the computation of the inner product

$$s_k = \text{fl}(x_1 y_1 + x_2 y_2 + \cdots + x_k y_k)$$

much as in Problem 1.25. Let

$$t_i = \text{fl}(x_i y_i) \qquad \text{for } i = 1, \ldots, k$$

and then set

$$s_1 = t_1 \qquad s_i = \text{fl}(s_{i-1} + t_i)$$

for $i = 1, \ldots, k$. This makes s_k the required inner product. Now find relations and estimates similar to those found in the earlier problem.

1.48. Using the conventions of Problem 1.17, interpret this floating-point symbol: 0100110011010. (This is as close as one can come to .1492 with only an 8-bit mantissa.)

1.49. Imitating Problem 1.21, imagine a floating-point system in which normalized mantissas have 4 bits and the exponents are -1, 0, and 1. Show that these numbers form three groups of eight, according to their exponents, one group falling in the interval $\frac{1}{4}$ to $\frac{1}{2}$, another in the interval $\frac{1}{2}$ to 1, and the third between 1 and 2. Which positive numbers will cause overflow? Underflow?

1.50. Complete the proof begun in Problem 1.31.

1.51. Complete Problem 1.33 by showing that the norm of the sum of two matrices does not exceed the sum of their norms.

1.52. By a suitable choice of unit vector (one component 1, the rest 0) show that the L_1 norm of a matrix A

can be computed as the maximum column sum of absolute elements. Compare with the related proof in Problem 1.36.

1.53. Show that for $A = \begin{bmatrix} a & b \\ b & a \end{bmatrix}$ the L_1, L_2, and L_∞ norms are equal.

1.54. Show that for $A = \begin{bmatrix} a & b \\ b & -a \end{bmatrix}$ the L_2 norm is $(a^2 + b^2)^{1/2}$.

1.55. Show that for $A = \begin{bmatrix} a & a \\ a & b \end{bmatrix}$ a vector V that maximizes $\|AV\|_2$ can be found in the form $(\cos t, \sin t)^T$ with $\cos 2t = 0$ in the case $b^2 = a^2$ while $\tan 2t = 2a/(a - b)$ otherwise.

1.56. It has been suggested that the following message be broadcast to outer space as a sign that this planet supports intelligent life. The idea is that any form of intelligent life elsewhere will surely comprehend its intellectual content and so deduce our own intelligent presence here. What is the meaning of the message?

$$11.001001000011111101110$$

1.57. If the vector V with components x, y is used to represent the point (x, y) of a plane, then points corresponding to unit vectors in the L_2 norm form the classic unit circle. As Fig. 1-1 shows, in the L_1 and L_∞ norms the "circle" takes a square shape. In a city of square blocks, which is the suitable norm for taxicab travel? (Find all the intersections at a given distance from a given intersection.) On a chessboard, why is the appropriate norm for the travels of the chess king the L_∞ norm?

Chapter 2

The Collocation Polynomial

APPROXIMATION BY POLYNOMIALS

Approximation by polynomials is one of the oldest ideas in numerical analysis, and still one of the most heavily used. A polynomial $p(x)$ is used as a substitute for a function $y(x)$, for any of a dozen or more reasons. Perhaps most important of all, polynomials are easy to compute, only simple integer powers being involved. But their derivatives and integrals are also found without much effort and are again polynomials. Roots of polynomial equations surface with less excavation than for other functions. The popularity of polynomials as substitutes is not hard to understand.

CRITERION OF APPROXIMATION

The difference $y(x)-p(x)$ is the error of the approximation, and the central idea is, of course, to keep this error reasonably small. The simplicity of polynomials permits this goal to be approached in various ways, of which we consider

1. collocation, 2. osculation, 3. least squares, 4. min.-max.

THE COLLOCATION POLYNOMIAL

The collocation polynomial is the target of this and the next few chapters. It coincides (collocates) with $y(x)$ at certain specified points. A number of properties of such polynomials, and of polynomials in general, play a part in the development.

1. **The existence and uniqueness theorem** states that there is exactly one collocation polynomial of degree n for arguments $x_0, \ldots, x_n$, that is, such that $y(x)=p(x)$ for these arguments. The existence will be proved by actually exhibiting such a polynomial in succeeeding chapters. The uniqueness is proved in the present chapter and is a consequence of certain elementary properties of polynomials.
2. **The division algorithm.** Any polynomial $p(x)$ may be expressed as

$$p(x)=(x-r)q(x)+R$$

   where r is any number, $q(x)$ is a polynomial of degree $n-1$, and R is a constant. This has two quick corollaries.
3. **The remainder theorem** states that $p(r)=R$.
4. **The factor theorem** states that if $p(r)=0$, then $x-r$ is a factor of $p(x)$.
5. **The limitation on zeros.** A polynomial of degree n can have at most n zeros, meaning that the equation $p(x)=0$ can have at most n roots. The uniqueness theorem is an immediate consequence, as will be shown.
6. **Synthetic division** is an economical procedure (or algorithm) for producing the $q(x)$ and R of the division algorithm. It is often used to obtain R, which by the remainder theorem equals $p(r)$. This path to $p(r)$ may be preferable to the direct computation of this polynomial value.
7. **The product** $\pi(x)=(x-x_0)(x-x_1)\cdots(x-x_n)$ plays a central role in collocation theory. Note that it vanishes at the arguments $x_0, x_1, \ldots, x_n$ which are our collocation arguments. The error of the collocation polynomial will be shown to be

$$y(x)-p(x)=\frac{y^{(n+1)}(\xi)\pi(x)}{(n+1)!}$$

where ξ depends upon x and is somewhere between the extreme points of collocation, provided x itself is. Note that this formula does reduce to zero at $x_0, x_1, \ldots, x_n$ so that $p(x)$ does collocate with $y(x)$ at those arguments. Elsewhere we think of $p(x)$ as an approximation to $y(x)$.

Solved Problems

2.1. Prove that any polynomial $p(x)$ may be expressed as

$$p(x) = (x-r)q(x) + R$$

where r is any number, $q(x)$ is a polynomial of degree $n-1$, and R is a constant.

This is an example of *the division algorithm*. Let $p(x)$ be of degree n.

$$p(x) = a_n x^n + a_{n-1}x^{n-1} + \cdots + a_0$$

Then
$$p(x) - (x-r)a_n x^{n-1} = q_1(x) = b_{n-1}x^{n-1} + \cdots$$

will be of degree $n-1$ or less. Similarly,

$$q_1(x) - (x-r)b_{n-1}x^{n-2} = q_2(x) = c_{n-2}x^{n-2} + \cdots$$

will be of degree $n-2$ or less. Continuing in this way, we eventually reach a polynomial $q_n(x)$ of degree zero, a constant. Renaming this constant R, we have

$$p(x) = (x-r)(a_n x^{n-1} + b_{n-1}x^{n-2} + \cdots) + R = (x-r)q(x) + R$$

2.2. Prove $p(r) = R$. This is called the **remainder theorem**.

Let $x = r$ in Problem 2.1. At once, $p(r) = 0 \cdot q(r) + R$.

2.3. Illustrate the "synthetic division" method for performing the division described in Problem 2.1, using $r = 2$ and $p(x) = x^3 - 3x^2 + 5x + 7$.

Synthetic division is merely an abbreviated version of the same operations described in Problem 2.1. Only the various coefficients appear. For the $p(x)$ and r above, the starting layout is

$r = 2$	1	−3	5	7 ← coefficients of $p(x)$
	1			

Three times we "multiply by r and add" to complete the layout.

$r = 2$	1	−3	5	7
		2	−2	6
	1	−1	3	13 ← the number R

(1, −1, 3: coefficients of $q(x)$)

Thus, $q(x) = x^2 - x + 3$ and $R = f(2) = 13$. This may be verified by computing $(x-r)q(x) + R$, which will be $p(x)$. It is also useful to find $q(x)$ by the "long division" method, starting from this familiar layout:

$$(x-2)\overline{\big)x^3 - 3x^2 + 5x + 7}$$

Comparing the resulting computation with the "synthetic" algorithm just completed, one easily sees the equivalence of the two.

2.4. Prove that if $p(r) = 0$, then $x - r$ is a factor of $p(x)$. This is the *factor theorem*. The other factor has degree $n - 1$.

If $p(r) = 0$, then $0 = 0q(x) + R$ making $R = 0$. Thus, $p(x) = (x - r)q(x)$.

2.5. Prove that a polynomial of degree n can have at most n zeros, meaning that $p(x) = 0$ can have at most n roots.

Suppose n roots exist. Call them $r_1, r_2, \ldots, r_n$. Then by n applications of the factor theorem,

$$p(x) = A(x - r_1)(x - r_2) \cdots (x - r_n)$$

where A has degree 0, a constant. This makes it clear that there can be no other roots. (Note also that $A = a_n$.)

2.6. Prove that at most one polynomial of degree n can take the specified values y_k at given arguments x_k, where $k = 0, 1, \ldots, n$.

Suppose there were two such polynomials, $p_1(x)$ and $p_2(x)$. Then the difference $p(x) = p_1(x) - p_2(x)$ would be of degree n or less, and would have zeros at all the arguments x_k: $p(x_k) = 0$. Since there are $n + 1$ such arguments this contradicts the result of the previous problem. Thus, at most one polynomial can take the specified values. The following chapters display this polynomial in many useful forms. It is called the **collocation polynomial**.

2.7. Suppose a polynomial $p(x)$ of degree n takes the same values as a function $y(x)$ for $x = x_0, x_1, \ldots, x_n$. [This is called collocation of the two functions and $p(x)$ is the collocation polynomial.] Obtain a formula for the difference between $p(x)$ and $y(x)$.

Since the difference is zero at the points of collocation, we anticipate a result of the form

$$y(x) - p(x) = C(x - x_0)(x - x_1) \cdots (x - x_n) = C\pi(x)$$

which may be taken as the definition of C. Now consider the following function $F(x)$:

$$F(x) = y(x) - p(x) - C\pi(x)$$

This $F(x)$ is zero for $x = x_0, x_1, \ldots, x_n$ and if we choose a new argument x_{n+1} and

$$C = \frac{y(x_{n+1}) - p(x_{n+1})}{\pi(x_{n+1})}$$

then $F(x_{n+1})$ will also be zero. Now $F(x)$ has $n + 2$ zeros at least. By Rolle's theorem $F'(x)$ then is guaranteed $n + 1$ zeros between those of $F(x)$, while $F''(x)$ is guaranteed n zeros between those of $F'(x)$. Continuing to apply Rolle's theorem in this way eventually shows that $F^{(n+1)}(x)$ has at least one zero in the interval from x_0 to x_n, say at $x = \xi$. Now calculate this derivative, recalling that the $(n + 1)$th derivative of $p(x)$ will be zero, and put x equal to ξ:

$$0 = y^{(n+1)}(\xi) - C(n + 1)!$$

This determines C, which may now be substituted back:

$$y(x_{n+1}) - p(x_{n+1}) = \frac{y^{(n+1)}(\xi)}{(n + 1)!}\pi(x_{n+1})$$

Since x_{n+1} can be any argument between x_0 and x_n except for $x_0, \ldots, x_n$ and since our result is clearly true for $x_0, \ldots, x_n$ also, we replace x_{n+1} by the simpler x:

$$y(x) - p(x) = \frac{y^{(n+1)}(\xi)}{(n + 1)!}\pi(x)$$

This result is often quite useful in spite of the fact that the number ξ is usually undeterminable, because we can estimate $y^{(n+1)}(\xi)$ independently of ξ.

2.8. Find a first-degree polynomial which takes the values $y(0)=1$ and $y(1)=0$, or in tabular form,

x_k	0	1
y_k	1	0

The result $p(x)=1-x$ is immediate either by inspection or by elementary geometry. This is the collocation polynomial for the meager data supplied.

2.9. The function $y(x)=\cos\frac{1}{2}\pi x$ also takes the values specified in Problem 2.8. Determine the difference $y(x)-p(x)$.

By Problem 2.7, with $n=1$,

$$y(x)-p(x)=-\frac{\pi^2\cos\frac{1}{2}\pi\xi}{8}x(x-1)$$

Even without determining ξ we can estimate this difference by

$$|y(x)-p(x)|\leqq\frac{\pi^2}{8}x(x-1)$$

Viewing $p(x)$ as a linear approximation to $y(x)$, this error estimate is simple, though generous. At $x=\frac{1}{2}$ it suggests an error of size roughly .3, while the actual error is approximately $\cos\frac{1}{4}\pi-(1-\frac{1}{2})=.2$.

2.10. As the degree n increases indefinitely, does the resulting sequence of collocation polynomials converge to $y(x)$?

The answer is slightly complicated. For carefully chosen collocation arguments x_k and reasonable functions $y(x)$, convergence is assured, as will appear later. But for the most popular case of equally spaced arguments x_k, divergence may occur. For some $y(x)$ the sequence of polynomials is convergent for all arguments x. For other functions, convergence is limited to a finite interval, with the error $y(x)-p(x)$ oscillating in the manner shown in Fig. 2-1. Within the interval of convergence the oscillation dies out and $\lim(y-p)=0$, but outside that interval $y(x)-p(x)$ grows arbitrarily large as n increases. The oscillation is produced by the $\pi(x)$ factor, the size being influenced by the derivatives of $y(x)$. This error behavior is a severe limitation on the use of high-degree collocation polynomials.

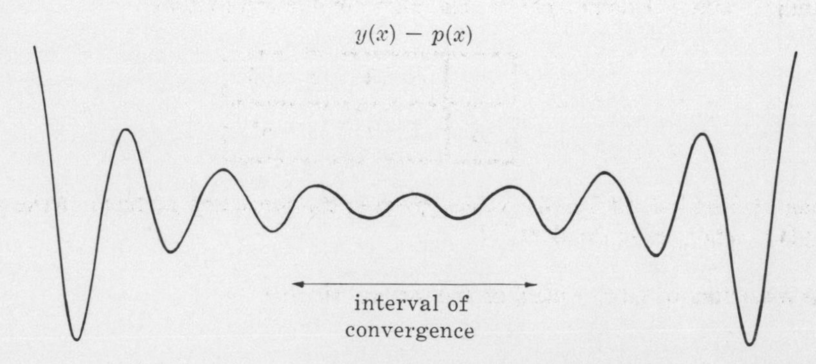

Fig. 2-1

Supplementary Problems

2.11. Apply synthetic division to divide $p(x)=x^3-x^2+x-1$ by $x-1$. Note that $R=f(1)=0$, so that $x-1$ is a factor of $p(x)$ and $r=1$ is a zero of $f(x)$.

2.12. Apply synthetic division to $p(x)=2x^4-24x^3+100x^2-168x+93$ to compute $p(1)$. (Divide by $x-1$ and take the remainder R.) Also compute $p(2)$, $p(3)$, $p(4)$, and $p(5)$.

2.13. To find a second-degree polynomial which takes the following values:

x_k	0	1	2
y_k	0	1	0

we could write $p(x)=A+Bx+Cx^2$ and substitute to find the conditions

$$0=A \qquad 1=A+B+C \qquad 0=A+2B+4C$$

Solve for A, B, and C and so determine this collocation polynomial. Theoretically the same procedure applies for higher-degree polynomials, but more efficient algorithms will be developed.

2.14. The function $y(x)=\sin\frac{1}{2}\pi x$ also takes the values specified in Problem 2.13. Apply Problem 2.7 to show that

$$y(x)-p(x)=-\frac{\pi^3\cos\frac{1}{2}\pi\xi}{48}x(x-1)(x-2)$$

where ξ depends on x.

2.15. Continuing Problem 2.14, show that

$$|y(x)-p(x)|\le\left|\frac{\pi^3}{48}x(x-1)(x-2)\right|$$

This estimates thc accuracy of the collocation polynomial $p(x)$ as an approximation to $y(x)$. Compute this estimate at $x=\frac{1}{2}$ and compare with the actual error.

2.16. Compare $y'(x)$ and $p'(x)$ for $x=\frac{1}{2}$.

2.17. Compare $y''(x)$ and $p''(x)$ for $x=\frac{1}{2}$.

2.18. Compare the integrals of $y(x)$ and $p(x)$ over the interval $(0,2)$.

2.19. Find the unique cubic polynomial $p(x)$ which takes the following values.

x_k	0	1	2	3
y_k	0	1	16	81

2.20. The function $y(x)=x^4$ also takes the values given in the preceding problem. Write a formula for the difference $y(x)-p(x)$, using Problem 2.7.

2.21. What is the maximum of $|y(x)-p(x)|$ on the interval $(0,3)$?

Chapter 3

Finite Differences

FINITE DIFFERENCES

Finite differences have had a strong appeal to mathematicians for centuries. Isaac Newton was an especially heavy user, and much of the subject originated with him. Given a discrete function, that is, a finite set of arguments x_k each having a mate y_k, and supposing the arguments equally spaced, so that $x_{k+1} - x_k = h$, the differences of the y_k values are denoted

$$\Delta y_k = y_{k+1} - y_k$$

and are called first differences. The differences of these first differences are denoted

$$\Delta^2 y_k = \Delta(\Delta y_k) = \Delta y_{k+1} - \Delta y_k = y_{k+2} - 2y_{k+1} + y_k$$

and are called second differences. In general,

$$\Delta^n y_k = \Delta^{n-1} y_{k+1} - \Delta^{n-1} y_k$$

defines the nth differences.

The *difference table* is the standard format for displaying finite differences. Its diagonal pattern makes each entry, except for the x_k, y_k, the difference of its two nearest neighbors to the left.

$$\begin{array}{cccccc}
x_0 & y_0 & & & & \\
 & & \Delta y_0 & & & \\
x_1 & y_1 & & \Delta^2 y_0 & & \\
 & & \Delta y_1 & & \Delta^3 y_0 & \\
x_2 & y_2 & & \Delta^2 y_1 & & \Delta^4 y_0 \\
 & & \Delta y_2 & & \Delta^3 y_1 & \\
x_3 & y_3 & & \Delta^2 y_2 & & \\
 & & \Delta y_3 & & & \\
x_4 & y_4 & & & &
\end{array}$$

Each differences proves to be a combination of the y values in column two. A simple example is $\Delta^3 y_0 = y_3 - 3y_2 + 3y_1 - y_0$. The general result is

$$\Delta^k y_0 = \sum_{i=0}^{k} (-1)^i \binom{k}{i} y_{k-1}$$

where $\binom{k}{i}$ is a binomial coefficient.

DIFFERENCE FORMULAS

Difference formulas for elementary functions somewhat parallel those of calculus. Examples include the following:

1. The **differences of a constant function** are zero. In symbols,

$$\Delta C = 0,$$

   where C denotes a constant (independent of k).

2. For a **constant times another function,** we have

$$\Delta(Cu_k) = C\Delta u_k$$

3. The **difference of a sum** of two functions is the sum of their differences:

$$\Delta(u_k + v_k) = \Delta u_k + \Delta v_k$$

4. The **linearity property** generalizes the two previous results to

$$\Delta(C_1 u_k + C_2 v_k) = C_1 \Delta u_k + C_2 \Delta v_k$$

where C_1 and C_2 are constants.

5. The **differences of a product** are given by the formula

$$\Delta(u_k v_k) = u_k \Delta v_k + v_{k+1} \Delta u_k$$

in which the argument $k+1$ should be noted.

6. The **differences of a quotient are**

$$\Delta\left(\frac{u_k}{v_k}\right) = \frac{v_k \Delta u_k - u_k \Delta v_k}{v_{k+1} v_k}$$

and again the argument $k+1$ should be noted.

7. The **differences of the power function** are given by

$$\Delta C^k = C^k (C - 1)$$

The special case $C = 2$ brings $\Delta y_k = y_k$.

8. The **differences of sine and cosine functions** are also reminiscent of corresponding results of calculus, but the details are not quite so attractive.

$$\Delta(\sin k) = 2 \sin \frac{1}{2} \cos \left(k + \frac{1}{2}\right)$$

$$\Delta(\cos k) = -2 \sin \frac{1}{2} \sin \left(k + \frac{1}{2}\right)$$

9. The **differences of the logarithm function** are a similar disappointment. With $x_k = x_0 + kh$, we have

$$\Delta(\log x_k) = \log\left(1 + \frac{h}{x_k}\right)$$

When h/x_k is very small this makes $\Delta(\log x_k)$ approximately h/x_k, but otherwise the reciprocal of x, which is so prominent in the calculus of logarithms, is quite remote.

10. The **unit error function,** for which $y_k = 1$ at a single argument and is otherwise zero, has a difference table consisting of the successive binomial coefficients with alternating signs. The detection of isolated errors in a table of y_k values can be based on this property of the unit error function.

11. The **oscillating error function,** for which $y_k = \pm 1$ alternately, has a difference table consisting of the successive powers of 2 with alternating signs.

12. **Other functions of special interest** will be studied in succeeding chapters, and the relationships between difference and differential calculus will be of continuing interest.

Solved Problems

3.1. Compute up through third differences of the discrete function displayed by the x_k y_k columns of Table 3.1. (The integer variable k also appears for convenience.)

The required differences appear in the remaining three columns. Table 3.1 is called a **difference table**. Its diagonal structure has become a standard format for displaying differences. Each entry in the difference columns is the difference of its two nearest neighbors to the left.

Table 3.1

k	x_k	y_k	Δy_k	$\Delta^2 y_k$	$\Delta^3 y_k$
0	1	1			
			7		
1	2	8		12	
			19		6
2	3	27		18	
			37		6
3	4	64		24	
			61		6
4	5	125		30	
			91		6
5	6	216		36	
			127		6
6	7	343		42	
			169		
7	8	512			

Any such table displays differences as shown in Table 3.2.

Table 3.2

0	x_0	y_0			
			Δy_0		
1	x_1	y_1		$\Delta^2 y_0$	
			Δy_1		$\Delta^3 y_0$
2	x_2	y_2		$\Delta^2 y_1$	
			Δy_2		$\Delta^3 y_1$
3	x_3	y_3		$\Delta^2 y_2$	
			Δy_3		
4	x_4	y_4			

For example,

$$\Delta y_0 = y_1 - y_0 = 8 - 1 = 7$$
$$\Delta^2 y_0 = \Delta y_1 - \Delta y_0 = 19 - 7 = 12 \quad \text{etc.}$$

3.2. What is true of all fourth and higher differences of the function of Problem 3.1?

Any such differences are zero. This is a special case of a result to be obtained shortly.

3.3. Prove that $\Delta^3 y_0 = y_3 - 3y_2 + 3y_1 - y_0$.

Either from Table 3.2 or by the definitions provided at the outset,

$$\Delta^3 y_0 = \Delta^2 y_1 - \Delta^2 y_0 = (y_3 - 2y_2 + y_1) - (y_2 - 2y_1 + y_0) = y_3 - 3y_2 + 3y_1 - y_0$$

3.4. Prove that $\Delta^4 y_0 = y_4 - 4y_3 + 6y_2 - 4y_1 + y_0$.

By definition, $\Delta^4 y_0 = \Delta^3 y_1 - \Delta^3 y_0$. Using the result of Problem 3.3 and the almost identical

$$\Delta^3 y_1 = y_4 - 3y_3 + 3y_2 - y_1$$

obtained by advancing all lower indices, the required result follows at once.

3.5. Prove that for any positive integer k,

$$\Delta^k y_0 = \sum_{i=0}^{k} (-1)^i \binom{k}{i} y_{k-i}$$

where the familiar symbol for binomial coefficients,

$$\binom{k}{i} = \frac{k!}{i!\,(k-i)!} = \frac{k(k-1)\cdots(k-i+1)}{i!}$$

has been used.

The proof will be by induction. For $k = 1, 2, 3,$ and 4, the result has already been established, by definition when k is 1. Assume it true when k is some particular integer p:

$$\Delta^p y_0 = \sum_{i=0}^{p} (-1)^i \binom{p}{i} y_{p-i}$$

By advancing all lower indices we have also

$$\Delta^p y_1 = \sum_{i=0}^{p} (-1)^i \binom{p}{i} y_{p-i+1}$$

and by a change in the summation index, namely, $i = j + 1$,

$$\Delta^p y_1 = y_{p+1} - \sum_{j=0}^{p-1} (-1)^j \binom{p}{j+1} y_{p-j}$$

It is also convenient to make a nominal change of summation index, $i = j$, in our other sum:

$$\Delta^p y_0 = \sum_{j=0}^{p-1} (-1)^j \binom{p}{j} y_{p-j} + (-1)^p y_0$$

Then $$\Delta^{p+1} y_0 = \Delta^p y_1 - \Delta^p y_0 = y_{p+1} - \sum_{j=0}^{p-1} (-1)^j \left[\binom{p}{j+1} + \binom{p}{j}\right] y_{p-j} - (-1)^p y_0$$

Now using $$\binom{p}{j+1} + \binom{p}{j} = \binom{p+1}{j+1}$$

(see Problem 4.5) and making a final change of summation index, $j + 1 = i$,

$$\Delta^{p+1} y_0 = y_{p+1} + \sum_{i=1}^{p} (-1)^i \binom{p+1}{i} y_{p+1-i} - (-1)^p y_0 = \sum_{i=0}^{p+1} (-1)^i \binom{p+i}{i} y_{p+1-i}$$

Thus our result is established when k is the integer $p + 1$. This completes the induction.

3.6. Prove that for a constant function all differences are zero.

Let $y_k = C$ for all k. This is a constant function. Then, for all k,

$$\Delta y_k = y_{k+1} - y_k = C - C = 0$$

3.7. Prove $\Delta(Cy_k) = C\Delta y_k$.

This is analogous to a result of calculus. $\Delta(Cy_k) = Cy_{k+1} - Cy_k = C\Delta y_k$.

Essentially this problem involves two functions defined for the same arguments x_k. One function has the values y_k, the other has values $z_k = Cy_k$. We have proved $\Delta z_k = C\Delta y_k$.

3.8. Consider two functions defined for the same set of arguments x_k. Call the values of these functions u_k and v_k. Also consider a third function with values

$$w_k = C_1 u_k + C_2 v_k$$

where C_1 and C_2 are two constants (independent of x_k). Prove

$$\Delta w_k = C_1 \Delta u_k + C_2 \Delta v_k$$

This is the **linearity property** of the difference operation.

The proof is direct from the definitions.

$$\Delta w_k = w_{k+1} - w_k = (C_1 u_{k+1} + C_2 v_{k+1}) - (C_1 u_k + C_2 v_k)$$
$$= C_1(u_{k+1} - u_k) + C_2(v_{k+1} - v_k) = C_1 \Delta u_k + C_2 \Delta v_k$$

Clearly the same proof would apply to sums of any finite length.

3.9. With the same symbolism as in Problem 3.8, consider the function with values $z_k = u_k v_k$ and prove $\Delta z_k = u_k \Delta v_k + v_{k+1} \Delta u_k$.

Again starting from the definitions,

$$\Delta z_k = u_{k+1} v_{k+1} - u_k v_k = u_{k+1} v_{k+1} - u_k v_{k+1} + u_k v_{k+1} - u_k v_k$$
$$= v_{k+1}(u_{k+1} - u_k) + u_k(v_{k+1} - v_k) = u_k \Delta v_k + v_{k+1} \Delta u_k$$

The result $\Delta z_k = u_{k+1} \Delta v_k + v_k \Delta u_k$ could also be proved.

3.10. Compute differences of the function displayed in the first two columns of Table 3.3. This may be viewed as a type of "error function," if one supposes that all its values should be zero but the single 1 is a *unit error.* How does this unit error affect the various differences?

Some of the required differences appear in the other columns of Table 3.3.

Table 3.3

x_0	0				
		0			
x_1	0		0		
		0		0	
x_2	0		0		1
		0		1	
x_3	0		1		−4
		1		−3	
x_4	1		−2		6
		−1		3	
x_5	0		1		−4
		0		−1	
x_6	0		0		1
		0		0	
x_7	0		0		
		0			
x_8	0				

This error influences a triangular portion of the difference table, increasing for higher differences and having a binomial coefficient pattern.

3.11. Compute differences for the function displayed in the first two columns of Table 3.4. This may be viewed as a type of error function, each value being a roundoff error of amount one unit. Show that the alternating ± pattern leads to serious error growth in the higher differences. Hopefully, roundoff errors will seldom alternate in just this way.

Some of the required differences appear in the other columns of Table 3.4. The error doubles for each higher difference.

Table 3.4

x_0	1						
		−2					
x_1	−1		4				
		2		−8			
x_2	1		−4		16		
		−2		8		−32	
x_3	−1		4		−16		64
		2		−8		32	
x_4	1		−4		16		
		−2		8			
x_5	−1		4				
		2					
x_6	1						

3.12. One number in this list is misprinted. Which one?

1 2 4 8 16 26 42 64 93

Calculating the first four differences, and displaying them horizontally for a change, we have

1 2 4 8 10 16 22 29

1 2 4 2 6 6 7

1 2 −2 4 0 1

1 −4 6 −4 1

and the impression is inescapable that these binomial coefficients arise from a data error of size 1 in the center entry 16 of the original list. Changing it to 15 brings the new list

1 2 4 8 15 26 42 64 93

from which we find the differences

1 2 4 7 11 16 22 29

1 2 3 4 5 6 7

which suggest a job well done. This is a very simple example of data smoothing, which we treat much more fully in a later chapter. There is always the possibility that data such as we have in our original list comes from a bumpy process, not from a smooth one, so that the bump (16 instead of 15) is real and not a misprint. The above analysis can then be viewed as bump detection, rather than as error correcting.

Supplementary Problems

3.13. Calculate up through fourth differences for the following y_k values. (Here it may be assumed that $x_k = k$.)

k	0	1	2	3	4	5	6
y_k	0	1	16	81	256	625	1296

3.14. Verify Problem 3.5 for $k=5$ by showing directly from the definition that

$$\Delta^5 y_0 = y_5 - 5y_4 + 10y_3 - 10y_2 + 5y_1 - y_0$$

3.15. Imitating Problem 3.9, prove that $\Delta \dfrac{u_k}{v_k} = \dfrac{v_k \Delta u_k - u_k \Delta v_k}{v_{k+1} v_k}$.

3.16. Calculate differences through the fifth order to observe the effect of adjacent "errors" of size 1.

k	0	1	2	3	4	5	6	7
y_k	0	0	0	1	1	0	0	0

3.17. Find and correct a single error in these y_k values.

k	0	1	2	3	4	5	6	7
y_k	0	0	1	6	24	60	120	210

3.18. Use the linearity property to show that if $y_k = k^3$, then

$$\Delta y_k = y_{k+1} - y_k = 3k^2 + 3k + 1 \qquad \Delta^2 y_k = \Delta y_{k+1} - \Delta y_k = 6k + 6 \qquad \Delta^3 y_k = \Delta^2 y_{k+1} - \Delta^2 y_k = 6$$

3.19. Show that if $y_k = k^4$, then $\Delta^4 y_k = 24$.

3.20. Show that if $y_k = 2^k$, then $\Delta y_k = y_k$.

3.21. Show that if $y_k = C^k$, then $\Delta y_k = C^k(C-1)$.

3.22. Compute the missing y_k values from the first differences provided

$$\begin{array}{rccccccccccccc}
y_k & 0 & & \cdot & & \cdot & & \cdot & & \cdot & & \cdot & & \cdot \\
\Delta y_k & & 1 & & 2 & & 4 & & 7 & & 11 & & 16 &
\end{array}$$

3.23. Compute the missing y_k and Δy_k values from the data provided.

$$\begin{array}{rccccccccccccc}
y_k & \cdot & & \cdot & & \cdot & & 6 & & \cdot & & \cdot & & \cdot \\
\Delta y_k & & \cdot & & \cdot & & 5 & & \cdot & & \cdot & & \cdot & \\
\Delta^2 y_k & & & 1 & & 4 & & 13 & & 18 & & 24 & &
\end{array}$$

3.24. Compute the missing y_k values from the data provided.

$$\begin{array}{rccccccccccccccccc}
y_k & 0 & & 0 & & 0 & & 6 & & 24 & & 60 & & \cdot & & \cdot & & \cdot \\
\Delta y_k & & 0 & & 0 & & 6 & & 18 & & 36 & & \cdot & & \cdot & & \cdot & \\
\Delta^2 y_k & & & 0 & & 6 & & 12 & & 18 & & \cdot & & \cdot & & \cdot & & \\
\Delta^3 y_k & & & & 6 & & 6 & & 6 & & 6 & & 6 & & 6 & & &
\end{array}$$

3.25. Find and correct a misprint in this data.

$$\begin{array}{rccccccccccc}
y_k & 1 & 3 & 11 & 31 & 69 & 113 & 223 & 351 & 521 & 739 & 1011
\end{array}$$

3.26. By advancing all subscripts in the formula $\Delta^2 y_0 = y_2 - 2y_1 + y_0$, write similar expansions for $\Delta^2 y_1$ and $\Delta^2 y_2$. Compute the sum of these second differences. It should equal $\Delta y_3 - \Delta y_0 = y_4 - y_3 - y_1 + y_0$.

3.27. Find a function y_k for which $\Delta y_k = 2y_k$.

3.28. Find a function y_k for which $\Delta^2 y_k = 9y_k$. Can you find two such functions?

3.29. Continuing the previous problem, find a function such that $\Delta^2 y_k = 9y_k$ and having $y_0 = 0,\ y_1 = 1$.

3.30. Prove $\Delta(\sin k) = 2 \sin \frac{1}{2} \cos (k + \frac{1}{2})$.

3.31. Prove $\Delta(\cos k) = -2 \sin \frac{1}{2} \sin (k + \frac{1}{2})$.

3.32. Prove $\Delta(\log x_k) = \log (1 + h/x_k)$ where $x_k = x_0 + kh$.

Chapter 4

Factorial Polynomials

FACTORIAL POLYNOMIALS

Factorial polynomials are defined by

$$y_k = k^{(n)} = k(k-1)\cdots(k-n+1)$$

where n is a positive integer. For example, $k^{(2)} = k(k-1) = k^2 - k$. These polynomials play a central role in the theory of finite differences because of their convenient properties. The various differences of a factorial polynomial are again factorial polynomials. More specifically, for the first difference,

$$\Delta k^{(n)} = nk^{(n-1)}$$

which is reminiscent of how *the powers of x* respond to differentiation. Higher differences then become further factorial polynomials of diminishing degree, until ultimately

$$\Delta^n k^{(n)} = n!$$

with all higher differences zero.

The *binomial coefficients* are related to factorial polynomials by

$$\binom{k}{n} = \frac{k^{(n)}}{n!}$$

and therefore share some of the properties of these polynomials, notably the famous recursion

$$\binom{k+1}{n+1} - \binom{k}{n+1} = \binom{k}{n}$$

which has the form of a finite difference formula.

The simple recursion

$$k^{(n+1)} = (k-n)k^{(n)}$$

follows directly from the definition of factorial polynomials. Rewriting it as

$$k^{(n)} = \frac{k^{(n+1)}}{k-n}$$

it may be used to extend the factorial idea successively to the integers $n = 0, -1, -2, \ldots$. The basic formula

$$\Delta k^{(n)} = nk^{(n-1)}$$

is then true for all integers n.

STIRLING'S NUMBERS

Stirling's numbers of the first kind appear when factorial polynomials are expressed in standard polynomial form. Thus

$$k^{(n)} = S_1^{(n)}k + \cdots + S_n^{(n)}k^n = \sum S_i^{(n)}k^i$$

the $S_i^{(n)}$ being the Stirling numbers. As an example,

$$k^{(3)} = 2k - 3k^2 + k^3$$

which makes $S_1^{(3)} = 2$, $S_2^{(3)} = -3$, and $S_3^{(3)} = 1$. The recursion formula

$$S_i^{(n+1)} = S_{i-1}^{(n)} - nS_i^{(n)}$$

permits rapid tabulation of these Stirling numbers.

Stirling's numbers of the second kind appear when the powers of k are represented as combinations of factorial polynomials. Thus

$$k^n = s_1^{(n)} k^{(1)} + \cdots + s_n^{(n)} k^{(n)} = \sum s_i^{(n)} k^{(i)}$$

the $s_i^{(n)}$ being the Stirling numbers. As an example,

$$k^3 = k^{(1)} + 3k^{(2)} + k^{(3)}$$

so that $s_1^{(3)} = 1$, $s_2^{(3)} = 3$, and $s_3^{(3)} = 1$. The recursion formula

$$s_i^{(n+1)} = s_{i-1}^{(n)} + is_i^{(n)}$$

permits rapid tabulation of these numbers. A basic theorem states that each power of k can have only one such representation as a combination of factorial polynomials. This assures the unique determination of the Stirling numbers of the second kind.

REPRESENTATION OF ARBITRARY POLYNOMIALS

The representation of arbitrary polynomials as combinations of factorial polynomials is a natural next step. Each power of k is so represented and the results are then combined. The representation is unique because of the basic theorem just quoted. For example,

$$k^2 + 2k + 1 = [k^{(2)} + k^{(1)}] + 2k^{(1)} + 1 = k^{(2)} + 3k^{(1)} + 1$$

Differences of arbitrary polynomials are conveniently found by first representing such polynomials as combinations of factorial polynomials and then applying our formula for differencing the separate factorial terms.

The principal theorem of the chapter is now accessible, and states that the difference of a polynomial of degree n is another polynomial, of degree $n-1$. This makes the nth differences of such a polynomial constant, and still higher differences zero.

Solved Problems

4.1. Consider the special function for which $y_k = k(k-1)(k-2)$ and prove $\Delta y_k = 3k(k-1)$.

$$\Delta y_k = y_{k+1} - y_k = (k+1)k(k-1) - k(k-1)(k-2) = [(k+1)-(k-2)]k(k-1) = 3k(k-1)$$

In tabular form this same result, for the first few integer values of k, is given in Table 4.1.

4.2. This generalizes Problem 4.1. Consider the special function

$$y_k = k(k-1)\cdots(k-n+1) = k^{(n)}$$

(Note that the upper index is not a power.) Prove, for $n > 1$,

$$\Delta y_k = nk^{(n-1)}$$

a result which is strongly reminiscent of the theorem on the derivative of the nth power function.

$$\begin{aligned}\Delta y_k = y_{k+1} - y_k &= [(k+1)\cdots(k-n+2)] - [k\cdots(k-n+1)] \\ &= [(k+1)-(k-n+1)]k(k-1)\cdots(k-n+2) = nk^{(n-1)}\end{aligned}$$

Table 4.1

k	y_k	Δy_k
0	0	
		0
1	0	
		0
2	0	
		6
3	6	
		18
4	24	
		36
5	60	

4.3. Prove that if $y_k = k^{(n)}$, then $\Delta^2 y_k = n(n-1)k^{(n-2)}$.

Problem 4.2 can be applied to Δy_k rather than to y_k.

$$\Delta^2 k^{(n)} = \Delta\Delta k^{(n)} = \Delta n k^{(n-1)} = n(n-1)k^{(n-2)}$$

Extensions to higher differences proceed just as with derivatives.

4.4. Prove $\Delta^n k^{(n)} = n!$ and $\Delta^{n+1} k^{(n)} = 0$.

After n applications of Problem 4.2, the first result follows. (The symbol $k^{(0)}$ can be interpreted as 1.) Since $n!$ is constant (independent of k) its differences are all 0.

4.5. The *binomial coefficients* are the integers

$$\binom{k}{n} = \frac{k^{(n)}}{n!} = \frac{k!}{n!\,(k-n)!}$$

Prove the recursion

$$\binom{k+1}{n+1} = \binom{k}{n+1} + \binom{k}{n}$$

Using factorial polynomials and applying Problem 4.2,

$$\binom{k+1}{n+1} - \binom{k}{n+1} = \frac{(k+1)^{(n+1)}}{(n+1)!} - \frac{k^{(n+1)}}{(n+1)!} = \frac{\Delta k^{(n+1)}}{(n+1)!} = \frac{(n+1)k^{(n)}}{(n+1)!} = \frac{k^{(n)}}{n!} = \binom{k}{n}$$

which transposes at once into what was to be proved. This famous result has already been used.

4.6. Use the recursion for the binomial coefficients to tabulate these numbers up through $k = 8$.

The first column of Table 4.2 gives $\binom{k}{0}$ which is defined to be 1. The diagonal, where $k = n$, is 1 by definition. The other entries result from the recursion. The table is easily extended.

4.7. Show that if k is a positive integer, then $k^{(n)}$ and $\binom{k}{n}$ are 0 for $n > k$. $\left[\text{For } n > k \text{ the symbol } \binom{k}{n} \text{ is defined as } k^{(n)}/n!\right]$

Table 4.2

$k \backslash n$	0	1	2	3	4	5	6	7	8
1	1	1							
2	1	2	1						
3	1	3	3	1					
4	1	4	6	4	1				
5	1	5	10	10	5	1			
6	1	6	15	20	15	6	1		
7	1	7	21	35	35	21	7	1	
8	1	8	28	56	70	56	28	8	1

Note that $k^{(k+1)} = k(k-1)\cdots 0$. For $n > k$ the factorial $k^{(n)}$ will contain this 0 factor, and so will $\binom{k}{n}$.

4.8. The binomial coefficient symbol and the factorial symbol are often used for nonintegral k. Calculate $k^{(n)}$ and $\binom{k}{n}$ for $k = \frac{1}{2}$ and $n = 2, 3$.

$$k^{(2)} = \left(\frac{1}{2}\right)^{(2)} = \frac{1}{2}\left(\frac{1}{2} - 1\right) = -\frac{1}{4} \qquad k^{(3)} = \left(\frac{1}{2}\right)^{(3)} = \frac{1}{2}\left(\frac{1}{2} - 1\right)\left(\frac{1}{2} - 2\right) = \frac{3}{8}$$

$$\binom{k}{2} = \frac{k^{(2)}}{2!} = \frac{1}{2}\left(-\frac{1}{4}\right) = -\frac{1}{8} \qquad \binom{k}{3} = \frac{k^{(3)}}{3!} = \frac{1}{6}\left(\frac{3}{8}\right) = \frac{1}{16}$$

4.9. The idea of factorial has also been extended to upper indices which are not positive integers. It follows from the definition that when n is a positive integer, $k^{(n+1)} = (k-n)k^{(n)}$. Rewriting this as

$$k^{(n)} = \frac{1}{k-n} k^{(n+1)}$$

and using it as a *definition* of $k^{(n)}$ for $n = 0, -1, -2, \ldots$, show that $k^{(0)} = 1$ and $k^{(-n)} = 1/(k+n)^{(n)}$.

With $n = 0$ the first result is instant. For the second we find successively

$$k^{(-1)} = \frac{1}{k+1} k^{(0)} = \frac{1}{k+1} = \frac{1}{(k+1)^{(1)}} \qquad k^{(-2)} = \frac{1}{k+2} k^{(-1)} = \frac{1}{(k+2)(k+1)} = \frac{1}{(k+2)^{(2)}}$$

and so on. An inductive proof is indicated but the details will be omitted. For $k = 0$ it is occasionally convenient to define $k^{(0)} = 1$ and to accept the consequences.

4.10. Prove that $\Delta k^{(n)} = nk^{(n-1)}$ for all integers n.

For $n > 1$, this has been proved in Problem 4.2. For $n = 1$ and 0, it is immediate. For n negative, say $n = -p$,

$$\Delta k^{(n)} = \Delta k^{(-p)} = \Delta \frac{1}{(k+p)^{(p)}} = \frac{1}{(k+1+p)\cdots(k+2)} - \frac{1}{(k+p)\cdots(k+1)}$$

$$= \frac{1}{(k+p)\cdots(k+2)}\left(\frac{1}{k+1+p} - \frac{1}{k+1}\right) = \frac{-p}{(k+1+p)\cdots(k+1)}$$

$$= \frac{n}{(k+1-n)^{(1-n)}} = nk^{(n-1)}$$

This result is analogous to the fact that the theorem of calculus "if $f(x) = x^n$, then $f'(x) = nx^{n-1}$" is also true for all integers.

4.11. Find $\Delta k^{(-1)}$.

By the previous problems, $\Delta k^{(-1)} = -k^{(-2)} = -1/(k+2)(k+1)$.

4.12. Show that $k^{(2)} = -k + k^2$, $k^{(3)} = 2k - 3k^2 + k^3$, $k^{(4)} = -6k + 11k^2 - 6k^3 + k^4$.

Directly from the definitions:

$$k^{(2)} = k(k-1) = -k + k^2$$
$$k^{(3)} = k^{(2)}(k-2) = 2k - 3k^2 + k^3$$
$$k^{(4)} = k^{(3)}(k-3) = -6k + 11k^2 - 6k^3 + k^4$$

4.13. Generalizing Problem 4.12, show that in the expansion of a factorial polynomial into standard polynomial form

$$k^{(n)} = S_1^{(n)}k + \cdots + S_n^{(n)}k^n = \sum S_i^{(n)}k^i$$

the coefficients satisfy the recursion

$$S_i^{(n+1)} = S_{i-1}^{(n)} - nS_i^{(n)}$$

These coefficients are called **Stirling's numbers of the first kind**.

Replacing n by $n+1$,

$$k^{(n+1)} = S_1^{(n+1)}k + \cdots + S_{n+1}^{(n+1)}k^{n+1}$$

and using the fact that $k^{(n+1)} = k^{(n)}(k-n)$, we find

$$S_1^{(n+1)}k + \cdots + S_{n+1}^{(n+1)}k^{n+1} = [S_1^{(n)}k + \cdots + S_n^{(n)}k^n](k-n)$$

Now compare coefficients of k^i on both sides. They are

$$S_i^{(n+1)} = S_{i-1}^{(n)} - nS_i^{(n)}$$

for $i = 2, \ldots, n$. The special cases $S_1^{(n+1)} = -nS_1^{(n)}$ and $S_{n+1}^{(n+1)} = S_n^{(n)}$ should also be noted, by comparing coefficients of k and k^{n+1}.

4.14. Use the formulas of Problem 4.13 to develop a brief table of Stirling's numbers of the first kind.

The special formula $S_1^{(n+1)} = -nS_1^{(n)}$ leads at once to column one of Table 4.3. For example, since $S_1^{(1)}$ is clearly 1,

$$S_1^{(2)} = -S_1^{(1)} = -1 \qquad S_1^{(3)} = -2S_1^{(2)} = 2$$

and so on. The other special formula fills the top diagonal of the table with 1s. Our main recursion then completes the table. For example,

$$S_2^{(3)} = S_1^{(2)} - 2S_2^{(2)} = (-1) - 2(1) = -3$$
$$S_2^{(4)} = S_1^{(3)} - 3S_2^{(3)} = (2) - 3(-3) = 11$$
$$S_3^{(4)} = S_2^{(3)} - 3S_3^{(3)} = (-3) - 3(1) = -6$$

and so on. Through $n = 8$ the table reads as follows:

Table 4.3

n \ i	1	2	3	4	5	6	7	8
1	1							
2	−1	1						
3	2	−3	1					
4	−6	11	−6	1				
5	24	−50	35	−10	1			
6	−120	274	−225	85	−15	1		
7	720	−1,764	1,624	−735	175	−21	1	
8	−5,040	13,068	−13,132	6,769	−1,960	322	−28	1

4.15. Use Table 4.3 to expand $k^{(5)}$.

Using row five of the table, $k^{(5)} = 24k - 50k^2 + 35k^3 - 10k^4 + k^5$.

4.16. Show that $k^2 = k^{(1)} + k^{(2)}$, $k^3 = k^{(1)} + 3k^{(2)} + k^{(3)}$, $k^4 = k^{(1)} + 7k^{(2)} + 6k^{(3)} + k^{(4)}$.

Using Table 4.3,

$$k^{(1)} + k^{(2)} = k + (-k + k^2) = k^2$$

$$k^{(1)} + 3k^{(2)} + k^{(3)} = k + 3(-k + k^2) + (2k - 3k^2 + k^3) = k^3$$

$$k^{(1)} + 7k^{(2)} + 6k^{(3)} + k^{(4)} = k + 7(-k + k^2) + 6(2k - 3k^2 + k^3) + (-6k + 11k^2 - 6k^3 + k^4) = k^4$$

4.17 As a necessary preliminary to the following problem, prove that a power of k can have only one representation as a combination of factorial polynomials.

Assume that two such representations exist for k^p.

$$k^p = A_1 k^{(1)} + \cdots + A_p k^{(p)} \qquad k^p = B_1 k^{(1)} + \cdots + B_p k^{(p)}$$

Subtracting leads to

$$0 = (A_1 - B_1)k^{(1)} + \cdots + (A_p - B_p)k^{(p)}$$

Since the right side is a polynomial, and no polynomial can be zero for all values of k, every power of k on the right side must have coefficient 0. But k^p appears only in the last term; hence A_p must equal B_p. And then k^{p-1} appears only in the last term remaining, which will be $(A_{p-1} - B_{p-1})k^{(p-1)}$; hence $A_{p-1} = B_{p-1}$. This argument prevails right back to $A_1 = B_1$.

This proof is typical of unique representation proofs which are frequently needed in numerical analysis. The analogous theorem, that two polynomials cannot have identical values without also having identical coefficients, is a classic result of algebra and has already been used in Problem 4.13.

4.18. Generalizing Problem 4.16, show that the powers of k can be represented as combinations of factorial polynomials

$$k^n = s_1^{(n)} k^{(1)} + \cdots + s_n^{(n)} k^{(n)} = \sum s_i^{(n)} k^{(i)}$$

and that the coefficients satisfy the recursion $s_i^{(n+1)} = s_{i-1}^{(n)} + i s_i^{(n)}$. These coefficients are called **Stirling's numbers of the second kind**.

We proceed by induction, Problem 4.16 already having established the existence of such representations for small k. Suppose

$$k^n = s_1^{(n)}k^{(1)} + \cdots + s_n^{(n)}k^{(n)}$$

and then multiply by k to obtain

$$k^{n+1} = ks_1^{(n)}k^{(1)} + \cdots + ks_n^{(n)}k^{(n)}$$

Now notice that $k \cdot k^{(i)} = (k-i)k^{(i)} + ik^{(i)} = k^{(i+1)} + ik^{(i)}$ so that

$$k^{n+1} = s_1^{(n)}(k^{(2)} + k^{(1)}) + \cdots + s_n^{(n)}(k^{(n+1)} + nk^{(n)})$$

This is already a representation of k^{n+1}, completing the induction, so that we may write

$$k^{n+1} = s_1^{(n+1)}k^{(1)} + \cdots + s_{n+1}^{(n+1)}k^{(n+1)}$$

By Problem 4.17, coefficients of $k^{(i)}$ in both these last lines must be the same, so that

$$s_i^{(n+1)} = s_{i-1}^{(n)} + is_i^{(n)}$$

for $i = 2, \ldots, n$. The special cases $s_1^{(n+1)} = s_1^{(n)}$ and $s_{n+1}^{(n+1)} = s_n^{(n)}$ should also be noted, by comparing coefficients of $k^{(1)}$ and $k^{(n+1)}$.

4.19. Use the formulas of Problem 4.18 to develop a brief table of Stirling's numbers of the second kind.

The special formula $s_1^{(n+1)} = s_1^{(n)}$ leads at once to column one of Table 4.4, since $s_1^{(1)}$ is clearly 1. The other special formula produces the top diagonal. Our main recursion then completes the table. For example,

$$s_2^{(3)} = s_1^{(2)} + 2s_2^{(2)} = (1) + 2(1) = 3 \qquad s_2^{(4)} = s_1^{(3)} + 2s_2^{(3)} = (1) + 2(3) = 7$$

$$s_3^{(4)} = s_2^{(3)} + 3s_3^{(3)} = (3) + 3(1) = 6$$

and so on. Through $n = 8$, the table reads as follows:

Table 4.4

n \ i	1	2	3	4	5	6	7	8
1	1							
2	1	1						
3	1	3	1					
4	1	7	6	1				
5	1	15	25	10	1			
6	1	31	90	65	15	1		
7	1	63	301	350	140	21	1	
8	1	127	966	1701	1050	266	28	1

4.20. Use Table 4.4 to expand k^5 in factorial polynomials.

Using row five of the table, $k^5 = k^{(1)} + 15k^{(2)} + 25k^{(3)} + 10k^{(4)} + k^{(5)}$.

4.21. Prove that the nth differences of a polynomial of degree n are equal, higher differences than the nth being zero.

Call the polynomial $P(x)$, and consider its values for a discrete set of equally spaced arguments $x_0, x_1, x_2, \ldots$. It is usually convenient to deal with the substitute integer argument k which we have used so frequently, related to x by $x_k - x_0 = kh$ where h is the uniform difference between consecutive x arguments. Denote the value of our polynomial for the argument k by the symbol P_k. Since the change of argument is linear, the polynomial has the same degree in terms of both x and k, and we may write it as

$$P_k = a_0 + a_1 k + a_2 k^2 + \cdots + a_n k^n$$

Problem 4.18 shows that each power of k can be represented as a combination of factorial polynomials, leading to a representation of P_k itself as such a combination.

$$P_k = b_0 + b_1 k^{(1)} + b_2 k^{(2)} + \cdots + b_n k^{(n)}$$

Applying Problem 4.2 and the linearity property

$$\Delta P_k = b_1 + 2b_2 k^{(1)} + \cdots + nb_n k^{(n-1)}$$

and reapplying Problem 4.2 leads eventually to $\Delta^n P_k = n!\, b_n$. So all the nth differences are this number. They do not vary with k and consequently higher differences are zero.

4.22. Assuming that the following y_k values belong to a polynomial of degree 4, compute the next three values.

k	0	1	2	3	4	5	6	7
y_k	0	1	2	1	0	·	·	·

A fourth-degree polynomial has constant fourth differences, according to Problem 4.21. Calculating from the given data, we obtain the entries to the left of the line in Table 4.5.

Table 4.5

1		1		−1		−1		5		21		51
	0		−2		0		6		16		30	
		−2		2		6		10		14		
			4		4		4		4			

Assuming the other fourth differences also to be 4 leads to the entires to the right of the line from which the missing entries may be predicted: $y_5 = 5,\ y_6 = 26,\ y_7 = 77$.

Supplementary Problems

4.23. Compute the factorials: $6^{(3)},\ 6^{(6)},\ 6^{(7)},\ (\frac{1}{3})^{(2)},\ (\frac{1}{3})^{(3)},\ (\frac{1}{3})^{(4)}$.

4.24. Compute the factorials: $6^{(-1)},\ 6^{(-2)},\ 6^{(-3)},\ (\frac{1}{3})^{(-1)},\ (\frac{1}{3})^{(-2)},\ (\frac{1}{3})^{(-3)}$.

4.25. Compute the binomial coefficients: $\binom{6}{3}, \binom{6}{6}, \binom{6}{7}, \binom{\frac{1}{3}}{2}, \binom{\frac{1}{3}}{3}, \binom{\frac{1}{3}}{4}$.

4.26. Compute differences through fourth order for these values of $y_k = k^{(4)}$.

k	0	1	2	3	4	5	6	7
y_k	0	0	0	0	24	120	360	840

4.27. Apply Problem 4.2 to express the first four differences of $y_k = k^{(4)}$ in terms of factorial polynomials.

4.28. Apply Problem 4.2 to express the first five differences of $y_k = k^{(5)}$ in terms of factorial polynomials.

4.29. Use Table 4.3 to express $y_k = 2k^{(3)} - k^{(2)} + 4k^{(1)} - 7$ as a conventional polynomial.

4.30. Use Table 4.3 to express $y_k = k^{(6)} + k^{(3)} + 1$ as a conventional polynomial.

4.31. Use Table 4.4 to express $y_k = \frac{1}{3}(2k^4 - 8k^2 + 3)$ as a combination of factorial polynomials.

4.32. Use Table 4.4 to express $y_k = 80k^3 - 30k^4 + 3k^5$ as a combination of factorial polynomials.

4.33. Use the result of the previous problem to obtain Δy_k in terms of factorial polynomials. Then apply Table 4.3 to convert the result to a conventional polynomial

4.34. Use the result of Problem 4.32 to obtain Δy_k and $\Delta^2 y_k$ in terms of factorial polynomials. Then apply Table 4.3 to convert both results to conventional polynomials.

4.35. Assuming that the following y_k values belong to a polynomial of degree 4, predict the next three values.

k	0	1	2	3	4	5	6	7
y_k	1	−1	1	−1	1			

4.36. Assuming that the following y_k values belong to a polynomial of degree 4, predict the next three values.

k	0	1	2	3	4	5	6	7
y_k	0	0	1	0	0			

4.37. What is the lowest degree possible for a polynomial which takes these values?

k	0	1	2	3	4	5
y_k	0	3	8	15	24	35

4.38. What is the lowest degree possible for a polynomial which takes these values?

k	0	1	2	3	4	5
y_k	0	1	1	1	1	0

4.39. Find a function y_k for which $\Delta y_k = k^{(2)} = k(k-1)$.

4.40. Find a function y_k for which $\Delta y_k = k(k-1)(k-2)$.

4.41. Find a function y_k for which $\Delta y_k = k^2 = k^{(2)} + k^{(1)}$.

4.42. Find a function y_k for which $\Delta y_k = k^3$.

4.43. Find a function y_k for which $\Delta y_k = 1(k+1)(k+2)$.

Chapter 5

Summation

Summation is the inverse operation to differencing, as integration is to differentiation. An extensive treatment appears in Chapter 17 but two elementary results are presented here.

1. **Telescoping sums** are sums of differences, and we have the simple but useful

$$\sum_{k=0}^{n-1} \Delta y_k = y_n - y_0$$

analogous to the integration of derivatives. Arbitrary sums may be converted to telescoping sums provided the equation $\Delta y_k = z_k$ can be solved for the function y_k. Then

$$\sum_{k=0}^{n-1} z_k = \sum_{k=0}^{n-1} \Delta y_k = y_n - y_0$$

Finite integration is the process of obtaining y_k from

$$\Delta y_k = z_k$$

where z_k is known. Since it obviously follows that

$$y_n = y_0 + \sum_{k=0}^{n-1} z_k$$

finite integration and summation are the same problem. As in integral calculus, however, there are times when explicit finite integrals (not involving Σ) are useful.

2. **Summation by parts** is another major result of summation calculus and involves the formula

$$\sum_{i=0}^{n-1} u\,\Delta v_i - u_n v_n - u_0 v_0 - \sum_{i-0}^{n-1} v\,\Delta u_i$$

which resembles the corresponding integration by parts formula.

Application of this formula involves exchanging one summation for a (presumably) simpler summation. If one of the Σ's is known, the formula serves to determine the other.

Infinite series may also be evaluated in certain cases where the partial sums respond to the telescoping or summation by parts methods.

Solved Problems

5.1. Prove $\sum_{k=0}^{n-1} \Delta y_k = y_n - y_0$.

This is a simple but useful result. Since it involves the summation of differences, it is usually compared with an analogous result of calculus involving the integration of a derivative. First notice that

$$\Delta y_0 = y_1 - y_0$$

$$\Delta y_0 + \Delta y_1 = (y_1 - y_0) + (y_2 - y_1) = y_2 - y_0$$

$$\Delta y_0 + \Delta y_1 + \Delta y_2 = (y_1 - y_0) + (y_2 - y_1) + (y_3 - y_2) = y_3 - y_0$$

which illustrate the sort of *telescoping sums* involved. In general,

$$\sum_{k=0}^{n-1} \Delta y_k = (y_1 - y_0) + (y_2 - y_1) + (y_3 - y_2) + \cdots + (y_n - y_{n-1}) = y_n - y_0$$

all other y values occurring both plus and minus. Viewed in a table of differences, this result looks even simpler. The sum of adjacent differences gives the difference of two entries in the row above.

$$\begin{array}{ccccccccccccccc} y_0 & & \cdot & & \cdot & & \cdot & & \cdot & & \cdot & & \cdot & & y_n \\ & \Delta y_0 & & \Delta y_1 & & \Delta y_2 & & \cdot & & \cdot & & \cdot & & \Delta y_{n-1} & \end{array}$$

Similar results hold elsewhere in the table.

5.2. Prove $1^2 + 2^2 + \cdots + n^2 = \sum_{i=1}^{n} i^2 = \dfrac{n(n+1)(2n+1)}{6}$.

We need a function for which $\Delta y_i = i^2$. This is similar to the integration problem of calculus. In this simple example, the y_i could be found almost by intuition, but even so we apply a method which handles harder problems just as well. First replace i^2 by a combination of factorial polynomials, using Stirling's numbers.

$$\Delta y_i = i^2 = i^{(2)} + i^{(1)}$$

A function having this difference is

$$y_i = \frac{1}{3} i^{(3)} + \frac{1}{2} i^{(2)}$$

as may easily be verified by computing Δy_i. Obtaining y_i from Δy_i is called **finite integration.** The resemblance to the integration of derivatives is obvious. Now rewrite the result of Problem 5.1 as $\sum_{i=1}^{n} \Delta y_i = y_{n+1} - y_1$ and substitute to obtain

$$\sum_{i=1}^{n} i^2 = \left[\frac{1}{3}(n+1)^{(3)} + \frac{1}{2}(n+1)^{(2)}\right] - \left[\frac{1}{3}(1)^{(3)} + \frac{1}{2}(1)^{(2)}\right]$$

$$= \frac{(n+1)n(n-1)}{3} + \frac{(n+1)n}{2} = \frac{n(n+1)(2n+1)}{6}$$

5.3. Evaluate the series $\sum_{i=0}^{\infty} \dfrac{1}{(i+1)(i+2)}$.

By an earlier result $\Delta i^{(-1)} = \dfrac{-1}{(i+1)(i+2)}$. Then, using Problem 4.9 to handle $0^{(-1)}$,

$$S_n = \sum_{i=0}^{n-1} \frac{1}{(i+1)(i+2)} = -\sum_{i=0}^{n-1} \Delta i^{(-1)} = -[n^{(-1)} - 0^{(-1)}] = 1 - \frac{1}{n+1}$$

The series is defined as $\lim S_n$ and is therefore equal to 1.

5.4. Consider two functions defined for the same set of arguments x_k, having values u_k and v_k. Prove

$$\sum_{i=0}^{n-1} u_i \, \Delta v_i = u_n v_n - u_0 v_0 - \sum_{i=0}^{n-1} v_{i+1} \, \Delta u_i$$

This is called **summation by parts** and is analogous to the result of calculus

$$\int_{x_0}^{x_n} u(x) v'(x)\, dx = u(x_n)v(x_n) - u(x_0)v(x_0) - \int_{x_0}^{x_n} v(x) u'(x)\, dx$$

The proof begins with the result of Problem 3.9, slightly rearranged.

$$u_i \, \Delta v_i = \Delta(u_i v_i) - v_{i+1} \, \Delta u_i$$

Sum from $i = 0$ to $i = n - 1$,

$$\sum_{i=0}^{n-1} u_i \, \Delta v_i = \sum_{i=0}^{n-1} \Delta(u_i v_i) - \sum_{i=0}^{n-1} v_{i+1} \, \Delta u_i$$

and then apply Problem 5.1 to the first sum on the right. The required result follows.

5.5. Evaluate the series $\sum_{i=0}^{\infty} iR^i$ where $-1 < R < 1$.

Since $\Delta R^i = R^{i+1} - R^i = R^i(R - 1)$, we may put $u_i = i$ and $v_i = R^i/(R - 1)$ and apply summation by parts. Take the finite sum

$$S_n = \sum_{i=0}^{n-1} iR^i = \sum_{i=0}^{n-1} u_i \, \Delta v_i = n \cdot \frac{R^n}{R-1} - 0 - \sum_{i=0}^{n-1} \frac{R^{i+1}}{R-1}$$

The last sum is *geometric* and responds to an elementary formula, making

$$S_n = \frac{nR^n}{R-1} + \frac{R(1-R^n)}{(1-R)^2}$$

Since nR^n and R^{n+1} both have limit zero, the value of the infinite series is $\lim S_n = R/(1-R)^2$.

5.6. A coin is tossed until heads first shows. A payoff is then made, equal to i dollars if heads first showed on the ith toss (one dollar if heads showed at once on the first toss, two dollars if the first head showed on the second toss, and so on). Probability theory leads to the series

$$1\left(\frac{1}{2}\right) + 2\left(\frac{1}{4}\right) + 3\left(\frac{1}{8}\right) + \cdots = \sum_{i=0}^{\infty} i\left(\frac{1}{2}\right)^i$$

for the average payoff. Use the previous problem to compute this series.

By Problem 5.5 with $R = \frac{1}{2}$, $\sum_{i=0}^{\infty} i(\frac{1}{2})^i = (\frac{1}{2})/(\frac{1}{4}) = 2$ dollars.

5.7. Apply summation by parts to evaluate the series $\sum_{i=0}^{\infty} i^2R^i$.

Putting $u_i = i^2$, $v_i = R^i/(R-1)$ we find $\Delta u_i = 2i + 1$ and so

$$S_n = \sum_{i=0}^{n-1} i^2R^i = \sum_{i=0}^{n-1} u_i \, \Delta v_i = n^2 \frac{R^n}{R-1} - 0 - \sum_{i=0}^{n-1} \frac{R^{i+1}}{R-1}(2i+1)$$

$$= n^2 \frac{R^n}{R-1} - \frac{2R}{R-1} \sum_{i=0}^{n-1} iR^i - \frac{R}{R-1} \sum_{i=0}^{n-1} R^i$$

The first of the two remaining sums was evaluated in Problem 5.5 and the second is geometric. So we come to

$$S_n = \frac{n^2R^n}{R-1} - \frac{2R}{R-1}\left[\frac{nR^n}{R-1} + \frac{R(1-R^n)}{(1-R)^2}\right] - \frac{R}{R-1} \cdot \frac{1-R^n}{1-R}$$

and letting $n \to \infty$ finally achieve $\lim S_n = (R + R^2)/(1-R)^3$.

5.8. A coin is tossed until heads first shows. A payoff is then made, equal to i^2 dollars if heads first showed on the ith toss. Probability theory leads to the series $\sum_{i=0}^{\infty} i^2(\frac{1}{2})^i$ for the average payoff. Evaluate the series.

By Problem 5.7 with $R = \frac{1}{2}$, $\sum_{i=0}^{\infty} i^2(\frac{1}{2})^i = (\frac{1}{2} + \frac{1}{4})/(\frac{1}{8}) = 6$ dollars.

Supplementary Problems

5.9. Use finite integration (as in Problem 5.2) to prove $\sum_{i=1}^{n} i = 1+2+\cdots+n = \frac{n(n+1)}{2}$.

5.10. Evaluate $\sum_{i=1}^{n} i^3$ by finite integration.

5.11. Show that $\sum_{i=0}^{n-1} A^i = \frac{A^n - 1}{A - 1}$ by using finite integration. (See Problem 3.21.) This is, of course, the *geometric* sum of elementary algebra.

5.12. Show that $\sum_{i=1}^{n-1} \binom{i}{k} = \binom{n}{k+1} - \binom{1}{k+1}$.

5.13. Evaluate by finite integration: $\sum_{i=0}^{\infty} \frac{1}{(i+1)(i+2)(i+3)}$.

5.14. Evaluate $\sum_{i=1}^{\infty} \frac{1}{i(i+2)}$.

5.15. Evaluate $\sum_{i=0}^{\infty} i^3 R^i$ for $-1 < R < 1$.

5.16. Alter Problem 5.8 so that the payoff is i^3. Use Problem 5.15 to evaluate the average payoff, which is $\sum_{i=0}^{\infty} i^3 (\frac{1}{2})^i$.

5.17. Alter Problem 5.8 so that the payoff is $+1$ when i is even and -1 when i is odd. The average payoff is $\sum_{i=1}^{\infty} (-1)^i (\frac{1}{2})^i$. Evaluate the series.

5.18. Evaluate $\sum_{i=1}^{n} \log\left(1 + \frac{1}{i}\right)$.

5.19. Evaluate $\sum_{i=1}^{N} i^n$ in terms of Stirling's numbers.

5.20. Evaluate $\sum_{i=1}^{\infty} \left[\frac{1}{i(i+n)}\right]$.

5.21. Evaluate $\sum_{i=0}^{\infty} i^n R^i$.

5.22. Express a finite integral of $\Delta y_k = 1/k$ in the form of a summation, avoiding $k = 0$.

5.23. Express a finite integral of $\Delta y_k = \log k$ in the form of a summation.

Chapter 6

The Newton Formula

The collocation polynomial can now be expressed in terms of finite differences and factorial polynomials. The summation formula

$$y_k = \sum_{i=0}^{k} \binom{k}{i} \Delta^i y_0$$

is proved first and leads directly to the **Newton formula** for the collocation polynomial, which can be written as

$$p_k = \sum_{i=0}^{n} \binom{k}{i} \Delta^i y_0$$

An alternative form of the Newton formula, in terms of the argument x_k, may be obtained using $x_k = x_0 + kh$, and proves to be

$$p(x_k) = y_0 + \left(\frac{\Delta y_0}{h}\right)(x_k - x_0) + \left(\frac{\Delta^2 y_0}{2!\,h^2}\right)(x_k - x_0)(x_k - x_1) + \cdots + \left(\frac{\Delta^n y_0}{n!\,h^n}\right)(x_k - x_0)\cdots(x_k - x_{n-1})$$

The points of collocation are $x_0, \ldots, x_n$. At these points (arguments) our polynomial takes the prescribed values $y_0, \ldots, y_n$.

Solved Problems

6.1. Prove that

$$y_1 = y_0 + \Delta y_0 \qquad y_2 = y_0 + 2\Delta y_0 + \Delta^2 y_0 \qquad y_3 = y_0 + 3\Delta y_0 + 3\Delta^2 y_0 + \Delta^3 y_0$$

and infer similar results such as

$$\Delta y_2 = \Delta y_0 + 2\Delta^2 y_0 + \Delta^3 y_0 \qquad \Delta^2 y_2 = \Delta^2 y_0 + 2\Delta^3 y_0 + \Delta^4 y_0$$

This is merely a preliminary to a more general result. The first result is obvious. For the second, with one eye on Table 6.1,

$$y_2 = y_1 + \Delta y_1 = (y_0 + \Delta y_0) + (\Delta y_0 + \Delta^2 y_0)$$

leading at once to the required result. Notice that this expresses y_2 in terms of entries in the top diagonal of Table 6.1. Notice also that almost identical computations produce

$$\Delta y_2 = \Delta y_0 + 2\Delta^2 y_0 + \Delta^3 y_0 \qquad \Delta^2 y_2 = \Delta^2 y_0 + 2\Delta^3 y_0 + \Delta^4 y_0$$

etc., expressing the entries on the "y_2 diagonal" in terms of those on the top diagonal. Finally,

$$y_3 = y_2 + \Delta y_2 = (y_0 + 2\Delta y_0 + \Delta^2 y_0) + (\Delta y_0 + 2\Delta^2 y_0 + \Delta^3 y_0)$$

leading quickly to the third required result. Similar expressions for Δy_3, $\Delta^2 y_3$, etc., can be written by simply raising the upper index on each Δ.

Table 6.1

x	y	Δy	$\Delta^2 y$	$\Delta^3 y$	$\Delta^4 y$
x_0	y_0				
		Δy_0			
x_1	y_1		$\Delta^2 y_0$		
		Δy_1		$\Delta^3 y_0$	
x_2	y_2		$\Delta^2 y_1$		$\Delta^4 y_0$
		Δy_2		$\Delta^3 y_1$	
x_3	y_3		$\Delta^2 y_2$		
		Δy_3			
x_4	y_4				

6.2. Prove that for any positive integer k, $y_k = \sum_{i=0}^{k} \binom{k}{i} \Delta^i y_0$. (Here $\Delta^0 y_0$ means simply y_0.)

The proof will be by induction. For $k = 1$, 2, and 3, see Problem 6.1. Assume the result true when k is some particular integer p.

$$y_p = \sum_{i=0}^{p} \binom{p}{i} \Delta^i y_0$$

Then, as suggested in the previous problem, the definition of our various differences makes

$$\Delta y_p = \sum_{i=0}^{p} \binom{p}{i} \Delta^{i+1} y_0$$

also true. We now find

$$\begin{aligned} y_{p+1} = y_p + \Delta y_p &= \sum_{j=0}^{p} \binom{p}{j} \Delta^j y_0 + \sum_{j=1}^{p+1} \binom{p}{j-1} \Delta^j y_0 \\ &= y_0 + \sum_{j=1}^{p} \left[\binom{p}{j} + \binom{p}{j-1} \right] \Delta^j y_0 + \Delta^{p+1} y_0 \\ &= y_0 + \sum_{j=1}^{p} \binom{p+1}{j} \Delta^j y_0 + \Delta^{p+1} y_0 = \sum_{j=0}^{p+1} \binom{p+1}{j} \Delta^j y_0 \end{aligned}$$

Problem 4.5 was used in the third step. The summation index may now be changed from j to i if desired. Thus our result is established when k is the integer $p + 1$, completing the induction.

6.3. Prove that the polynomial of degree n,

$$\begin{aligned} p_k &= y_0 + k\ \Delta y_0 + \frac{1}{2!} k^{(2)} \Delta^2 y_0 + \cdots + \frac{1}{n!} k^{(n)} \Delta^n y_0 \\ &= \sum_{i=0}^{n} \frac{1}{i!} k^{(i)} \Delta^i y_0 = \sum_{i=0}^{n} \binom{k}{i} \Delta^i y_0 \end{aligned}$$

takes the values $p_k = y_k$ for $k = 0, 1, \ldots, n$. This is Newton's formula.

Notice first that when k is 0 only the y_0 term on the right contributes, all others being 0. When k is 1 only the first two terms on the right contribute, all others being 0. When k is 2 only the first three terms contribute. Thus, using Problem 6.1,

$$p_0 = y_0 \qquad p_1 = y_0 + \Delta y_0 = y_1 \qquad p_2 = y_0 + 2\Delta y_0 + \Delta^2 y_0 = y_2$$

and the nature of our proof is indicated. In general, if k is any integer from 0 to n, then $k^{(i)}$ will be 0 for $i > k$. (It will contain the factor $k - k$.) The sum abbreviates to

$$p_k = \sum_{i=0}^{k} \frac{1}{i!} k^{(i)} \Delta^i y_0$$

and by Problem 6.2 this reduces to y_k. The polynomial of this problem therefore takes the same values as our y_k function for the integer arguments $k = 0, \ldots, n$. (The polynomial is, however, defined for any argument k.)

6.4. Express the result of Problem 6.3 in terms of the argument x_k, where $x_k = x_0 + kh$.

Notice first that

$$k = \frac{x_k - x_0}{h} \qquad k - 1 = \frac{x_{k-1} - x_0}{h} = \frac{x_k - x_1}{h} \qquad k - 2 = \frac{x_{k-2} - x_0}{h} = \frac{x_k - x_2}{h}$$

and so on. Using the symbol $p(x_k)$ instead of p_k, we now find

$$p(x_k) = y_0 + \frac{\Delta y_0}{h}(x_k - x_0) + \frac{\Delta^2 y_0}{2!\,h^2}(x_k - x_0)(x_k - x_1) + \cdots + \frac{\Delta^n y_0}{n!\,h^n}(x_k - x_0)\cdots(x_k - x_{n-1})$$

which is Newton's formula in its alternative form.

6.5. Find the polynomial of degree three which takes the four values listed in the y_k column below at the corresponding arguments x_k.

The various differences needed appear in the remaining columns of Table 6.2.

Table 6.2

k	x_k	y_k	Δy_k	$\Delta^2 y_k$	$\Delta^3 y_k$
0	④	①			
			②		
1	⑥	3		③	
			5		④
2	⑧	8		7	
			12		
3	10	20			

Substituting the circled numbers in their places in Newton's formula,

$$p(x_k) = 1 + \frac{2}{2}(x_k - 4) + \frac{3}{8}(x_k - 4)(x_k - 6) + \frac{4}{48}(x_k - 4)(x_k - 6)(x_k - 8)$$

which can be simplified to

$$p(x_k) = \frac{1}{24}(2x_k^3 - 27x_k^2 + 142x_k - 240)$$

though often in applications the first form is preferable.

6.6 Express the polynomial of Problem 6.5 in terms of the argument k.

Directly from Problem 6.3,

$$p_k = 1 + 2k + \frac{3}{2}k^{(2)} + \frac{4}{6}k^{(3)}$$

which is a convenient form for computing p_k values and so could be left as is. It can also be rearranged into

$$p_k = 1 + \frac{11}{6}k - \frac{1}{2}k^2 + \frac{2}{3}k^3$$

6.7. Apply Newton's formula to find a polynomial of degree four or less which takes the y_k values of Table 6.3.

The needed differences are circled. Substituting the circled entries into their places in Newton's formula,

$$p_k = 1 - 2k + \frac{4}{2}k^{(2)} - \frac{8}{6}k^{(3)} + \frac{16}{24}k^{(4)}$$

which is also

$$p_k = \frac{1}{3}(2k^4 - 16k^3 + 40k^2 - 32k + 3)$$

Table 6.3

k	x_k	y_k	Δ	Δ^2	Δ^3	Δ^4
0	1	(1)				
			(−2)			
1	2	−1		(4)		
			2		(−8)	
2	3	1		−4		(16)
			−2		8	
3	4	−1		4		
			2			
4	5	1				

Since $k = x_k - 1$, this result can also be written as

$$p(x_k) = \frac{1}{3}(2x_k^4 - 24x_k^3 + 100x_k^2 - 168x_k + 93)$$

Supplementary Problems

6.8. Find a polynomial of degree four which takes these values.

x_k	2	4	6	8	10
y_k	0	0	1	0	0

6.9. Find a polynomial of degree two which takes these values.

$k = x_k$	0	1	2	3	4	5	6	7
y_k	1	2	4	7	11	16	22	29

6.10. Find a polynomial of degree three which takes these values.

x_k	3	4	5	6
y_k	6	24	60	120

6.11. Find a polynomial of degree five which takes these values.

$k = x_k$	0	1	2	3	4	5
y_k	0	0	1	1	0	0

6.12. Find the cubic polynomial which includes these values.

$k = x_k$	0	1	2	3	4	5
y_k	1	2	4	8	15	26

(See also Problem 3.12.)

6.13. Expressing a polynomial of degree n in the form

$$p_k = a_0 + a_1 k^{(1)} + a_2 k^{(2)} + \cdots + a_n k^{(n)}$$

calculate $\Delta p_k, \Delta^2 p_k, \ldots, \Delta^n p_k$. Then show that the requirement

$$p_k = y_k \qquad k = 0, \ldots, n$$

leads to $\Delta p_0 = \Delta y_0$, $\Delta^2 p_0 = \Delta^2 y_0$, etc. Next deduce

$$a_0 = y_0 \qquad a_1 = \Delta y_0 \qquad a_2 = \frac{1}{2}\Delta^2 y_0 \qquad \cdots \qquad a_n = \frac{1}{n!}\Delta^n y_0$$

and substitute these numbers to obtain once again Newton's formula.

6.14. Find a quadratic polynomial which collocates with $y(x) = x^4$ at $x = 0, 1, 2$.

6.15. Find a cubic polynomial which collocates with $y(x) = \sin(\pi x/2)$ at $x = 0, 1, 2, 3$. Compare the two functions at $x = 4$. Compare them at $x = 5$.

6.16. Is there a polynomial of degree four which collocates with $y(x) = \sin(\pi x/2)$ at $x = 0, 1, 2, 3, 4$?

6.17. Is there a polynomial of degree two which collocates with $y(x) = x^3$ at $x = -1, 0, 1$?

6.18. Find a polynomial of degree four which collocates with $y(x) = |x|$ at $x = -2, -1, 0, 1, 2$. Where is the polynomial greater than $y(x)$, and where less?

6.19. Find a polynomial of degree two which collocates with $y(x) = \sqrt{x}$ at $x = 0, 1, 4$. Why is Newton's formula not applicable?

6.20. Find a solution of $\Delta^3 y_k = 1$ for all integers k with $y_0 = \Delta y_0 = \Delta^2 y_0 = 0$.

Chapter 7

Operators and Collocation Polynomials

OPERATORS

Operators are used here and there in numerical analysis, in particular for simplifying the development of complicated formulas. Some of the most interesting applications are carried out in a spirit of optimism, without excessive attention to logical precision, the results subject to verification by other methods, or checked experimentally.

A number of the formulas to be derived in this chapter are, in part, of historical interest, providing a view of the numerical priorities of an earlier time. The attached names, such as Newton and Gauss, indicate their importance in those times. Changes in computing hardware have reduced their range of application, a point that will be repeated in Chapter 12 where certain classic applications will be offered.

The specific operator concepts now to be used are these:

1. The **operator** Δ is defined by

$$\Delta y_k = y_{k+1} - y_k$$

We now think of Δ as an operator which when offered y_k as an input produces $y_{k+1} - y_k$ as an output, for all k values under consideration.

$$\boxed{y_k} \longrightarrow \boxed{\Delta} \longrightarrow \boxed{y_{k+1} - y_k}$$

The analogy between operator and an algorithm (as described in Chapter 1) is apparent.

2. The **operator** E is defined by

$$Ey_k = y_{k+1}$$

Here the input to the operator is again y_k. The output is y_{k+1}.

$$\boxed{y_k} \longrightarrow \boxed{E} \longrightarrow \boxed{y_{k+1}}$$

Both Δ and E have the linearity property, that is,

$$\Delta(C_1 y_k + C_2 z_k) = C_1 \Delta y_k + C_2 \Delta z_k$$
$$E(C_1 y_k + C_2 z_k) = C_1 E y_k + C_2 E z_k$$

where C_1 and C_2 are any constants (independent of k). All the operators to be introduced will have this property.

3. **Linear combinations of operators.** Consider two operators, call them L_1 and L_2, which produce outputs $L_1 y_k$ and $L_2 y_k$ from the input y_k. Then the sum of these operators is defined as the operator which outputs $L_1 y_k + L_2 y_k$.

$$\boxed{y_k} \longrightarrow \boxed{L_1 + L_2} \longrightarrow \boxed{L_1 y_k + L_2 y_k}$$

A similar definition introduces the difference of two operators.

More generally, if C_1 and C_2 are constants (independent of k) the operator $C_1L_1 + C_2L_2$ produces the output $C_1L_1y_k + C_2L_2y_k$.

$$\boxed{y_k} \longrightarrow \boxed{C_1L_1 + C_2L_2} \longrightarrow \boxed{C_1L_1y_k + C_2L_2y_k}$$

4. The **product of operators** L_1 and L_2 is defined as the operator which outputs $L_1L_2y_k$. A diagram makes this clearer.

$$\boxed{y_k} \longrightarrow \boxed{L_2} \longrightarrow \boxed{L_2y_k} \longrightarrow \boxed{L_1} \longrightarrow \boxed{L_1L_2y_k}$$

The operator L_1 is applied to the output produced by L_2. The center three parts of the diagram together represent the operator L_1L_2.

$$\boxed{y_k} \longrightarrow \boxed{L_1L_2} \longrightarrow \boxed{L_1L_2y_k}$$

With this definition of product, numbers such as the C_1 and C_2 above may also be thought of as operators. For instance, C being any number, the operator C performs a multiplication by the number C.

5. **Equality of operators.** Two operators L_1 and L_2 are called equal if they produce identical outputs for all inputs under consideration. In symbols,

$$L_1 = L_2 \qquad \text{if} \qquad L_1y_k = L_2y_k$$

for all arguments k under consideration. With this definition a comparison of outputs shows at once that for any operators L_1, L_2, and L_3,

$$L_1 + L_2 = L_2 + L_1$$
$$L_1 + (L_2 + L_3) = (L_1 + L_2) + L_3$$
$$L_1(L_2L_3) = (L_1L_2)L_3$$
$$L_1(L_2 + L_3) = L_1L_2 + L_1L_3$$

but the commutative law of multiplication is not always true:

$$L_1L_2 \neq L_2L_1$$

If either operator is a number C, however, equality is obvious from a comparison of outputs,

$$CL_1 = L_1C$$

6. **Inverse operators.** For many of the other operators we shall use, commutativity will also be true. As a special case, L_1 and L_2 are called inverse operators if

$$L_1L_2 = L_2L_1 = 1$$

In such a case we use the symbols

$$L_1 = L_2^{-1} = \frac{1}{L_2} \qquad L_2 = L_1^{-1} = \frac{1}{L_1}$$

The operator 1 is known as the **identity operator** and it is easy to see that it makes $1 \cdot L = L \cdot 1$ for any operator L.

7. **Simple equations relating Δ and E** include, among others,

$$E = 1 + \Delta \qquad \Delta^2 = E^2 - 2E + 1$$
$$E\Delta = \Delta E \qquad \Delta^3 = E^3 - 3E^2 + 3E - 1$$

Two related theorems, already proved earlier by other means, appear as follows in operator symbolism:

$$\Delta^k = \sum_{i=0}^{k} (-1)^i \binom{k}{i} E^{k-i} \qquad E^k = \sum_{i=0}^{k} \binom{k}{i} \Delta^i$$

8. The **backward difference operator** ∇ is defined by

$$\nabla y_k = y_k - y_{k-1}$$

and it is then easy to verify that

$$\nabla E = E\nabla = \Delta$$

The relationship between ∇ and E^{-1} proves to be

$$E^{-1} = 1 - \nabla$$

and leads to the expansion

$$y_k = y_0 + \sum_{i=1}^{-k} \frac{k(k+1)\cdots(k+i-1)}{i!} \nabla^i y_0$$

for negative integers k.

9. The **central difference operator** is defined by

$$\delta = E^{1/2} - E^{-1/2}$$

It follows that $\delta E^{1/2} = \Delta$. In spite of the fractional arguments this is a heavily used operator. It is closely related to the following operator.

10. The **averaging operator** is defined by

$$\mu = \frac{1}{2}(E^{1/2} + E^{-1/2})$$

and is the principal mechanism by which fractional arguments can be eliminated from central difference operations.

COLLOCATION POLYNOMIALS

The collocation polynomial can now be expressed in a variety of alternative forms, all equivalent to the Newton formula of Chapter 6, but each suited to somewhat different circumstances. We discuss the following, which find use beginning with Chapter 12.

1. **Newton's backward formula**

$$p_k = y_0 + k\,\nabla y_0 + \frac{k(k+1)}{2!}\nabla^2 y_0 + \cdots + \frac{k\cdots(k+n-1)}{n!}\nabla^n y_0$$

represents the collocation polynomial which takes the values y_k for $k = 0, -1, \ldots, -n$.

2. The **Gauss forward formula** may be obtained by developing the relationship between E and δ and reads

$$p_k = y_0 + \sum_{i=1}^{n} \left[\binom{k+i-1}{2i-1} \delta^{2i-1} y_{1/2} + \binom{k+i-1}{2i} \delta^{2i} y_0 \right]$$

if the polynomial is of even degree $2n$ and collocation is at $k=-n,\ldots,n$. It becomes

$$p_k=\sum_{i=0}^{n}\left[\binom{k+i-1}{2i}\delta^{2i}y_0+\binom{k+1}{2i+1}\delta^{2i+1}y_{1/2}\right]$$

if the polynomial is of odd degree $2n+1$ and collocation is at $k=-n,\ldots,n+1$.

3. The **Gauss backward formula** may be derived in a similar way. For even degree it takes the form

$$p_k=y_0+\sum_{i=1}^{n}\left[\binom{k+i-1}{2i-1}\delta^{2i-1}y_{-1/2}+\binom{k+i}{2i}\delta^{2i}y_0\right]$$

with collocation again at $k=-n,\ldots,n$. One principal use of the two formulas of Gauss is in deriving Stirling's formula.

4. **Stirling's formula** is one of the most heavily applied forms of the collocation polynomial. It reads

$$p_k=y_0+\binom{k}{1}\delta\mu y_0+\left(\frac{k}{2}\right)\binom{k}{1}\delta^2y_0+\binom{k+1}{3}\delta^3\mu y_0$$
$$+\left(\frac{k}{4}\right)\binom{k+1}{3}\delta^4y_0+\cdots+\binom{k+n-1}{2n-1}\delta^{2n-1}\mu y_0+\left(\frac{k}{2n}\right)\binom{k+n-1}{2n-1}\delta^{2n}y_0$$

and is a very popular formula. Needless to say, collocation is at $k=-n,\ldots,n$.

5. **Everett's formula** takes the form

$$p_k=\binom{k}{1}y_1+\binom{k+1}{3}\delta^2y_1+\binom{k+2}{5}\delta^4y_1+\cdots+\binom{k+n}{2n+1}\delta^{2n}y_1$$
$$-\binom{k-1}{1}y_0-\binom{k}{3}\delta^2y_0-\binom{k+1}{5}\delta^4y_0-\cdots-\binom{k+n-1}{2n+1}\delta^{2n}y_0$$

and may be obtained by rearranging the ingredients of the Gauss forward formula of odd degree. Collocation is at $k=-n,\ldots,n+1$. Note that only even differences appear.

6. **Bessel's formula** is a rearrangement of Everett's and can be written as

$$p_k=\mu y_{1/2}+\left(k-\frac{1}{2}\right)\delta y_{1/2}+\binom{k}{2}\mu\delta^2y_{1/2}+\frac{1}{3}\left(k-\frac{1}{2}\right)\binom{k}{2}\delta^3y_{1/2}$$
$$+\cdots+\binom{k+n-1}{2n}\mu\delta^{2n}y_{1/2}+\left(\frac{1}{2n+1}\right)\left(k-\frac{1}{2}\right)\binom{k+n-1}{2n}\delta^{2n+1}y_{1/2}$$

Solved Problems

7.1. Prove $E=1+\Delta$.

By definition of E, $Ey_k=y_{k+1}$; and by definition of $1+\Delta$,

$$(1+\Delta)=1\cdot y_k+\Delta y_k=y_k+(y_{k+1}-y_k)=y_{k+1}$$

Having identical outputs for all arguments k, the operators E and $1+\Delta$ are equal. This result can also be written as $\Delta=E-1$.

7.2. Prove $E\Delta=\Delta E$.

$$E\Delta y_k=E(y_{k+1}-y_k)=y_{k+2}-y_{k+1}\qquad\text{and}\qquad\Delta Ey_k=\Delta y_{k+1}=y_{k+2}-y_{k+1}$$

The equality of outputs makes the operators equal. This is an example in which the commutative law of multiplication is true.

7.3. Prove $\Delta^2 = E^2 - 2E + 1$.

Using various operator properties,

$$\Delta^2 = (E-1)(E-1) = E^2 - 1 \cdot E - E \cdot 1 + 1 = E^2 - 2E + 1$$

7.4. Apply the binomial theorem to prove $\Delta^k y_0 = \sum_{i=0}^{k} (-1)^i \binom{k}{i} y_{k-i}$.

The binomial theorem, $(a+b)^k = \sum_{i=0}^{k} \binom{k}{i} a^{k-i} b^i$, is valid as long as a and b (and therefore $a+b$) commute in multiplication. In the present situation these elements will be E and -1 and these do commute. Thus,

$$\Delta^k = (E-1)^k = \sum_{i=0}^{k} (-1)^i \binom{k}{i} E^{k-i}$$

Noticing that $Ey_0 = y_1$, $E^2 y_0 = y_2$, etc., we have finally

$$\Delta^k y_0 = \sum_{i=0}^{k} (-1)^i \binom{k}{i} y_{k-i}$$

which duplicates the result of Problem 3.5.

7.5. Prove $y_k = \sum_{i=0}^{k} \binom{k}{i} \Delta^i y_0$.

Since $E = 1 + \Delta$, the binomial theorem produces $E^k = (1+\Delta)^k = \sum_{i=0}^{k} \binom{k}{i} \Delta^i$. Applying this operator to y_0, and using the fact that $E^k y_0 = y_k$, produces the required result at once. Note that this duplicates Problem 6.2.

7.6. The **backward difference** is defined by $\nabla y_k = y_k - y_{k-1} = \Delta y_{k-1}$. Clearly it involves assigning a new symbol to $y_k - y_{k-1}$. Show that $\nabla E = E \nabla = \Delta$, $E^{-1} = 1 - \nabla$.

$$\nabla E y_k = \nabla y_{k+1} = y_{k+1} - y_k = \Delta y_k \qquad E \nabla y_k = E(y_k - y_{k-1}) = y_{k+1} - y_k = \Delta y_k$$

Since these are true for all arguments k, we have $\nabla E = E \nabla = \Delta = E - 1$.

Using the symbol E^{-1} for the operator defined by $E^{-1} y_k = y_{k-1}$, we see that $EE^{-1} y_k$ and $E^{-1} E y_k$ are both y_k. In operator language this means that these two operators are inverses: $EE^{-1} = E^{-1} E = 1$. Finally, as an exercise with operator calculations,

$$\nabla = E^{-1} E \nabla = E^{-1} \Delta = E^{-1}(E-1) = 1 - E^{-1} \qquad \text{and} \qquad E^{-1} = 1 - \nabla$$

7.7. Backward differences are normally applied only at the bottom of a table, using negative k arguments as shown in Table 7.1. Using the symbols $\nabla^2 y_k = \nabla \nabla y_k$, $\nabla^3 y_k = \nabla \nabla^2 y_k$, etc., show that $\Delta^n y_k = \nabla^n y_{k+n}$.

Since $\Delta = E \nabla$, we have $\Delta^n = (E \nabla)^n$. But E and ∇ commute, so the $2n$ factors on the right side may be rearranged to give $\Delta^n = \nabla^n E^n$. Applying this to y_k, $\Delta^n y_k = \nabla^n E^n y_k = \nabla^n y_{k+n}$.

Table 7.1

k	x	y				
-4	x_{-4}	y_{-4}				
			∇y_{-3}			
-3	x_{-3}	y_{-3}		$\nabla^2 y_{-2}$		
			∇y_{-2}		$\nabla^3 y_{-1}$	
-2	x_{-2}	y_{-2}		$\nabla^2 y_{-1}$		$\nabla^4 y_0$
			∇y_{-1}		$\nabla^3 y_0$	
-1	x_{-1}	y_{-1}		$\nabla^2 y_0$		
			∇y_0			
0	x_0	y_0				

7.8. Prove that

$$y_{-1} = y_0 - \nabla y_0 \qquad y_{-2} = y_0 - 2\nabla y_0 + \nabla^2 y_0 \qquad y_{-3} = y_0 - 3\nabla y_0 + 3\nabla^2 y_0 - \nabla^3 y_0$$

and that in general for k a negative integer, $y_k = y_0 + \sum_{i=1}^{-k} \frac{k(k+1)\cdots(k+i-1)}{i!}\nabla^i y_0$.

Take the general case at once: $y_k = E^k y_0 = (E^{-1})^{-k} y_0 = (1-\nabla)^{-k} y_0$. With k a negative integer the binomial theorem applies, making

$$y_k = \sum_{i=0}^{-k} (-1)^i \binom{-k}{i} \nabla^i y_0 = y_0 + \sum_{i=1}^{-k} (-1)^i \frac{(-k)(-k-1)\cdots(-k-i+1)}{i!}\nabla^i y_0$$
$$= y_0 + \sum_{i=1}^{-k} \frac{k(k+1)\cdots(k+i-1)}{i!}\nabla^i y_0$$

The special cases now follow for $k = -1, -2, -3$ by writing out the sum.

7.9. Prove that the polynomial of degree n which has values defined by the following formula reduces to $p_k = y_k$ when $k = 0, -1, \ldots, -n$. (This is Newton's backward difference formula.)

$$p_k = y_0 + k\,\nabla y_0 + \frac{k(k+1)}{2!}\nabla^2 y_0 + \cdots + \frac{k\cdots(k+n-1)}{n!}\nabla^n y_0$$
$$= y_0 + \sum_{i=1}^{n} \frac{k(k+1)\cdots(k+i-1)}{i!}\nabla^i y_0$$

The proof is very much like the one in Problem 6.3. When k is 0, only the first term on the right side contributes. When k is -1, only the first two terms contribute, all others being zero. In general, if k is any integer from 0 to $-n$, then $k(k+1)\cdots(k+i-1)$ will be 0 for $i > -k$. The sum abbreviates to

$$p_k = y_0 + \sum_{i=1}^{-k} \frac{k(k+1)\cdots(k+i-1)}{i!}\nabla^i y_0$$

and by Problem 7.8 this reduces to y_k. The polynomial of this problem therefore agrees with our y_k function for $k = 0, -1, \ldots, -n$.

7.10. Find the polynomial of degree three which takes the four values listed as y_k in Table 7.2 at the corresponding x_k arguments.

The differences needed appear in the remaining columns of Table 7.2.

Table 7.2

k	x_k	y_k	∇y_k	$\nabla^2 y_k$	$\nabla^3 y_k$
−3	4	1			
			2		
−2	6	3		3	
			5		(4)
−1	8	8		(7)	
			(12)		
0	10	(20)			

Substituting the circled numbers in their places in Newton's backward difference formula,

$$p_k = 20 + 12k + \frac{7}{2}k(k+1) + \frac{4}{6}k(k+1)(k+2)$$

Notice that except for the arguments k this data is the same as that of Problem 6.5. Eliminating k by the relation $x_k = 10 + 2k$, the formula found in that problem

$$p(x_k) = \frac{1}{24}(2x_k^3 - 27x_k^2 + 142x_k - 240)$$

is again obtained. Newton's two formulas are simply rearrangements of the same polynomial. Other rearrangements now follow.

7.11. The central difference operator δ is defined by $\delta = E^{1/2} - E^{-1/2}$ so that $\delta y_{1/2} = y_1 - y_0 = \Delta y_0 = \nabla y_1$, and so on. Observe that $E^{1/2}$ and $E^{-1/2}$ are inverses and that $(E^{1/2})^2 = E$, $(E^{-1/2})^2 = E^{-1}$. Show that $\Delta^n y_k = \delta^n y_{k+n/2}$.

From the definition of δ, we have $\delta E^{1/2} = E - 1 = \Delta$ and $\Delta^n = \delta^n E^{n/2}$. Applied to y_k, this produces the required result.

7.12. In δ notation, the usual difference table may be rewritten as in Table 7.3.

Table 7.3

k	y_k	δ	δ^2	δ^3	δ^4
−2	y_{-2}				
		$\delta y_{-3/2}$			
−1	y_{-1}		$\delta^2 y_{-1}$		
		$\delta y_{-1/2}$		$\delta^3 y_{-1/2}$	
0	y_0		$\delta^2 y_0$		$\delta^4 y_0$
		$\delta y_{1/2}$		$\delta^3 y_{1/2}$	
1	y_1		$\delta^2 y_1$		
		$\delta y_{3/2}$			
2	y_2				

Express $\delta y_{1/2}$, $\delta^2 y_0$, $\delta^2 y_{1/2}$, and $\delta^4 y_0$ using the Δ operator.

By Problem 7.11, $\delta y_{1/2} = \Delta y_0$, $\delta^2 y_0 = \Delta^2 y_{-1}$, $\delta^3 y_{1/2} = \Delta^3 y_{-1}$, $\delta^4 y_0 = \Delta^4 y_{-2}$.

7.13. The **averaging operator** μ is defined by $\mu = \frac{1}{2}(E^{1/2} + E^{-1/2})$ so that $\mu y_{1/2} = \frac{1}{2}(y_1 + y_0)$, and so on. Prove $\mu^2 = 1 + \frac{1}{4}\delta^2$.

First we compute $\delta^2 = E - 2 + E^{-1}$. Then $\mu^2 = \frac{1}{4}(E + 2 + E^{-1}) = \frac{1}{4}(\delta^2 + 4) = 1 + \frac{1}{4}\delta^2$.

7.14. Verify the following for the indicated arguments k:

$k = 0, 1$ $\qquad y_k = y_0 + \binom{k}{1}\delta y_{1/2}$

$k = -1, 0, 1$ $\qquad y_k = y_0 + \binom{k}{1}\delta y_{1/2} + \binom{k}{2}\delta^2 y_0$

$k = -1, 0, 1, 2$ $\qquad y_k = y_0 + \binom{k}{1}\delta y_{1/2} + \binom{k}{2}\delta^2 y_0 + \binom{k+1}{3}\delta^3 y_{1/2}$

$k = -2, -1, 0, 1, 2$ $\qquad y_k = y_0 + \binom{k}{1}\delta y_{1/2} + \binom{k}{2}\delta^2 y_0 + \binom{k+1}{3}\delta^3 y_{1/2} + \binom{k+1}{4}\delta^4 y_0$

For $k = 0$ only the y_0 terms on the right contribute. When $k = 1$ all right sides correspond to the operator

$$1 + \delta E^{1/2} = 1 + (E - 1) = E$$

which does produce y_1. For $k = -1$ the last three formulas lead to

$$1 - \delta E^{1/2} + \delta^2 = 1 - (E - 1) + (E - 2 + E^{-1}) = E^{-1}$$

which produces y_{-1}. When $k = 2$ the last two formulas bring

$$1 + 2\delta E^{1/2} + \delta^2 + \delta^3 E^{1/2} = 1 + 2(E - 1) + (E - 2 + E^{-1})(1 + E - 1) = E^2$$

producing y_2. Finally when $k = -2$ the last formula involves

$$1 - 2\delta E^{1/2} + 3\delta^2 - \delta^3 E^{1/2} + \delta^4 = 1 - 2(E - 1) + (E - 2 + E^{-1})[3 - (E - 1) + (E - 2 + E^{-1})] = E^{-2}$$

leading to y_{-2}.

The formulas of this problem generalize to form the **Gauss forward formula**. It represents a polynomial either of degree $2n$

$$p_k = y_0 + \sum_{i=1}^{n}\left[\binom{k+i-1}{2i-1}\delta^{2i-1}y_{1/2} + \binom{k+i-1}{2i}\delta^{2i}y_0\right]$$

taking the values $p_k = y_k$ for $k = -n, \ldots, n$, or of degree $2n + 1$

$$p_k = \sum_{i=0}^{n}\left[\binom{k+i-1}{2i}\delta^{2i}y_0 + \binom{k+i}{2i+1}\delta^{2i+1}y_{1/2}\right]$$

taking the values $p_k = y_k$ for $k = -n, \ldots, n + 1$. (In special cases the degree may be lower.)

7.15. Apply Gauss' formula with $n = 2$ to find a polynomial of degree four or less which takes the y_k values in Table 7.4.

The differences needed are listed as usual. This resembles a function used in illustrating the two Newton formulas, with a shift in the argument k and an extra number pair added at the top. Since the fourth difference is 0 in this example, we anticipate a polynomial of degree three. Substituting the circled entries into their places in Gauss' formula,

$$p_k = 3 + 5k + \frac{3}{2}k(k - 1) + \frac{4}{6}(k + 1)k(k - 1)$$

If k is eliminated by the relation $x_k = 6 + 2k$, the cubic already found twice before appears once again.

Table 7.4

k	x_k	y_k				
−2	2	−2				
			3			
−1	4	1		−1		
			2		4	
0	6	(3)		(3)		(0)
			(5)		(4)	
1	8	8		7		
			12			
2	10	20				

7.16. Apply Gauss' forward formula to find a polynomial of degree four or less which takes the y_k values in Table 7.5.

The needed differences are circled.

Table 7.5

k	x_k	y_k				
−2	1	1				
			−2			
−1	2	−1		4		
			2		−8	
0	3	(1)		(−4)		(16)
			(−2)		(8)	
1	4	−1		4		
			2			
2	5	1				

Substituting into their places in the Gauss formula,

$$p_k = 1 - 2k - 4\frac{k(k-1)}{2} + 8\frac{(k+1)k(k-1)}{6} + 16\frac{(k+1)k(k-1)(k-2)}{24}$$

which simplifies to

$$p_k = \frac{1}{3}(2k^4 - 8k^2 + 3)$$

Since $k = x_k - 3$, this result can also be written as

$$p(x_k) = \frac{1}{3}(2x_k^4 - 24x_k^3 + 100x_k^2 - 168x_k + 93)$$

agreeing, of course, with the polynomial found earlier by Newton's formula.

7.17. Verify that, for $k = -1, 0, 1$,

$$y_k = y_0 + \binom{k}{1}\delta y_{-1/2} + \binom{k+1}{2}\delta^2 y_0$$

and, for $k = -2, -1, 0, 1, 2$,

$$y_k = y_0 + \binom{k}{1}\delta y_{-1/2} + \binom{k+1}{2}\delta^2 y_0 + \binom{k+1}{3}\delta^3 y_{-1/2} + \binom{k+2}{4}\delta^4 y_0$$

For $k=0$, only the y_0 terms on the right contribute. When $k=1$ both formulas involve the operator

$$1+\delta E^{-1/2}+\delta^2=1+(1-E^{-1})+(E-2+E^{-1})=E$$

which does produce y_1. For $k=-1$ both formulas involve

$$1-\delta E^{-1/2}=1-(1-E^{-1})=E^{-1}$$

which does produce y_{-1}. Continuing with the second formula, we find, for $k=2$,

$$1+2\delta E^{-1/2}+3\delta^2+\delta^3E^{-1/2}+\delta^4=1+2(1-E^{-1})+(E-2+E^{-1})(3+1-E^{-1}+E-2+E^{-1})=E^2$$

and, for $k=-2$,

$$1-2\delta E^{-1/2}+\delta^2-\delta^3E^{-1/2}=1-2(1-E^{-1})+(E-2+E^{-1})(1-1+E^{-1})=E^{-2}$$

as required.

The formulas of this problem can be generalized to form the **Gauss backward formula**. It represents the same polynomial as the Gauss forward formula of even order and can be verified as above.

$$p_k=y_0+\sum_{i=1}^{n}\left[\binom{k+i-1}{2i-1}\delta^{2i-1}y_{-1/2}+\binom{k+i}{2i}\delta^{2i}y_0\right]$$

7.18. Prove $\binom{k+i}{2i}+\binom{k+i-1}{2i}=\frac{k}{i}\binom{k+i-1}{2i-1}$.

From the definitions of binomial coefficients,

$$\binom{k+i}{2i}+\binom{k+i-1}{2i}=\binom{k+i-1}{2i-1}[(k+i)+(k-i)]\frac{1}{2i}$$

as required.

7.19. Deduce *Stirling's formula*, given below, from the Gauss formulas.

Adding the Gauss formulas for degree $2n$ term by term, dividing by two, and using Problem 7.18,

$$\begin{aligned}p_k&=y_0+\sum_{i=1}^{n}\left[\binom{k+i-1}{2i-1}\delta^{2i-1}\mu y_0+\frac{k}{2i}\binom{k+i-1}{2i-1}\delta^{2i}y_0\right]\\&=y_0+\binom{k}{1}\delta\mu y_0+\frac{k}{2}\binom{k}{1}\delta^2y_0+\binom{k+1}{3}\delta^3\mu y_0+\frac{k}{4}\binom{k+1}{3}\delta^4y_0\\&\qquad+\cdots+\binom{k+n-1}{2n-1}\delta^{2n-1}\mu y_0+\frac{k}{2n}\binom{k+n-1}{2n-1}\delta^{2n}y_0\end{aligned}$$

This is Stirling's formula.

7.20. Apply Stirling's formula with $n=2$ to find a polynomial of degree four or less which takes the y_k values in Table 7.6.

The differences needed are again listed. Substituting the circled entries into their places in Stirling's formula,

$$p_k=3+\frac{2+5}{2}k+3\frac{k^2}{2}+\frac{4+4}{2}\frac{(k+1)k(k-1)}{6}$$

which is easily found to be a minor rearrangement of the result found by the Gauss forward formula.

7.21. Prove $\binom{k+i-1}{2i}\delta^{2i}y_0+\binom{k+i}{2i+1}\delta^{2i+1}y_{1/2}=\binom{k+i}{2i+1}\delta^{2i}y_1-\binom{k+i-1}{2i+1}\delta^{2i}y_0$.

Table 7.6

k	x_k	y_k	δ	δ^2	δ^3	δ^4
−2	2	−2				
			3			
−1	4	1		−1		
			②		④	
0	6	③		③		⓪
			⑤		④	
1	8	8		7		
			12			
2	10	20				

The left side becomes (using Problem 4.5)

$$\left[\binom{k+i}{2i+1}-\binom{k+i-1}{2i+1}\right]\delta^{2i}y_0+\binom{k+i}{2i+1}\delta^{2i+1}y_{1/2}=\binom{k+i}{2i+1}[\delta^{2i}(1+\delta E^{1/2})y_0]-\binom{k+i-1}{2i+1}\delta^{2i}y_0$$

$$=\binom{k+i}{2i+1}\delta^{2i}y_1-\binom{k+i-1}{2i+1}\delta^{2i}y_0$$

where in the last step we used $1+\delta E^{1/2}=E$.

7.22. Deduce Everett's formula from the Gauss forward formula of odd degree.

Using Problem 7.21, we have at once

$$p_k=\sum_{i=0}^{n}\left[\binom{k+i}{2i+1}\delta^{2i}y_1-\binom{k+i-1}{2i+1}\delta^{2i}y_0\right]$$

$$=\binom{k}{1}y_1+\binom{k+1}{3}\delta^2y_1+\binom{k+2}{5}\delta^4y_1+\cdots+\binom{k+n}{2n+1}\delta^{2n}y_1$$

$$-\binom{k-1}{1}y_0-\binom{k}{3}\delta^2y_0-\binom{k+1}{5}\delta^4y_0-\cdots-\binom{k+n-1}{2n+1}\delta^{2n}y_0$$

which is **Everett's formula**. Since it is a rearrangement of the Gauss formula it is the same polynominal of degree $2n+1$, satisfying $p_k=y_k$ for $k=-n,\ldots,n+1$. It is a heavily used formula because of its simplicity, only even differences being involved.

7.23. Apply Everett's formula with $n=2$ to find a polynomial of degree five or less which takes the y_k values of Table 7.7.

The needed differences are circled.

Table 7.7

k	x_k	y_k	δ	δ^2	δ^3	δ^4
−2	0	0				
			−1			
−1	1	−1		10		
			9		108	
0	2	(8)		(118)		(216)
			127		324	
1	3	(135)		(442)		(336)
			569		660	
2	4	704		1102		
			1671			
3	5	2375				

Substituting the circled entries into their places in Everett's formula,

$$p_k = 135k + 442\frac{(k+1)k(k-1)}{6} + 336\frac{(k+2)(k+1)k(k-1)(k-2)}{120} - 8(k-1) - 118\frac{k(k-1)(k-2)}{6} - 216\frac{(k+1)k(k-1)(k-2)(k-3)}{120}$$

which can be simplified, using $x_k = k + 2$, to $p(x_k) = x_k^5 - x_k^4 - x_k^3$.

7.24. Show that

$$\binom{k+i-1}{2i}\mu\delta^{2i}y_{1/2} + \frac{k-\frac{1}{2}}{2i+1}\binom{k+i-1}{2i}\delta^{2i+1}y_{1/2} = \binom{k+i}{2i+1}\delta^{2i}y_1 - \binom{k+i-1}{2i+1}\delta^{2i}y_0$$

The left side corresponds to the operator

$$\delta^{2i}\binom{k+i-1}{2i}\frac{1}{2}\left[E+1+\frac{2k-1}{2i+1}(E-1)\right] = \delta^{2i}\binom{k+i-1}{2i}\left(\frac{k+i}{2i+1}E - \frac{k-i-1}{2i+1}\right)$$

The right side corresponds to the operator

$$\delta^{2i}\left[\binom{k+i}{2i+1}E - \binom{k+i-1}{2i+1}\right] = \delta^{2i}\binom{k+i-1}{2i}\left(\frac{k+i}{2i+1}E - \frac{k-i-1}{2i+1}\right)$$

so that both sides are the same.

7.25. Show that *Bessel's formula* is a rearrangement of Everett's formula.

Bessel's formula is

$$p_k = \sum_{i=0}^{n}\left[\binom{k+i-1}{2i}\mu\delta^{2i}y_{1/2} + \frac{1}{2i+1}\left(k-\frac{1}{2}\right)\binom{k+i-1}{2i}\delta^{2i+1}y_{1/2}\right]$$

$$= \mu y_{1/2} + \left(k-\frac{1}{2}\right)\delta y_{1/2} + \binom{k}{2}\mu\delta^2 y_{1/2} + \frac{1}{3}\left(k-\frac{1}{2}\right)\binom{k}{2}\delta^3 y_{1/2} + \cdots + \binom{k+n-1}{2n}\mu\delta^{2n}y_{1/2} + \frac{1}{2n+1}\left(k-\frac{1}{2}\right)\binom{k+n-1}{2n}\delta^{2n+1}y_{1/2}$$

By the previous problem it reduces at once to Everett's.

7.26. Apply Bessel's formula with $n = 1$ to find a polynomial of degree three or less which takes the y_k values in Table 7.8.

Table 7.8

k	x_k	y_k			
−1	4	1			
			2		
0	6	③		③	
			⑤		④
1	8	⑧		⑦	
			12		
2	10	20			

The needed differences are circled and have been inserted into their places in Bessel's formula. Needless to say, the resulting polynomial is the same one already found by other formulas.

$$p_k = \frac{3+8}{2} + 5\left(k - \frac{1}{2}\right) + \frac{3+7}{2}\frac{k(k-1)}{2} + \frac{1}{3}(4)\left(k - \frac{1}{2}\right)\frac{k(k-1)}{2}$$

This can be verified to be equivalent to earlier results.

Supplementary Problems

7.27. Prove $\nabla = \delta E^{-1/2} = 1 - E^{-1} = 1 - (1 + \Delta)^{-1}$.

7.28. Prove $\sqrt{1 + \delta^2\mu^2} = 1 + \frac{1}{2}\delta^2$.

7.29. Prove $E^{1/2} = \mu + \frac{1}{2}\delta$ and $E^{-1/2} = \mu - \frac{1}{2}\delta$.

7.30. Two operators L_1 and L_2 commute if $L_1L_2 = L_2L_1$. Show that μ, δ, E, Δ, and ∇ all commute with one another.

7.31. Prove $\mu\delta = \frac{1}{2}\Delta E^{-1} + \frac{1}{2}\Delta$.

7.32. Prove $\Delta = \frac{1}{2}\delta^2 + \delta\sqrt{1 + \frac{1}{4}\delta^2}$.

7.33. Apply Newton's backward formula to the following data, to obtain a polynomial of degree four in the argument k:

k	-4	-3	-2	-1	0
x_k	1	2	3	4	5
y_k	1	-1	1	-1	1

Then use $x_k = k + 5$ to convert to a polynomial in x_k. Compare the final result with that of Problem 6.7.

7.34. Apply Newton's backward formula to find a polynomial of degree three which includes the following x_k, y_k pairs:

x_k	3	4	5	6
y_k	6	24	60	120

Using $x_k = k + 6$, convert to a polynomial in x_k and compare with the result of Problem 6.10.

7.35. Show that the change of argument $x_k = x_0 + kh$ converts Newton's backward formula into

$$p(x_k) = y_0 + \frac{\nabla y_0}{h}(x - x_0) + \frac{\nabla^2 y_0}{2!h^2}(x - x_0)(x - x_{-1}) + \cdots + \frac{\nabla^n y_0}{n!h^n}(x - x_0)\cdots(x - x_{-n+1})$$

7.36. Apply Problem 7.35 to the data of Problem 7.34 to produce the cubic polynomial directly in the argument x_k.

7.37. Apply the Gauss forward formula to the data below and compare the result with that of Problem 6.8.

k	-2	-1	0	1	2
x_k	2	4	6	8	10
y_k	0	0	1	0	0

7.38. Apply the Gauss backward formula to the data of Problem 7.37.

7.39. Apply the Gauss backward formula to the data of Problem 7.34, with the argument k shifted so that $k = 0$ at $x = 6$.

7.40. Apply the Gauss forward formula to the data below and compare the result with that of Problem 6.11.

k	-2	-1	0	1	2	3
x_k	0	1	2	3	4	5
y_k	0	0	1	1	0	0

7.41. Verify that for $k = -1, 0$

$$y_k = y_0 + \binom{k}{1} \delta y_{-1/2}$$

and that for $k = -2, -1, 0, 1$

$$y_k = y_0 + \binom{k}{1} \delta y_{-1/2} + \binom{k+1}{2} \delta^2 y_0 + \binom{k+1}{3} \delta^3 y_{-1/2}$$

These can also be considered forms of the Gauss backward formula, the degree of these polynomials being odd rather than even.

7.42. Apply Stirling's formula to the data of Problem 7.37.

7.43. Apply Stirling's formula to the data of Problem 6.9. Choose any three equally spaced arguments and let them correspond to $k = -1, 0, 1$.

7.44. Apply Everett's formula to the data of Problem 7.34, with the center pair of arguments corresponding to $k = 0$ and 1.

7.45. Apply Everett's formula to the data of Problem 7.40.

7.46. Apply Everett's formula to the data of Problem 6.9.

7.47. Apply Bessel's formula to the data of Problem 7.44.

7.48. Apply Bessel's formula to the data of Problem 7.40.

7.49. Prove $E^{1/2} = \frac{1}{2}\delta + \mu = (1 + \frac{1}{4}\delta^2)^{1/2} + \frac{1}{2}\delta = 1 + \frac{1}{2}\delta + \frac{1}{8}\delta^2 + \cdots$.

7.50. Show that $\mu^{-1} = 1 - \frac{1}{8}\delta^2 + \frac{3}{128}\delta^4 - \frac{5}{1024}\delta^6 + \cdots$.

7.51. Prove $\delta(f_k g_k) = \mu f_k \delta g_k + \mu g_k \delta f_k$.

Chapter 8

Unequally Spaced Arguments

The collocation polynomial for unequally spaced arguments $x_0, \ldots, x_n$ may be found in several ways. The methods of Lagrange, determinants, and divided differences will be presented in this chapter.

1. **Lagrange's formula** is

$$p(x) = \sum_{i=0}^{n} L_i(x) y_i$$

where $L_i(x)$ is the Lagrange multiplier function

$$L_i(x) = \frac{(x-x_0)(x-x_1)\cdots(x-x_{i-1})(x-x_{i+1})\cdots(x-x_n)}{(x_i-x_0)(x_i-x_1)\cdots(x_i-x_{i-1})(x_i-x_{i+1})\cdots(x_i-x_n)}$$

having the properties

$$L_i(x_k) = 0 \qquad \text{for } k \neq i \qquad L_i(x_i) = 1$$

Lagrange's formula does represent the collocation polynomial, that is, $p(x_k) = y_k$ for $k = 0, \ldots, n$. The function

$$\pi(x) = (x-x_0)\cdots(x-x_n) = \prod_{i=0}^{n} (x-x_i)$$

may be used to express the Lagrange multiplier function in the more compact form

$$L_i(x) = \frac{\pi(x)}{(x-x_i)\pi'(x_i)}$$

The closely related function

$$F_k(x) = \prod_{i \neq k} (x-x_i)$$

leads to a second compact representation of the Lagrange multiplier function,

$$L_i(x) = \frac{F_i(x)}{F_i(x_i)}$$

2. A **determinant form** of the collocation polynomial $p(x)$ is

$$\begin{vmatrix} p(x) & 1 & x & x^2 & \cdots & x^n \\ y_0 & 1 & x_0 & x_0^2 & \cdots & x_0^n \\ y_1 & 1 & x_1 & x_1^2 & \cdots & x_1^n \\ \cdots & \cdots & \cdots & \cdots & \cdots & \cdots \\ y_n & 1 & x_n & x_n^2 & \cdots & x_n^n \end{vmatrix} = 0$$

since $p(x_k) = y_k$ for $k = 0, \ldots, n$. It finds occasional use, mostly in theoretical work.

3. The **first divided difference** between x_0 and x_1 is defined as

$$y(x_0, x_1) = \frac{y_1 - y_0}{x_1 - x_0}$$

with a similar formula applying between other argument pairs.

Then **higher divided differences** are defined in terms of lower divided differences. For example,

$$y(x_0, x_1, x_2) = \frac{y(x_1, x_2) - y(x_0, x_1)}{x_2 - x_0}$$

is a second difference, while

$$y(x_0, x_1, \ldots, x_n) = \frac{y(x_1, \ldots, x_n) - y(x_0, \ldots, x_{n-1})}{x_n - x_0}$$

is an nth difference. In many ways these differences play roles equivalent to those of the simpler differences used earlier.

A *difference table* is again a convenient device for displaying differences, the standard diagonal form being used.

x_0	y_0				
		$y(x_0, x_1)$			
x_1	y_1		$y(x_0, x_1, x_2)$		
		$y(x_1, x_2)$		$y(x_0, x_1, x_2, x_3)$	
x_2	y_2		$y(x_1, x_2, x_3)$		$y(x_0, x_1, x_2, x_3, x_4)$
		$y(x_2, x_3)$		$y(x_1, x_2, x_3, x_4)$	
x_3	y_3		$y(x_2, x_3, x_4)$		
		$y(x_3, x_4)$			
x_4	y_4				

The **representation theorem**

$$y(x_0, x_1, \ldots, x_n) = \sum_{i=0}^{n} \frac{y_i}{F_i^n(x_i)}$$

shows how each divided difference may be represented as a combination of y_k values. This should be compared with a corresponding theorem in Chapter 3.

The *symmetry property* of divided differences states that such differences are invariant under all permutations of the arguments x_k, provided the y_k values are permuted in the same way. This very useful result is an easy consequence of the representation theorem.

Divided differences and derivatives are related by

$$y(x, x_0, \ldots, x_n) = \frac{y^{(n+1)}(\xi)}{(n+1)!}$$

In the case of equally spaced arguments, divided differences reduce to ordinary finite differences; specifically,

$$y(x_0, x_1, \ldots, x_n) = \frac{\Delta^n y_0}{n!\, h^n}$$

A useful property of ordinary finite differences may be obtained in this way, namely,

$$\Delta^n y_0 = y^{(n)}(\xi) h^n$$

For a function $y(x)$ with bounded derivatives, all $y^n(x)$ having a bound independent of n, it follows that, for small h,

$$\lim \Delta^n y_0 = 0$$

for increasing n. This generalizes the result found earlier for polynomials and explains why the higher differences in a table are often found to tend toward zero.

The collocation polynomial may now be obtained in terms of divided differences. The classic result is **Newton's divided difference formula,**

$$p(x) = y_0 + (x - x_0)y(x_0, x_1) + (x - x_0)(x - x_1)y(x_0, x_1, x_2)$$
$$+ \cdots + (x - x_0)(x - x_1) \cdots (x - x_{n-1})y(x_0, \ldots, x_n)$$

the arguments x_k not being required to have equal spacing. This generalizes the Newton formula of Chapter 6, and in the case of equal spacing reduces to it.

The error $y(x) - p(x)$, where $y(x)$ and $p(x)$ collocate at the arguments $x_0, \ldots, x_n$, is still given by the formula obtained earlier,

$$y(x) - p(x) = \frac{y^{(n+1)}(\xi)\pi(x)}{(n+1)!}$$

since we are still discussing the same collocation polynomial $p(x)$. An alternative form of this error, using divided differences, is

$$y(x) - p(x) = y(x, x_0, \ldots, x_n)(x - x_0) \cdots (x - x_n)$$

Solved Problems

8.1. What values does the Lagrange multiplier function

$$L_i(x) = \frac{(x - x_0)(x - x_1) \cdots (x - x_{i-1})(x - x_{i+1}) \cdots (x - x_n)}{(x_i - x_0)(x_i - x_1) \cdots (x_i - x_{i-1})(x_i - x_{i+1}) \cdots (x_i - x_n)}$$

take at the data points $x = x_0, x_1, \ldots, x_n$?

First notice that the numerator factors guarantee $L_i(x_k) = 0$ for $k \neq i$, and then the denominator factors guarantee that $L_i(x_i) = 1$.

8.2. Verify that the polynomial $p(x) = \sum_{i=0}^{n} L_i(x)y_i$ takes the value y_k at the argument x_k, for $k = 0, \ldots n$. This is *Lagrange's formula* for the collocation polynomial.

By Problem 8.1, $p(x_k) = \sum_{i=0}^{n} L_i(x_k)y_i = L_k(x_k)y_k = y_k$ so that Lagrange's formula does provide the collocation polynomial.

8.3. With $\pi(x)$ defined as the product $\pi(x) = \prod_{i=0}^{n} (x - x_i)$, show that

$$L_k(x) = \frac{\pi(x)}{(x - x_k)\pi'(x_k)}$$

Since $\pi(x)$ is the product of $n + 1$ factors, the usual process of differentiation produces $\pi'(x)$ as the sum of $n + 1$ terms, in each of which one factor has been differentiated. If we define

$$F_k(x) = \prod_{i \neq k} (x - x_i)$$

to be the same as $\pi(x)$ except that the factor $x - x_k$ is omitted, then

$$\pi'(x) = F_0(x) + \cdots + F_n(x)$$

But then at $x = x_k$ all terms are zero except $F_k(x_k)$, since this is the only term not containing $x - x_k$. Thus

$$\pi'(x_k) = F_k(x_k) = (x_k - x_0) \cdots (x_k - x_{k-1})(x_k - x_{k+1}) \cdots (x_k - x_n)$$

and
$$\frac{\pi(x)}{(x - x_k)\pi'(x_k)} = \frac{F_k(x)}{\pi'(x_k)} = \frac{F_k(x)}{F_k(x_k)} = L_k(x)$$

8.4. Show that the determinant equation

$$\begin{vmatrix} p(x) & 1 & x & x^2 & \cdots & x^n \\ y_0 & 1 & x_0 & x_0^2 & \cdots & x_0^n \\ y_1 & 1 & x_1 & x_1^2 & \cdots & x_1^n \\ \cdots & \cdots & \cdots & \cdots & \cdots & \cdots \\ y_n & 1 & x_n & x_n^2 & \cdots & x_n^n \end{vmatrix} = 0$$

also provides the collocation polynomial $p(x)$.

Expansion of this determinant using minors of the first-row elements would clearly produce a polynomial of degree n. Substituting $x = x_k$ and $p(x) = y_k$ makes two rows identical so that the determinant is zero. Thus $p(x_k) = y_k$ and this polynomial is the collocation polynomial. As attractive as this result is, it is not of much use due to the difficulty of evaluating determinants of large size.

8.5. Find the polynomial of degree three which takes the values prescribed below.

x_k	0	1	2	4
y_k	1	1	2	5

The polynomial can be written directly.

$$p(x) = \frac{(x-1)(x-2)(x-4)}{(0-1)(0-2)(0-4)}1 + \frac{x(x-2)(x-4)}{1(1-2)(1-4)}1 + \frac{x(x-1)(x-4)}{2(2-1)(2-4)}2 + \frac{x(x-1)(x-2)}{4(4-1)(4-2)}5$$

It can be rearranged into $p(x) = \frac{1}{12}(-x^3 + 9x^2 - 8x + 12)$.

8.6. Compute divided differences through the third for the y_k values in Table 8.1.

The differences are listed in the last three columns.

Table 8.1

x_k	y_k			
0	1			
		0		
1	1		$\frac{1}{2}$	
		1		$-\frac{1}{12}$
2	2		$\frac{1}{6}$	
		$\frac{3}{2}$		
4	5			

For example,

$$y(2, 4) = \frac{5-2}{4-2} = \frac{3}{2} \qquad y(1, 2, 4) = \frac{\frac{3}{2}-1}{4-1} = \frac{1}{6}$$

$$y(0, 1, 2) = \frac{1-0}{2-0} = \frac{1}{2} \qquad y(0, 1, 2, 4) = \frac{\frac{1}{6}-\frac{1}{2}}{4-0} = -\frac{1}{12}$$

8.7. Prove $y(x_0, x_1) = y(x_1, x_0)$. This is called **symmetry** of the first divided difference.

This is obvious from the definition but can also be seen from the fact that

$$y(x_0, x_1) = \frac{y_0}{x_0 - x_1} + \frac{y_1}{x_1 - x_0}$$

since interchanging x_0 with x_1 and y_0 with y_1 here simply reverses the order of the two terms on the right. This procedure can now be applied to higher differences.

8.8. Prove $y(x_0, x_1, x_2)$ is symmetric.

Rewrite this difference as

$$y(x_0, x_1, x_2) = \frac{y(x_1, x_2) - y(x_0, x_1)}{x_2 - x_0} = \frac{1}{x_2 - x_0}\left(\frac{y_2 - y_1}{x_2 - x_1} - \frac{y_1 - y_0}{x_1 - x_0}\right)$$

$$= \frac{y_0}{(x_0 - x_1)(x_0 - x_2)} + \frac{y_1}{(x_1 - x_0)(x_1 - x_2)} + \frac{y_2}{(x_2 - x_0)(x_2 - x_1)}$$

Interchanging any two arguments x_j and x_k and the corresponding y values now merely interchanges the y_j and y_k terms on the right, leaving the overall result unchanged. Since any permutation of the arguments x_k can be effected by successive interchanges of pairs, the divided difference is invariant under all permutations (of both the x_k and y_k numbers).

8.9. Prove that, for any positive integer n,

$$y(x_0, x_1, \ldots, x_n) = \sum_{i=0}^{n} \frac{y_i}{F_i^n(x_i)}$$

where $F_i^n(x_i) = (x_i - x_0)(x_i - x_1) \cdots (x_i - x_{i-1})(x_i - x_{i+1}) \cdots (x_i - x_n)$. This generalizes the results of the previous two problems.

The proof is by induction. We already have this result for $n = 1$ and 2. Suppose it true for $n = k$. Then by definition,

$$y(x_0, x_1, \ldots, x_{k+1}) = \frac{y(x_1, \ldots, x_{k+1}) - y(x_0, \ldots, x_k)}{x_{k+1} - x_0}$$

Since we have assumed our result true for differences of order k, the coefficient of y_k on the right, for $i = 1, 2, \ldots, k$, will be

$$\frac{1}{x_{k+1} - x_0}\left[\frac{1}{(x_i - x_1) \cdots (x_i - x_{k+1})} - \frac{1}{(x_i - x_0) \cdots (x_i - x_k)}\right]$$

where it is understood that the factor $(x_i - x_i)$ is not involved in the denominator products. But this coefficient reduces to

$$\frac{1}{(x_i - x_0) \cdots (x_i - x_{k+1})} = \frac{1}{F_i^{k+1}(x_i)}$$

as claimed. For $i = 0$ or $i = k + 1$ the coefficient of y_i comes in one piece instead of two, but in both cases is easily seen to be what is claimed in the theorem with $n = k + 1$, that is,

$$\frac{1}{(x_0 - x_1) \cdots (x_0 - x_{k+1})} \qquad \frac{1}{(x_{k+1} - x_0) \cdots (x_{k+1} - x_k)}$$

This completes the induction and proves the theorem.

8.10. Prove that the nth divided difference is symmetric.

This follows at once from the previous problem. If any pair of arguments are interchanged, say x_j and x_k, the terms involving y_j and y_k on the right are interchanged and nothing else changes.

8.11. Evaluate the first few differences of $y(x)=x^2$ and x^3.

Take $y(x)=x^2$ first. Then

$$y(x_0, x_1)=\frac{x_1^2-x_0^2}{x_1-x_0}=x_1+x_0 \qquad y(x_0, x_1, x_2)=\frac{(x_2+x_1)-(x_1+x_0)}{x_2-x_0}=1$$

Higher differences will clearly be 0. Now take $y(x)=x^3$.

$$y(x_0, x_1)=\frac{x_1^3-x_0^3}{x_1-x_0}=x_1^2+x_1x_0+x_0^2$$

$$y(x_0, x_1, x_2)=\frac{(x_2^2+x_2x_1+x_1^2)-(x_1^2+x_1x_0+x_0^2)}{x_2-x_0}=x_0+x_1+x_2$$

$$y(x_0, x_1, x_2, x_3)=\frac{(x_1+x_2+x_3)-(x_0+x_1+x_2)}{x_3-x_0}=1$$

Again higher differences are clearly zero. Notice that in both cases all the differences are symmetric polynomials.

8.12. Prove that the kth divided difference of a polynomial of degree n is a polynomial of degree $n-k$ if $k \leqq n$, and is zero if $k>n$.

Call the polynomial $p(x)$. A typical divided difference is

$$p(x_0, x_1)=\frac{p(x_1)-p(x_0)}{x_1-x_0}$$

Thinking of x_0 as fixed and x_1 as the argument, the various parts of this formula can be viewed as functions of x_1. In particular, the numerator is a polynomial in x_1, of degree n, with a zero at $x_1=x_0$. By the factor theorem the numerator contains x_1-x_0 as a factor and therefore the quotient, which is $p(x_0, x_1)$, is a polynomial in x_1 of degree $n-1$. By the symmetry of $p(x_0, x_1)$ it is therefore also a polynomial in x_0 of degree $n-1$. The same argument may now be repeated. A typical second difference is

$$p(x_0, x_1, x_2)=\frac{p(x_1, x_2)-p(x_0, x_1)}{x_2-x_0}$$

Thinking of x_0 and x_1 as fixed, and x_2 as the argument, the numerator is a polynomial in x_2, of degree $n-1$, with a zero at $x_2=x_0$. By the factor theorem $p(x_0, x_1, x_2)$ is therefore a polynomial in x_2 of degree $n-2$. By the symmetry of $p(x_0, x_1, x_2)$ it is also a polynomial in either x_0 or x_1, again of degree $n-2$. Continuing in this way the required result is achieved. An induction is called for, but it is an easy one and the details are omitted.

8.13. Prove that *Newton's divided difference formula*

$$p(x)=y_0+(x-x_0)y(x_0, x_1)+(x-x_0)(x-x_1)y(x_0, x_1, x_2) + \cdots + (x-x_0)(x-x_1)\cdots(x-x_{n-1})y(x_0, \ldots, x_n)$$

represents the collocation polynomial. That is, it takes the values $p(x_k)=y_k$ for $k=0, \ldots, n$.

The fact that $p(x_0)=y_0$ is obvious. Next, from the definition of divided differences, and using symmetry,

$$y_k=y_0+(x_k-x_0)y(x_0, x_k)$$

$$y(x_0, x_k)=y(x_0, x_1)+(x_k-x_1)y(x_0, x_1, x_k)$$

$$y(x_0, x_1, x_k)=y(x_0, x_1, x_2)+(x_k-x_2)y(x_0, x_1, x_2, x_k)$$

$$\cdots\cdots\cdots\cdots\cdots\cdots\cdots\cdots\cdots\cdots\cdots\cdots$$

$$y(x_0, \ldots, x_{n-2}, x_k)=y(x_0, \ldots, x_{n-1})+(x_k-x_{n-1})y(x_0, \ldots, x_{n-1}, x_k)$$

For example, the second line follows from

$$y(x_0, x_1, x_k) = y(x_1, x_0, x_k) = \frac{y(x_0, x_k) - y(x_1, x_0)}{x_k - x_1}$$

For $k = 1$ the first of these proves $p(x_1) = y_1$. Substituting the second into the first brings

$$y_k = y_0 + (x_k - x_0)y(x_0, x_1) + (x_k - x_0)(x_k - x_1)y(x_0, x_1, x_k)$$

which for $k = 2$ proves $p(x_2) = y_2$. Successive substitutions verify $p(x_k) = y_k$ for each x_k in its turn until finally we reach

$$y_n = y_0 + (x_n - x_0)y(x_0, x_1) + (x_n - x_0)(x_n - x_1)y(x_0, x_1, x_2) + \cdots + (x_n - x_0)(x_n - x_1) \cdots (x_n - x_{n-1})y(x_0, \ldots, x_{n-1}, x_n)$$

which proves $p(x_n) = y_n$.

Since this Newton formula represents the same polynomial as the Lagrange formula, the two are just rearrangements of each other.

8.14. Find the polynomial of degree three which takes the values given in Table 8.1.

Using Newton's formula, which involves the differences on the top diagonal of Table 8.1,

$$p(x) = 1 + (x - 0)0 + (x - 0)(x - 1)\frac{1}{2} + (x - 0)(x - 1)(x - 2)\left(-\frac{1}{12}\right)$$

which simplifies to $p(x) = \frac{1}{12}(-x^3 + 9x^2 - 8x + 12)$, the same result as found by Lagrange's formula.

Supplementary Problems

8.15. Use Lagrange's formula to produce a cubic polynomial which includes the following x_k, y_k number pairs. Then evaluate this polynomial for $x = 2, 3, 5$.

x_k	0	1	4	6
y_k	1	−1	1	−1

8.16. Use Lagrange's formula to produce a fourth-degree polynomial which includes the following x_k, y_k number pairs. Then evaluate the polynomial for $x = 3$.

x_k	0	1	2	4	5
y_k	0	16	48	88	0

8.17. Deduce Lagrange's formula by determining the coefficients a_i in the partial fractions expansion

$$\frac{p(x)}{\pi(x)} = \sum_{i=0}^{n} \frac{a_i}{x - x_i}$$

[Multiply both sides by $x - x_i$ and let x approach x_i as limit, remembering that $p(x_i) = y_i$ for collocation.] The result is $a_i = \dfrac{y_i}{\pi'(x_i)}$.

8.18. Apply Problem 8.17 to express $\dfrac{3x^2+x+1}{x^3-6x^2+11x-6}$ as a sum of partial fractions

$$\frac{a_0}{x-x_0}+\frac{a_1}{x-x_1}+\frac{a_2}{x-x_2}$$

[*Hint.* Think of the denominator as $\pi(x)$ for some x_0, x_1, x_2 and then find the corresponding y_0, y_1, y_2. This amounts to regarding $p(k)$ as a collocation polynomial.]

8.19. Express $\dfrac{x^2+6x+1}{(x^2-1)(x-4)(x-6)}$ as a sum of partial fractions.

8.20. Show that

$$L_0(x)=1+\frac{x-x_0}{x_0-x_1}+\frac{(x-x_0)(x-x_1)}{(x_0-x_1)(x_0-x_2)}+\cdots+\frac{(x-x_0)\cdots(x-x_{n-1})}{(x_0-x_1)\cdots(x_0-x_n)}$$

Similar expansions can be written by symmetry for the other coefficients.

8.21. Write the three-point Lagrange formula for arguments x_0, $x_0+\epsilon$ and x_1 and then consider the limit as ε tends to 0. Show that

$$p(x)=\frac{(x_1-x)(x+x_1-2x_0)}{(x_1-x_0)^2}y(x_0)+\frac{(x-x_0)(x_1-x)}{(x_1-x_0)}y'(x_0)+\frac{(x-x_0)^2}{(x_1-x_0)^2}y(x_1)+\frac{1}{6}(x-x_0)^2(x-x_1)y'''(\xi)$$

This determines a quadratic polynomial in terms of $y(x_0)$, $y'(x_0)$, and $y(x_1)$.

8.22. Proceed as in the previous problem, beginning with the Lagrange formula for arguments $x_0, x_0+\epsilon$, $x_1-\epsilon$, x_1 to represent a cubic polynomial in terms of $y(x_0)$, $y'(x_0)$, $y(x_1)$, and $y'(x_1)$.

8.23. Calculate divided differences through third order for the following x_k, y_k pairs:

x_k	0	1	4	6
y_k	1	−1	1	−1

8.24. Find the collocation polynomial of degree three for the x_k, y_k pairs of Problem 8.23. Use Newton's formula. Compare your result with that obtained by the Lagrange formula.

8.25. Rearrange the number pairs of Problem 8.23 as follows:

x_k	4	1	6	0
y_k	1	−1	−1	1

Compute the third divided difference again. It should be the same number as before, illustrating the symmetry property.

8.26. Calculate a fourth divided difference for the following y_k values:

x_k	0	1	2	4	5
y_k	0	16	48	88	0

8.27. Apply Newton's formula to find the collocation polynomial for the data of Problem 8.26. What value does this polynomial take at $x = 3$?

8.28. Show that

$$y(x_0, x_1) = \frac{\begin{vmatrix} 1 & y_0 \\ 1 & y_1 \end{vmatrix}}{\begin{vmatrix} 1 & x_0 \\ 1 & x_1 \end{vmatrix}} \qquad y(x_0, x_1, x_2) = \frac{\begin{vmatrix} 1 & x_0 & y_0 \\ 1 & x_1 & y_1 \\ 1 & x_2 & y_2 \end{vmatrix}}{\begin{vmatrix} 1 & x_0 & x_0^2 \\ 1 & x_1 & x_1^2 \\ 1 & x_2 & x_2^2 \end{vmatrix}}$$

8.29. For $y(x) = (x - x_0)(x - x_1) \cdots (x - x_n) = \pi(x)$, prove that

$$y(x_0, x_1, \ldots, x_p) = 0 \qquad \text{for } p = 0, 1, \ldots, n$$

$$y(x_0, x_1, \ldots, x_n, x) = 1 \qquad \text{for all } x$$

$$y(x_0, x_1, \ldots, x_n, x, z) = 0 \qquad \text{for all } x, z$$

8.30. Show that

$$p(x) = y_0 + \frac{y(x_1, x_0) + y(x_0, x_{-1})}{2}(x - x_0) + y(x_1, x_0, x_{-1})(x - x_0)\left(x - \frac{x_1 + x_{-1}}{2}\right)$$

$$+ \frac{y(x_2, x_1, x_0, x_{-1}) + y(x_1, x_0, x_{-1}, x_{-2})}{2}(x - x_1)(x - x_0)(x - x_{-1})$$

$$+ y(x_2, x_1, x_0, x_{-1}, x_{-2})(x - x_0)(x - x_1)(x - x_{-1})\left(x - \frac{x_2 + x_{-2}}{2}\right)$$

is another way of writing the collocation polynomial, by verifying

$$p(x_k) = y_k \qquad \text{for } k = -2, -1, 0, 1, 2$$

This is a generalization of Stirling's formula for unequal spacing. It can be extended to higher degree. Bessel's formula and others can also be generalized.

8.31. Show that for arguments which are equally spaced, so that $x_{k+1} - x_k = h$, we have

$$y(x_0, x_1, \ldots, x_n) = \frac{\Delta^n y_0}{n!\, h^n}$$

8.32. Divided differences with two or more arguments equal can be defined by limiting processes. For example, $y(x_0, x_0)$ can be defined as $\lim y(x, x_0)$, where $\lim x = x_0$. This implies that

$$y(x_0, x_0) = \lim \frac{y(x) - y_0}{x - x_0} = y'(x_0)$$

Verify this directly when $y(x) = x^2$ by showing that in this case $y(x, x_0) = x + x_0$ so that $\lim y(x, x_0) = y'(x_0) = 2x_0$. Also verify it directly when $y(x) = x^3$ by showing first that in this case $y(x, x_0) = x^2 + xx_0 + x_0^2$.

8.33. In the second divided difference

$$y(x_0, x, x_2) = \frac{y(x, x_2) - y(x_0, x_2)}{x - x_0}$$

the right side may be viewed as having the form $\dfrac{f(x) - f(x_0)}{x - x_0}$ with x_2 considered a constant. If $\lim x = x_0$, we define

$$y(x_0, x_0, x_2) = \lim y(x_0, x, x_2)$$

This implies that

$$y(x_0, x_0, x_2) = y'(x, x_2) \mid x = x_0$$

Verify this directly when $y(x) = x^3$ by showing first that in this case

$$y(x_0, x, x_2) = x + x_0 + x_2 \qquad \text{while} \qquad y(x, x_2) = x^2 + xx_2 + x_2^2$$

Chapter 9

Splines

Instead of using a single polynomial, presumably of high degree, to represent a given function over an interval to achieve a required accuracy, we may instead join together several polynomial segments, each of low degree. The classic example is, of course, a set of line segments, each fit to the given data over a subinterval. Such an approximation is continuous but has a first derivative with discontinuities at the interval ends, the corners (Fig. 9-1). It is the basis for elementary interpolations in tables and for the trapezoidal rule for numerical integration. The implicit assumption that between data points the given function is almost linear may be reasonable if the points are close enough together.

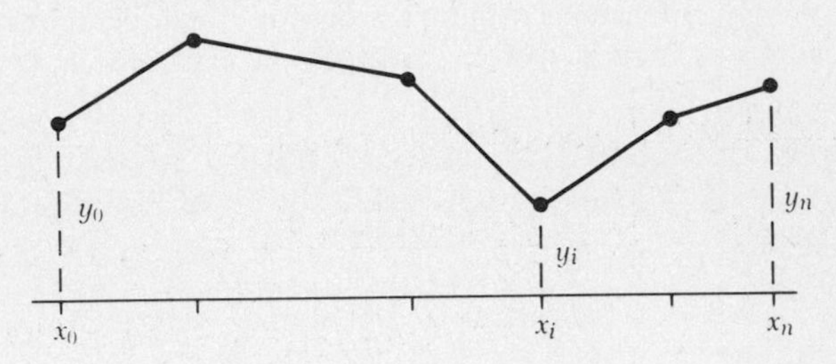

Fig. 9-1 A primitive spline.

In Chapter 14 we will fit parabolic segments (quadratic polynomials) together to develop Simpson's rule for numerical integration. Other examples using slightly higher degree polynomials will also be given. In all these cases there will be corners where the segments are joined.

We now consider a method in which cubic segments are pieced together in such a way that the corners are rounded, both the first and second derivatives of the approximation being continuous. High-degree polynomials have an oscillatory character. One of degree n can have as many as $n-1$ turning points. When such a polynomial represents a given function accurately, it is usually by oscillating back and forth across the function. This has undesirable side effects, poor approximation of the derivative to mention only one. The *spline* approximation now to be derived avoids such oscillations, because it consists of low-degree segments. The word spline is borrowed from the drafting instrument of the same name, a flexible strip used in drawing curves.

Given an interval $(a, b) = I$ divided into n subintervals by the points $x_0 = a, x_1, x_2, \ldots, x_n = b$, a cubic segment is to be fit on each subinterval, taking specified values y_i at the points x_i, with first and second derivatives on adjacent subintervals agreeing in value at the join. The points x_1 to x_{n-1} are called the **nodes**, or **knots**, of the spline (Fig. 9-2). Details of the development of these spline segments will be worked out in the solved problems, and examples will be provided.

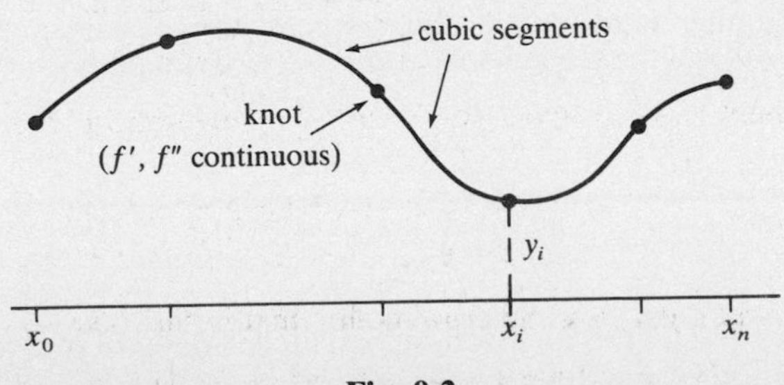

Fig. 9-2

Solved Problems

9.1. A polynomial of degree three, a cubic, has four coefficients. In a common representation

$$p(x) = c_0 + c_1 x + c_2 x^2 + c_3 x^3$$

With the conventions of Fig. 9-2, the n cubic segments together will involve $4n$ coefficients. How does this compare with the number of conditions being imposed upon the spline?

The point is that ordinarily we expect $4n$ coefficients to be well determined by $4n$ conditions. Here we have four conditions to be met at each of the knots x_1 to x_{n-1}, namely, the segment on either side must reach this point, and the first two derivatives have to agree. This comes to $4n-4$ conditions. At the two endpoints we are asking only for collocation, two more conditions, making a grand total of $4n-2$. The spline is, therefore, not completely defined by the specifications given. Two degrees of freedom remain. Sometimes these are used to make the second derivative zero at the endpoints, leading to what is called the natural spline. Alternatively one may require the end segments to match the end derivative values of the given function, if these are known or can be approximated. A third option, of reducing the specifications at knots x_1 and x_{n-1}, will also be explored.

9.2. Let the subintervals of Fig. 9-2 be called I_1 to I_n, so that $I_i = (x_{i-1}, x_i)$. Also define $h_i = x_i - x_{i-1}$, noting that the subintervals need not be of equal length. If $S_i(x)$ is the spline segment on I_i, show that

$$S_i''(x) = C_{i-1}\frac{x_i - x}{h_i} + C_i \frac{x - x_{i-1}}{h_i}$$

for constants C_i and $i = 1, \ldots, n$.

On I_i the spline segment is cubic, so its first derivative will be quadratic and the second derivative linear. It remains to verify the continuity at each knot x_k for $k = 1, \ldots, n-1$. The segment S_k touches this knot at its right end while S_{k+1} touches it at its left end. The required derivatives are thus

$$S_k''(x_k^-) = C_{k-1}\frac{x_k - x_k}{h_k} + C_k \frac{x_k - x_{k-1}}{h_k}$$

and

$$S_{k+1}''(x_k^+) = C_k \frac{x_{k+1} - x_k}{h_{k+1}} + C_{k+1}\frac{x_k - x_k}{h_{k+1}}$$

both of which reduce to C_k. Continuity is thus assured and we discover that the constants C_k are in fact the common values of spline second derivatives.

9.3. Integrate the result of the preceding problem twice to obtain the spline segments and then impose the requirement that segments pass through appropriate knots to determine the constants of integration.

The two integrations manage

$$S_i(x) = C_{i-1}\frac{(x_i - x)^3}{6h_i} + C_i \frac{(x - x_{i-1})^3}{6h_i} + c_i(x_i - x) + d_i(x - x_{i-1})$$

the last two terms being the linear function introduced by the constants of integration. For collocation at the knots, we must have $S_i(x_{i-1}) = y_{i-1}$ and $S_i(x_i) = y_i$. These conditions fix c_i and d_i and lead to

$$S_i(x) = C_{i-1}\frac{(x_i - x)^3}{6h_i} + C_i\frac{(x - x_{i-1})^3}{6h_i} + \left(y_{i-1} - \frac{C_{i-1}h_i^2}{6}\right)\frac{x_i - x}{h_i} + \left(y_i - \frac{C_i h_i^2}{6}\right)\frac{x - x_{i-1}}{h_i}$$

as may be verified by inserting x_{i-1} and x_i.

9.4. It remains to ensure the continuity of the first derivatives. To arrange this, differentiate the result of the preceding problem and compare adjoining values as in Problem 9.2.

Differentiating

$$S_i'(x) = -C_{i-1}\frac{(x_i - x)^2}{2h_i} + C_i\frac{(x - x_{i-1})^2}{2h_i} + \frac{y_i - y_{i-1}}{h_i} - \frac{C_i - C_{i-1}}{6}h_i$$

so the required derivatives at knot x_k are

$$S_k'(x_k^-) = \frac{h_k}{6}C_{k-1} + \frac{h_k}{3}C_k + \frac{y_k - y_{k-1}}{h_k}$$

and

$$S_{k+1}'(x_k^+) = -\frac{h_{k+1}}{3}C_k - \frac{h_{k+1}}{6}C_{k+1} + \frac{y_{k+1} - y_k}{h_{k+1}}$$

Since these are to be equal, we have, for $k = 1, \ldots, n-1$,

$$\frac{h_k}{6}C_{k-1} + \frac{h_k + h_{k+1}}{3}C_k + \frac{h_{k+1}}{6}C_{k+1} = \frac{y_{k+1} - y_k}{h_{k+1}} - \frac{y_k - y_{k-1}}{h_k}$$

which is a linear system of $n-1$ equations for the constants C_0 to C_n. As observed earlier, the system is underdetermined. We are two equations short.

There is an interesting way to include two additional equations in the linear system, keeping our options open and preserving the general character of the matrix. First let

$$\alpha_i = \frac{h_{i+1}}{h_i + h_{i+1}}$$

$$\beta_i = 1 - \alpha_i = \frac{h_i}{h_i + h_{i+1}}$$

$$d_i = \frac{6}{h_i + h_{i+1}}\left(\frac{y_{i+1} - y_i}{h_{i+1}} - \frac{y_i - y_{i-1}}{h_i}\right)$$

for $i = 1, \ldots, n-1$. The system can then be rewritten, still for $i = 1, \ldots, n-1$, as

$$\beta_i C_{i-1} + 2C_i + \alpha_i C_{i+1} = d_i$$

Now take two additional conditions in the form

$$2C_0 + \alpha_0 C_1 = d_0 \qquad \beta_n C_{n-1} + 2C_n = d_n$$

with α_0, d_0, β_n, and d_n at our disposal. The combined system then takes this shape:

$$\begin{bmatrix} 2 & \alpha_0 & 0 & & & & \\ \beta_1 & 2 & \alpha_1 & & & & \\ 0 & \beta_2 & 2 & & & & \\ & & & \ddots & & & \\ & & & & 2 & \alpha_{n-2} & 0 \\ & & & & \beta_{n-1} & 2 & \alpha_{n-1} \\ & & & & 0 & \beta_n & 2 \end{bmatrix} \begin{bmatrix} C_0 \\ C_1 \\ C_2 \\ \vdots \\ C_{n-2} \\ C_{n-1} \\ C_n \end{bmatrix} = \begin{bmatrix} d_0 \\ d_1 \\ d_2 \\ \vdots \\ d_{n-2} \\ d_{n-1} \\ d_n \end{bmatrix}$$

The coefficient matrix is triple diagonal, all other elements being zero.

9.5. How can the linear system of the preceding problem be used to find a natural spline?

Choose α_0, d_0, β_n, and d_n as zero. The top and bottom equations then force C_0 and C_n to be zero also and this is what identifies the natural spline. The system is reduced to order $n-1$ for determining the remaining C_1 to C_{n-1}.

9.6. Similarly, how can we arrange that the end conditions

$$S_1'(x_0) = y_0' \qquad S_n'(x_n) = y_n'$$

be met?

Borrowing appropriate formulas from Problem 9.4, we have

$$S_1'(x_0^+) = -\frac{h_1}{3}C_0 - \frac{h_1}{6}C_1 + \frac{y_1 - y_0}{h_1} = y_0'$$

and

$$S_n'(x_n^-) = \frac{h_n}{6}C_{n-1} + \frac{h_n}{3}C_n + \frac{y_n - y_{n-1}}{h_n} = y_n'$$

which are easily converted to

$$2C_0 + C_1 = \frac{6}{h_1}\left(\frac{y_1 - y_0}{h_1} - y_0'\right)$$

and

$$C_{n-1} + 2C_n = \frac{6}{h_n}\left(y_n' - \frac{y_n - y_{n-1}}{h_n}\right)$$

Now comparing with the first and last equations of the linear system, namely $2C_0 + \alpha_0 C_1 = d_0$ and $\beta_n C_{n-1} + 2C_n = d_n$, suggests the choices

$$\alpha_0 = 1 = \beta_n \qquad d_0 = \frac{6}{h_1}\left(\frac{y_1 - y_0}{h_1} - y_0'\right) \qquad d_n = \frac{6}{h_n}\left(y_n' - \frac{y_n - y_{n-1}}{h_n}\right)$$

which will, in fact, provide the required end values.

9.7. Fit cubic spline segments to the function $f(x) = \sin x$ on the interval $(0, \pi)$. Use just the two interior points $\pi/3$ and $2\pi/3$.

The corresponding data set is

x_i	0	$\pi/3$	$2\pi/3$	π
y_i	0	$\sqrt{3}/2$	$\sqrt{3}/2$	0

with $i = 0, \ldots, 3$ and all $h_i = \pi/3$. There are three cubic segments to find. The uniform h_i values at once make α_1, α_2, β_1, and β_2 all equal to $\frac{1}{2}$. Then

$$d_1 = \frac{3}{h}\left(\frac{0}{h} - \frac{\sqrt{3}/2}{h}\right) = -\frac{27\sqrt{3}}{2\pi^2}$$

with the same result for d_2. This leads us to the equations

$$\frac{1}{2}C_0 + 2C_1 + \frac{1}{2}C_2 \qquad = \frac{-27\sqrt{3}}{2\pi^2}$$

$$\frac{1}{2}C_1 + 2C_2 + \frac{1}{2}C_3 = \frac{-27\sqrt{3}}{2\pi^2}$$

and to the matter of end conditions. The natural spline is certainly appropriate here because the sine function does have zero second derivatives at the endpoints. So we set C_0 and C_3 to zero. The remaining system then quickly yields $C_1 = C_2 = -27\sqrt{3}/5\pi^2$. Substituting into the formulas of Problem 9.3 finally produces the spline segments, which after simplifications are these:

$$S_1(x) = \left(\frac{-27\sqrt{3}}{10\pi^3}\right)x^3 + \left(\frac{9\sqrt{3}}{5\pi}\right)x$$

$$S_2(x) = \left(\frac{-27\sqrt{3}}{10\pi^3}\right)\left(\frac{2\pi}{3} - x\right)^3 + \left(x - \frac{\pi}{3}\right)^3 + \frac{3\sqrt{3}}{5}$$

$$S_3(x) = \left(\frac{-27\sqrt{3}}{10\pi^3}\right)(\pi - x)^3 + \left(\frac{9\sqrt{3}}{5\pi}\right)(\pi - x)$$

Problem 9.19 asks that these cubics be verified by checking all the conditions imposed upon them. The simplicity of the example has allowed exact values to be carried throughout. Notice also that the central "cubic" segment is actually quadratic.

9.8. Again fit cubic segments to the sine function, this time asking that endpoint first derivatives equal sine derivatives.

The new endpoint conditions are $S_1'(0)=1$ and $S_3'(\pi)=-1$. From Problem 9.6 we find

$$\alpha_0=\beta_n=1 \qquad d_0=d_3=\left(\frac{18}{\pi}\right)\left(\frac{3\sqrt{3}}{2\pi}-1\right)$$

so the new linear system is

$$\begin{aligned}
2C_0+C_1 &= \left(\frac{18}{\pi}\right)\left(\frac{3\sqrt{3}}{2\pi}-1\right)\\
\frac{1}{2}C_0+2C_1+\frac{1}{2}C_2 &= \frac{-27\sqrt{3}}{2\pi^2}\\
\frac{1}{2}C_1+2C_2+\frac{1}{2}C_3 &= \frac{-27\sqrt{3}}{2\pi^2}\\
C_2+2C_3 &= \left(\frac{18}{\pi}\right)\left(\frac{3\sqrt{3}}{2\pi}-1\right)
\end{aligned}$$

and has this solution:

$$C_0=C_3=\frac{18\sqrt{3}}{\pi^2}-\frac{10}{\pi}$$

$$C_1=C_2=\frac{2}{\pi}-\frac{9\sqrt{3}}{\pi^2}$$

Substituting into the $S_i(x)$ formulas of Problem 9.3, we again have the cubic segments. Verification that these segments meet all conditions imposed upon them is requested as Problem 9.20, where it may also be found that the end values of $S''(x)$ are not zero.

9.9. A third way to obtain a well-determined system for spline approximation is to relax our requirements slightly. For example, omitting the segments $S_1(x)$ and $S_n(x)$, we can ask $S_2(x)$ and $S_{n-1}(x)$ to take care of the endpoint collocations. This also eliminates continuity requirements at x_1 and x_{n-1}, which are no longer knots. Show that the resulting problem will have just as many conditions to be met as coefficients available to meet them.

There will now be $n-2$ instead of n cubic segments, with $4n-8$ coefficients available. But there will be only $n-3$ rather than $n-1$ knots. With four requirements per knot, this makes $4n-12$ conditions to be satisfied. Since collocation is also required at x_0, x_1, x_{n-1}, and x_n the count of conditions climbs to $4n-8$.

9.10. Modify the developments in Problems 9.2 to 9.4 to meet the requirements suggested in Problem 9.9.

A careful rereading of the problems mentioned will show that a great deal can be saved. The center $n-3$ equations of our linear system, as presented in Problem 9.4, are still valid because they refer to knots x_2 to x_{n-2} where no changes are being made. These already provide $n-3$ equations for the $n-1$ coefficients C_1 to C_{n-1}. The other two needed equations will make $S_2(x_0)=y_0$ and $S_{n-1}(x_n)=y_n$. Returning to the $S_i(x)$ formula given in Problem 9.3, these conditions can be implemented. After some algebraic manipulation they can be induced to take the form

$$2C_1+\alpha_1C_2=d_1 \qquad \beta_{n-1}C_{n-2}+2C_{n-1}=d_{n-1}$$

with the following definitions:

$$\alpha_1 = \frac{2(h_1h_2^2 - h_1^3)}{(h_1+h_2)^3 - (h_1+h_2)h_2^2}$$

$$\beta_{n-1} = \frac{2(h_{n-1}^2 h_n - h_n^3)}{(h_{n-1}+h_n)^3 - (h_{n-1}+h_n)h_{n-1}^2}$$

$$d_1 = \frac{12h_2\left(y_0 - \dfrac{h_1+h_2}{h_2}y_1 + \dfrac{h_1}{h_2}y_2\right)}{(h_1+h_2)^3 - (h_1+h_2)h_2^2}$$

$$d_{n-1} = \frac{12h_{n-1}\left(y_n - \dfrac{h_{n-1}+h_n}{h_{n-1}}y_{n-1} + \dfrac{h_n}{h_{n-1}}y_{n-2}\right)}{(h_{n-1}+h_n)^3 - (h_{n-1}+h_n)h_{n-1}^2}$$

The final form of the system is then

$$\begin{bmatrix} 2 & \alpha_1 & 0 & & & & \\ \beta_2 & 2 & \alpha_2 & & & & \\ 0 & \beta_3 & 2 & & & & \\ & & & \ddots & & & \\ & & & 2 & \alpha_{n-3} & 0 \\ & & & \beta_{n-2} & 2 & \alpha_{n-2} \\ & & & 0 & \beta_{n-1} & 2 \end{bmatrix} \begin{bmatrix} C_1 \\ C_2 \\ C_3 \\ \vdots \\ C_{n-3} \\ C_{n-2} \\ C_{n-1} \end{bmatrix} = \begin{bmatrix} d_1 \\ d_2 \\ d_3 \\ \vdots \\ d_{n-3} \\ d_{n-2} \\ d_{n-1} \end{bmatrix}$$

again triple diagonal, all other elements being zero.

9.11. Apply the method just developed to $f(x) = \sin x$ on the interval $(0, \pi)$ using three equally spaced interior points.

There are four subintervals, with spline segments to be found for the inner two. The one knot will be at $x_2 = \pi/2$. This makes it clear why we are not continuing the earlier example, which had one fewer interval. There would be no knots at all and a single cubic would interpolate the four given points. The present data set is

x_i	0	$\pi/4$	$\pi/2$	$3\pi/4$	π
y_i	0	$\sqrt{2}/2$	1	$\sqrt{2}/2$	0

with all $h_i = \pi/4$. The formulas for α_i and β_i now apply only at the knot x_2 and yield $\alpha_2 = \beta_2 = \frac{1}{2}$. We also find $d_2 = 48(\sqrt{2}-2)/\pi^2$ and then the one equation

$$\frac{1}{2}C_1 + 2C_2 + \frac{1}{2}C_3 = \frac{48(\sqrt{2}-2)}{\pi^2}$$

Turning to more recent formulas, $\alpha_1 = 0$, $\beta_3 = 0$, and

$$d_1 = d_3 = \frac{32(1-\sqrt{2})}{\pi^2}$$

so our linear system is the following:

$$\begin{aligned} 2C_1 \qquad\qquad &= d_1 \\ \frac{1}{2}C_1 + 2C_2 + \frac{1}{2}C_3 &= \left(\frac{3\sqrt{2}}{2}\right)d_1 \\ 2C_3 &= d_1 \end{aligned}$$

Solving, and appealing again to Problem 9.3, we come to these two spline segments:

$$S_2(x) = \frac{16(1-\sqrt{2})(\pi-2x)^3 + (4\sqrt{2}-7)(4x-\pi)^3}{12\pi^3} + \frac{(8\sqrt{2}-2)(2\pi-4x) + (19-4\sqrt{2})(4x-\pi)}{12\pi}$$

$$S_3(x) = \frac{16(1-\sqrt{2})(2x-\pi)^3 + (4\sqrt{2}-7)(3\pi-4x)^3}{12\pi^3} + \frac{(8\sqrt{2}-2)(4x-2\pi) + (19-4\sqrt{2})(3\pi-4x)}{12\pi}$$

With a little patience it can be verified that S_2 joins the first three points, S_3 the last three, and that they make a proper knot at x_2. This is all that was required. Bonuses such as $S_2'(0) = 1$ or $S_2''(\pi/2) = -1$ would have been nice, but there is no point in being greedy. The approximations 1.05 and -1.09 will have to do.

9.12. What is the error of a spline approximation?

It can be shown that

$$\max |f(x) - S(x)| \leqq \frac{5}{384} \max |f^{(4)}(x)|\, H^4$$

where H is the largest of the h_i and the maxima are on the interval I.

9.13. Apply the error bound of Problem 9.12 to the spline of Problem 9.7.

The fourth derivative of $\sin x$ is, of course, bounded by 1 and $H = \pi/3$. Thus

$$\max |\sin x - S(x)| \leqq \frac{5}{384} \frac{\pi^4}{81} = .016$$

9.14. How well does a spline approximate the derivative $f'(x)$?

It can be shown that

$$\max |f'(x) - S'(x)| \leqq \frac{\max |f^{(4)}(x)|\, H^3}{24}$$

9.15. Apply the formula of Problem 9.14 to the spline of Problem 9.12.

We find $H^3/24 = .05$ approximately. Generally speaking, splines are quite good approximations to derivatives.

9.16. What is meant by saying that a spline is a global approximation to $f(x)$?

The segments of the spline are not determined independently of each other. Each is linked with all the others. The set of coefficients C_i which identify the segments is determined by one linear system. By way of contrast, one could fit a cubic polynomial to the first four points, x_0 to x_3, then another to set x_3 to x_6, and so on across the interval I. Each segment would then be found independently of the others, but the continuity properties of the spline at knots would almost surely be absent.

9.17. Show that the natural spline on (a, b) uniquely minimizes

$$\int_a^b f''(x)^2\, dx$$

among all functions $f(x)$ which have continuous second derivatives and satisfy $f(x_i) = y_i$ at the knots.

First note that

$$\int_a^b f''(x)^2\,dx - \int_a^b S''(x)^2\,dx = \int_a^b [f''(x) - S''(x)]^2\,dx + 2\int_a^b S''(x)[f''(x) - S''(x)]\,dx$$

with $S(x)$ the cubic spline. Integration by parts over each subinterval converts the last integral as follows:

$$\int_{x_{i-1}}^{x_i} S_i''(x)[f''(x) - S_i''(x)]\,dx = S_i''(x)[f'(x) - S_i'(x)]\Big|_{x_{i-1}}^{x_i} - \int_{x_{i-1}}^{x_i} [f'(x) - S_i'(x)]S_i^{(3)}(x)\,dx$$

$$= S_i''(x)[f'(x) - S_i'(x)]\Big|_{x_{i-1}}^{x_i} - S_i^{(3)}(x)[f(x) - S_i(x)]\Big|_{x_{i-1}}^{x_i} + \int_{x_{i-1}}^{x_i} [f(x) - S_i(x)]S_i^{(4)}(x)\,dx$$

The last two terms vanish since $f(x)$ equals $S_i(x)$ at the knots and $S_i^{(4)}(x)$ is zero. Summing what is left for $i = 1, \ldots, n$ there is cancellation of all interior values leaving

$$S''(b)[f'(b) - S'(b)] - S''(a)[f'(a) - S'(a)]$$

which also vanishes since S is the natural spline. Notice that this remnant would still vanish if we assumed instead that f' and S' agree at the endpoints. In either case, reordering the original equation just slightly,

$$\int_a^b S''(x)^2\,dx = \int_a^b f''(x)^2\,dx - \int_a^b [f''(x) - S''(x)]^2\,dx$$

which does make the first integral smaller than the second.

9.18. Fit a cubic spline to this data.

x_i	0	2	2.5	3	3.5	4	4.5	5	6
y_i	0	2.9	3.5	3.8	3.5	3.5	3.5	2.6	0

Choosing the natural spline, the system of Problem 9.4 provides seven equations for the seven interior C_i. Their solution, rounded to two places, follows:

i	1	2	3	4	5	6	7
C_i	−.23	−.72	−4.08	2.65	.69	−5.40	−.70

A plot of the nine data points and the spline segments appears as Fig. 9-3. Recalling that the C_i are the second-derivative values at the data points, with C_0 and C_8 zero, it is reassuring to observe their behavior across the interval, particularly the large values more or less where expected.

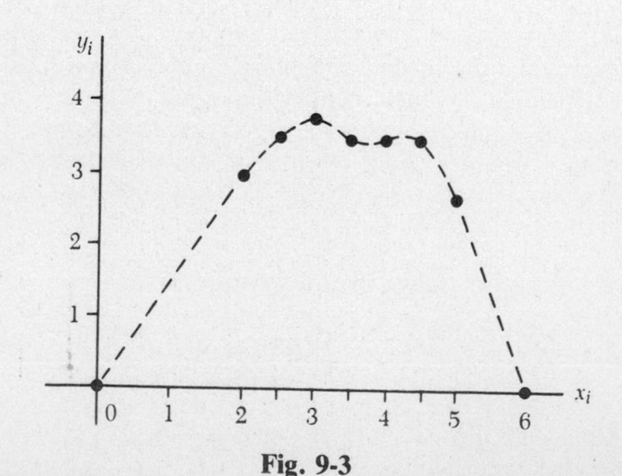

Fig. 9-3

Supplementary Problems

9.19. Verify that the spline of Problem 9.7 meets all the conditions imposed upon it.

9.20. Verify that the first cubic segment in Problem 9.8 is

$$S_1(x) = \frac{C_0}{2}\left(\frac{\pi}{3} - x\right)^3 + \frac{C_1}{2}x^3 - \frac{C_0\pi^2}{54}\left(1 - \frac{3x}{\pi}\right) + \left(\frac{\sqrt{3}}{2} - \frac{C_1\pi^2}{54}\right)\frac{3x}{\pi}$$

and find the other two segments. Verify that they meet the requirements imposed upon them.

9.21. Verify the details given in Problem 9.10.

9.22. Find the natural spline that passes through these points.

x_i	0	1	2	3	4
y_i	0	0	1	0	0

9.23. Apply the procedure of Problem 9.10 to the preceding data, finding a spline of two segments on the center two subintervals. The only knot will be at $x = 2$, but the spline must also, of course, pass through the two endpoints.

9.24. The case in which all data points fall on a straight line is hardly one that calls for a spline, but it is worth a moment's attention. Recall that the constants C_i are values of the second derivative and in this case must all be zero. How does our linear system manage this?

9.25. What happens to our linear system if all data points fall on a parabola?

Chapter 10

Osculating Polynomials

Osculating polynomials not only agree in value with a given function at specified arguments, which is the idea of collocation, but their derivatives up to some order also match the derivatives of the given function, usually at the same arguments. Thus for the simplest osculation, we require

$$p(x_k) = y(x_k) \qquad p'(x_k) = y'(x_k)$$

for $k = 0, 1, \ldots, n$. In the language of geometry, this makes the curves representing our two functions tangent to each other at these $n+1$ points. Higher-order osculation would also require $p''(x_k) = y''(x_k)$, and so on. The corresponding curves then have what is called contact of higher order. The existence and uniqueness of osculating polynomials can be proved by methods resembling those used with the simpler collocation polynomials.

Hermite's formula, for example, exhibits a polynomial of degree $2n+1$ or less which has first-order osculation. It has the form

$$p(x) = \sum_{i=0}^{n} U_i(x)y_i + \sum_{i=0}^{n} V_i(x)y_i'$$

where y_i and y_i' are the values of the given function and its derivative at x_i. The functions $U_i(x)$ and $V_i(x)$ are polynomials having properties similar to those of the Lagrange multipliers $L_i(x)$ presented earlier. In fact,

$$U_i(x) = [1 - 2L_i'(x_i)(x - x_i)][L_i(x)]^2$$
$$V_i(x) = (x - x_i)[L_i(x)]^2$$

The *error of Hermite's formula* can be expressed in a form resembling that of the collocation error but with a higher-order derivative, an indication of the greater accuracy obtainable by osculation. The error is

$$y(x) - p(x) = \frac{y^{(2n+2)}(\xi)}{(2n+2)!}[\pi(x)]^2$$

A *method of undetermined coefficients* may be used to obtain polynomials having higher-order osculation. For example, taking $p(x)$ in standard form

$$p(x) = c_0 + c_1x + c_2x^2 + \cdots + c_{3n+2}x^{3n+2}$$

and requiring $p(x_k) = y_k$, $p'(x_k) = y_k'$, $p''(x_k) = y_k''$ for the arguments $x_0, \ldots, x_n$ leads to $3n+3$ equations for the $3n+3$ coefficients c_i. Needless to say, for large n this will be a large system of equations. The methods of a later chapter may be used to solve such a system. In certain cases special devices may be used to effect simplifications.

Solved Problems

10.1. Verify that $p(x) = \sum_{i=0}^{n} U_i(x)y_i + \sum_{i=0}^{n} V_i(x)y_i'$ will be a polynomial of degree $2n+1$ or less, satisfying $p(x_k) = y_k$, $p'(x_k) = y_k'$ provided

(*a*) $U_i(x)$ and $V_i(x)$ are polynomials of degree $2n+1$.

(*b*) $U_i(x_k) = \delta_{ik}$, $V_i(x_k) = 0$.

(c) $U_i'(x_k)=0,\ V_i'(x_k)=\delta_{ik}$.

where $\delta_{ik}=\begin{cases}0 & \text{for } i\neq k\\ 1 & \text{for } i=k\end{cases}$.

The degree issue is obvious, since an additive combination of polynomials of given degree is a polynomial of the same or lower degree. Substituting $x=x_k$ we have

$$p(x_k)=U_k(x_k)y_k+0=y_k$$

and similarly substituting $x=x_k$ into $p'(x)$,

$$p'(x_k)=V_k'(x_k)y_k'=y_k'$$

all other terms being zero.

10.2. Recalling that the Lagrangian multiplier $L_i(x)$ satisfies $L_i(x_k)=\delta_{ik}$, show that

$$U_i(x)=[1-2L_i'(x_i)(x-x_i)][L_i(x)]^2 \qquad V_i(x)=(x-x_i)[L_i(x)]^2$$

meet the requirements listed in Problem 10.1.

Since $L_i(x)$ is of degree n, its square has degree $2n$ and both $U_i(x)$ and $V_i(x)$ are of degree $2n+1$. For the second requirement we note that $U_i(x_k)=V_i(x_k)=0$ for $k\neq i$, since $L_i(x_k)=0$. Also, substituting $x=x_i$,

$$U_i(x_i)=[L_i(x_i)]^2=1 \qquad V_i(x_i)=0$$

so that $U_i(x_k)=\delta_{ik}$ and $V_i(x_k)=0$. Next calculate the derivatives

$$U_i'(x)=[1-2L_i'(x_i)(x-x_i)]2L_i'(x)L_i(x)-2L_i'(x_i)[L_i(x)]^2$$
$$V_i'(x)=(x-x_i)2L_i(x)L_i'(x)+[L_i(x)]^2$$

At once $U_i'(x_k)=0$ and $V_i'(x_k)=0$ for $k\neq i$ because of the $L_i(x_k)$ factor. And for $x=x_i$, $U_i'(x_i)=2L_i'(x_i)-2L_i'(x_i)=0$ since $L_i(x_i)=1$. Finally, $V_i'(x_i)=[L_i(x_i)]^2=1$. The Hermite formula is therefore

$$p(x)=\sum_{i=0}^{n}[1-2L_i'(x_i)(x-x_i)][L_i(x)]^2y_i+(x-x_i)[L_i(x)]^2y_i'$$

10.3. A switching path between parallel railroad tracks is to be a cubic polynomial joining positions $(0,0)$ and $(4,2)$ and tangent to the lines $y=0$ and $y=2$, as shown in Fig. 10-1. Apply Hermite's formula to produce this polynomial.

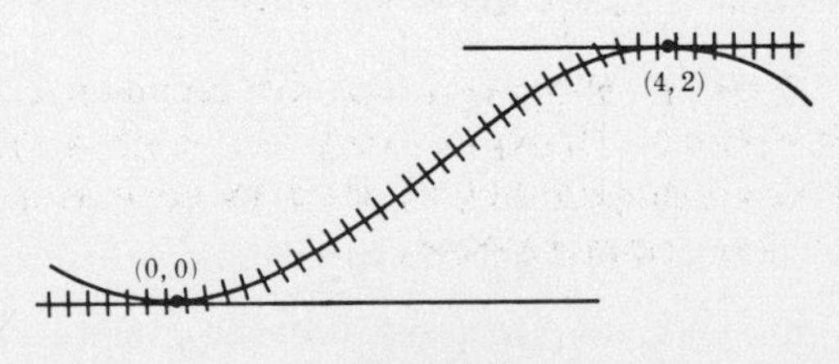

Fig. 10-1

The specifications ask for a cubic polynomial matching this data.

x_k	y_k	y_k'
0	0	0
4	2	0

With $n=1$, we have

$$L_0(x)=\frac{x-x_1}{x_0-x_1} \qquad L_1(x)=\frac{x-x_0}{x_1-x_0} \qquad L_0'(x)=\frac{1}{x_0-x_1} \qquad L_1'(x)=\frac{1}{x_1-x_0}$$

and substituting into Hermite's formula (only the y_1 term need be computed since $y_0=y_0'=y_1'=0$),

$$p(x)=\left(1-2\frac{x-4}{4-0}\right)\left(\frac{x-0}{4-0}\right)^2\cdot 2=\frac{1}{16}(6-x)x^2$$

The significance of this switching path is, of course, that it provides a smooth journey. Being tangent to both of the parallel tracks, there are no sudden changes of direction, no corners. Since $p''(0)$ and $p''(4)$ are not zero, there are, however, discontinuities in curvature. (But see Problem 10.7.)

10.4. Obtain a formula for the difference between $y(x)$ and its polynomial approximation $p(x)$.

The derivation is very similar to that for the simpler collocation polynomial. Since $y(x)=p(x)$ and $y'(x)=p'(x)$ at the arguments $x_0,\ldots,x_n$ we anticipate a result of the form

$$y(x)-p(x)=C[\pi(x)]^2$$

where $\pi(x)=(x-x_0)\cdots(x-x_n)$ as before. Accordingly we define the function

$$F(x)=y(x)-p(x)-C[\pi(x)]^2$$

which has $F(x_k)=F'(x_k)=0$ for $k=0,\ldots,n$. By choosing any new argument x_{n+1} in the interval between x_0 and x_n, and making

$$C=\frac{y(x_{n+1})-p(x_{n+1})}{[\pi(x_{n+1})]^2}$$

we also make $F(x_{n+1})=0$. Since $F(x)$ now has $n+2$ zeros at least, $F'(x)$ will have $n+1$ zeros at intermediate points. It also has zeros at $x_0,\ldots,x_n$, making $2n+2$ zeros in all. This implies that $F''(x)$ has $2n+1$ zeros at least. Successive applications of Rolle's theorem now show that $F^{(3)}(x)$ has $2n$ zeros at least, $F^{(4)}(x)$ has $2n-1$ zeros, and so on to $F^{(2n+2)}(x)$ which is guaranteed at least one zero in the interval between x_0 and x_n, say at $x=\xi$. Calculating this derivative, we get

$$F^{(2n+2)}(\xi)=y^{(2n+2)}(\xi)-C(2n+2)!=0$$

which can be solved for C. Substituting back,

$$y(x_{n+1})-p(x_{n+1})=\frac{y^{(2n+2)}(\xi)}{(2n+2)!}[\pi(x_{n+1})]^2$$

Recalling that x_{n+1} can be any argument other than $x_0,\ldots,x_n$ and noticing that this result is even true for $x_0,\ldots,x_n$ (both sides being zero), we replace x_{n+1} by the simpler x:

$$y(x)-p(x)=\frac{y^{(2n+2)}(\xi)}{(2n+2)!}[\pi(x)]^2$$

10.5. Prove that only one polynomial can meet the specifications of Problem 10.1.

Suppose there were two. Since they must share common y_k and y_k' values at the arguments x_k, we may choose one of them as the $p(x)$ of Problem 10.4 and the other as the $y(x)$. In other words, we may view one polynomial as an approximation to the other. But since $y(x)$ is now a polynomial of degree $2n+1$, it follows that $y^{(2n+2)}(\xi)$ is zero. Thus $y(x)$ is identical with $p(x)$, and our two polynomials are actually one and the same.

10.6. How can a polynomial be found which matches the following data?

x_0	y_0	y_0'	y_0''
x_1	y_1	y_1'	y_1''

In other words, at two arguments the values of the polynomial and its first two derivatives are specified.

Assume for simplicity that $x_0 = 0$. If this is not true, then a shift of argument easily achieves it. Let

$$p(x) = y_0 + xy_0' + \frac{1}{2}x^2y_0'' + Ax^3 + Bx^4 + Cx^5$$

with A, B, and C to be determined. At $x = x_0 = 0$ the specifications have already been met. At $x = x_1$ they require

$$Ax_1^3 + Bx_1^4 + Cx_1^5 = y_1 - y_0 - x_1y_0' - \frac{1}{2}x_1^2y_0''$$

$$3Ax_1^2 + 4Bx_1^3 + 5Cx_1^4 = y_1' - y_0' - x_1y_0''$$

$$6Ax_1 + 12Bx_1^2 + 20Cx_1^3 = y_1'' - y_0''$$

These three equations determine A, B, C uniquely.

10.7. A switching path between parallel railroad tracks is to join positions $(0, 0)$ and $(4, 2)$. To avoid discontinuities in both direction and curvature the following specifications are made:

x_k	y_k	y_k'	y_k''
0	0	0	0
4	2	0	0

Find a polynomial which meets these specifications.

Applying the procedure of Problem 10.6,

$$p(x) = Ax^3 + Bx^4 + Cx^5$$

the quadratic portion vanishing entirely. At $x_1 = 4$ we find

$$64A + 256B + 1024C = 2 \qquad 48A + 256B + 1280C = 0 \qquad 24A + 192B + 1280C = 0$$

from which $A = \frac{40}{128}$, $B = -\frac{15}{128}$, $C = \frac{3}{256}$. Substituting, $p(x) = \frac{1}{256}(80x^3 - 30x^4 + 3x^5)$.

Supplementary Problems

10.8. Apply Hermite's formula to find a cubic polynomial which meets these specifications.

x_k	y_k	y_k'
0	0	0
1	1	1

This can be viewed as a switching path between nonparallel tracks.

10.9. Apply Hermite's formula to find a polynomial which meets these specifications.

x_k	y_k	y_k'
0	0	0
1	1	0
2	0	0

10.10. Apply the method of Problem 10.6 to find a fifth-degree polynomial which meets these specifications.

x_k	y_k	y'_k	y''_k
0	0	0	0
1	1	1	0

This is a smoother switching path than that of Problem 10.8.

10.11. Find two second-degree polynomials, one having $p_1(0) = p'_1(0) = 0$, the other having $p_2(4) = 2$, $p'_2(4) = 0$, both passing through $(2, 1)$, as shown in Fig. 10-2. Show that $p'_1(2) = p'_2(2)$ so that a pair of parabolic arcs also serves as a switching path between parallel tracks, as well as the cubic of Problem 10.3.

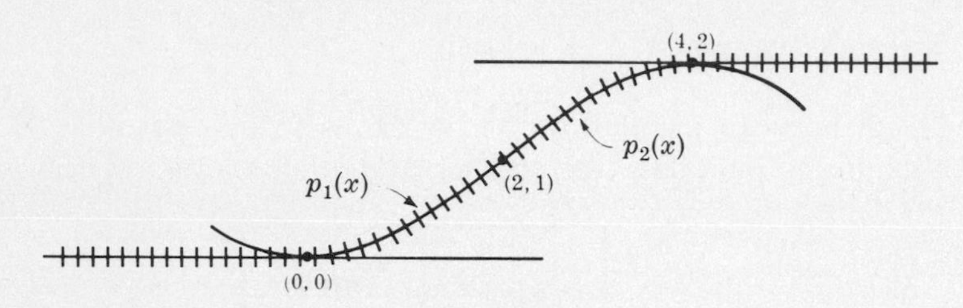

Fig. 10-2

10.12. Find two fourth-degree polynomials, one having $p_1(0) = p'_1(0) = p''_1(0) = 0$, the other having $p_2(4) = 2$, $p'_2(4) = p''_2(4) = 0$, both passing through $(2, 1)$ with $p''_1(2) = p''_2(2) = 0$. This is another switching path for which direction and curvature are free of discontinuities, like the fifth-degree polynomial of Problem 10.7. Verify this by showing that first and second derivatives agree on both sides of $(0, 0)$, $(2, 1)$, and $(4, 2)$ where the four pieces of track are butted together.

10.13. From Hermite's formula for two-point osculation derive the midpoint formula

$$p_{1/2} = \frac{1}{2}(y_0 + y_1) + \frac{1}{8}L(y'_0 - y'_1)$$

where $L = x_1 - x_0$.

10.14. Show that the error of the formula in Problem 10.13 is $L^4 y^{(4)}(\xi)/384$.

10.15. Find a polynomial of degree four which meets the following conditions:

x_k	y_k	y'_k
0	1	0
1	0	—
2	9	24

Note that one of the y'_k values is not available.

10.16. Find a polynomial of degree four which meets these conditions.

x_k	y_k	y'_k	y''_k
0	1	−1	0
1	2	7	—

10.17. Find a polynomial of degree three which meets these conditions.

x_k	y_k	y''_k
0	1	−2
1	1	4

Chapter 11

The Taylor Polynomial

TAYLOR POLYNOMIAL

The Taylor polynomial is the ultimate in osculation. For a single argument x_0 the values of the polynomial and its first n derivatives are required to match those of a given function $y(x)$. That is,

$$p^{(i)}(x_0) = y^{(i)}(x_0) \qquad \text{for } i = 0, 1, \ldots, n$$

The existence and uniqueness of such a polynomial will be proved, and are classical results of analysis. The **Taylor formula** settles the existence issue directly by exhibiting such a polynomial in the form

$$p(x) = \sum_{i=0}^{n} \frac{y^{(i)}(x_0)}{i!}(x - x_0)^i$$

The *error of the Taylor polynomial,* when viewed as an approximation to $y(x)$, can be expressed by the *integral formula*

$$y(x) - p(x) = \frac{1}{n!}\int_{x_0}^{x} y^{(n+1)}(x_0)(x - x_0)^n \, dx_0$$

Lagrange's error formula may be deduced by applying a mean value theorem to the integral formula. It is

$$y(x) - p(x) = \frac{y^{(n+1)}(\xi)}{(n+1)!}(x - x_0)^{n+1}$$

and clearly resembles our error formulas of collocation and osculation.

If the derivatives of $y(x)$ are bounded independently of n, then either error formula serves to estimate the degree n required to reduce $|y(x) - p(x)|$ below a prescribed tolerance over a given interval of arguments x.

Analytic functions have the property that, for n tending to infinity, the above error of approximation has limit zero for all arguments x in a given interval. Such functions are then represented by the Taylor series

$$y(x) = \sum_{i=0}^{\infty} \frac{y^{(i)}(x_0)}{i!}(x - x_0)^i$$

The **binomial series** is an especially important case of the Taylor series. For $-1 < x < 1$ we have

$$(1 + x)^p = \sum_{i=0}^{\infty} \binom{p}{i} x^i$$

DIFFERENTIATION OPERATOR D

The **differentiation operator** D *is defined by*

$$D = h\frac{d}{dx}$$

The **exponential operator** may then be defined by

$$e^{kD} = \sum_{i=0}^{\infty} \frac{k^i D^i}{i!}$$

and the **Taylor series in operator form** becomes

$$y(x_k) = e^{kD} y_0(x_0)$$

The relationship between D and Δ may be expressed in either of the forms

$$\Delta + 1 = e^D \qquad D = \Delta - \frac{1}{2}\Delta^2 + \frac{1}{3}\Delta^3 - \cdots$$

both of which involve "infinite series" operators.

The *Euler transformation* is another useful relationship between infinite series operators. It may be written as

$$(1+E)^{-1} = \frac{1}{2}\left[1 - \frac{1}{2}\Delta + \frac{1}{4}\Delta^2 - \frac{1}{8}\Delta^3 + \cdots\right]$$

by using the binomial series.

The **Bernoulli numbers** B_i are defined by

$$\frac{x}{e^x - 1} = \sum_{i=0}^{\infty} \frac{1}{i!} B_i x^i$$

Actually expanding the left side into its Taylor series we shall find $B_0 = 1$, $B_1 = -\frac{1}{2}$, $B_2 = \frac{1}{6}$, and so on. These numbers occur in various operator equations. For example, the indefinite summation operator Δ^{-1} is defined by

$$\Delta F_k = y_k \qquad F_k = \Delta^{-1} y_k$$

and is related to D by

$$\Delta^{-1} = D^{-1} \sum_{i=0}^{\infty} \frac{1}{i!} B_i D^i$$

where the B_i are Bernoulli numbers. The operator D^{-1} is the familiar indefinite integral operator.

The **Euler-Maclaurin formula** may be deduced from the previous relationship,

$$\sum_{i=0}^{n-1} y_i = \int_0^n y_k \, dk - \frac{1}{2}(y_n - y_0) + \frac{h}{12}(y'_n - y'_0) + \cdots$$

and is often used for the evaluation of either sums or integrals.

The powers of D may be expressed in terms of the central difference operator δ by using Taylor series. Some examples are the following:

$$D = \mu\left(\delta - \frac{1^2}{3!}\delta^3 + \frac{1^2 \cdot 2^2}{5!}\delta^5 - \frac{1^2 \cdot 2^2 \cdot 3^2}{7!}\delta^7 + \cdots\right)$$

$$D^2 = \delta^2 - \frac{1}{12}\delta^4 + \frac{1}{90}\delta^6 - \frac{1}{560}\delta^8 + \frac{1}{3150}\delta^{10} - \cdots$$

Solved Problems

11.1. Find the polynomial $p(x)$, of degree n or less, which together with its first n derivatives takes the values $y_0, y_0^{(1)}, y_0^{(2)}, \ldots, y_0^{(n)}$ for the argument x_0.

A polynomial of degree n can be written

$$p(x) = a_0 + a_1(x - x_0) + \cdots + a_n(x - x_0)^n$$

Successive differentiations produce

$$p^{(1)}(x) = a_1 + 2a_2(x - x_0) + \cdots + na_n(x - x_0)^{n-1}$$
$$p^{(2)}(x) = 2a_2 + 3 \cdot 2a_3(x - x_0) + \cdots + n(n-1)a_n(x - x_0)^{n-2}$$
$$\cdots\cdots\cdots\cdots\cdots\cdots\cdots$$
$$p^{(n)}(x) = n!\, a_n$$

The specifications then require

$$p(x_0) = a_0 = y_0 \qquad p^{(1)}(x_0) = a_1 = y_0^{(1)} \qquad p^{(2)}(x_0) = 2a_2 = y_0^{(2)} \qquad \cdots \qquad p^{(n)}(x_0) = n!\, a_n = y_0^{(n)}$$

Solving for the a_n coefficients and substituting

$$p(x) = y_0 + y_0^{(1)}(x - x_0) + \cdots + \frac{1}{n!} y_0^{(n)}(x - x_0)^n = \sum_{i=0}^{n} \frac{1}{i!} y_0^{(i)}(x - x_0)^i$$

11.2. **Find a polynomial $p(x)$ of degree n, such that, at $x_0 = 0$, $p(x)$ and e^x agree in value together with their first n derivatives.**

Since for e^x derivatives of all orders are also e^x,

$$y_0 = y_0^{(1)} = y_0^{(2)} = \cdots = y_0^{(n)} = 1$$

The Taylor polynomial can then be written

$$p(x) = \sum_{i=0}^{n} \frac{1}{i!} x^n = 1 + x + \frac{1}{2}x^2 + \frac{1}{6}x^3 + \cdots + \frac{1}{n!}x^n$$

11.3. **Consider a second function $y(x)$ also having the specifications of Problem 11.1. We shall think of $p(x)$ as a polynomial approximation to $y(x)$. Obtain a formula for the difference $y(x) - p(x)$ in integral form, assuming $y^{(n+1)}(x)$ continuous between x_0 and x.**

Here it is convenient to use a different procedure from that which led us to error estimates for the collocation and osculating polynomials. We start by temporarily calling the difference R,

$$R = y(x) - p(x)$$

or in full detail

$$R(x, x_0) = y(x) - y(x_0) - y'(x_0)(x - x_0) - \frac{1}{2} y''(x_0)(x - x_0)^2 - \cdots - \frac{1}{n!} y^{(n)}(x_0)(x - x_0)^n$$

This actually defines R as a function of x and x_0. Calculating the derivative of R relative to x_0, holding x fixed, we find

$$R'(x, x_0) = -y'(x_0) + y'(x_0) - y''(x_0)(x - x_0) + y''(x_0)(x - x_0)$$
$$- \frac{1}{2} y^{(3)}(x_0)(x - x_0)^2 + \cdots - \frac{1}{n!} y^{(n+1)}(x_0)(x - x_0)^n$$
$$= -\frac{1}{n!} y^{(n+1)}(x_0)(x - x_0)^n$$

since differentiation of the second factor in each product cancels the result of differentiating the first factor in the previous product. Only the very last term penetrates through. Having differentiated relative to x_0, we reverse direction and integrate relative to x_0 to recover R.

$$R(x, x_0) = -\frac{1}{n!} \int_x^{x_0} y^{(n+1)}(u)(x - u)^n \, du + \text{constant}$$

By the original definition of R, $R(x_0, x_0) = 0$ and the constant of integration is 0. Reversing the limits,

$$R(x, x_0) = \frac{1}{n!} \int_{x_0}^{x} y^{(n+1)}(u)(x - u)^n \, du$$

which is known as an integral form of the error.

11.4. Obtain Lagrange's form of the error from the integral form.

Here we use the mean value theorem of calculus, which says that if $f(x)$ is continuous and $w(x)$ does not change sign in the interval (a, b) then

$$\int_a^b f(x)w(x)\,dx = f(\xi)\int_a^b w(x)\,dx$$

where ξ is between a and b. Choosing $w(x) = (x - x_0)^n$, we easily get

$$R(x, x_0) = \frac{1}{(n+1)!}y^{(n+1)}(\xi)(x - x_0)^{n+1}$$

where ξ is between x_0 and x but otherwise unknown. This form of the error is very popular because of its close resemblance to the terms of the Taylor polynomial. Except for a ξ in place of an x_0 it would be the term which produced the Taylor polynomial of next higher degree.

11.5. Estimate the degree of a Taylor polynomial for the function $y(x) = e^x$, with $x_0 = 0$, which guarantees approximations correct to three decimal places for $-1 < x < 1$. To six decimal places.

By the Lagrange formula for the error,

$$|e^x - p(x)| = |R| \leqq \frac{e}{(n+1)!}$$

For three-place accuracy this should not exceed .0005, a condition which is satisfied for $n = 7$ or higher. The polynomial

$$p(x) = \sum_{i=0}^{7} \frac{1}{i!}x^i$$

is therefore adequate. Similarly, for six-place accuracy $|R|$ should not exceed .0000005, which will be true for $n = 10$.

11.6. The *operator* D is defined by $D = h\dfrac{\partial}{\partial x}$. What is the result of applying the successive powers of D to $y(x)$?

We have at once $D^i y(x) = h^i y^{(i)}(x)$.

11.7. Express the Taylor polynomial in operator symbolism.

Let $x - x_0 = kh$. This is the symbolism we have used earlier, with x_k now abbreviated to x. Then direct substitution into the Taylor polynomial of Problem 11.1 brings

$$p(x) = \sum_{i=0}^{n} \frac{1}{i!} y_0^{(i)}(x - x_0)^i = \sum_{i=0}^{n} \frac{1}{i!} y_0^{(i)} k^i h^i = \sum_{i=0}^{n} \frac{1}{i!} k^i D^i y(x_0)$$

A common way of rewriting this result is

$$p(x) = \left(\sum_{i=0}^{n} \frac{1}{i!} k^i D^i\right) y(x_0)$$

or in terms of the integer variable k alone as

$$p_k = \left(\sum_{i=0}^{n} \frac{1}{i!} k^i D^i\right) y_0$$

where as usual $p(x_k) = p_k$.

11.8. A function $y(x)$ is called *analytic* on the interval $|x - x_0| \leqq r$ if as $n \to \infty$,

$$\lim R(x, x_0) = 0$$

for all arguments x in the interval. It is then customary to write $y(x)$ as an infinite series, called a **Taylor series**

$$y(x) = \lim p(x) = \sum_{i=0}^{\infty} \frac{1}{i!} y_0^{(i)}(x - x_0)^i$$

Express this in operator form.

Proceeding just as in Problem 11.7, we find $y(x_k) = \left(\sum_{i=0}^{\infty} \frac{1}{i!} k^i D^i \right) y_0$. This is our first "infinite series operator." The arithmetic of such operators is not so easy to justify as was the case with the simpler operators used earlier.

11.9. The operator e^{kD} is defined by $e^{kD} = \sum_{i=0}^{\infty} \frac{1}{i!} k^i D^i$. Write the Taylor series using this operator.

We have at once $y(x_k) = e^{kD} y_0$.

11.10. Prove $e^D = E$.

By Problem 11.9 with $k = 1$ and the definition of E, $y(x_1) = y_1 = Ey_0 = e^D y_0$ making $E = e^D$.

11.11. Develop the Taylor series for $y(x) = \ln(1 + x)$, using $x_0 = 0$.

The derivatives are $y^{(i)}(x) = (-1)^{i+1}(i-1)!/(1+x)^i$ so that $y^{(i)}(0) = (-1)^{i+1}(i-1)!$. Since $y(0) = \ln 1 = 0$, we have

$$y(x) = \ln(1 + x) = \sum_{i=1}^{\infty} \frac{(-1)^{i+1}}{i} x^i = x - \frac{1}{2}x^2 + \frac{1}{3}x^3 - \frac{1}{4}x^4 + \cdots$$

The familiar ratio test shows this to be convergent for $-1 < x < 1$. It does not, however, prove that the series equals $\ln(1 + x)$. To prove this let $p(x)$ represent the Taylor polynomial, of degree n. Then by the Lagrange formula for the error,

$$|\ln(1 + x) - p(x)| \leqq \frac{1}{(n+1)!} \cdot \frac{n!}{(1+\xi)^{n+1}} \cdot x^{n+1}$$

For simplicity consider only the interval $0 \leqq x < 1$. The series is applied mostly to this interval anyway. Then the error can be estimated by replacing ξ by 0 and x by 1 to give $|\ln(1 + x) - p(x)| \leqq 1/(n + 1)$ and this does have limit 0. Thus $\lim p(x) = \ln(1 + x)$, which was our objective.

11.12. Estimate the degree of a Taylor polynomial for the function $y(x) = \ln(1 + x)$, with $x_0 = 0$, which guarantees three-decimal-place accuracy for $0 < x < 1$.

By the Lagrange formula for the error,

$$|\ln(1 + x) - p(x)| \leqq \frac{1}{(n+1)!} \cdot \frac{n!}{(1+\xi)^n} \cdot x^{n+1} \leqq \frac{1}{n+1}$$

Three-place accuracy requires that this not exceed .0005, which is satisfied for $n = 2000$ or higher. A polynomial of degree 2000 would be needed! This is an example of a slowly convergent series.

11.13. Express the operator D in terms of the operator Δ.

From $e^D = E$ we find $D = \ln E = \ln(1 + \Delta) = \Delta - \frac{1}{2}\Delta^2 + \frac{1}{3}\Delta^3 - \frac{1}{4}\Delta^4 + \cdots$.

The validity of this calculation is surely open to suspicion, and any application of it must be carefully checked. It suggests that the final series operator will produce the same result as the operator D.

11.14. Express $y(x) = (1+x)^p$ as a Taylor series.

For p a positive integer this is the binomial theorem of algebra. For other values of p it is the *binomial series*. Its applications are extensive. We easily find

$$y^{(i)}(x) = p(p-1)\cdots(p-i+1)(1+x)^{p-i} = p^{(i)}(1+x)^{p-i}$$

where $p^{(i)}$ is again the factorial polynomial. Choosing $x_0 = 0$

$$y^{(i)}(0) = p^{(i)}$$

and substituting into the Taylor series,

$$y(x) = \sum_{i=0}^{\infty} \frac{p^{(i)}}{i!} x^i = \sum_{i=0}^{\infty} \binom{p}{i} x^i$$

where $\binom{p}{i}$ is the generalized binomial coefficient. The convergence of this series to $y(x)$ for $-1 < x < 1$ can be demonstrated.

11.15. Use the binomial series to derive the *Euler transformation*.

The **Euler transformation** is an extensive rearrangement of the alternating series $S = a_0 - a_1 + a_2 - a_3 + \cdots$ which we rewrite as

$$S = (1 - E + E^2 - E^3 + \cdots)a_0 = (1+E)^{-1}a_0$$

by the binomial theorem with $p = -1$. The operator $(1+E)^{-1}$ may be interpreted as the inverse operator of $1 + E$. A second application of the binomial theorem now follows.

$$S = (1+E)^{-1}a_0 = (2+\Delta)^{-1}a_0 = \frac{1}{2}\left(1 + \frac{\Delta}{2}\right)^{-1} a_0$$

$$= \frac{1}{2}\left(1 - \frac{\Delta}{2} + \frac{\Delta^2}{4} - \frac{\Delta^3}{8} + \cdots\right)a_0 = \frac{1}{2}\left(a_0 - \frac{1}{2}\Delta a_0 + \frac{1}{4}\Delta^2 a_0 - \frac{1}{8}\Delta^3 a_0 + \cdots\right)$$

Our derivation of this formula has been a somewhat optimistic application of operator arithmetic. No general, easy-to-apply criterion for ensuring its validity exists.

11.16. The **Bernoulli numbers** are defined to be the numbers B_i in the following series:

$$y(x) = \frac{x}{e^x - 1} = \sum_{i=0}^{\infty} \frac{1}{i!} B_i x^i$$

Find $B_0, \ldots, B_{10}$.

The Taylor series requires that $y^{(i)}(0) = B_i$, but it is easier in this case to proceed differently. Multiplying by $e^x - 1$ and using the Taylor series for e^x, we get

$$x = \left(x + \frac{1}{2}x^2 + \frac{1}{6}x^3 + \cdots\right)\left(B_0 + B_1 x + \frac{1}{2}B_2 x^2 + \frac{1}{6}B_3 x^3 + \cdots\right)$$

Now comparing the coefficients of the successive powers of x,

$$B_0 = 1 \qquad B_1 = -\frac{1}{2} \qquad B_2 = \frac{1}{6} \qquad B_3 = 0 \qquad B_4 = -\frac{1}{30} \qquad B_5 = 0$$

$$B_6 = \frac{1}{42} \qquad B_7 = 0 \qquad B_8 = -\frac{1}{30} \qquad B_9 = 0 \qquad B_{10} = \frac{5}{66}$$

The process could be continued in an obvious way.

11.17. Suppose $\Delta F_k = y_k$. Then an *inverse operator* Δ^{-1} can be defined by

$$F_k = \Delta^{-1} y_k$$

This inverse operator is "indefinite" in that for given y_k the numbers F_k are determined except for an arbitrary additive constant. For example, in the following table the numbers y_k are listed as first differences. Show that the number F_0 can be chosen arbitrarily and that the other F_k numbers are then determined.

F_k	F_0		·		·		·		·		·		·		·
y_k		y_0		y_1		y_2		y_3		y_4		·		·	·

We have at once

$$F_1 = F_0 + y_0 \qquad F_2 = F_1 + y_1 = F_0 + y_0 + y_1 \qquad F_3 = F_2 + y_2 = F_0 + y_0 + y_1 + y_2$$

and in general $F_k = F_0 + \sum_{i=0}^{k-1} y_i$. The requirements plainly hold for an arbitrary F_0, and the analogy with indefinite integration is apparent.

11.18. Obtain a formula for Δ^{-1} in terms of the operator D.

The result $e^D = 1 + \Delta$ suggests

$$\Delta^{-1} = (e^D - 1)^{-1} = D^{-1}[D(e^D - 1)^{-1}]$$

where D^{-1} is an *indefinite integral operator,* an inverse of D. From the definition of Bernoulli numbers,

$$\Delta^{-1} = D^{-1} \sum_{i=0}^{\infty} \frac{1}{i!} B_i D^i$$

$$= D^{-1}\left(1 - \frac{1}{2} D + \frac{1}{12} D^2 - \frac{1}{720} D^4 + \cdots\right) = D^{-1} - \frac{1}{2} + \frac{1}{12} D - \frac{1}{720} D^3 + \cdots$$

As always with the indefinite integral (and here we also have an indefinite summation) the presence of an additive constant may be assumed.

11.19. Derive the Euler–Maclaurin formula operationally.

Combining the results of the previous two problems, we have

$$F_k = \Delta^{-1} y_k = F_0 + \sum_{i=0}^{k-1} y_i$$

$$= \left(D^{-1} - \frac{1}{2} + \frac{1}{12} D - \frac{1}{720} D^3 + \cdots\right) y_k$$

From the first of these,

$$F_n - F_0 = \sum_{i=0}^{n-1} y_i$$

while from the second,

$$F_n - F_0 = \frac{1}{h} \int_{x_0}^{x_n} y(x)\, dx - \frac{1}{2}(y_n - y_0) + \frac{h}{12}(y'_n - y'_0) - \frac{h^3}{720}(y_n^{(3)} - y_0^{(3)}) + \cdots$$

so that finally,

$$\sum_{i=0}^{n-1} y_i = \frac{1}{h} \int_{x_0}^{x_n} y(x)\, dx - \frac{1}{2}(y_n - y_0) + \frac{h}{12}(y'_n - y'_0) + \cdots$$

which is the Euler-Maclaurin formula. The operator arithmetic used in this derivation is clearly in need of supporting logic, but the result is useful in spite of its questionable pedigree and in spite of the fact that the series obtained is usually *not convergent.*

Supplementary Problems

11.20. Find the Taylor polynomials of degree n for $\sin x$ and $\cos x$, using $x_0 = 0$.

11.21. Express the error term in Lagrange's form, for both $\sin x$ and $\cos x$. Show that as $n \to \infty$ this error has limit 0 for any argument x.

11.22. For what value of n will the Taylor polynomial approximate $\sin x$ correctly to three decimal places for $0 < x < \pi/2$?

11.23. For what value of n will the Taylor polynomial approximate $\cos x$ correctly to three decimal places for $0 < x < \pi/2$? To six decimal places?

11.24. Express the operator Δ as a series operator in D.

11.25. The functions $\sinh x$ and $\cosh x$ are defined by

$$\sinh x = \frac{e^x - e^{-x}}{2} \qquad \cosh x = \frac{e^x + e^{-x}}{2}$$

Show that their Taylor series are

$$\sinh x = \sum_{i=0}^{\infty} \frac{1}{(2i+1)!} x^{2i+1} \qquad \cosh x = \sum_{i=0}^{\infty} \frac{1}{(2i)!} x^{2i}$$

11.26. Show by operator arithmetic that $\delta = 2 \sinh \frac{1}{2}D$, $\mu = \cosh \frac{1}{2}D$.

11.27. Use the binomial series to express $\Delta = \frac{1}{2}\delta^2 + \delta\sqrt{1 + \frac{1}{4}\delta^2}$ as a series in powers of δ, through the term in δ^7.

11.28. Combine the results of Problems 11.13 and 11.27 to express D as a series in powers of δ, verifying these terms through δ^7.

$$D = \delta - \frac{1^2}{2^2 \cdot 3!}\delta^3 + \frac{1^2 \cdot 3^2}{2^4 \cdot 5!}\delta^5 - \frac{1^2 \cdot 3^2 \cdot 5^2}{2^6 \cdot 7!}\delta^7 + \cdots$$

11.29. Verify these terms of a Taylor series for D^2:

$$D^2 = \delta^2 - \frac{1}{12}\delta^4 + \frac{1}{90}\delta^6 - \frac{1}{560}\delta^8 + \frac{1}{3150}\delta^{10} - \cdots$$

by squaring the result of Problem 11.28 and collecting the various powers of δ.

Chapter 12

Interpolation

HISTORICAL PLACE

Previous chapters have consisted almost entirely of supporting theory. That theory will now be used in several ways, beginning with the classic problem of interpolation. Interpolation is the familiar process of estimating the values of a function $y(x)$ for arguments between $x_0, \ldots, x_n$ at which the values $y_0, \ldots, y_n$ are known. Inverse interpolation simply proceeds in the opposite direction. Subtabulation is the systematic interpolation of many values between each pair of arguments x_i, x_{i+1} and so reducing the spacing of a table of values, perhaps from h to $h/10$. Prediction requires estimating a value $y(x)$ for x outside the interval in which the data arguments fall.

All these operations were much more pressing before the arrival of high-speed computers, which now calculate values of all the familiar functions by series or other nontabular ways. The formulas of this chapter bear the names of prominent mathematicians of a century and more ago, when tables of functions were indispensable. Their place in our subject is partly, but not entirely, historical. It is interesting to see how the computational hurdles of an earlier time were surmounted, but important to note that tables of special functions are still constructed so that some of this work continues to have a useful role.

METHODS OF SOLUTION

The methods of interpolation involve substituting for $y(x)$ some more easily computed function, often a polynomial, and simplest of all a straight line. The values $y_0, \ldots, y_n$ may be introduced into any of our polynomial formulas (Newton, Everett, ...) which then becomes an algorithm for interpolation, the output being an approximation to $y(x)$. It was realized that using data from both sides of the interpolation argument x "made sense" and led to better values or briefer computations. The formulas of Stirling, Bessel, and Everett were motivated by this reasoning and a study of the errors involved provides logical support. At the ends of a table this could not be done and the Newton forward and backward formulas had their turn. It was unnecessary to choose the degree of the approximating polynomial in advance, simply to continue fitting differences from the table into appropriate places as long as the results seemed to warrant. It was also realized that a point of diminishing returns occurs, where results deteriorate instead of improve, and that this point depends upon the accuracy of the tabulated values.

The alternative procedure of Lagrange fits the polynomial to the data without using finite differences. The degree has to be chosen in advance, but the method has compensating advantages. Aitken's method is another variant, not requiring equal spacing of tabular arguments or of the polynomial's degree at the outset.

Osculating polynomials and the Taylor polynomial also find application to interpolation problems in special circumstances.

INPUT AND ALGORITHM ERRORS

Input and algorithm errors occur in all these applications. Their impact on the completed outputs can be estimated only up to a point. It is customary to identify three main error sources.

1. **Input errors** arise when the given values $y_0, \ldots, y_n$ are inexact, as experimental or computed values usually are.
2. **Truncation error** is the difference $y(x) - p(x)$, which we accept the moment we decide to

use a polynomial approximation. This error has been found earlier to be

$$y(x) - p(x) = \frac{\pi(x)}{(n+1)!} y^{(n+1)}(\xi)$$

Though ξ is unknown, this formula can still be used at times to obtain error bounds. Truncation error is one type of algorithm error. In prediction problems this error can be substantial, since the factor $\pi(x)$ becomes extremely large outside of the interval in which the data arguments $x_0, \ldots, x_n$ fall.

3. **Roundoff errors** occur since computers operate with a fixed number of digits and any excess digits produced in multiplications or divisions are lost. They are another type of algorithm error.

Solved Problems

12.1. Predict the two missing values of y_k.

$k = x_k$	0	1	2	3	4	5	6	7
y_k	1	2	4	8	15	26		

This is a simple example, but it will serve to remind us that the basis on which applications are to be made is polynomial approximation. Calculate some differences.

$$\begin{array}{ccccc} 1 & 2 & 4 & 7 & 11 \\ & 1 & 2 & 3 & 4 \\ & & 1 & 1 & 1 \end{array}$$

Presumably the missing y_k values might be any numbers at all, but the evidence of these differences points strongly toward a polynomial of degree three, suggesting that the six y_k values given and the two to be predicted all belong to such a polynomial. Accepting this as the basis for prediction, it is not even necessary to find this collocation polynomial. Adding two more 1s to the row of third differences, we quickly supply a 5 and 6 to the row of second differences, a 16 and 22 as new first differences, and then predict $y_6 = 42$, $y_7 = 64$. This is the same data used in Problem 6.12 where the cubic collocation polynomial was found.

12.2. Values of $y(x) = \sqrt{x}$ are listed in Table 12.1, rounded off to four decimal places, for arguments $x = 1.00(.01)1.06$. (This means that the arguments run from 1.00 to 1.06 and are equally spaced with $h = .01$.) Calculate differences to Δ^6 and explain their significance.

The differences are also listed in Table 12.1.

For simplicity, leading zeros are often omitted in recording differences. In this table all differences are in the fourth decimal place. Though the square root function is certainly not linear, the first differences are almost constant, suggesting that over the interval tabulated and to four-place accuracy this function may be accurately approximated by a linear polynomial. The entry Δ^2 is best considered a unit roundoff error, and its effect on higher differences follows the familiar binomial coefficient pattern observed in Problem 3.10. In this situation one would ordinarily calculate only the first differences. Many familiar functions such as $\sqrt{x}$, $\log x$, $\sin x$, etc., have been tabulated in this way, with arguments so tightly spaced that first differences are almost constant and the function can be accurately approximated by a linear polynomial.

12.3. Apply Newton's forward formula with $n = 1$ to interpolate for $\sqrt{1.005}$.

Table 12.1

x	$y(x)=\sqrt{x}$	Δ	Δ^2	Δ^3	Δ^4	Δ^5	Δ^6
1.00	1.0000						
		50					
1.01	1.0050		0				
		50		−1			
1.02	1.0100		−1		2		
		49		1		−3	
1.03	1.0149		0		−1		4
		49		0		1	
1.04	1.0198		0		0		
		49		0			
1.05	1.0247		0				
		49					
1.06	1.0296						

Newton's formula reads

$$p_k = y_0 + \binom{k}{1}\Delta y_0 + \binom{k}{2}\Delta^2 y_0 + \cdots + \binom{k}{n}\Delta^n y_0$$

Choosing $n = 1$ for a linear approximation we find, with $k = \dfrac{x - x_0}{h} = \dfrac{1.005 - 1.00}{.01} = \dfrac{1}{2}$,

$$p_k = 1.0000 + \frac{1}{2}(.0050) = 1.0025$$

This is hardly a surprise. Since we have used a linear collocation polynomial, matching our $y = \sqrt{x}$ values at arguments 1.00 and 1.01, we could surely have anticipated this midway result.

12.4. What would be the effect of using a higher-degree polynomial for the interpolation of Problem 12.3?

An easy computation shows the next several terms of the Newton formula, beginning with the second difference term, to be approximately .00001. They would not affect our result at all.

12.5. Values of $y(x) = \sqrt{x}$ are listed in Table 12.2, rounded off to five decimal places, for arguments $x = 1.00(.05)1.30$. Calculate differences to Δ^6 and explain their significance.

The differences are listed in Table 12.2.

Table 12.2

x	$y(x)=\sqrt{x}$	Δ	Δ^2	Δ^3	Δ^4	Δ^5	Δ^6
1.00	1.00000						
		2470					
1.05	1.02470		−59				
		2411		5			
1.10	1.04881		−54		−1		
		2357		4		−1	
1.15	1.07238		−50		−2		4
		2307		2		3	
1.20	1.09544		−48		1		
		2259		3			
1.25	1.11803		−45				
		2214					
1.30	1.14017						

Here the error pattern is more confused but the fluctuations of + and − signs in the last three columns are reminiscent of the effects produced in Problems 3.10 and 3.11. It may be best to view these three columns as error effects, not as useful information for computing the square root function.

12.6. Use the data of Problem 12.5 to interpolate for $\sqrt{1.01}$.

Newton's forward formula is convenient for interpolations near the top of a table. With $k=0$ at the top entry $x_0=1.00$, this choice usually leads to diminishing terms and makes the decision of how many terms to use almost automatic. Substituting into the formula as displayed in Problem 12.3, with $k=(x-x_0)/h=(1.01-1.00)/.05=\frac{1}{5}$, we find

$$p_k = 1.00000 + \frac{1}{5}(.02470) - \frac{2}{25}(-.00059) + \frac{6}{125}(.00005)$$

stopping with this term since it will not affect the fifth decimal place. Notice that this last term uses the highest-order difference which we felt, in Problem 12.5, to be significant for square root computations. We have not trespassed into columns which were presumably only error effects. The value p_k reduces to

$$p_k = 1.000000 + .004940 + .000048 + .000002 = 1.00499$$

which is correct to five places. (It is a good idea to carry an extra decimal place during computations, if possible, to control "algorithm errors" described in Chapter 1. In machine computations, of course, the number of digits is fixed anyway, so this remark would not apply.)

12.7. Use the data of problem 12.5 to interpolate for $\sqrt{1.28}$.

Here Newton's backward formula is convenient and most of the remarks made in Problem 12.6 again apply. With $k=0$ at the bottom entry $x_0=1.30$, we have $k=(x-x_0)/h=(1.28-1.30)/.05=-\frac{2}{5}$. Substituting into the backward formula (Problem 7.9)

$$p_k = y_0 + k\,\nabla y_0 = \frac{k(k+1)}{2}\nabla^2 y_0 + \frac{k(k+1)(k+2)}{3!}\nabla^3 y_0 + \cdots + \frac{k(k+1)\cdots(k+n-1)}{n!}\nabla^n y_0$$

we obtain $$p_k = 1.14017 + \left(-\frac{2}{5}\right)(.02214) + \left(-\frac{3}{25}\right)(-.00045) + \left(-\frac{8}{125}\right)(.00003)$$

$$= 1.140170 - .008856 + .000054 - .000002 = 1.13137$$

which is correct to five places.

12.8. The previous two problems have treated special cases of the interpolation problem, working near the top or near the bottom of a table. This problem is more typical in that data will be available on both sides of the point of interpolation. Interpolate for $\sqrt{1.12}$ using the data of Problem 12.5.

The central difference formulas are now convenient since they make it easy to use data more or less equally from both sides. In Problem 12.15 we will see that this also tends to keep the truncation error small. Everett's formula will be used.

$$p_k = \binom{k}{1}y_1 + \binom{k+1}{3}\delta^2 y_1 + \binom{k+2}{5}\delta^4 y_1 + \cdots - \binom{k-1}{1}y_0 - \binom{k}{3}\delta^2 y_0 - \binom{k+1}{5}\delta^4 y_0 - \cdots$$

where higher-order terms have been omitted since we will not need them in this problem. Choosing $k=0$ at $x_0=1.10$, we have $k=(x-x_0)/h=(1.12-1.10)/.05=\frac{2}{5}$. Substituting into Everett's formula,

$$p_k = \left(\frac{2}{5}\right)(1.07238) + \left(-\frac{7}{125}\right)(-.00050) + \left(\frac{168}{5^6}\right)(-.00002)$$

$$-\left(-\frac{3}{5}\right)(1.04881) - \left(\frac{8}{125}\right)(-.00054) - \left(-\frac{182}{5^6}\right)(-.00001)$$

$$= .428952 + .000028 + .629286 + .000035$$

the two highest-order terms contributing nothing (as we hoped, since these are drawn from the error effects columns). Finally $p_k = 1.05830$, which is correct to five places. Notice that the three interpolations made in Table 12.2 have all been based on collocation polynomials of degree three.

12.9. The laboratory's newest employee has been asked to "look up" the value $y(.3333)$ in table NBS-AMS 52 of the National Bureau of Standards Applied Mathematics Series. On the appropriate page of this extensive volume he finds abundant information, a small part of which is reproduced in Table 12.3. Apply Everett's formula for the needed interpolation.

Table 12.3

x	$y(x)$	δ^2
.31	.1223 4609	2392
.32	.1266 9105	2378
.33	.1310 5979	2365
.34	.1354 5218	2349
.35	.1398 6806	2335

Choosing $x = 0$ at $x_0 = .33$, we have $k = (x - x_0)/h = (.3333 - .33)/.01 = .33$. Writing Everett's formula through second differences in the form

$$p_k = ky_1 + (1-k)y_0 + E_1\delta^2 y_1 - E_0\delta^2 y_0$$

where $E_1 = \binom{k+1}{3}$ and $E_0 = \binom{k}{3}$, the interpolator will find all ingredients available in tables. For $k = .33$, we find $E_1 = -.0490105$, $E_0 = .0615395$. Then

$$p_k = (.33)(.13545218) + (.67)(.13105979) + (-.0490105)(.00002349) - (.0615395)(.00002365)$$
$$= .13250667$$

This table was prepared with Everett's formula in mind.

12.10. Apply the Lagrange formula to obtain $\sqrt{1.12}$ from the data of Table 12.2.

The Lagrange formula does not require equally spaced arguments. It can of course be applied to such arguments as a special case, but there are difficulties. The degree of the collocation polynomial must be chosen at the outset. With the Newton, Everett, or other difference formulas the degree can be determined by computing terms until they no longer appear significant. Each term is an additive correction to terms already accumulated. But with the Lagrange formula a change of degree involves a completely new computation, of all terms. In Table 12.2 the evidence is strong that a third-degree polynomial is suitable. On this basis we may proceed to choose $x_0 = 1.05, \ldots, x_3 = 1.20$ and substitute into

$$p = \frac{(x-x_1)(x-x_2)(x-x_3)}{(x_0-x_1)(x_0-x_2)(x_0-x_3)}y_0 + \frac{(x-x_0)(x-x_2)(x-x_3)}{(x_1-x_0)(x_1-x_2)(x_1-x_3)}y_1$$
$$+ \frac{(x-x_0)(x-x_1)(x-x_3)}{(x_2-x_0)(x_2-x_1)(x_2-x_3)}y_2 + \frac{(x-x_0)(x-x_1)(x-x_2)}{(x_3-x_0)(x_3-x_1)(x_3-x_2)}y_3$$

to produce

$$p = \frac{-8}{125}(1.02470) + \frac{84}{125}(1.04881) + \frac{56}{125}(1.07238) + \frac{-7}{125}(1.09544) = 1.05830$$

This agrees with the result of Problem 12.8.

12.11. The problem of *inverse interpolation* reverses the roles of x_k and y_k. We may view the y_k numbers as arguments and the x_k as values. Clearly the new arguments are not usually equally spaced. Given that $\sqrt{x} = 1.05$, use the data of Table 12.2, to find x.

Since we could easily find $x = (1.05)^2 = 1.1025$ by a simple multiplication, this is plainly another "test case" of our available algorithms. Since it applies to unequally spaced arguments, suppose we use Lagrange's formula. Interchanging the roles of x and y,

$$p = \frac{(y-y_1)(y-y_2)(y-y_3)}{(y_0-y_1)(y_0-y_2)(y_0-y_3)}x_0 + \frac{(y-y_0)(y-y_2)(y-y_3)}{(y_1-y_0)(y_1-y_2)(y_1-y_3)}x_1$$
$$+ \frac{(y-y_0)(y-y_1)(y-y_3)}{(y_2-y_0)(y_2-y_1)(y_2-y_3)}x_2 + \frac{(y-y_0)(y-y_1)(y-y_2)}{(y_3-y_0)(y_3-y_1)(y_3-y_2)}x_3$$

With the same four x_k, y_k pairs used in Problem 12.10, this becomes

$$p = (-.014882)1.05 + (.97095)1.10 + (.052790)1.15 + (-.008858)1.20 = 1.1025$$

as expected.

12.12. Apply Everett's formula to the inverse interpolation problem just solved.

Since the Everett formula requires equally spaced arguments, we return x and y to their original roles. Writing Everett's formula as

$$1.05 = k(1.07238) + \binom{k+1}{3}(-.00050) + \binom{k+2}{5}(-.00002)$$
$$+ (1-k)(1.04881) - \binom{k}{3}(-.00054) - \binom{k+1}{5}(-.00001)$$

we have a fifth-degree polynomial equation in k. This is a problem treated extensively in a later chapter. Here a simple, iterative procedure can be used. First neglect all differences and obtain a first approximation by solving

$$1.05 = k(1.07238) + (1-k)(1.04881)$$

The result of this linear inverse interpolation is $k = .0505$. Insert this value into the δ^2 terms, still neglecting the δ^4 terms, and obtain a new approximation from

$$1.05 = k(1.07238) + \binom{1.0505}{3}(-.00050) + (1-k)(1.04881) - \binom{.0505}{3}(.00054)$$

This proves to be $k = .0501$. Inserting this value into both the δ^2 and δ^4 terms then produces $k = .0500$. Reintroduced into the δ^2 and δ^4 terms this last value of k reproduces itself, so we stop. The corresponding value of x is 1.1025 to four places.

12.13. Interpolate for $\sqrt{1.125}$ and $\sqrt{1.175}$ in Table 12.2.

For these arguments which are midway between tabulated arguments, Bessel's formula has a strong appeal. First choose $k = 0$ at $x_0 = 1.10$, making $k = (1.125 - 1.10)/.05 = \frac{1}{2}$. The Bessel formula (Problem 7.25) is

$$p_k = \mu y_{1/2} + \binom{k}{2}\mu\delta^2 y_{1/2} + \binom{k+1}{4}\mu\delta^4 y_{1/2}$$

if we stop at degree four. The odd difference terms disappear entirely because of the factor $k - \frac{1}{2}$. Substituting,

$$p_k = 1.06060 + \left(-\frac{1}{8}\right)(-.00052) + \left(\frac{3}{128}\right)(-.000015) = 1.06066$$

with the δ^4 term again making no contribution. Similarly in the second case, with $k = 0$ now at $x_0 = 1.15$, we again have $k = \frac{1}{2}$ and find $p_k = 1.08397$. By finding all such midway values, the size of a table may be doubled. This is a special case of the problem of *subtabulation*.

12.14. In using a collocation polynomial $p(x)$ to compute approximations to a function $y(x)$, we accept what is called a *truncation error*, $y(x) - p(x)$. Estimate this error for our interpolations in Table 12.1.

The formula for truncation error of a collocation polynomial was derived in Chapter 2 and is

$$y(x) - p(x) = \frac{\pi(x)}{(n+1)!} y^{(n+1)}(\xi)$$

when the polynomial approximation is of degree n. For Table 12.1 we found $n = 1$ suitable. The collocation points may be called x_0 and x_1, leading to this error estimate for linear interpolation:

$$y(x) - p(x) = \frac{(x - x_0)(x - x_1)}{2} y^{(2)}(\xi) = \frac{k(k-1)}{2} h^2 y^{(2)}(\xi)$$

Since $h = .01$ and $y^{(2)}(x) = -\frac{1}{4}x^{-3/2}$, we have

$$|y(x) - p(x)| \leqq \frac{k(k-1)}{8}(.0001)$$

For k between 0 and 1, which we arrange for any interpolation by our choice of x_0, the quadratic $k(k-1)$ has a maximum size of $\frac{1}{4}$ at the midpoint $k = \frac{1}{2}$ (see Fig. 12-1). This allows us to complete our truncation error estimate,

$$|y(x) - p(x)| \leqq \frac{1}{32}(.0001)$$

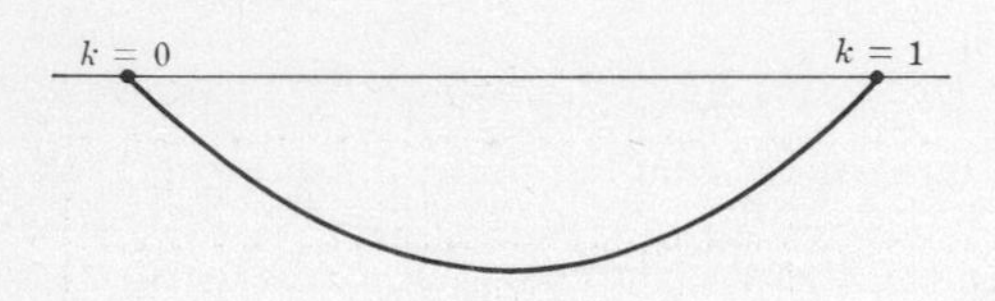

Fig. 12-1

and we discover that it cannot affect the fourth decimal place. Table 12.1 was prepared with linear interpolation in mind. The interval $h = .01$ was chosen to keep truncation error this small.

12.15. Estimate truncation errors for our computations in Table 12.2.

Here for the most part we used Everett's formula for a cubic polynomial. For other cubic formulas the same error estimate follows. Assuming equally spaced collocation arguments x_{-1}, x_0, x_1, and x_2,

$$y(x) - p(x) = \frac{(x - x_{-1})(x - x_0)(x - x_1)(x - x_2)}{4!} y^{(4)}(\xi)$$

$$= \frac{(k+1)k(k-1)(k-2)h^4 y^{(4)}\xi}{24}$$

The polynomial $(k+1)k(k-1)(k-2)$ has the general shape of Fig. 12-2. Outside the interval $-1 < k < 2$ it climbs sensationally. Inside $0 < k < 1$ it does not exceed $\frac{9}{16}$ and this is the appropriate part for interpolation. We now have, for the maximum error in cubic interpolation,

$$|y(x) - p(x)| \leqq \frac{9}{16} \cdot \frac{1}{24} h^4 |y^{(4)}(\xi)| = \frac{3}{128} h^4 |y^{(4)}(\xi)|$$

For this example $h = .05$ and $y^{(4)}(x) = -\frac{15}{16}x^{-7/2}$, and hence $|y(x) - p(x)| \leqq \frac{1}{64}(.00005)$ so that truncation error has not affected our five-decimal calculations.

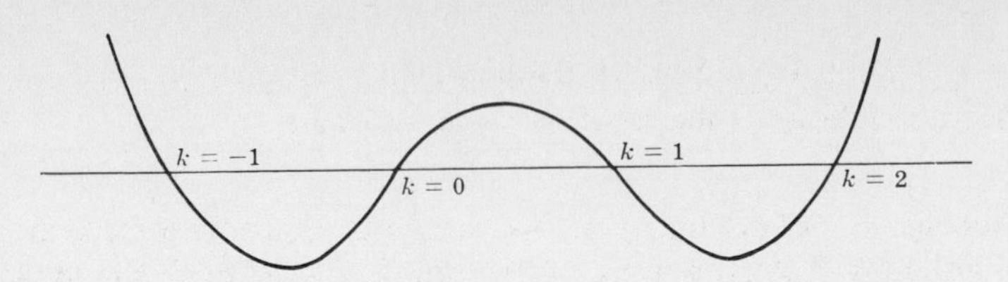

Fig. 12-2

12.16. How large could the interval length h be made in a table of $\sqrt{x}$ with a cubic formula still giving five-place accuracy? (Assume $1 \leqq x$.)

This sort of question is naturally of interest to table makers. Our truncation error formula can be written as

$$|y(x) - p(x)| \leqq \frac{9}{16} h^4 \left(\frac{15}{16}\right)\left(\frac{1}{24}\right)$$

To keep this less than .000005 requires $h^4 < .000228$, or very closely $h < \frac{1}{8}$. This is somewhat larger than the $h = .05$ used in Table 12.1, but other errors enter our computations and it pays to be on the safe side.

12.17. The previous problem suggests that Table 12.2 may be abbreviated to half length, if Everett's cubic polynomial is to be used for interpolations. Find the second differences needed in this Everett formula.

The result is Table 12.4, in which first differences may be ignored.

Table 12.4

x_k	y_k	δ	δ^2
1.00	1.00000		
		4881	
1.10	1.04881		−217
		4664	
1.20	1.09544		−191
		4473	
1.30	1.14017		

12.18. Use Table 12.4 to interpolate for $y(1.15)$.

With Everett's formula and $k = \frac{1}{2}$,

$$p_k = \frac{1}{2}(1.09544) - \frac{1}{16}(-.00191) + \frac{1}{2}(1.04881) - \frac{1}{16}(-.00217) = 1.07238$$

as listed in Table 12.2. This confirms Problem 12.16 in this instance.

12.19. Estimate the truncation error for a fifth-degree formula.

Assume the collocation arguments equally spaced and at $k = -2, -1, \ldots, 3$ as in Everett's formula. (The position is actually immaterial.)

$$y(x) - p(x) = \frac{\pi(x)}{(n+1)!} y^{(n+1)}(\xi) = \frac{(k+2)(k+1)k(k-1)(k-2)(k-3)}{720} h^6 y^{(6)}(\xi)$$

The numerator factor, for $0<k<1$, takes a maximum absolute value of $\frac{225}{64}$ at $k=\frac{1}{2}$, as may easily be verified, making

$$|y(x)-p(x)| \leqq \frac{1}{720} \cdot \frac{225}{64} \cdot h^6 |y^{(6)}(\xi)|$$

12.20. For the function $y(x)=\sqrt{x}$, and $1 \leqq x$, how large an interval h is consistent with five-place accuracy if Everett's fifth-degree formula is to be used in interpolations?

For this function, $y^{(6)}(x)=\frac{945}{64}x^{-11/2} \leqq \frac{945}{64}$. Substituting this into the result of the previous problem and requiring five-place accuracy,

$$\frac{1}{720} \cdot \frac{225}{64} \cdot h^6 \cdot \frac{945}{64} \leqq .000005$$

leading to $h \leqq \frac{1}{5}$ approximately. Naturally the interval permitted with fifth-degree interpolation exceeds that for third-degree interpolation.

12.21. For the function $y(x)=\sin x$, how large an interval h is consistent with five-place accuracy if Everett's fifth-degree formula is to be used in interpolations?

For this function $y^{(6)}(x)$ is bounded absolutely by 1, so we need $\frac{1}{720} \cdot \frac{225}{64} \cdot h^6 \leqq .000005$, leading to $h \leqq .317$. This is the equivalent of 18° intervals, and means that only four values of the sine function, besides sin 0 and sin 90° are needed to cover this entire basic interval!

12.22. A second source of error in the use of our formulas for the collocation polynomial (the first source being truncation error) is the presence of *inaccuracies in the data values*. The numbers y_k, for example, if obtained by physical measurement will contain inaccuracy due to the limitations imposed by equipment, and if obtained by computations probably contain roundoff errors. Show that linear interpolation does not magnify such errors.

The linear polynomial may be written in Lagrangian form,

$$p = ky_1 + (1-k)y_0$$

where the y_k are as usual the actual data values. Suppose these values are inaccurate. With Y_1 and Y_0 denoting the exact but unknown values, we may write

$$Y_0 = y_0 + e_0 \qquad Y_1 = y_1 + e_1$$

where the numbers e_0 and e_1 are the errors. The exact result desired is therefore

$$P = kY_1 + (1-k)Y_0$$

making the error of our computed result

$$P - p = ke_1 + (1-k)e_0$$

If the errors e_k do not exceed E in magnitude, then

$$|P-p| \leqq kE + (1-k)E = E$$

for $0<k<1$. This means that the error in the computed value p does not exceed the maximum data error. No magnification of error has occurred.

12.23. Estimate the magnification of data inaccuracies due to cubic interpolation.

Again using the Lagrangian form but assuming equally spaced arguments at $k=-1, 0, 1, 2$, the cubic can be written as

$$p = \frac{k(k-1)(k-2)}{-6}y_{-1} + \frac{(k+1)(k-1)(k-2)}{2}y_0 + \frac{(k+1)k(k-2)}{-2}y_1 + \frac{(k+1)k(k-1)}{6}y_2$$

As in Problem 12.22, we let $Y_k = y_k + e_k$, with Y_k denoting the exact data values. If P again stands for the exact result desired, then the error is

$$P - p = \frac{k(k-1)(k-2)}{-6}e_{-1} + \frac{(k+1)(k-1)(k-2)}{2}e_0 + \frac{(k+1)k(k-2)}{-2}e_1 + \frac{(k+1)k(k-1)}{6}e_2$$

Notice that for $0 < k < 1$ the errors e_{-1} and e_2 have negative coefficients while the other two have positive coefficients. This means that if the errors do not exceed E in magnitude,

$$|P - p| \leqq E\left[\frac{k(k-1)(k-2)}{6} + \frac{(k+1)(k-1)(k-2)}{2} + \frac{(k+1)k(k-2)}{-2} + \frac{(k+1)k(k-1)}{-6}\right]$$

which simplifies to
$$|P - p| \leqq (-k^2 + k + 1)E = m_k E$$

Not surprisingly the quadratic magnification factor m_k takes its maximum at $k = \frac{1}{2}$ (Fig. 12-3) and so $|P - p| \leqq \frac{5}{4}E$. The data error E may be magnified by as much as $\frac{5}{4}$. This is, of course, a pessimistic estimate. In certain cases errors may even annul one another, making the computed value p more accurate than the data y_k.

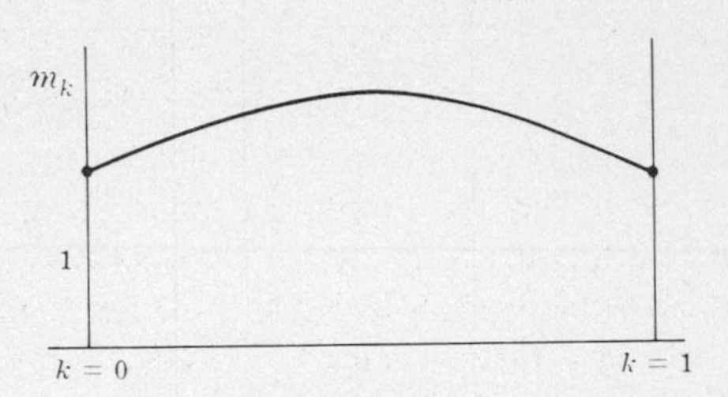

Fig. 12-3

12.24. What other source of error is there in an interpolation?

One source which is very important to keep in mind, even though it is often entirely out of one's control, is the continual necessity to make roundoffs during the carrying out of the algorithm. Working to a limited number of digits, this cannot be avoided. Our various formulas, even they represent exactly the same collocation polynomial, process the data involved in differing ways. In other words, they represent different algorithms. Such formulas accept the same input error (data inaccuracies) and may have the same truncation error but still differ in the way algorithm roundoffs develop.

12.25. Describe how Taylor's series may be used for interpolation.

Consider the function $y = e^x$. But Taylor's series,

$$e^{x+t} = e^x \cdot e^t = e^x(1 + t + \tfrac{1}{2}t^2 + \cdots)$$

Assume the factor e^x known. Truncating the series after the t^2 term means an error (inside the parentheses) of at most $\frac{1}{6}(h/2)^3$ where h is the interval at which arguments are spaced in the table. This assumes that interpolation will always be based on the nearest tabular entry. If $h = .05$ this error is $(\frac{125}{48})10^{-6}$, or $(2.6)10^{-6}$. This means that, stopping at the t^2 term, accuracy to five digits (not decimal places) will be obtained in the computed value of e^{x+t}. For example, using the data of Table 12.5 the interpolation for $e^{2.718}$ runs as follows. With $t = .018$, $1 + t + \frac{1}{2}t^2 = 1.01816$ and

$$e^{2.718} = e^{2.70}(1.01816) = (14.880)(1.01816) = 15.150$$

which is correct to its full five digits. Our collocation polynomials would also produce this result.

Table 12.5

x	2.60	2.65	2.70	2.75	2.80
$y = e^x$	13.464	14.154	14.880	15.643	16.445

12.26. How can Taylor series interpolation be used for the function $y(x) = \sin x$?

Since $\sin x$ and $\cos x$ are usually tabulated together, we may express

$$\sin(x \pm t) = \sin x \pm t \cos x - \tfrac{1}{2}t^2 \sin x$$

Here, of course, t is measured in radians. If the tabular interval is $h = .0001$, as it is NBS-AMS 36, of which Table 12.6 is a brief extract, then the above formula will give accuracy to nine digits, since $\frac{1}{6}(h/2)^3$ is out beyond the twelfth place.

Table 12.6

x	$\sin x$	$\cos x$
1.0000	.8414 70985	.5403 02306
1.0001	.8415 25011	.5402 18156
1.0002	.8415 79028	.5401 34001
1.0003	.8416 33038	.5400 49840

12.27. Compute sin 1.00005 by the Taylor series interpolation.

With $x = 1$ and $t = .00005$,

$$\sin 1.00005 = .8414\,70985 + (.00005)(.5403\,02306) - \left(\frac{1}{8}\right)(10^{-8})(.8414\,70985) = .8414\,97999$$

12.28. Apply Newton's backward formula to the *prediction* of $\sqrt{1.32}$ in Table 12.2.

With $k = 0$ at $x_0 = 1.30$ we find $k = (1.32 - 1.30)/.05 = .4$. Substituting into the Newton formula,

$$p = 1.14017 + (.4)(.02214) + (.28)(-.00045) + (.224)(.00003) = 1.14891$$

which is correct as far as it goes. Newton's backward formula seems the natural choice for such prediction problems, since the supply of available dfferences is greatest for this formula and one may introduce difference terms until they do not contribute to the decimal places retained. This allows the degree of the approximating polynomial to be chosen as the computation progresses.

12.29. Analyze the truncation error in prediction.

The truncation error of the collocation polynomial can be expressed as

$$\frac{k(k+1)\cdots(k+n)}{(n+1)!}h^{n+1}y^{(n+1)}(\xi)$$

where the collocation points are at $k = 0, -1, \ldots, -n$ as is the case when Newton's backward formula is used. For prediction, k is positive. The numerator factor grows rapidly with increasing k, more rapidly for large n, as Fig. 12-4 suggests. This indicates that truncation error will not be tolerable beyond a certain point, and that prediction far beyond the end of a table is dangerous, as might be anticipated. The truncation error of a collocation polynomial is oscillatory between the points of collocation, but once outside the interval of these points it becomes explosive.

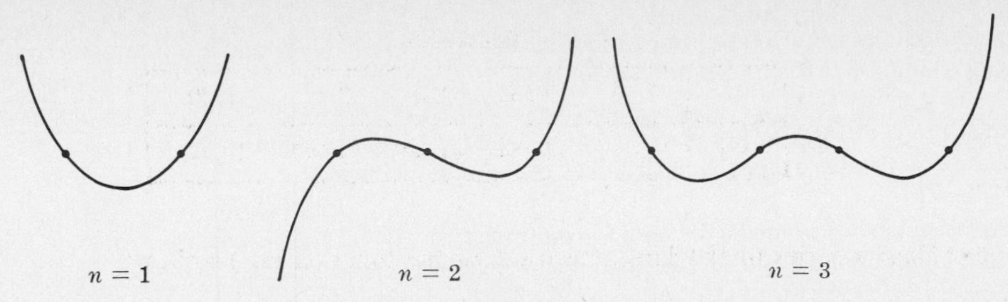

Fig. 12-4

12.30. Predict $\sqrt{1.50}$ from the data of Table 12.2.

With $k = (1.50 - 1.30)/.05 = 4$,

$$p = 1.14017 + (4)(.02214) + (10)(-.00045) + (20)(.00003) = 1.22483$$

while the correct result is 1.22474. Note also that higher difference terms, which we believe to be error effects anyway, would only make the result worse because they are positive.

Supplementary Problems

12.31. From the data of Table 12.1 obtain $\sqrt{1.012}$ and $\sqrt{1.017}$ by linear interpolation, to four decimal places. Would the second difference term affect the result? Would higher-order terms?

12.32. From the data of Table 12.1 obtain $\sqrt{1.059}$ by linear interpolation. Note that if Newton's forward formula is used (with $k = 0$ at $x = 1.05$) no second difference would be available in this case.

12.33. Interpolate for $\sqrt{1.03}$ in Table 12.2.

12.34. Interpolate for $\sqrt{1.26}$ in Table 12.2.

12.35. Apply Stirling's formula to obtain $\sqrt{1.12}$ from the data of Table 12.2. Does the result agree with that of Problem 12.8?

12.36. Apply Everett's formula to Table 12.3, obtaining $y(.315)$.

12.37. Apply the Lagrange formula to interpolate for $y(1.50)$ using some of the following values of the normal error function, $y(x) = e^{-x^2/2}/\sqrt{2\pi}$.

x_k	1.00	1.20	1.40	1.60	1.80	2.00
y_k	.2420	.1942	.1497	.1109	.0790	.0540

The correct result is .1295.

12.38. Use Lagrange's formula to inverse interpolate for the number x corresponding to $y = .1300$ in the data of Problem 12.37.

12.39. Apply the method of Problem 12.12 to the inverse interpolation of Problem 12.38.

12.40. Apply Bessel's formula to obtain $y(1.30)$, $y(1.50)$, and $y(1.70)$ for the data of Problem 12.37.

12.41. In a table of the function $y(x) = \sin x$ to four decimal places, what is the largest interval h consistent with linear interpolation? (Keep truncation error well below .00005.)

12.42. In a table of $y(x) = \sin x$ to five places, what is the largest interval h consistent with linear interpolation? Check these estimates against familiar tables of the sine function.

12.43. If Everett's cubic polynomial were used for interpolations, rather than a linear polynomial, how large an interval h could be used in a four-decimal-place table of $y(x) = \sin x$? In a five-place table?

12.44. In quadratic approximation with Newton's formula, the function $k(k-1)(k-2)$ appears in the truncation error estimate. Show that this function has the shape indicated in Fig. 12-5 and that for $0 < k < 2$ it does not exceed $2\sqrt{3}/9$ in absolute value.

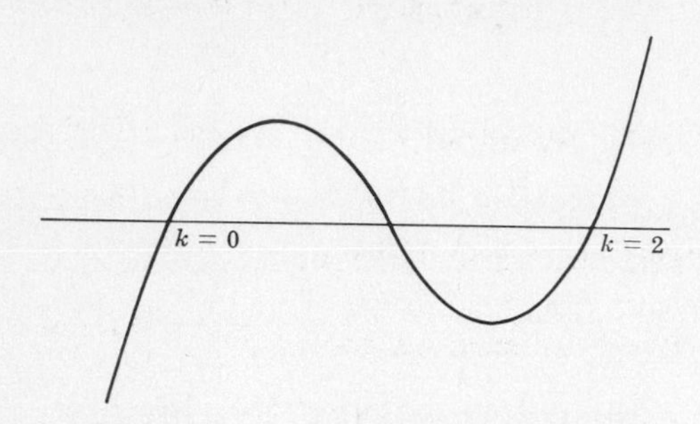

Fig. 12-5

12.45. The function $k(k^2-1)(k^2-4)$ appears in the truncation error estimate for Stirling's formula. Diagram this for $-2 < k < 2$ and estimate its maximum absolute value for $-\frac{1}{4} < k < \frac{1}{4}$, which is the interval to which use of this formula is usually limited.

12.46. Show that the relative maxima and minima of the polynomials

$$k(k^2-1)(k^2-4) \qquad k(k^2-1)(k^2-4)(k^2-9)$$

increase in magnitude as their distance from the interval $-1 < k < 1$ increases. These polynomials appear in the truncation error for Stirling's formula. The implication is that this formula is most accurate in the center of the range of collocation.

12.47. Show that the relative maxima and minima of the polynomials

$$(k+1)k(k-1)(k-2) \qquad (k+2)(k+1)k(k-1)(k-2)(k-3)$$

increase in magnitude with distance from the interval $0 < k < 1$. These polynomials appear in the truncation error for Everett's or Bessel's formula. The implication is that these formulas are most accurate over this central interval.

12.48. How large an interval h is consistent with interpolation by Everett's fifth-degree formula if the function is $y(x) = \log x$ and five-place accuracy is required?

12.49. Estimate the magnification of data inaccuracies due to second-degree interpolation. Follow the argument of Problems 12.22 and 12.23, with $0 < k < 1$.

12.50. Estimate the magnification of data inaccuracies due to fourth-degree interpolation, again for $0 < k < 1$.

12.51. Apply Stirling's formula to compute $y(2.718)$ from the data of Table 12.5.

12.52. Compute $\sin 1.00015$ from the data provided in Table 12.6.

12.53. Show that the Taylor series interpolation

$$\log(x+t) = \log x + \log\left(1+\frac{t}{x}\right) = \log x + \frac{t}{x} - \frac{t^2}{2x^2} + \cdots$$

may be truncated after the t^2 term with six-decimal-place accuracy for $1<x$, provided the tabular spacing is $h=.01$.

12.54. Use Newton's backward formula to predict $\sqrt{1.35}$, $\sqrt{1.40}$, and $\sqrt{1.45}$ from the data of Table 12.2.

12.55. Predict $\sqrt{1.40}$ and $\sqrt{1.50}$ from the data of Table 12.4.

12.56. Diagram the error of the quadratic polynomial of Problem 6.14. Show that the error equals zero at $x=-3$ as well as at the points of collocation. How can this be explained in terms of our collocation error formula $\pi(x)y^{(3)}(\xi)/3!$?

12.57. In Problem 6.15 how can the zero error at $x=4$ be explained in terms of the error formula $\pi(x)y^{(4)}(\xi)/4!$?

12.58. Use the result of Problem 10.15 to estimate the missing $y'(1)$.

12.59. Use the result of Problem 10.16 to estimate the missing $y''(1)$.

12.60. Use the result of Problem 10.17 to estimate the missing $y'(0)$ and $y'(1)$.

Chapter 13

Numerical Differentiation

APPROXIMATE DERIVATIVES

Approximate derivatives of a function $y(x)$ may be found from a polynomial approximation $p(x)$ simply by accepting $p', p^{(2)}, p^{(3)}, \ldots$ in place of $y', y^{(2)}, y^{(3)}, \ldots$. Our collocation polynomials lead to a broad variety of useful formulas of this sort. The three well-known formulas

$$y(x) \simeq \frac{y(x+h)-y(x)}{h} \qquad y'(x) \simeq \frac{y(x+h)-y(x-h)}{2h} \qquad y'(x) \simeq \frac{y(x)-y(x-h)}{h}$$

follow by differentiation of the Newton forward, Stirling, and Newton backward formulas, respectively, in each case only one term being used. More complicated formulas are available simply by using more terms. Thus

$$y'(x) \simeq \frac{1}{h}\left[\Delta y_0 + \left(k-\frac{1}{2}\right)\Delta^2 y_0 + \frac{3k^2-6k+2}{6}\Delta^3 y_0 + \cdots\right]$$

comes from the Newton formula, while

$$y'(x) \simeq \frac{1}{h}\left(\delta\mu y_0 + k\delta^2 y_0 + \frac{3k^2-1}{6}\delta^3\mu y_0 + \cdots\right)$$

results from differentiating Stirling's. Other collocation formulas produce similar approximations. For second derivatives one popular result is

$$y^{(2)}(x) \simeq \frac{1}{h^2}\left(\delta^2 y_0 + k\delta^3\mu y_0 + \frac{6k^2-1}{12}\delta^4 y_0 + \cdots\right)$$

and comes from the Stirling formula. Retaining only the first term, we have the familiar

$$y^{(2)}(x) \simeq \frac{y(x+h)-2y(x)+y(x-h)}{h^2}$$

SOURCES OF ERROR IN APPROXIMATE DIFFERENTIATION

The study of test cases suggests that approximate derivatives obtained from collocation polynomials be viewed with skepticism unless very accurate data are available. Even then the accuracy diminishes with increasing order of the derivatives.

The basic difficulty is that $y(x)-p(x)$ may be very small while $y'(x)-p'(x)$ is very large. In geometrical language, two curves may be close together but still have very different slopes. All the other familiar sources of error are also present, including input errors in the y_i values, truncation errors such as $y'-p'$, $y^{(2)}-p^{(2)}$, etc., and internal roundoffs.

The dominant error source is the input errors themselves. These are critical, even when small, because the algorithms magnify them enormously. A crucial factor in this magnification is the reciprocal power of h which occurs in the formulas, multiplying both the true values and the errors which are blended together to make the y_i data. An optimum choice of the interval h may sometimes be made. Since truncation error depends directly on h, while input error magnification depends inversely, the usual method of calculus may be used to minimize the combination.

Large errors should be anticipated in approximate derivatives based on collocation polynomials. Error bounds should be obtained whenever possible. Alternative methods for approximate differentiation may be based upon polynomials obtained by least squares or min-max procedures rather than by collocation. (See Chapters 21 and 22.) Since these methods also smooth the given data, they are usually more satisfactory. Trigonometric approximation (Chapter 24) provides still another alternative.

Solved Problems

13.1. Differentiate Newton's forward formula,

$$p_k = y_0 + \binom{k}{1}\Delta y_0 + \binom{k}{2}\Delta^2 y_0 + \binom{k}{3}\Delta^3 y_0 + \binom{k}{4}\Delta^4 y_0 + \cdots$$

The Stirling numbers may be used to express the factorials as powers, after which an easy computation produces derivatives relative to k. With the operator D continuing to represent such derivatives, Dp_k, $D^2p_k, \ldots$, we use the familiar $x = x_0 + kh$ to obtain derivatives relative to the argument x.

$$p'(x) = \frac{Dp_k}{h} \qquad p^{(2)}(x) = \frac{D^2 p_k}{h^2} \qquad \cdots$$

The results are

$$p'(x) = \frac{1}{h}\left[\Delta y_0 + \left(k - \frac{1}{2}\right)\Delta^2 y_0 + \frac{3k^2 - 6k + 2}{6}\Delta^3 y_0 + \frac{2k^3 - 9k^2 + 11k - 3}{12}\Delta^4 y_0 + \cdots\right]$$

$$p^{(2)}(x) = \frac{1}{h^2}\left(\Delta^2 y_0 + (k-1)\Delta^3 y_0 + \frac{6k^2 - 18k + 11}{12}\Delta^4 y_0 + \cdots\right)$$

$$p^{(3)}(x) = \frac{1}{h^3}\left(\Delta^3 y_0 + \frac{2k-3}{2}\Delta^4 y_0 + \cdots\right)$$

$$p^{(4)}(x) = \frac{1}{h^4}(\Delta^4 y_0 + \cdots) \qquad \text{and so on}$$

13.2. Apply the formulas of Problem 13.1 to produce $p'(1)$, $p^{(2)}(1)$, and $p^{(3)}(1)$ from the data of Table 13.1. (This is the same as Table 12.2 with the differences beyond the third suppressed. Recall that those differences were written off as error effects. The table is reproduced here for convenience.)

Table 13.1

x	$y(x) = \sqrt{x}$			
1.00	1.00000			
		2470		
1.05	1.02470		−59	
		2411		5
1.10	1.04881		−54	
		2357		4
1.15	1.07238		−50	
		2307		2
1.20	1.09544		−48	
		2259		3
1.25	1.11803		−45	
		2214		
1.30	1.14017			

With $h = .05$, and $k = 0$ at $x_0 = 1.00$, our formulas produce

$$p'(1) = 20(.02470 + .000295 + .000017) = .50024$$

$$p^{(2)}(1) = 400(-.00059 - .00005) = -.256$$

$$p^{(3)}(1) = 8000(.00005) = .4$$

The correct results are, since $y(x) = \sqrt{x}$, $y'(1) = \frac{1}{2}$, $y^{(2)}(1) = -\frac{1}{4}$, and $y^{(3)}(1) = \frac{3}{8}$.

Though the input data are accurate to five decimal places, we find $p'(1)$ correct to only three places, $p^{(2)}(1)$ not quite correct to two places, and $p^{(3)}(1)$ correct to only one. Obviously, algorithm errors are prominent.

13.3. Differentiate Stirling's formula,

$$p_k = y_0 + \binom{k}{1}\delta\mu y_0 + \frac{k}{2}\binom{k}{1}\delta^2 y_0 + \binom{k+1}{3}\delta^3\mu y_0 + \frac{k}{4}\binom{k+1}{3}\delta^4 y_0 + \cdots$$

Proceeding as in Problem 13.1, we find

$$p'(x) = \frac{1}{h}\left(\delta\mu y_0 + k\delta^2 y_0 + \frac{3k^2-1}{6}\delta^3\mu y_0 + \frac{2k^3-k}{12}\delta^4 y_0 + \cdots\right)$$

$$p^{(2)}(x) = \frac{1}{h^2}\left(\delta^2 y_0 + k\delta^3\mu y_0 + \frac{6k^2-1}{12}\delta^4 y_0 + \cdots\right)$$

$$p^{(3)}(x) = \frac{1}{h^3}(\delta^3\mu y_0 + k\delta^4 y_0 + \cdots)$$

$$p^{(4)}(x) = \frac{1}{h^4}(\delta^4 y_0 + \cdots) \qquad \text{and so on}$$

13.4. Apply the formulas of Problem 13.3 to produce $p'(1.10)$, $p^{(2)}(1.10)$, and $p^{(3)}(1.10)$ from the data of Table 13.1.

With $k = 0$ at $x_0 = 1.10$, our formulas produce

$$p'(1.10) = 20\left[\frac{.02411 + .02357}{2} + 0 - \frac{1}{6}\left(\frac{.00005 + .00004}{2}\right)\right] = .4766$$

$$p^{(2)}(1.10) = 400(-.00054 + 0) = -.216$$

$$p^{(3)}(1.10) = 8000(.000045) = .360$$

The correct results are $y'(1.10) = .47674$, $y^{(2)}(1.10) = -.2167$, and $y^{(3)}(1.10) = .2955$.

The input data were correct to five places, but our approximations to these first three derivatives are correct to roughly four, three, and one place, respectively.

13.5. The previous problems suggest that approximate differentiation is an inaccurate affair. Illustrate this further by comparing the function $y(x) = e \sin(x/e^2)$ with the polynomial approximation $p(x) = 0$.

The two functions collocate at the equally spaced arguments $x = ie^2\pi$ for integers i. For a very small number e, the approximation is extremely accurate, $y(x) - p(x)$ never exceeding e. However, since $y'(x) = (1/e)\cos(x/e^2)$ and $p'(x) = 0$, the difference in derivatives is enormous. This example shows that accurate approximation of a function should not be expected to mean accurate approximation of its derivative. See Fig. 13-1.

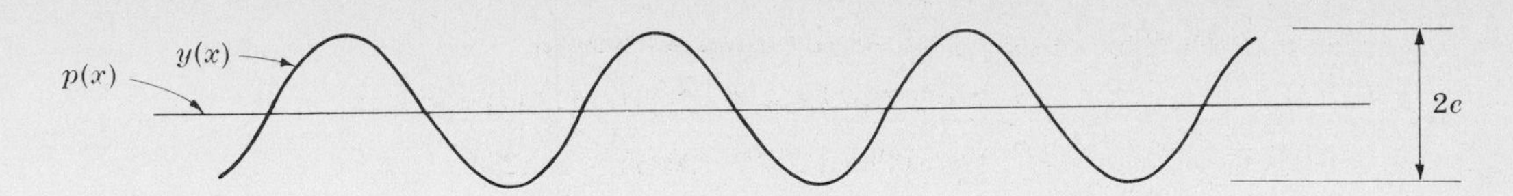

Fig. 13-1

13.6. Problems 13.1, 13.3, and 13.23 suggest three approximations to $y'(x_0)$ using only first differences,

$$\frac{y_1 - y_0}{h} \qquad \frac{y_1 - y_{-1}}{2h} \qquad \frac{y_0 - y_{-1}}{h}$$

Interpreted geometrically, these are the slopes of the three lines shown in Fig. 13-2. The tangent line at x_0 is also shown. It would appear that the middle approximation is closest to the slope of the tangent line. Confirm this by computing the truncation errors of the three formulas.

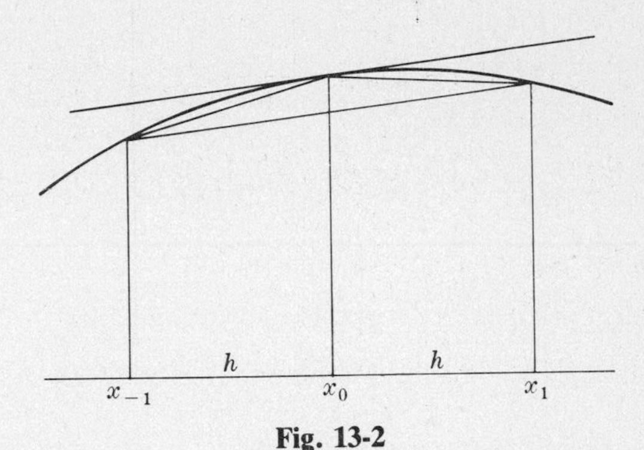

Fig. 13-2

Newton's forward formula, truncated after the first difference term, leaves the truncation error

$$y(x) - p(x) = \frac{h^2}{2}[k(k-1)y^{(2)}(\xi)]$$

with $x = x_0 + kh$ as usual. It is helpful here to consider k as a continuous argument, no longer restricting it to integer values. Assuming $y^{(2)}(\xi)$ continuous, we then find the error of our derivative formula (by the chain rule) for $k = 0$.

$$y'(x_0) - p'(x_0) = -\frac{h}{2}y^{(2)}(\xi_0)$$

Note that for $k = 0$ the derivative of the troublesome $y^{(2)}(\xi)$ factor is not involved. Similarly for Newton's backward formula,

$$y'(x_0) - p'(x_0) = \frac{h}{2}y^{(2)}(\xi_0)$$

With Stirling's formula we receive an unexpected bonus. Retaining even the *second* difference term in our approximation we find that at $k = 0$ it disappears from $p'(x)$. (See Problem 13.3.) Thus we may consider the middle approximation under discussion as arising from a second-degree polynomial approximation. The truncation error is then

$$y(x) - p(x) = \frac{h^2}{6}[(k+1)k(k-1)y^{(3)}(\xi)]$$

leading to
$$y'(x_0) - p'(x_0) = \frac{-h^2}{6}y^{(3)}(\xi)$$

It is true that the symbol ξ probably represents three distinct unknown numbers in these three computations. But since h is usually small, the appearance of h^2 in the last result, compared with h in the others, suggests that this truncation error is the smallest, by an "order of magnitude." This confirms the geometrical evidence.

13.7. Apply the middle formula of Problem 13.6 to approximate $y'(1.10)$ for the data of Table 13.1. Find the actual error of this result and compare with the truncation error estimate of Problem 13.6.

This approximation is actually the first term computed in Problem 13.4: $y'(1.10) \simeq .4768$. The actual error is, to five places,

$$y'(1.10) - .4768 = .47674 - .47680 = -.00006$$

The estimate obtained in Problem 13.6 was $-h^2 y^{(3)}(\xi)/6$. Since $y^{(3)}(x) = \frac{3}{8}x^{-5/2}$ we exaggerage only slightly by replacing the unknown ξ by 1, obtaining $-h^2 y^{(3)}(\xi)/6 \simeq -(.05)^2(\frac{1}{16}) = -.00016$. This estimate is generous, though not unrealistic.

13.8. Convert the formula for $p'(x_0)$ obtained in Problem 13.3 to a form which exhibits the y_k values used rather than the differences.

We have $k = 0$ for this case, making

$$p'(x_0) = \frac{1}{h}\left[\frac{1}{2}(y_1 - y_{-1}) - \frac{1}{12}(y_2 - 2y_1 + 2y_{-1} - y_{-2}\right] = \frac{1}{12h}(y_{-2} - 8y_{-1} + 8y_1 - y_2)$$

13.9. Estimate the truncation error in the formula of Problem 13.8.

Since the formula was based on Stirling's fourth-degree polynomial,

$$y(x) - p(x) = \frac{h^5(k^2-4)(k^2-1)ky^{(5)}(\xi)}{120}$$

Differentiating as in Problem 13.6 and putting $k = 0$, $y'(x_0) - p'(x_0) = h^4 y^{(5)}(\xi)/30$.

13.10. Compare the estimate of Problem 13.9 with the actual error of the computed result in Problem 13.4.

To five places the actual error is

$$y'(1.10) - p'(1.10) = .47674 - .47660 = .00014$$

while the formula of Problem 13.9, with $y^{(5)}(1)$ substituting for the unknown $y^{(5)}(\xi)$ and causing a slight exaggeration, yields

$$\frac{h^4 y^{(5)}(\xi)}{30} \simeq (.05)^4\left(\frac{7}{64}\right) = .0000007$$

Surely this is disappointing! Though the truncation error has been essentially eliminated by using differences of higher order, the actual error is greater. Clearly another source of error is dominant in these algorithms. It proves to be the input errors of the y_i values, and how the algorithm magnifies them. For brevity we shall include this in the term roundoff error.

13.11. Estimate the roundoff error behavior for the formula $(y_1 - y_{-1})/2h$.

As before, let Y_1 and Y_{-1} be the exact (unknown) data values. Then $Y_1 = y_1 + e_1$ and $Y_{-1} = y_{-1} + e_{-1}$ with e_1 and e_{-1} representing data errors. The difference

$$\frac{Y_1 - Y_{-1}}{2h} - \frac{y_1 - y_{-1}}{2h} = \frac{e_1 - e_{-1}}{2h}$$

is then the error in our output due to input inaccuracies. If e_1 and e_{-1} do not exceed E in magnitude, then this output error is at worst $2E/2h$, making the maximum roundoff error E/h.

13.12. Apply the estimate of Problem 13.11 to the computation of Problem 13.7.

Here $h = .05$ and $E = .000005$, making $E/h = .00010$. Thus roundoff error in the algorithm may influence the fourth place slightly.

13.13. Estimate roundoff error behavior for the formula of Problem 13.8.

Proceeding just as in Problem 13.10, we find $(1/12h)(e_{-2} - 8e_{-1} + 8e_1 - e_2)$ for the error in the output due to input inaccuracies. If the e_k do not exceed E in magnitude, then this output error is at worst $18E/12h$, i.e., maximum roundoff error $= (3/2h)E$. The factor $(3/2h)$ is the magnification factor, as $(1/h)$ was in Problem 13.11. Note that for small h, which we generally associate with high accuracy, this factor is large and roundoff errors in the input information become strongly magnified.

13.14. Apply the estimate of Problem 13.13 to the computation of Problem 13.4. Then compare the various errors associated with our efforts to compute $y'(1.10)$.

With $h = .05$ and $E = .000005$, $(3/2h)E = .00015$. The various errors are grouped in Table 13.2.

Table 13.2

Formula	Actual error	Est. trunc. error	Max. R.O. error
$(y_1 - y_{-1})/2h$	−.00006	−.00016	±.00010
$(y_{-2} - 8y_{-1} + 8y_1 - y_2)/12h$	.00014	.0000007	±.00015

In the first case roundoff error has helped, but in the second case it has hurt. Plainly, the high magnification of such errors makes low truncation errors pointless, except for extremely accurate data.

13.15. Estimate the truncation error of the formula

$$y^{(2)}(x_0) \simeq \frac{1}{h^2}\delta^2 y_0 = \frac{1}{h^2}(y_1 - 2y_0 + y_{-1})$$

obtainable from Problem 13.3 by stopping after the second difference term.

Here it may be convenient to follow a different route to the truncation error, using Taylor series. In particular

$$y_1 = y_0 + hy_0' + \frac{1}{2}h^2 y_0^{(2)} + \frac{1}{6}h^3 y_0^{(3)} + \frac{1}{24}h^4 y^{(4)}(\xi_1)$$

$$y_{-1} = y_0 - hy_0' + \frac{1}{2}h^2 y_0^{(2)} - \frac{1}{6}h^3 y_0^{(3)} + \frac{1}{24}h^4 y^{(4)}(\xi_2)$$

so that adding these up and then subtracting $2y_0$ we find

$$\delta^2 y_0 = h^2 y_0^{(2)} + \frac{1}{24}h^4[y^{(4)}(\xi_1) + y^{(4)}(\xi_2)]$$

Unfortunately ξ_1 is probably not the same as ξ_2, but for an estimate of truncation error suppose we replace both fourth derivatives by a number $y^{(4)}$ which remains open for our choice. For complete safety we could choose $y^{(4)} = \max|y^{(4)}(x)|$ over the interval involved, leading to an upper bound for the magnitude of truncation error, but conceivably other choices might be possible. We now have

$$\text{Truncation error} = y_0^{(2)} - \frac{1}{h^2}\delta^2 y_0 = -\frac{h^2}{12}y^{(4)}$$

13.16. Apply the estimate in Problem 13.15 to the computation of Problem 13.4.

The computation of $p^{(2)}(1.10)$ in Problem 13.4 was actually made by the formula

$$p^{(2)}(1.10) = \frac{\delta^2 y_0}{h^2} = -.21600$$

since higher difference terms contributed nothing. The result has already been compared with the correct $y''(1.10) = -.21670$. The truncation error estimate of Problem 13.15, with

$$y^{(4)}(x) = -\frac{15}{16}x^{-7/2} \simeq -\frac{15}{16}$$

suggests a slight exaggeration

$$\text{Truncation error} \simeq \frac{1}{5120} = .00020$$

The actual error is $-.00070$, again indicating that truncation is not the major error source.

13.17. Estimate the roundoff error of the formula $\delta^2 y_0/h^2$.

Proceeding as before, we find the output error due to input inaccuracies to be $(1/h^2)(e_1 - 2e_0 + e_{-1})$ where the e_k are the input errors. If these do not exceed E in magnitude, then this can be at worst $(4/h^2)E$; thus the maximum roundoff error $= (4/h^2)E$.

13.18. Apply the formula of Problem 13.17 to the computation of Problem 13.4 and compare the actual error of our approximation to $y^{(2)}(1.10)$ with truncation and roundoff estimates.

As before $h = .05$ and $E = .000005$, making $(4/h^2)E = .00800$.

The magnification factor $(4/h^2)$ has a powerful effect. Our results confirm that roundoff has been the principal error source in our approximation of $y^{(2)}(1.10)$, and it has contributed only about 90 of a potential 800 units.

Actual error	Est. truncation error	Max. R.O. error
−.00070	.00020	±.00800

13.19. Apply the splines of Problems 9.7 and 9.8 to find approximate derivatives of the sine function.

In Problem 9.7 we found the natural spline, having zero second derivatives at the endpoints. Since the sine itself has these end derivatives, the natural spline is appropriate in this case. Taking the center point first, we find the derivative of the center spline segment S_2 to be

$$S_2'(x) = -\frac{27\sqrt{3}}{10\pi^3}(2\pi x - \pi^2)$$

which is precisely zero at $x = \pi/2$. Clearly the symmetry has been helpful. A fairer test may be made at $x = \pi/3$ which was one of the knots, where we find S_2' to be .496. The error of .4 percent may be judged keeping in mind that only three spline segments were used over the interval $(0, \pi)$.

In Problem 9.8 we found the spline that matched the endpoint first derivatives of the sine function. For the center section we found

$$S_2'(x) = \frac{2\pi - 9\sqrt{3}}{2\pi^3}(2\pi x - \pi^2)$$

which is again zero at $x = \pi/2$. At $x = \pi/3$, it manages $(9\sqrt{3} - 2\pi)/6\pi$ or .494.

For the second derivative the anticipated deterioration again appears. The natural spline predicts $S_2'' = -.948$ for the entire center interval, where the true second derivative ranges from $-.866$ to -1.

13.20. How can the Richardson extrapolation method be applied to numerical differentiation?

As usual, information about the error in an approximation formula is used to make a correction. As an illustration take the central formula

$$y'(x) = \frac{y(x+h) - y(x-h)}{2h} + T$$

where T is the truncation error. An easy calculation using Taylor series finds

$$T = a_1h^2 + a_2h^4 + a_3h^6 + \cdots$$

Making two applications, using h and $h/2$, we have

$$y'(x) = F(h) + a_1h^2 + a_2h^4 + \cdots$$

$$y'(x) = F\left(\frac{h}{2}\right) + \frac{a_1h^2}{4} + \frac{a_2h^4}{16} + \cdots$$

with $F(h)$ and $F(h/2)$ denoting the approximate derivatives, and where we assume that the a_i do not change much for small h. Eliminating the a_1 terms leads to

$$y'(x) = \frac{4F(h/2) - F(h)}{3} + b_1h^4 + O(h^6)$$

so that in

$$F_1\left(\frac{h}{2}\right) - \frac{4F(h/2) - F(h)}{3}$$

we have an approximate differentiation formula of fourth-order accuracy, obtained by combining two results from a formula of second-order accuracy.

The argument can now be repeated, beginning with

$$y'(x) = F_1\left(\frac{h}{2}\right) + b_1h^4 + O(h^6)$$

$$y'(x) = F_1\left(\frac{h}{4}\right) + \frac{b_1h^4}{16} + O(h^6)$$

and eliminating the b_1 term to produce an approximation

$$F_2\left(\frac{h}{2}\right) = \frac{16F_1(h/4) - F_1(h/2)}{15}$$

with sixth-order accuracy. Clearly further repetitions are possible, the overall process being known as extrapolation to the limit.

The set of approximations calculated during an extrapolation to the limit is usually displayed as follows:

	F	F_1	F_2	F_3
h	$F(h)$			
$h/2$	$F(h/2)$	$F_1(h/2)$		
$h/4$	$F(h/4)$	$F_1(h/4)$	$F_2(h/4)$	
$h/8$	$F(h/8)$	$F_1(h/8)$	$F_2(h/8)$	$F_3(h/8)$

more entries being added as needed. The general formula is this:

$$F_m\left(\frac{h}{2^k}\right) = F_{m-1}\left(\frac{h}{2^k}\right) + \frac{F_{m-1}(h/2^k) - F_{m-1}(h/2^{k-1})}{2^{2m} - 1}$$

It is not hard to modify the process just sketched so that the step size is reduced in some other way, perhaps $h_i = r^{i-1}h_1$ with h_1 the initial h. An arbitrary sequence of h_i could even be handled at little cost. Examples exist to show that sometimes these variations can be profitable.

13.21. Apply Richardson extrapolation to the function $y(x) = -1/x$ to find $y'(.05)$. The exact value is 400.

The computations are summarized in Table 13.3 and were carried out on an eight-digit computer. The original formula of Problem 13.20 produced the column headed F (all table entries being reduced by 400) so its best effort, for $h = .0001$, was off in the third decimal place. After that roundoff error took over. Looking elsewhere in the table one sees that values almost correct to five places appear.

Table 13.3†

h	F	F_1	F_2	F_3
.0128	28.05289			
.0064	6.66273	−.46732		
.0032	1.64515	−.02737	.00196	
.0016	.41031	−.00130	.00043	.00041
.0008	.10250	−.00010	−.00002	−.00002
.0004	.02625	.00084	.00090	.00091
.0002	.00750	.00125	.00127	.00127
.0001	.00500	.00417	.00436	.00441
.00005	.01000	.01166	.01215	.01227

† Entries reduced by 400

Supplementary Problems

13.22. Differentiate Bessel's formula, obtaining derivatives up to $p^{(5)}(x)$ in terms of differences through the fifth.

13.23. Apply the results of the previous problem to produce p', $p^{(2)}$, and $p^{(3)}$ at $x = 1.125$ from the data of Table 13.1.

13.24. Find the truncation error of the formula for $p'(x)$ obtained in Problem 13.22 using $k = \frac{1}{2}$. Estimate it by using $\xi = 1$. Compare with the actual error.

13.25. Find the maximum possible roundoff error of the formula of the previous problem. Compare the actual error with the truncation and roundoff error estimates.

13.26. Show that Stirling's formula of degree six produces

$$p'(x_0) = \frac{1}{h}\left(\delta\mu y_0 - \frac{1}{6}\delta^3\mu y_0 + \frac{1}{30}\delta^5\mu y_0\right)$$

Show that the truncation error of this formula is $-h^6 y^{(7)}(\xi)/140$.

13.27. Convert the formula of the previous problem to the form

$$p'(x_0) = \frac{1}{60h}(-y_{-3} + 9y_{-2} - 45y_{-1} + 45y_1 - 9y_2 + y_3)$$

and prove that the maximum roundoff error is $11E/6h$.

13.28. Find the argument corresponding to $y' = 0$ in Table 13.4 by inverse cubic interpolation, using either the Lagrange or Everett formula. (See again Problems 12.11 and 12.12.) Then find the corresponding y value by direct interpolation.

Table 13.4

x	y	y'
1.4	.98545	.16997
1.5	.99749	.07074
1.6	.99957	−.02920
1.7	.99166	−.12884

13.29. Ignoring the top and bottom lines of Table 13.4, apply Hermite's formula to find a cubic polynomial fitting the remaining data. Where does the derivative of this cubic equal zero? Compare with the previous problem. [Here the data correspond to $y(x) = \sin x$ and so the correct argument is $\pi/2$.]

13.30. The normal distribution function $y(x) = (1/\sqrt{2\pi})e^{-x^2/2}$ has an inflection point exactly at $x = 1$. How closely could this be determined from each of the following four-place data tables independently?

x	y	x	y
.50	.3521	.98	.2468
.75	.3011	.99	.2444
1.00	.2420	1.00	.2420
1.25	.1827	1.01	.2396
1.50	.1295	1.02	.2371

13.31. From Problems 13.9 and 13.13 we find the combined truncation and roundoff errors of the approximation

$$y'(x_0) \simeq \frac{1}{12h}(y_{-2} - 8y_{-1} + 8y_1 - y_2)$$

to have the form $Ah^4 + 3E/2h$ where $A = |y^{(5)}(\xi)/30|$. For what interval h will this be a minimum? Compute your result for the square root function and five-place accuracy.

13.32. Show that the truncation error of the formula $y^{(4)}(x_0) = \delta^4 y_0/h^4$ is $h^2 y^{(6)}(\xi)/6$.

13.33. Show that the maximum roundoff error of the formula in Problem 13.38 is $16E/h^4$.

Chapter 14

Numerical Integration

The importance of numerical integration may be appreciated by noting how frequently the formulation of problems in applied analysis involves derivatives. It is then natural to anticipate that the solutions of such problems will involve integrals. For most integrals no representation in terms of elementary functions is possible, and approximation becomes necessary.

POLYNOMIAL APPROXIMATION

Polynomial approximation serves as the basis for a broad variety of integration formulas, the main idea being that if $p(x)$ is an approximation to $y(x)$, then

$$\int_a^b p(x)\,dx \simeq \int_a^b y(x)\,dx$$

and on the whole this approach is very successful. In numerical analysis integration is the "easy" operation and differentiation the "hard" one, while the reverse is more or less true in elementary analysis. The best-known examples are the following:

1. **Integrating Newton's forward formula** of degree n between x_0 and x_n (the full range of collocation) leads to several useful formulas, including

$$\int_{x_0}^{x_1} p(x)\,dx = \frac{h}{2}(y_0 + y_1)$$

$$\int_{x_0}^{x_2} p(x)\,dx = \frac{h}{3}(y_0 + 4y_1 + y_2)$$

$$\int_{x_0}^{x_3} p(x)\,dx = \frac{3h}{8}(y_0 + 3y_1 + 3y_2 + y_3)$$

for $n = 1$, 2, and 3. The truncation error of any such formula is

$$\int_{x_0}^{x_n} y(x)\,dx - \int_{x_0}^{x_n} p(x)\,dx$$

and may be estimated in various ways. A Taylor series argument, for example, shows this error to be approximately $-h^3 y^{(2)}(\xi)/12$ when $n = 1$, and approximately $-h^5 y^{(4)}(\xi)/90$ when $n = 2$.

2. **Composite formulas** are obtained by applying the simple formulas just exhibited repeatedly to cover longer intervals. This amounts to using several connected line segments or parabolic segments, etc., and has advantages in simplicity over the use of a single high-degree polynomial.

3. The **trapezoidal rule,**

$$\int_{x_0}^{x_n} y(x)\,dx \simeq \frac{1}{2}h(y_0 + 2y_1 + \cdots + 2y_{n-1} + y_n)$$

is an elementary, but typical, composite formula. It, of course, uses connected line segments as the approximation to $y(x)$. Its truncation error is approximately $-(x_n - x_0)h^2 y^{(2)}(\xi)/12$.

4. **Simpson's rule,** $\displaystyle\int_{x_0}^{x_n} y(x)\,dx \simeq \frac{h}{3}(y_0 + 4y_1 + 2y_2 + 4y_3 + \cdots + 2y_{n-2} + 4y_{n-1} + y_n)$

is also a composite formula, and comes from using connected parabolic segments as the approximation to $y(x)$. It is one of the most heavily used formulas for approximate integration. The truncation error is about $-(x_n - x_0)h^4 y^{(4)}(\xi)/180$.

5. **Romberg's method** is based upon the fact that the truncation error of the trapezoidal rule is nearly proportional to h^2. Halving h and reapplying the rule thus reduces the error by a factor of $\frac{1}{4}$. Comparing the two results leads to an estimate of the error remaining. This estimate may then be used as a correction. Romberg's method is a systematic refinement of this simple idea.

6. **More complex formulas** may be obtained by integrating collocation polynomials over less than the full range of collocation. For example, *Simpson's rule with correction terms* may be derived by integrating Stirling's formula of degree six, which provides collocation at $x_{-3}, \ldots, x_3$, over just the center two intervals x_{-1} to x_1, and then using the result to develop a composite formula. The result is

$$\int_{x_0}^{x_n} y(x)\,dx \simeq \frac{h}{3}(y_0 + 4y_1 + 2y_2 + \cdots + y_n) - \frac{h}{90}(\delta^4 y_1 + \delta^4 y_3 + \cdots + \delta^4 y_{n-1}) + \frac{h}{756}(\delta^6 y_1 + \delta^6 y_3 + \cdots + \delta^6 y_{n-1})$$

the first part of which is Simpson's rule.

7. **Gregory's formula** takes the form of the trapezoidal rule with correction terms. It may be derived from the Euler–Maclaurin formula by expressing all derivatives as suitable combinations of differences to obtain

$$\int_{x_0}^{x_n} y(x)\;dx \simeq \frac{h}{2}(y_0 + 2y_1 + \cdots + 2y_{n-1} + y_n) - \frac{h}{12}(\nabla y_n - \Delta y_0) - \frac{h}{24}(\nabla^2 y_n + \Delta^2 y_0) - \frac{19h}{720}(\nabla^3 y_n - \Delta^3 y_0) - \cdots$$

and again the first part is the trapezoidal rule. The Euler–Maclaurin formula itself may be used as an approximate integration formula.

8. **Taylor's theorem** may be applied to develop the integrand as a power series, after which term-by-term integration sometimes leads to a feasible computation of the integral. More sophisticated ways of using this theorem have also been developed.

9. **The method of undetermined coefficients** may be used to generate integration formulas of a wide variety of types for special purposes.

10. **Adaptive integration** covers the many methods that have been devised to deal with the fact that most functions are harder to integrate accurately over certain intervals than over others. A particularly difficult section might, for example, force the use of a very small h value in Simpson's rule and lead to a great deal of unnecessary computation. Adaptive methods use finer subdivisions only where they are actually needed. One systematic way of doing this will be illustrated.

ERROR SOURCES

The usual error sources are present. However, input errors in the data values $y_0, \ldots, y_n$ are not magnified by most integration formulas, so this source of error is not nearly so troublesome as it is in numerical differentiation. The truncation error, which is

$$\int_a^b [y(x) - p(x)]\,dx$$

for our simplest formulas, and a composite of similar pieces for most of the others, is now the major contributor. A wide variety of efforts to estimate this error have been made. A related question is that of *convergence*. This asks whether, as continually higher degree polynomials are used, or as continually smaller intervals h_n between data points are used with $\lim h_n = 0$, a sequence of approximations is produced for which the limit of truncation error is zero. In many cases, the trapezoidal and Simpson rules being excellent examples, convergence can be proved. Roundoff errors also have a strong effect. A small interval h means substantial computation and much rounding off.

These algorithm errors ultimately obscure the convergence which should theoretically occur, and it is found in practice that decreasing h below a certain level leads to larger errors rather than smaller. As truncation error becomes negligible, roundoff errors accumulate, limiting the accuracy obtainable by a given method.

Solved Problems

14.1. Integrate Newton's formula for a collocation polynomial of degree n. Use the limits x_0 and x_n which are the outside limits of collocation. Assume equally spaced arguments.

This involves integrating a linear function from x_0 to x_1, or a quadratic from x_0 to x_2, and so on. See Fig. 14-1.

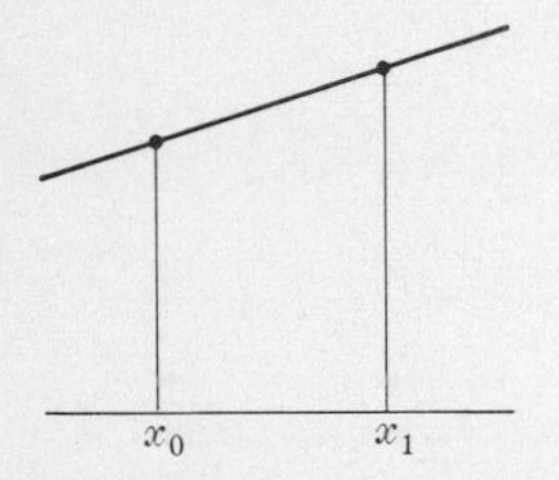

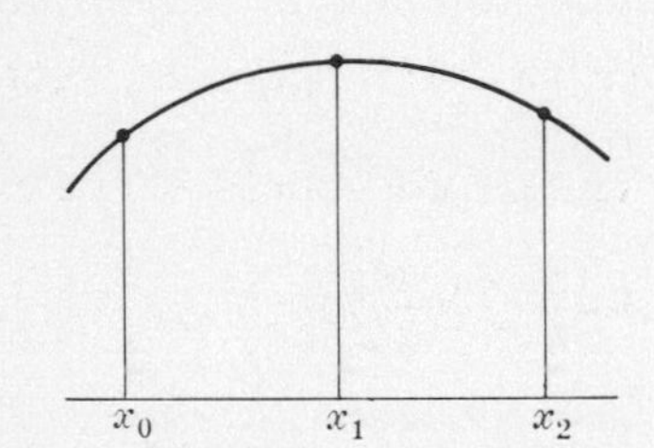

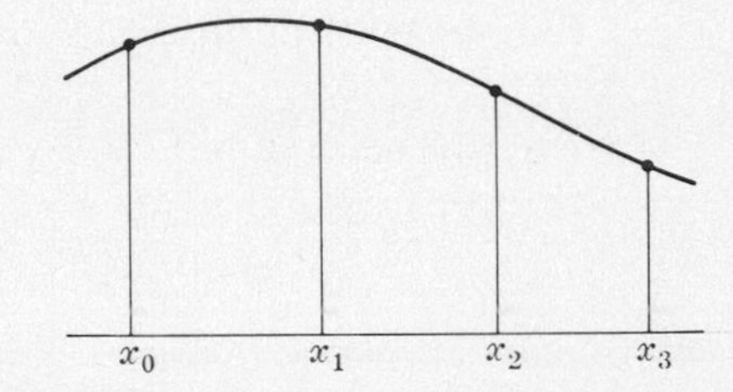

Fig. 14-1

The linear function certainly leads to $\frac{1}{2}h(y_0 + y_1)$. For the quadratic

$$p_k = y_0 + k\Delta y_0 + \frac{1}{2}k(k-1)\Delta^2 y_0$$

and easy computation produces, since $x = x_0 + kh$,

$$\int_{x_0}^{x_2} p(x)\,dx = h\int_0^2 p_k\,dk = h\left(2y_0 + 2\Delta y_0 + \frac{1}{3}\Delta^2 y_0\right) = \frac{h}{3}(y_0 + 4y_1 + y_2)$$

For the cubic polynomial a similar calculation produces

$$\int_{x_0}^{x_1} p(x)\,dx = h\int_0^3 p_k\,dk = h\int_0^3 \left[y_0 + k\Delta y_0 + \binom{k}{2}\Delta^2 y_0 + \binom{k}{3}\Delta^3 y_0\right] dk$$

$$= h\left(3y_0 + \frac{9}{2}\Delta y_0 + \frac{9}{4}\Delta^2 y_0 + \frac{3}{8}\Delta^3 y_0\right) = \frac{3h}{8}(y_0 + 3y_1 + 3y_2 + y_3)$$

Results for higher-degree polynomials can also be obtained in the same form

$$\int_{x_0}^{x_n} p(x)\,dx = Ch(c_0 y_0 + \cdots + c_n y_n)$$

and values of C and c_i for the first few values of n are given in Table 14.1. Such formulas are called the **Cotes formulas**.

Table 14.1

n	C	c_0	c_1	c_2	c_3	c_4	c_5	c_6	c_7	c_8
1	1/2	1	1							
2	1/3	1	4	1						
3	3/8	1	3	3	1					
4	2/45	7	32	12	32	7				
6	1/140	41	216	27	272	27	216	41		
8	4/14,175	989	5888	−928	10,496	−4540	10,496	−928	5888	989

Higher-degree formulas are seldom used, partly because simpler and equally accurate formulas are available, and partly because of the somewhat surprising fact that higher-degree polynomials do not always mean improved accuracy.

14.2. Estimate the truncation error of the $n = 1$ formula.

For this simple case we can integrate the formula

$$y(x) - p(x) = \frac{1}{2}(x - x_0)(x - x_1)y^{(2)}(\xi)$$

directly and apply the mean value theorem as follows, obtaining the exact error:

$$\int_{x_0}^{x_1} y(x)\,dx - \frac{1}{2}h(y_0 + y_1) = \int_{x_0}^{x_1} \frac{1}{2}(x - x_0)(x - x_1)y^{(2)}(\xi)\,dx$$

$$= y^{(2)}(\xi)\int_{x_0}^{x_1} \frac{1}{2}(x - x_0)(x - x_1)\,dx = -\frac{1}{12}h^3y^{(2)}(\xi)$$

where $h = x_1 - x_0$. The application of the mean value theorem is possible because $(x - x_0)(x - x_1)$ does not change sign in (x_0, x_1). The continuity of $y^{(2)}(\xi)$ is also involved. For $n > 1$ a sign change prevents a similar application of the mean value theorem and many methods have been devised to estimate truncation error, most having some disadvantages. We now illustrate one of the oldest methods, using the Taylor series, for the present simple case $n = 1$. First we have

$$\frac{1}{2}h(y_0 + y_1) = \frac{1}{2}h\left[y_0 + \left(y_0 + hy_0' + \frac{1}{2}h^2y_0^{(2)} + \cdots\right)\right]$$

Using an indefinite integral $F(x)$, where $F'(x) = y(x)$, we can also find

$$\int_{x_0}^{x_1} y(x)\,dx = F(x_1) - F(x_0) = hF'(x_0) + \frac{1}{2}h^2F^{(2)}(x_0) + \frac{1}{6}h^3F^{(3)}(x_0) + \cdots = hy_0 + \frac{1}{2}h^2y_0' + \frac{1}{6}h^3y_0^{(2)} + \cdots$$

and subtracting,

$$\int_{x_0}^{x_1} y(x)\,dx - \frac{1}{2}h(y_0 + y_1) = -\frac{h^3}{12}y_0^{(2)} + \cdots$$

presenting the truncation error in series form. The first term may be used as an error estimate. It should be compared with the actual error as given by $-(h^3/12)y^{(2)}(\xi)$ where $x_0 < \xi < x_1$.

14.3. Estimate the truncation error of the $n = 2$ formula.

Proceeding as in the previous problem, we find first

$$\frac{1}{3}h(y_0+4y_1+y_2)=\frac{1}{3}h\Big[y_0+4\Big(y_0+hy_0'+\frac{1}{2}h^2y_0^{(2)}+\frac{1}{6}h^3y_0^{(3)}+\frac{1}{24}h^2y_0^{(4)}+\cdots\Big)$$

$$+\Big(y_0+2hy_0'+2h^2y_0^{(2)}+\frac{4}{3}h^3y_0^{(3)}+\frac{2}{3}h^4y_0^{(4)}+\cdots\Big)\Big]$$

$$=\frac{1}{3}h\Big(6y_0+6hy_0'+4h^2y_0^{(2)}+2h^3y_0^{(3)}+\frac{5}{6}h^4y_0^{(4)}+\cdots\Big)$$

The integral itself is

$$\int_{x_0}^{x_2} y(x)\,dx = F(x_2)-F(x_0)$$

$$=2hF'(x_0)+\frac{1}{2}(2h)^2F^{(2)}(x_0)+\frac{1}{6}(2h)^3F^{(3)}(x_0)+\frac{1}{24}(2h)^4F^{(4)}(x_0)+\frac{1}{120}(2h)^5F^{(5)}(x_0)+\cdots$$

$$=2hy_0+2h^2y_0+\frac{4}{3}h^3y_0^{(2)}+\frac{2}{3}h^4y_0^{(3)}+\frac{4}{15}h^5y_0^{(4)}+\cdots$$

and subtracting,

$$\int_{x_0}^{x_2} y(x)\,dx-\frac{1}{3}h(y_0+4y_1+y_2)=-\frac{1}{90}h^5y_0^{(4)}+\cdots$$

we again have the truncation error in series form. The first term will be used as an approximation. It can also be shown that the error is given by $-(h^5/90)y^{(4)}(\xi)$ where $x_0<\xi<x_2$. (See Problem 14.65.)

A similar procedure applies to the other formulas. Results are presented in Table 14.2, the first term only being shown.

Table 14.2

n	**Truncation error**	n	**Truncation error**
1	$-(h^3/12)y^{(2)}$	4	$-(8h^7/945)y^{(6)}$
2	$-(h^5/90)y^{(4)}$	6	$-(9h^9/1400)y^{(8)}$
3	$-(3h^5/80)y^{(4)}$	8	$-(2368h^{11}/467{,}775)y^{(10)}$

Notice that formulas for odd n are comparable with those for the next smaller integer. (Of course, such formulas do cover one more interval of length h, but this does not prove to be significant. The even formulas are superior.)

14.4. Derive the *trapezoidal rule*.

This ancient formula stills finds application and illustrates very simply how the formulas of Problem 14.1 may be stretched to cover many intervals. The trapezoidal rule applies our $n=1$ formula to successive intervals up to x_n.

$$\frac{1}{2}h(y_0+y_1)+\frac{1}{2}h(y_1+y_2)+\frac{1}{2}h(y_2+y_3)+\cdots+\frac{1}{2}h(y_{n-1}+y_n)$$

This leads to the formula

$$\int_{x_0}^{x_n} y(x)\,dx \simeq \frac{1}{2}h(y_0+2y_1+\cdots+2y_{n-1}+y_n)$$

which is the trapezoidal rule.

14.5. Apply the trapezoidal rule to the integration of $\sqrt{x}$ between the arguments 1.00 and 1.30. Use the data of Table 13.1. Compare with the correct value of the integral.

We easily find

$$\int_{1.00}^{1.30} \sqrt{x}\,dx \simeq \frac{.05}{2}[1 + 2(1.02470 + \cdots + 1.11803) + 1.14017] = .32147$$

The correct value is $\frac{2}{3}[(1.3)^{3/2} - 1] = .32149$ to five places, making the actual error .00002.

14.6. Derive an estimate of the truncation error of the trapezoidal rule.

The result of Problem 14.2 may be applied to each interval, producing a total truncation error of about

$$-\frac{h^3}{12}[y_0^{(2)} + y_1^{(2)} + \cdots + y_{n-1}^{(2)}]$$

Assuming the second derivative bounded, $m < y^{(2)} < M$, the sum in brackets will be between nm and nM. Also assuming this derivative continuous allows the sum to be written as $ny^{(2)}(\xi)$ where $x_0 < \xi < x_n$. This is because $y^{(2)}(\xi)$ then assumes all values intermediate to m and M. It is also convenient to call the ends of the interval of integration $x_0 = a$ and $x_n = b$, making $b - a = nh$. Putting all this together, we have

$$\text{Truncation error} \simeq -\frac{(b-a)h^2}{12}y^{(2)}(\xi)$$

14.7. Apply the estimate of Problem 14.6 to our square root integral.

With $h = .05$, $b - a = .30$, and $y^{(2)}(x) = -x^{-3/2}/4$, truncation error $\simeq .000016$ which is slightly less than the actual error of .00002. However, rounding to five places and adding this error estimate to our computed result does produce .32149, the correct result.

14.8. Estimate the effect of inaccuracies in the y_k values on results obtained by the trapezoidal rule.

With Y_k denoting the true values, as before, we find $\frac{1}{2}h(e_0 + 2e_1 + \cdots + 2e_{n-1} + e_n)$ as the error due to inaccuracies $e_k = Y_k - y_k$. If the e_k do not exceed E in magnitude, this output error is bounded by $\frac{1}{2}h[E + 2(n-1)E + E) = (b-a)E$.

14.9. Apply the above to the square root integral of Problem 14.5.

We have $(b - a)E = (.30)(.000005) = .0000015$, so that this source of error is negligible.

14.10. Derive *Simpson's rule*.

This may be the most popular of all integration formulas. It involves applying our $n = 2$ formula to successive pairs of intervals up to x_n, obtaining the sum

$$\frac{h}{3}(y_0 + 4y_1 + y_2) + \frac{h}{3}(y_2 + 4y_3 + y_4) + \cdots + \frac{h}{3}(y_{n-2} + 4y_{n-1} + y_n)$$

which simplifies to

$$\frac{h}{3}(y_0 + 4y_1 + 2y_2 + 4y_3 + \cdots + 2y_{n-2} + 4y_{n-1} + y_n)$$

This is Simpson's rule. It requires n to be an even integer.

14.11. Apply Simpson's rule to the integral of Problem 14.5.

$$\int_{1.00}^{1.30} \sqrt{x}\,dx = \frac{.05}{3}[1.0000 + 4(1.02470 + 1.07238 + 1.11803) + 2(1.04881 + 1.09544) + 1.14017] = .32149$$

which is correct to five places.

14.12. Estimate the truncation error of Simpson's rule.

The result of Problem 14.3 may be applied to each pair of intervals, producing a total truncation error of about

$$-\frac{h^5}{90}(y_0^{(4)} + y_2^{(4)} + \cdots + y_{n-2}^{(4)})$$

Assuming the fourth derivative continuous allows the sum in brackets to be written as $(n/2)y^{(4)}(\xi)$ where $x_0 < \xi < x_n$. (The details are almost the same as in Problem 14.6.) Since $b - a = nh$,

$$\text{Truncation error} \simeq -\frac{(b-a)h^4}{180}y^{(4)}(\xi)$$

14.13. Apply the estimate of Problem 14.12 to our square root integral.

Since $y^{(4)}(x) = -\frac{15}{16}x^{-7/2}$, truncation error $\simeq .00000001$ which is minute.

14.14. Estimate the effect of data inaccuracies on results computed by Simpson's rule.

As in Problem 14.8, this error is found to be

$$\frac{1}{3}h(e_0 + 4e_1 + 2e_2 + 4e_3 + \cdots + 2e_{n-2} + 4e_{n-1} + e_n)$$

and if the data inaccuracies e_k do not exceed E in magnitude, this output error is bounded by

$$\frac{1}{3}hE\left[1 + 4\left(\frac{1}{2}n\right) + 2\left(\frac{1}{2}n - 1\right) + 1\right] = (b-a)E$$

exactly as for the trapezoidal rule. Applying this to the square root integral of Problem 14.11 we obtain the same .0000015 as in Problem 14.9, so that once again this source of error is negligible.

14.15. Compare the results of applying Simpson's rule with intervals $2h$ and h and obtain a new estimate of truncation error.

Assuming data errors negligible, we compare the two truncation errors. Let E_1 and E_2 denote these errors for the intervals $2h$ and h, respectively. Then

$$E_1 = -\frac{(b-a)(2h)^4}{180}y^{(4)}(\xi_1) \qquad E_2 = -\frac{(b-a)h^4}{180}y^{(4)}(\xi_2)$$

so that $E_2 \simeq E_1/16$. The error is reduced by a factor of 16 by halving the interval h. This may now be used to get another estimate of the truncation error of Simpson's rule. Call the correct value of the integral I, and the two Simpson approximations A_1 and A_2. Then

$$I = A_1 + E_1 = A_2 + E_2 \simeq A_1 + 16E_2$$

Solving for E_2, the truncation error associated with interval h is $E_2 \simeq (A_2 - A_1)/15$.

14.16. Use the estimate of Problem 14.15 to correct the Simpson's rule approximation.

This is an elementary but very useful idea. We find

$$I = A_2 + E_2 \simeq A_2 + \frac{A_2 - A_1}{15} = \frac{16A_2 - A_1}{15}$$

14.17. Apply the trapezoidal, Simpson, and $n = 6$ formulas to compute the integral of $\sin x$ between 0 and $\pi/2$ from the seven values provided in Table 14.3. Compare with the correct value of 1.

Table 14.3

x	0	$\pi/12$	$2\pi/12$	$3\pi/12$	$4\pi/12$	$5\pi/12$	$\pi/2$
$\sin x$	.00000	.25882	.50000	.70711	.86603	.96593	1.00000

The trapezoidal rule produces .99429. Simpson manages 1.00003. The $n = 6$ formula leads to

$$\frac{\pi}{140(12)}[41(0) + 216(.25882) + 27(.5) + 272(.70711) + 27(.86603) + 216(.96593) + 41(1)] = 1.000003$$

Clearly the $n = 6$ rule performs best for this fixed data supply.

14.18. Show that to obtain the integral of the previous problem correct to five places by using the trapezoidal rule would require an interval h of approximately .006 radian. By contrast, Table 14.3 has $h = \pi/12 \simeq .26$.

The truncation error of Problem 14.6 suggests that we want

$$\frac{(b-a)h^2}{12} y^{(2)}(\xi) \leqq \frac{(\pi/2)h^2}{12} < .000005$$

which will occur provided $h < .006$.

14.19. What interval h would be required to obtain the integral of Problem 14.17 correct to five places using Simpson's rule?

The truncation error of Problem 14.12 suggests

$$\frac{(b-a)h^4}{180} y^{(4)}(\xi) \leqq \frac{(\pi/2)h^4}{180} < .000005$$

or $h < .15$ approximately.

14.20. Prove that the trapezoidal and Simpson's rules are *convergent.*

If we assume truncation to be the only source of error, then in the case of the trapezoidal rule

$$I - A = -\frac{(b-a)h^2}{12} y^{(2)}(\xi)$$

where I is the exact integral and A the approximation. (Here we depend upon the exact representation of truncation error mentioned at the end of Problem 14.2.) If $\lim h = 0$ then assuming $y^{(2)}$ bounded, $\lim (I - A) = 0$. (This is the definition of convergence.)

For Simpson's rule we have the similar result

$$I - A = -\frac{(b-a)h^4}{180} y^{(4)}(\xi)$$

If $\lim h = 0$ then assuming $y^{(4)}$ bounded, $\lim (I - A) = 0$. Multiple use of higher-degree formulas also leads to convergence.

14.21. Apply Simpson's rule to the integral $\int_0^{\pi/2} \sin x\, dx$, continually halving the interval h in the search for greater accuracy.

Machine computations, carrying eight digits, produce the results in Table 14.4.

Table 14.4

h	Approx. integral	h	Approx. integral
$\pi/8$	1.0001344	$\pi/128$	.99999970
$\pi/16$	1.0000081	$\pi/256$	.99999955
$\pi/32$	1.0000003	$\pi/512$	.99999912
$\pi/64$	.99999983 (best)	$\pi/1024$	.99999870

14.22. The computations of Problem 14.21 indicate a durable error source which does not disappear as h diminishes, actually increases as work continues. What is this error source?

For very small intervals h the truncation error is small and, as seen earlier, data inaccuracies have little impact on Simpson's rule for any interval h. But small h means much computing, with the prospect of numerous computational roundoffs. This error source has not been a major factor in the much briefer algorithms encountered in interpolation and approximate differentiation. Here it has become dominant and limits the accuracy obtainable, even though our algorithm is convergent (Problem 14.20) and the effect of data inaccuracies small (we are saving eight decimal places). This problem emphasizes the importance of continuing search for briefer algorithms.

14.23. Develop the idea of Problems 14.15 and 14.16 into Romberg's method of approximate integration.

Suppose that the error of an approximate formula is proportional to h^n. Then two applications of the formula, with intervals h and $2h$, involve errors

$$E_1 \simeq C(2h)^n \qquad E_2 \simeq Ch^n$$

making $E_2 \simeq E_1/2^n$. With $I = A_1 + E_1 = A_2 + E_2$ as before, we soon find the new approximation

$$I \simeq A_2 + \frac{A_2 - A_1}{2^n - 1} = \frac{2^n A_2 - A_1}{2^n - 1}$$

For $n = 4$ this duplicates Problem 14.16. For $n = 2$ it applies to the trapezoidal rule in which the truncation error is proportional to h^2. It is not hard to verify that for $n = 2$ our last formula duplicates Simpson's rule, and that for $n = 4$ it duplicates the Cotes $n = 4$ formula. It can be shown that the error in this formula is proportional to h^{n+2} and this suggests a recursive computation. Apply the trapezoidal rule several times, continually halving h. Call the results $A_1, A_2, A_3, \ldots$. Apply our formula above with $n = 2$ to each pair of consecutive A_i. Call the results $B_1, B_2, B_3, \ldots$. Since the error is now proportional to h^4 we may reapply the formula, with $n = 4$, to the B_i. The results may be called $C_1, C_2, C_3, \ldots$. Continuing in this fashion an array of results is obtained.

$$\begin{array}{ccccc} A_1 & A_2 & A_3 & A_4 & \cdots \\ & B_1 & B_2 & B_3 & \cdots \\ & & C_1 & C_2 & \cdots \\ & & & D_1 & \cdots \end{array}$$

The computation is continued until entries at the lower right of the array agree within the required tolerance.

14.24. Apply Romberg's method to the integral of Problem 14.21.

The various results are as follows:

Points used	4	8	16	32
Trapezoidal result	.987116	.996785	.999196	.999799
		1.000008	1.000000	1.000000
			1.000000	1.000000
				1.000000

Convergence to the correct value of 1 is apparent.

14.25. More accurate integration formulas may be obtained by integrating a polynomial over less than the full range of collocation. Integrate Stirling's formula over the two center intervals.

Up through sixth differences Stirling's formula is

$$p_k = y_0 + k\mu\delta y_0 + \frac{1}{2}k^2\delta^2 y_0 + \frac{k(k^2-1)}{6}\mu\delta^3 y_0 + \frac{k^2(k^2-1)}{24}\delta^4 y_0 + \frac{k(k^2-1)(k^2-4)}{120}\mu\delta^5 y_0 + \frac{k^2(k^2-1)(k^2-4)}{720}\delta^6 y_0$$

Integration brings, since $x - x_0 = kh$ and $dx = h\,dk$,

$$\int_{x_0-h}^{x_0+h} p(x)\,dx = h\int_{-1}^{1} p_k\,dk = h\left(2y_0 + \frac{1}{3}\delta^2 y_0 - \frac{1}{90}\delta^4 y_0 + \frac{1}{756}\delta^6 y_0\right)$$

More terms are clearly available by increasing the degree of the polynomial. Stopping with the second difference term leaves us once again with the starting combination of Simpson's rule, in the form $(h/3)(y_{-1} + 4y_0 + y_1)$. In this case the integration has extended over the full range of collocation, as in Problem 14.1. With the fourth difference term we integrate over only half the range of collocation (Fig. 14-2).

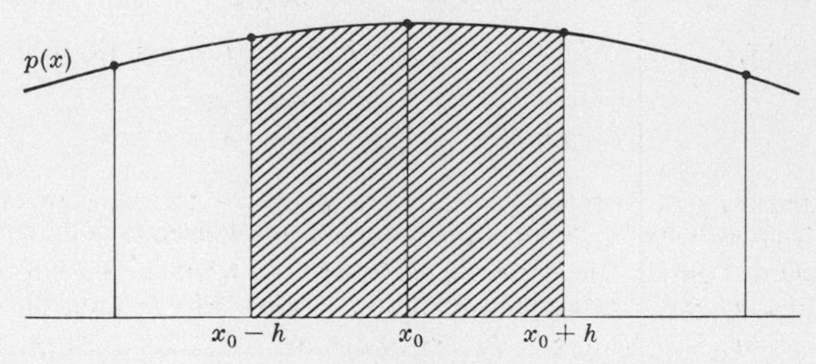

Fig. 14-2

As more differences are used $y(x)$ and $p(x)$ collocate at additional arguments, but the integration is extended over the center two intervals only. Since these are the intervals where Stirling's formula has the smallest truncation error (Problem 12.64), it can be anticipated that an integration formula obtained in this way will be more accurate. This extra accuracy is, however, purchased at a price; in application such formulas require y_k values outside the interval of integration.

The truncation error of this formula may be estimated by the Taylor series method used in Problem 14.6, and proves to be approximately $-\dfrac{23h^9}{113{,}400}y_0^{(8)} + \cdots$.

14.26. Use the result of Problem 14.25 to develop Simpson's rule with correction terms.

We make $n/2$ applications centered at $x_1, x_3, \ldots, x_{n-1}$, where n is even. The result is

$$\int_{x_0}^{x_n} p(x)\,dx = \frac{h}{3}(y_0 + 4y_1 + 2y_2 + \cdots + 4y_{n-1} + y_n)$$

$$-\frac{h}{90}(\delta^4 y_1 + \delta^4 y_3 + \cdots + \delta^4 y_{n-1}) + \frac{h}{756}(\delta^6 y_1 + \delta^6 y_3 + \cdots + \delta^6 y_{n-1})$$

This can be extended to higher differences if desired.

The truncation error of the result will be approximately $n/2$ times that of the previous problem and can be written as $-\dfrac{23(x_n - x_0)h^8}{226{,}800} y_0^{(8)} + \cdots$.

14.27. Develop the idea of adaptive integration.

The essential idea is to subdivide each part of the interval of integration just finely enough for it to contribute only its proportion of the overall error. There are many ways of doing this. Suppose the overall permissible error is E. Select an integration formula and apply it to the interval. Apply an error estimator. If the error is less than E, we are finished. If not, apply the formula to the left half of the interval. If the new error estimate is less than $E/2$, we are finished with that half interval. If not, this interval is halved and the process goes on. Eventually an interval of length $(b-a)/2^k$ is reached, (a, b) being the original interval, where the formula in use produces an acceptable result, the error being less than $E/2^k$. The process then resumes, beginning at the right edge of the accepted interval.

As the basic integration formula, Simpson's rule

$$A_2 = \frac{h}{3}(y_0 + 4y_1 + 2y_2 + 4y_3 + y_4)$$

might be chosen. As error measure, the doubled interval rule

$$A_1 = \frac{2h}{3}(y_0 + 4y_2 + y_4)$$

is then convenient, since Problem 14.15 then estimates the error as $(A_2 - A_1)/15$. The approximation A_2 is then accepted whenever $A_2 - A_1 \leqq 15E/2^k$ and is accumulated into the sum of other accepted results to its left. Clearly, the process ends when the accepted fragments cover (a, b).

14.28. Apply the adaptive integration method of the preceding problem to this integral:

$$\int_0^8 x^5\,dx$$

A few runs were made with different tolerances and slight changes in the upper limit. The following abbreviated output is typical. Note especially the values of k, which started at 1 (not printed) and rose to 7. An effort to increase the upper limit further found k skyrocketing.

x	$x^6/6$	Computed	k
2	10.667	10.667	4
4	682.667	682.667	5
6	7,776.000	7,775.99	6
8	43,690.67	43,690.58	7

14.29. Apply adaptive integration to the arcsine integral

$$\int_0^1 \frac{dx}{\sqrt{1-x^2}}$$

Tension is generated by the infinite discontinuity at the upper limit, which suggests a diminishing step size near this end much as in the preceding problem. Values of k climbed steadily as the computation progressed and reached 15 with this result:

$$\text{Upper limit} = .9999$$

$$\text{Integral} = 1.5573$$

At this point the correct arcsine value is 1.5575.

14.30. Derive Gregory's formula.

This is a form of the trapezoidal rule with correction terms and can be derived in many ways. One way begins with the Euler–Maclaurin formula (Problem 11.19) in the form

$$\int_{x_0}^{x_n} y(x)\,dx = \frac{h}{2}(y_0 + 2y_1 + \cdots + 2y_{n-1} + y_n) - \frac{h^2}{12}(y_n' - y_0') + \frac{h^4}{720}(y_n^{(3)} - y_0^{(3)}) - \frac{h^6}{30,240}(y_n^{(5)} - y_0^{(5)})$$

more terms being available if needed. Now express the derivatives at x_n in terms of backward differences and the derivatives at x_0 in terms of forward differences (Problem 13.1).

$$hy_0' = \left(\Delta - \frac{1}{2}\Delta^2 + \frac{1}{3}\Delta^3 - \frac{1}{4}\Delta^4 + \frac{1}{5}\Delta^5 - \cdots\right)y_0$$

$$hy_n' = \left(\nabla + \frac{1}{2}\nabla^2 + \frac{1}{3}\nabla^3 + \frac{1}{4}\nabla^4 + \frac{1}{5}\nabla^5 + \cdots\right)y_n$$

$$h^3y_0^{(3)} = \left(\Delta^3 - \frac{3}{2}\Delta^4 + \frac{7}{4}\Delta^5 - \cdots\right)y_0$$

$$h^3y_n^{(3)} = \left(\nabla^3 + \frac{3}{2}\nabla^4 + \frac{7}{4}\nabla^5 + \cdots\right)y_n$$

$$h^5y_0^{(5)} = (\Delta^5 - \cdots)y_0$$

$$h^5y_n^{(5)} = (\nabla^5 + \cdots)y_n$$

The result of substituting these expressions is

$$\int_{x_0}^{x_n} p(x)\,dx = \frac{h}{2}(y_0 + 2y_1 + \cdots + 2y_{n-1} + y_n) - \frac{h}{12}(\nabla y_n - \Delta y_0)$$

$$- \frac{h}{24}(\nabla^2 y_n + \Delta^2 y_0) - \frac{19h}{720}(\nabla^3 y_n - \Delta^3 y_0) - \frac{3h}{160}(\nabla^4 y_n + \Delta^4 y_0) - \frac{863h}{60,480}(\nabla^5 y_n - \Delta^5 y_0)$$

and again more terms can be computed if needed. This is Gregory's formula. It does not require y_k values outside the interval of integration.

14.31. Apply Taylor's theorem to evaluate the error function integral

$$H(x) = \frac{2}{\sqrt{\pi}}\int_0^x e^{-t^2}\,dt$$

for $x = .5$ and $x = 1$, correct to four decimal places.

The series $e^{-t^2} = 1 - t^2 + \frac{t^4}{2} - \frac{t^6}{6} + \frac{t^8}{24} - \frac{t^{10}}{120} + \cdots$ leads to

$$H(x) = \frac{2}{\sqrt{\pi}}\left(x - \frac{x^3}{3} + \frac{x^5}{10} - \frac{x^7}{42} + \frac{x^9}{216} - \frac{x^{11}}{1320} + \cdots\right)$$

For $x = .5$ this produces .5205, and for $x = 1$ we find .8427. The character of this series assures that the error made in truncating it does not exceed the last term used, so we can be confident in our results. The series method has performed very well here, but it becomes clear that if more decimal places are wanted or if larger upper limits x are to be used, then many more terms of this series will become involved. In such cases it is usually more convenient to proceed as in the next problem.

14.32. Tabulate the error function integral for $x = 0(.1)4$ to six decimal places.

$$H(x) = \frac{2}{\sqrt{\pi}} \int_0^x e^{-t^2}\, dt$$

We adopt the method which was used to prepare the fifteen-place table of this function, NBS-AMS 41. The derivatives needed are

$$H'(x) = \frac{2}{\sqrt{\pi}} e^{-x^2} \qquad H^{(2)}(x) = -2xH'(x) \qquad H^{(3)}(x) = -2xH^{(2)}(x) - 2H'(x)$$

and in general
$$H^{(n)}(x) = -2xH^{(n-1)}(x) - 2(n-2)H^{(n-2)}(x)$$

The Taylor series may be written as

$$H(x+h) = H(x) + hH'(x) + \cdots + \frac{h^n}{n!} H^{(n)}(x) + R$$

where the remainder is the usual $R = h^{n+1}H^{(n+1)}(\xi)/(n+1)!$. Notice that if M denotes the sum of even power terms and N the sum of odd power terms, then

$$H(x+h) = M + N \qquad H(x-h) = M - N$$

For six-place accuracy we use terms of the Taylor series which affect the eighth place, because the length of the task ahead makes substantial roundoff error growth a possibility. With $H(0) = 0$, the computation begins with

$$H(.1) = \frac{2}{\sqrt{\pi}}(.1) - \frac{2}{3\sqrt{\pi}}(.1)^3 + \frac{1}{5\sqrt{\pi}}(.1)^5 = .11246291$$

only the odd powers contributing. Next we put $x = .1$ and find

$$H'(.1) = \frac{2}{\sqrt{\pi}} e^{-.01} = 1.1171516$$

$$H^{(2)}(.1) = -.2H'(.1) = -.22343032$$

$$H^{(3)}(.1) = -.2H^{(2)}(.1) - 2H'(.1) = -2.1896171$$

$$H^{(4)}(.1) = -.2H^{(3)}(.1) - 4H^{(2)}(.1) = 1.3316447$$

$$H^{(5)}(.1) = -.2H^{(4)}(.1) - 6H^{(3)}(.1) = 12.871374$$

$$H^{(6)}(.1) = -.2H^{(5)}(.1) - 8H^{(4)}(.1) = -13.227432$$

leading to

$$M = .11246291 - .00111715 + .00000555 - .00000002 = .11135129$$

$$N = .11171516 - .00036494 + .00000107 = .11135129$$

Since $H(x-h) = M - N$, we rediscover $H(0) = 0$ which serves as a check on the correctness of the computation. We also obtain

$$H(.2) = H(x+h) = M + N = .22270258$$

The process is now repeated to obtain a check on $H(.1)$ and a prediction of $H(.3)$. Continuing in this way one eventually reaches $H(4)$. The last two decimal places can then be rounded off. Correct values to six places are given in Table 14.15 for $x = 0(.5)4$. In NBS-AMS 41 computations were carried to 25 places, then rounded to 15. Extensive subtabulations were then made for small x arguments.

Table 14.5

x	.5	1.0	1.5	2.0	2.5	3.0	3.5	4.0
$H(x)$	.520500	.842701	.966105	.995322	.999593	.999978	.999999	1.000000

14.33. Illustrate the *method of undetermined coefficients* for deriving approximate integration formulas, by applying it to the derivation of Simpson's rule.

In this method we aim directly for a formula of a preselected type. For Simpson's rule the choice

$$\int_{-h}^{h} y(x)\,dx = h(c_{-1}y_{-1} + c_0 y_0 + c_1 y_1)$$

is convenient. The selection of the coefficients c_k can proceed in many ways, but for Simpson's rule the choice is made on the basis that the resulting formula be exact when $y(x)$ is any of the first three powers of x. Taking $y(x)=1$, x, and x^2 in turn, we are led to the conditions

$$2 = c_{-1} + c_0 + c_1 \qquad 0 = -c_{-1} + c_1 \qquad \frac{2}{3} = c_{-1} + c_1$$

which yield $c_{-1} = c_1 = \frac{1}{3}$, $c_0 = \frac{4}{3}$ making

$$\int_{-h}^{h} y(x)\,dx = \frac{h}{3}(y_{-1} + 4y_0 + y_1)$$

Applying this result to successive pairs of intervals between x_0 and x_n again generates Simpson's rule.

As a bonus, this result also proves to be exact for $y(x)=x^3$, as is easily seen from the symmetries. This means by addition that it is also exact for any polynomial of degree three or less. For higher-degree polynomials there is an error term.

14.34. Apply the method of undetermined coefficients to derive a formula of the type

$$\int_0^h y(x)\,dx = h(a_0 y_0 + a_1 y_1) + h^2(b_0 y_0' + b_1 y_1')$$

With four coefficients available, we try to make the formula exact when $y(x) = 1$, x, x^2, and x^3. This leads to the four conditions

$$\begin{aligned} 1 &= a_0 + a_1 \\ \frac{1}{2} &= a_1 + b_0 + b_1 \\ \frac{1}{3} &= a_1 + 2b_1 \\ \frac{1}{4} &= a_1 + 3b_1 \end{aligned}$$

which yield $a_0 = a_1 = \frac{1}{2}$, $b_0 = -b_1 = \frac{1}{12}$. The resulting formula is

$$\int_0^h y(x)\,dx = \frac{h}{2}(y_0 + y_1) + \frac{h^2}{12}(y_0' - y_1')$$

which reproduces the first terms of the Euler–Maclaurin formula. A great variety of formulas may be generated by this method of undetermined coefficients. As in the examples just offered, a little preliminary planning and use of symmetry can often simplify the system of equations which ultimately determines the coefficients.

Supplementary Problems

14.35 Integrate Newton's formula for a collocation polynomial of degree four and so verify the $n=4$ row of Table 14.1.

14.36. Verify the $n=6$ row of Table 14.1.

14.37. Use the Taylor series method to obtain the truncation error estimate for the $n=3$ formula as listed in Table 14.2.

14.38. Use the Taylor series method to verify the truncation error estimate for the $n=4$ formula.

14.39. Apply various formulas to the following limited data supply to approximate the integral of $y(x)$:

x	1.0	1.2	1.4	1.6	1.8	2.0
$y(x)$	1.0000	.8333	.7143	.6250	.5556	.5000

Use the trapezoidal rule, applying correction terms. How much confidence do you place in your result? Does it appear correct to four places? (See the next problem.)

14.40. The data of Problem 14.39 actually belong to the function $y(x)=1/x$. The correct integral is, therefore, to four places, $\ln 2 = .6931$. Has any approximate method produced this?

14.41. Use the truncation error estimate for the trapezoidal rule to predict how tightly values of $y(x)$ must be packed (what interval h) for the trapezoidal rule itself to achieve a correct result to four places for $\int_1^2 dx/x$.

14.42. Suppose the data of Problem 14.39 augmented by the inclusion of these new number pairs:

x	1.1	1.3	1.5	1.7	1.9
$y(x)$	.9091	.7692	.6667	.5882	.5263

Reapply the trapezoidal rule to the full data supply. Use this result as A_2, the corresponding result in Problem 14.39 as A_1, and the formula of Problem 14.23 to obtain still another approximation to I. Is it correct to four places?

14.43. Apply the trapezoidal rule with correction terms to the full data supply now available for $y(x)=1/x$.

14.44. Apply Simpson's rule to the data of Problem 14.39. Will correction terms as in Problem 14.26 be needed? If so, apply them.

14.45. Use the truncation error estimate for Simpson's rule to predict how many values of $y(x)$, or how small an interval h, will be needed for this rule to produce $\ln 2$ correct to four places.

14.46. How small an interval h would be required to obtain $\ln 2$ correct to eight places using the trapezoidal rule? Using Simpson's rule?

14.47. Apply the Euler–Maclaurin formula (Problem 14.30) up through the fifth-derivative terms to evaluate $\ln 2$ to eight decimal places. The correct value is .69314718. (Try $h=.1$.)

14.48. From the following data estimate $\int_0^2 y(x)\,dx$ as well as you can.

x	0	.25	.50	.75	1.00	1.25	1.50	1.75	2
$y(x)$	1.000	1.284	1.649	2.117	2.718	3.490	4.482	5.755	7.389

How much confidence do you place in your results? Do you believe them correct to three places?

14.49. The data of Problem 14.48 were taken from the exponential function $y(x) = e^x$. The correct integral is, therefore, to three places, $\int_0^2 e^x\,dx = e^2 - 1 = 6.389$. Were any of our formulas able to produce this result?

14.50. From the following data, estimate $\int_1^5 y(x)\,dx$ as well as you can.

x	1	1.5	2	2.5	3	3.5	4	4.5	5
$y(x)$	0	.41	.69	.92	1.10	1.25	1.39	1.50	1.61

How much confidence do you place in your results?

14.51. The data of Problem 14.50 corresponds to $y(x) = \log x$. The correct integral is, therefore, to two places, $\int_1^5 \log x\,dx = 5\log 5 - 4 = 4.05$. Were any of our formulas able to produce this result?

14.52. Calculate $\displaystyle\int_0^1 \frac{dx}{1+x^2}$ correct to seven places by adaptive integration. The correct value is $\pi/4$, or to seven places .7853982.

14.53. Calculate $\int_0^{\pi/2} \sqrt{1 - \frac{1}{4}\sin^2 t}\,dt$ to four decimal places. This is called an elliptic integral. Its correct value is 1.4675. Use adaptive integration.

14.54. Show that to four places $\int_0^{\pi/2} \sqrt{1 - \frac{1}{2}\sin^2 t}\,dt = 1.3506$.

14.55. Use adaptive integration to verify

$$\int_0^{\pi/2} \frac{dx}{\sin^2 x + \frac{1}{4}\cos^2 x} = 3.1415927$$

the exact value being π.

14.56. Apply the Taylor series method as in Problem 14.31, to compute the sine integral

$$\text{Si}(x) = \int_0^x \frac{\sin t}{t}\,dt$$

for $x = 0(.1)1$, to five decimal places. The refined procedure used in Problem 14.32 is not necessary here. [The last result should be Si (1) = .94608.]

14.57. Apply the Taylor series method as in Problem 14.32 to compute the sine integral for $x = 0(.5)15$, to five decimal places. The final result should be Si (15) = 1.61819.

14.58. Apply the Taylor series method to compute $\int_0^1 \sqrt{x}\sin x\,dx$ to eight decimal places.

14.59. Apply the Taylor series method to compute $\int_0^1 (1/\sqrt{1+x^4})\,dx$ to four decimal places.

14.60. Compute the total arc length of the ellipse $x^2 + y^2/4 = 1$ to six decimal places.

14.61. By adding $(h/140)\delta^6 y_3$ to the $n=6$ formula of Table 14.1, derive Weddle's rule,

$$\int_{x_0}^{x_6} y(x)\,dx = \frac{3h}{10}(y_0 + 5y_1 + y_2 + 6y_3 + y_4 + 5y_5 + y_6)$$

14.62. Use the method of undetermined coefficients to derive a formula of the form

$$\int_{-h}^{h} y(x)\,dx = h(a_{-1}y_{-1} + a_0 y_0 + a_1 y_1) + h^2(b_{-1}y'_{-1} + b_0 y'_0 + b_1 y'_1)$$

which is exact for polynomials of as high a degree as possible.

14.63. Use the method of undetermined coefficients to derive the formula

$$\int_0^h y(x)\,dx = \frac{h}{2}(y_0 + y_1) - \frac{h^3}{24}(y_0^{(2)} + y_1^{(2)})$$

proving it exact for polynomials of degree up through three.

14.64. Use the method of undetermined coefficients to derive

$$\int_0^h y(x)\,dx = \frac{h}{2}(y_0 + y_1) + \frac{h^2}{10}(y'_0 - y'_1) + \frac{h^3}{120}(y_0^{(2)} + y_1^{(2)})$$

proving it exact for polynomials of degree up through five.

14.65. Derive an exact expression for the truncation error of our $n=2$ formula by the following method. Let

$$F(h) = \int_{-h}^{h} y(x)\,dx - \frac{h}{3}[y(-h) + 4y(0) + y(h)]$$

Differentiate three times relative to h, using the theorem on "differentiating under the integral sign"

$$\frac{d}{dh}\int_{a(h)}^{b(h)} y(x, h)\,dx = \int_a^b \frac{\partial y}{\partial h}\,dx + y(b, h)b'(h) - y(a, h)a'(h)$$

to obtain

$$F^{(3)}(h) = -\frac{h}{3}[y^{(3)}(h) - y^{(3)}(-h)]$$

Notice that $F'(0) = F^{(2)}(0) = F^{(3)}(0) = 0$. Assuming $y^{(4)}(x)$ continuous, the mean value theorem now produces

$$F^{(3)}(h) = -\frac{2}{3}h^2 y^{(4)}(\theta h)$$

where θ depends on h and falls between -1 and 1. We now reverse direction and recover $F(h)$ by integration. It is convenient to replace h by t (making θ a function of t). Verify that

$$F(h) = -\frac{1}{3}\int_0^h (h-t)^2 t^2 y^{(4)}(\theta t)\,dt$$

by differentiating three times relative to h to recover the above $F^{(3)}(h)$. Since this formula also makes $F(0) = F'(0) = F^{(2)}(0)$, it is the original $F(h)$. Next apply the mean value theorem

$$\int_a^b f(t)g(t)\,dt = g(\xi)\int_a^b f(t)\,dt$$

with $a < \xi < b$, which is valid for continuous functions provided $f(t)$ does not change sign between a and b. These conditions do hold here with $f(t) = -t^2(h-t)^2/3$. The result is

$$F(h) = y^{(4)}(\xi)\int_0^h f(t)\,dt = -\frac{h^5}{90}y^{(4)}(\xi)$$

This is the result mentioned in Problem 14.3. The early parts of this proof, in which we maneuver from $F(h)$ to its third derivative and back again, have as their goal a representation of $F(h)$ to which the mean value theorem can be applied. [Recall that $f(t)$ did not change sign in the interval of integration.] This is often the central difficulty in obtaining a truncation error formula of the sort just achieved.

14.66. Modify the argument of Problem 14.65 to obtain the formula given at the end of Problem 14.2,

$$\text{Truncation error} = -\frac{h^3}{12}y^{(2)}(\xi)$$

for the $n = 1$ formula.

14.67. Evaluate $\int_0^1 e^{-x^3}\,dx$ correct to six places.

Chapter 15

Gaussian Integration

CHARACTER OF A GAUSSIAN FORMULA

The main idea behind Gaussian integration is that in the selection of a formula

$$\int_a^b y(x)\,dx \simeq \sum_{i=1}^n A_i y(x_i)$$

it may be wise not to specify that the arguments x_i be equally spaced. All the formulas of the preceding chapter assume equal spacing, and if the values $y(x_i)$ are obtained experimentally this will probably be true. Many integrals, however, involve familiar analytic functions which may be computed for any argument and to great accuracy. In such cases it is useful to ask what choice of the x_i and A_i together will bring maximum accuracy. It proves to be convenient to discuss the slightly more general formula

$$\int_a^b w(x)y(x)\,dx \simeq \sum_{i=1}^n A_i y(x_i)$$

in which $w(x)$ is a weighting function to be specified later. When $w(x) = 1$ we have the original, simpler formula.

One approach to such Gaussian formulas is to ask for perfect accuracy when $y(x)$ is one of the power functions $1, x, x^2, \ldots, x^{2n-1}$. This provides $2n$ conditions for determining the $2n$ numbers x_i and A_i. In fact,

$$A_i = \int_a^b w(x)L_i(x)\,dx$$

where $L_i(x)$ is the Lagrange multiplier function introduced in Chapter 8. The arguments $x_1, \ldots, x_n$ are the zeros of the nth-degree polynomial $p_n(x)$ belonging to a family having the orthogonality property

$$\int_a^b w(x)p_n(x)p_m(x)\,dx = 0 \qquad \text{for } m \neq n$$

These polynomials depend upon $w(x)$. The weighting function therefore influences both the A_i and the x_i but does not appear explicitly in the Gaussian formula.

Hermite's formula for an osculating polynomial provides another approach to Gaussian formulas. Integrating the osculating polynomial leads to

$$\int_a^b w(x)y(x)\,dx \simeq \sum_{i=1}^n [A_i y(x_i) + B_i y'(x_i)]$$

but the choice of the arguments x_i as the zeros of a member of an orthogonal family makes all $B_i = 0$. The formula then reduces to the prescribed type. This suggests, and we proceed to verify, that a simple collocation polynomial at these unequally spaced arguments would lead to the same result.

Orthogonal polynomials therefore play a central role in Gaussian integration. A study of their main properties forms a substantial part of this chapter.

The *truncation error* of the Gaussian formula is

$$\int_a^b w(x)y(x)\,dx - \sum_{i=1}^n A_i y(x_i) = \frac{y^{(2n)}(\xi)}{(2n)!}\int_a^b w(x)[\pi(x)]^2\,dx$$

where $\pi(x) = (x - x_1)\cdots(x - x_n)$. Since this is proportional to the $(2n)$th derivative of $y(x)$, such formulas are exact for all polynomials of degree $2n - 1$ or less. In the formulas of the previous chapter it is $y^{(n)}(\xi)$ which appears in this place. In a sense our present formulas are *twice as accurate* as those based on equally spaced arguments.

PARTICULAR TYPES OF GAUSSIAN FORMULAS

Particular types of Gaussian formulas may be obtained by choosing $w(x)$ and the limits of integration in various ways. Occasionally one may also wish to impose constraints, such as specifying certain x_i in advance. A number of particular types are presented.

1. **Gaussian–Legendre formulas** occur when $w(x) = 1$. This is the prototype of the Gaussian method and we discuss it in more detail than the other types. It is customary to normalize the interval (a, b) to $(-1, 1)$. The orthogonal polynomials are then the Legendre polynomials

$$P_n(x) = \frac{1}{2^n n!} \frac{d^n}{dx^n} (x^2 - 1)^n$$

with $P_0(x) = 1$. The x_i are the zeros of these polynomials and the coefficients are

$$A_i = \frac{2(1 - x_i^2)}{n^2[P_{n-1}(x_i)]^2}$$

Tables of the x_i and A_i are available to be substituted directly into the Gauss–Legendre formula

$$\int_a^b y(x)\,dx \simeq \sum_{i=1}^n A_i y(x_i)$$

Various properties of Legendre polynomials are required in the development of these results, including the following:

$$\int_{-1}^1 x^k P_n(x)\,dx = 0 \qquad \text{for } k = 0, \ldots, n-1$$

$$\int_{-1}^1 x^n P_n(x)\,dx = \frac{2^{n+1}(n!)^2}{(2n+1)!}$$

$$\int_{-1}^1 [P_n(x)]^2\,dx = \frac{2}{2n+1}$$

$$\int_{-1}^1 P_m(x)P_n(x)\,dx = 0 \qquad \text{for } m \neq n$$

$$P_n(x) \text{ has } n \text{ real zeros in } (-1, 1)$$

$$(n+1)P_{n+1}(x) = (2n+1)xP_n(x) - nP_{n-1}(x)$$

$$(t - x)\sum_{i=0}^n (2i+1)P_i(x)P_i(t) = (n+1)[P_{n+1}(t)P_n(x) - P_n(t)P_{n+1}(x)]$$

$$\int_{-1}^1 \frac{P_n(x)}{x - x_k}\,dx = \frac{-2}{(n+1)P_{n+1}(x_k)}$$

$$(1 - x^2)P_n'(x) + nxP_n(x) = nP_{n-1}(x)$$

Lanczos' estimate of truncation error for Gauss–Legendre formulas takes the form

$$E \simeq \frac{1}{2n+1}\left[y(1) + y(-1) - I - \sum_{i=1}^n A_i x_i y'(x_i)\right]$$

where I is the approximate integral obtained by the Gaussian n-point formula. Note that the Σ term involves applying this same formula to the function $xy'(x)$. This error estimate seems to be fairly accurate for smooth functions.

2. **Gauss–Laguerre formulas** take the form

$$\int_0^\infty e^{-x}y(x)\,dx \simeq \sum_{i=1}^{n} A_i y(x_i)$$

the arguments x_i being the zeros of the nth Laguerre polynomial

$$L_n(x) = e^x \frac{d^n}{dx^n}(e^{-x}x^n)$$

and the coefficients A_i being

$$A_i = \frac{(n!)^2}{x_i[L_n'(x_i)]^2}$$

The numbers x_i and A_i are available in tables.

The derivation of Gauss–Laguerre formulas parallels that of Gauss–Legendre very closely, using properties of the Laguerre polynomials.

3. **Gauss–Hermite formulas** take the form

$$\int_{-\infty}^{\infty} e^{-x^2}y(x)\,dx \simeq \sum_{i=1}^{n} A_i y(x_i)$$

the arguments x_i being the zeros of the nth Hermite polynomial

$$H_n(x) = (-1)^n e^{x^2}\frac{d^n}{dx^n}(e^{-x^2})$$

and the coefficients A_i being

$$A_i = \frac{2^{n+1}n!\sqrt{\pi}}{[H_n'(x_i)]^2}$$

The numbers x_i and A_i are available in tables.

4. **Gauss–Chebyshev formulas** take the form

$$\int_{-1}^{1} \frac{y(x)}{\sqrt{1-x^2}}\,dx \simeq \frac{\pi}{n}\sum_{i=1}^{n} y(x_i)$$

the arguments x_i being the zeros of the nth Chebyshev polynomial $T_n(x) = \cos(n \arccos x)$.

Solved Problems

THE GAUSSIAN METHOD

15.1. Integrate Hermite's formula for an osculating polynomial approximation to $y(x)$ at arguments x_1 to x_n.

Here it is convenient to delete the argument x_0 in our osculating polynomial. This requires only minor changes in our formulas of Chapter 10. The Hermite formula itself becomes

$$p(x) = \sum_{i=1}^{n} [1 - 2L_i'(x_i)(x - x_i)][L_i(x)]^2 y_i + (x - x_i)[L_i(x)]^2 y_i'$$

where $L_i(x) = F_i(x)/F_i(x_i)$ is the Lagrange multiplier function, $F_i(x)$ being the product $F_i(x) = \prod_{k\neq i}(x - x_k)$.

Integrating, we find

$$\int_a^b w(x)p(x)\,dx = \sum_{i=1}^{n} (A_i y_i + B_i y_i')$$

where $$A_i = \int_a^b w(x)[1 - 2L_i'(x_i)(x - x_i)][L_i(x)]^2\,dx \qquad B_i = \int_a^b w(x)(x - x_i)[L_i(x)]^2\,dx$$

15.2. Find the truncation error of the formula in Problem 15.1.

Surprisingly enough, this comes more easily than for formulas obtained from simple collocation polynomials, because the mean value theorem applies directly. The error of Hermite's formula (Problem 10.4), with n in place of $n+1$ because we have deleted one argument, becomes

$$y(x) - p(x) = \frac{y^{(2n)}(\xi)}{(2n)!}[\pi(x)]^2$$

Multiplying by $w(x)$ and integrating,

$$\int_a^b w(x)[y(x) - p(x)]\,dx = \int_a^b w(x)\frac{y^{(2n)}(\xi)}{(2n)!}[\pi(x)]^2\,dx$$

Since $w(x)$ is to be chosen a nonnegative function and $[\pi(x)]^2$ is surely positive, the mean value theorem at once yields

$$E = \int_a^b w(x)[y(x) - p(x)]\,dx = \frac{y^{(2n)}(\theta)}{(2n)!}\int_a^b w(x)[\pi(x)]^2\,dx$$

for the truncation error. Here $a < \theta < b$, but as usual θ is not otherwise known. Notice that if $y(x)$ were a polynomial of degree $2n-1$ or less, this error term would be exactly 0. Our formula will be exact for all such polynomials.

15.3. Show that all the coefficients B_i will be 0 if

$$\int_a^b w(x)\pi(x)x^k\,dx = 0 \qquad \text{for } k = 0, 1, \ldots, n-1$$

By Problem 8.3 $(x - x_i)L_i(x) = \pi(x)/\pi'(x_i)$. Substituting this into the formula for B_i,

$$B_i = \frac{1}{\pi'(x_i)}\int_a^b w(x)\pi(x)L_i(x)\,dx$$

But $L_i(x)$ is a polynomial in x of degree $n-1$ and so

$$B_i = \frac{1}{\pi'(x_i)}\int_a^b w(x)\pi(x)\sum_{k=0}^{n-1}\alpha_k x^k\,dx = \frac{1}{\pi'(x_i)}\sum_{k=0}^{n-1}\alpha_k\int_a^b w(x)\pi(x)x^k\,dx = 0$$

15.4. Define *orthogonal functions* and restate the result of Problem 15.3 in terms of orthogonality.

Functions $f_1(x)$ and $f_2(x)$ are called orthogonal on the interval (a, b) with weight function $w(x)$ if

$$\int_a^b w(x)f_1(x)f_2(x)\,dx = 0$$

The coefficients B_i of our formula will be zero if $\pi(x)$ is orthogonal to x^p for $p = 0, 1, \ldots, n-1$. By addition $\pi(x)$ will then be orthogonal to any polynomial of degree $n-1$ or less, including the Lagrange multiplier functions $L_i(x)$. Such orthogonality depends upon and determines our choice of the collocation arguments x_k and is assumed for the remainder of this chapter.

15.5. Prove that with all the $B_i = 0$, the coefficients A_i reduce to $A_i = \int_a^b w(x)[L_i(x)]^2\,dx$ and are therefore positive numbers.

$$A_i = \int_a^b w(x)[L_i(x)]^2\,dx - 2L_i'(x_i)B_i \quad \text{reduces to the required form when } B_i = 0$$

15.6. Derive the simpler formula $A_i = \int_a^b w(x)L_i(x)\,dx$.

The result follows if we can show that $\int_a^b w(x)L_i(x)[L_i(x) - 1]\,dx = 0$.
But $L_i(x) - 1$ must contain $(x - x_i)$ as a factor, because $L_i(x_i) - 1 = 1 - 1 = 0$. Therefore

$$L_i(x)[L_i(x) - 1] = \frac{\pi(x)}{\pi'(x_i)(x - x_i)}[L_i(x) - 1] = \pi(x)p(x)$$

with $p(x)$ of degree $n - 1$ at most. Problem 15.3 then guarantees that the integral is zero.

15.7. The integration formula of this section can now be written as

$$\int_a^b w(x)y(x)\,dx \simeq \sum_{i=1}^n A_i y(x_i)$$

where $A_i = \int_a^b w(x)L_i(x)\,dx$ and the arguments x_i are to be chosen by the orthogonality requirements of Problem 15.3. This formula was obtained by integration of an osculating polynomial of degree $2n - 1$ determined by the y_i and y_i' values at arguments x_i. Show that the same formula is obtained by integration of the simpler collocation polynomial of degree $n - 1$ determined by the y_i values alone. (This is one way of looking at Gaussian formulas; they extract high accuracy from polynomials of relatively low degree.)

The collocation polynomial is $p(x) = \sum_{i=1}^n L_i(x)y(x_i)$ so that integration produces

$$\int_a^b w(x)p(x)\,dx = \sum_{i=1}^n A_i y(x_i)$$

as suggested. Here $p(x)$ represents the collocation polynomial. In Problem 15.1 it stood for the more complicated osculating polynomial. Both lead to the same integration formula. (For a specific example of this, see Problem 15.25.)

GAUSS–LEGENDRE FORMULAS

15.8. The special case $w(x) = 1$ leads to Gauss–Legendre formulas. It is the custom to use the interval of integration $(-1, 1)$. As a preliminary exercise, determine the arguments x_k directly from the conditions of Problem 15.3

$$\int_{-1}^1 \pi(x)x^k\,dx = 0 \qquad k = 0, 1, \ldots, n - 1$$

for the value $n = 3$.

The polynomial $\pi(x)$ is then cubic, say $\pi(x) = a + bx + cx^2 + x^3$. Integrations produce

$$2a + \frac{2}{3}c = 0 \qquad \frac{2}{3}b + \frac{2}{5} = 0 \qquad \frac{2}{3}a + \frac{2}{5}c = 0$$

which lead quickly to $a = c = 0$, $b = -\frac{3}{5}$. This makes

$$\pi(x) = x^3 - \frac{3}{5}x = \left(x + \sqrt{\frac{3}{5}}\right)x\left(x - \sqrt{\frac{3}{5}}\right)$$

The collocation arguments are therefore $x_k = -\sqrt{\frac{3}{5}}, 0, \sqrt{\frac{3}{5}}$.

Theoretically this procedure would yield the x_k for any value of n but it is quicker to use a more sophisticated approach.

15.9. The Legendre polynomial of degree n is defined by

$$P_n(x) = \frac{1}{2^n n!} \frac{d^n}{dx^n}(x^2-1)^n$$

with $P_0(x) = 1$. Prove that for $k = 0, 1, \ldots, n-1$

$$\int_{-1}^{1} x^k P_n(x)\,dx = 0$$

making $P_n(x)$ also orthogonal to any polynomial of degree less than n.

Apply integration by parts k times.

$$\int_{-1}^{1} x^k \frac{d^n}{dx^n}(x^2-1)^n\,dx = \underbrace{\left[x^k \frac{d^{n-1}}{dx^{n-1}}(x^2-1)^n\right]_{-1}^{1}}_{=0} - \int_{-1}^{1} kx^{k-1}\frac{d^{n-1}}{dx^{n-1}}(x^2-1)^n\,dx$$

$$= \cdots = (-1)^k k! \int_{-1}^{1} \frac{d^{n-k}}{dx^{n-k}}(x^2-1)^n\,dx = 0$$

15.10. Prove $\displaystyle\int_{-1}^{1} x^n P_n(x)\,dx = \frac{2^{n+1}(n!)^2}{(2n+1)!}$.

Taking $k = n$ in the preceding problem,

$$\int_{-1}^{1} x^n \frac{d^n}{dx^n}(x^2-1)^n\,dx = (-1)^n n! \int_{-1}^{1} (x^2-1)^n\,dx$$

$$= 2n! \int_0^1 (1-x^2)^n\,dx = 2n! \int_0^{\pi/2} \cos^{2n+1} t\,dt$$

This last integral responds to the treatment

$$\int_0^{\pi/2} \cos^{2n+1} t\,dt = \underbrace{\left[\frac{\cos^{2n} t \sin t}{2n+1}\right]_0^{\pi/2}}_{=0} + \frac{2n}{2n+1}\int_0^{\pi/2} \cos^{2n-1} t\,dt$$

$$= \cdots = \frac{2n(2n-2)\cdots 2}{(2n+1)(2n-1)\cdots 3}\int_0^{\pi/2} \cos t\,dt$$

so that

$$\int_{-1}^{1} x^n \frac{d^n}{dx^n}(x^2-1)^n\,dx = 2n!\frac{2n(2n-2)\cdots 2}{(2n+1)(2n-1)\cdots 3}$$

Now multiply top and bottom by $2n(2n-2)\cdots 2 = 2^n n!$ and recall the definition of $P_n(x)$ to obtain, as required,

$$\int_{-1}^{1} x^n P_n(x)\,dx = \frac{1}{2^n n!} 2n! \frac{2^n n!\, 2^n n!}{(2n+1)!} = \frac{2^{n+1}(n!)^2}{(2n+1)!}$$

15.11. Prove $\displaystyle\int_{-1}^{1} [P_n(x)]^2\,dx = \frac{2}{2n+1}$.

Splitting off the highest power of x in one $P_n(x)$ factor,

$$\int_{-1}^{1} [P_n(x)]^2\,dx = \int_{-1}^{1} \left[\frac{1}{2^n n!}\frac{(2n)!}{n!}x^n + \cdots\right] P_n(x)\,dx$$

Powers below x^n make no contribution, by Problem 15.9. Using the preceding problem, we have

$$\int_{-1}^{1} [P_n(x)]^2\,dx = \frac{(2n)!}{2^n (n!)^2}\frac{2^{n+1}(n!)^2}{(2n+1)!} = \frac{2}{2n+1}$$

15.12. Prove that for $m \neq n$, $\int_{-1}^{1} P_m(x)P_n(x)\,dx = 0$.

Writing out the lower-degree polynomial, we find each power in it orthogonal to the higher-degree polynomial. In particular with $m = 0$ and $n \neq 0$ we have the special case $\int_{-1}^{1} P_n(x)\,dx = 0$.

15.13. Prove that $P_n(x)$ has n real zeros between -1 and 1.

The polynomial $(x^2-1)^n$ is of degree $2n$ and has multiple zeros at ± 1. Its derivative therefore has one interior zero, by Rolle's theorem. This first derivative is also zero at ± 1, making three zeros in all. The second derivative is then guaranteed two interior zeros by Rolle's theorem. It also vanishes at ± 1, making four zeros in all. Continuing in this way we find that the nth derivative is guaranteed n interior zeros, by Rolle's theorem. Except for a constant factor, this derivative is the Legendre polynomial $P_n(x)$.

15.14. Show that for the weight function $w(x) = 1$, $\pi(x) = [2^n(n!)^2/(2n)!]P_n(x)$.

Let the n zeros of $P_n(x)$ be called $x_1, \ldots, x_n$. Then

$$\left[\frac{2^n n!^2}{(2n)!}\right] P_n(x) = (x - x_1) \cdots (x - x_n)$$

The only other requirement on $\pi(x)$ is that it be orthogonal to x^k for $k = 0, 1, \ldots, n-1$. But this follows from Problem 15.9.

15.15. Calculate the first several Legendre polynomials directly from the definition, noticing that only even or only odd powers can occur in any such polynomial.

$P_0(x)$ is defined to be 1. Then we find

$$P_1(x) = \frac{1}{2}\frac{d}{dx}(x^2-1) = x \qquad P_3(x) = \frac{1}{48}\frac{d^3}{dx^3}(x^2-1)^3 = \frac{1}{2}(5x^3 - 3x)$$

$$P_2(x) = \frac{1}{8}\frac{d^2}{dx^2}(x^2-1)^2 = \frac{1}{2}(3x^2-1) \qquad P_4(x) = \frac{1}{16 \cdot 24}\frac{d^4}{dx^4}(x^2-1)^4 = \frac{1}{8}(35x^4 - 30x^2 + 3)$$

Similarly,

$$P_5(x) = \frac{1}{8}(63x^5 - 70x^3 + 15x) \qquad P_7(x) = \frac{1}{16}(429x^7 - 693x^5 + 315x^3 - 35x)$$

$$P_6(x) = \frac{1}{16}(231x^6 - 315x^4 + 105x^2 - 5) \qquad P_8(x) = \frac{1}{128}(6435x^8 - 12{,}012x^6 + 6930x^4 - 1260x^2 + 35)$$

and so on. Since $(x^2-1)^n$ involves only even powers of x, the result of differentiating n times will contain only even or only odd powers.

15.16. Show that x^n can be expressed as a combination of Legendre polynomials up through $P_n(x)$. The same is then true of any polynomial of degree n.

Solving in turn for successive powers, we find

$$1 = P_0(x) \qquad x = P_1(x) \qquad x^2 = \frac{1}{3}[2P_2(x) + P_0(x)]$$

$$x^3 = \frac{1}{5}[2P_3(x) + 3P_1(x)] \qquad x^4 = \frac{1}{35}[8P_4(x) + 20P_2(x) + 7P_0(x)]$$

and so on. The fact that each $P_k(x)$ begins with a nonzero term in x^k allows this procedure to continue indefinitely.

15.17. Prove the recursion for Legendre polynomials,

$$(n+1)P_{n+1}(x) = (2n+1)xP_n(x) - nP_{n-1}(x)$$

The polynomial $xP_n(x)$ is of degree $n+1$, and so can be expressed as the combination (see Problem 15.16)

$$xP_n(x) = \sum_{i=0}^{n+1} c_i P_i(x)$$

Multiply by $P_k(x)$ and integrate to find

$$\int_{-1}^{1} xP_k(x)P_n(x)\,dx = c_k \int_{-1}^{1} P_k^2(x)\,dx$$

all other terms on the right vanishing since Legendre polynomials of different degrees are orthogonal. But for $k < n-1$ we know $P_n(x)$ is also orthogonal to $xP_k(x)$, since this product then has degree at most $n-1$. (See Problem 15.9.) This makes $c_k = 0$ for $k < n-1$ and

$$xP_n(x) = c_{n+1}P_{n+1}(x) + c_nP_n(x) + c_{n-1}P_{n-1}(x)$$

Noticing that, from the definition, the coefficient of x^n in $P_n(x)$ will be $(2n)!/2^n(n!)^2$, we compare coefficients of x^{n+1} in the above to find

$$\frac{(2n)!}{2^n(n!)^2} = c_{n+1}\frac{(2n+2)!}{2^{n+1}[(n+1)!]^2}$$

from which $c_{n+1} = (n+1)/(2n+1)$ follows. Comparing the coefficients of x^n, and remembering that only alternate powers appear in any Legendre polynomial, brings $c_n = 0$. To determine c_{n-1} we return to our integrals. With $k = n-1$ we imagine $P_k(x)$ written out as a sum of powers. Only the term in x^{n-1} need be considered, since lower terms, even when multiplied by x, will be orthogonal to $P_n(x)$. This leads to

$$\frac{(2n-2)!}{2^{n-1}[(n-1)!]^2}\int_{-1}^{1} x^nP_n(x)\,dx = c_{n-1}\int_{-1}^{1} P_{n-1}^2(x)\,dx$$

and using the results of Problems 15.10 and 15.11 one easily finds $c_{n-1} = n/(2n+1)$. Substituting these coefficients into our expression for $xP_n(x)$ now brings the required recursion. As a bonus we also have the integral

$$\int_{-1}^{1} xP_{n-1}(x)P_n(x)\,dx = \frac{n}{2n+1}\frac{2}{2n-1} = \frac{2n}{4n^2-1}$$

15.18. Illustrate the use of the recursion formula.

Taking $n = 5$, we find

$$P_6(x) = \frac{11}{6}xP_5(x) - \frac{5}{6}P_4(x) = \frac{1}{16}(231x^6 - 315x^4 + 105x^2 - 5)$$

and with $n = 6$,

$$P_7(x) = \frac{13}{7}xP_6(x) - \frac{6}{7}P_5(x) = \frac{1}{16}(429x^7 - 693x^5 + 315x^3 - 35x)$$

confirming the results obtained in Problem 15.15. The recursion process is well suited to automatic computation of these polynomials, while the differentiation process of Problem 15.15 is not.

15.19. Derive Christoffel's identity,

$$(t-x)\sum_{i=0}^{n}(2i+1)P_i(x)P_i(t) = (n+1)[P_{n+1}(t)P_n(x) - P_n(t)P_{n+1}(x)]$$

The recursion formula of Problem 15.17 can be multiplied by $P_i(t)$ to obtain

$$(2i+1)xP_i(x)P_i(t) = (i+1)P_{i+1}(x)P_i(t) + iP_{i-1}(x)P_i(t)$$

Writing this also with arguments x and t reversed (since it is true for any x and t) and then subtracting, we have

$$(2i+1)(t-x)P_i(x)P_i(t) = (i+1)[P_{i+1}(t)P_i(x) - P_i(t)P_{i+1}(x)] - i[P_i(t)P_{i-1}(x) - P_{i-1}(t)P_i(x)]$$

Summing from $i = 1$ to $i = n$, and noticing the "telescoping effect" on the right, we have

$$(t-x)\sum_{i=1}^{n}(2i+1)P_i(x)P_i(t) = (n+1)[P_{n+1}(t)P_n(x) - P_n(t)P_{n+1}(x)] - (t-x)$$

The last term may be transferred to the left side where it may be absorbed into the sum as an $i = 0$ term. This is the Christoffel identity.

15.20. Use the Christoffel identity to evaluate the integration coefficients for the Gauss–Legendre case, proving $A_k = \dfrac{2}{nP'_n(x_k)P_{n-1}(x_k)}$.

Let x_k be a zero of $P_n(x)$. Then the preceding problem, with t replaced by x_k, makes

$$\frac{(n+1)P_{n+1}(x_k)P_n(x)}{x - x_k} = -\sum_{i=0}^{n}(2i+1)P_i(x)P_i(x_k)$$

Now integrate from -1 to 1. By a special case of Problem 15.12 only the $i = 0$ term survives on the right, and we have

$$\int_{-1}^{1}\frac{P_n(x)}{x - x_k}\,dx = \frac{-2}{(n+1)P_{n+1}(x_k)}$$

The recursion formula with $x = x_k$ makes $(n+1)P_{n+1}(x_k) = -nP_{n-1}(x_k)$ which allows us the alternative

$$\int_{-1}^{1}\frac{P_n(x)}{x - x_k}\,dx = \frac{2}{nP_{n-1}(x_k)}$$

By Problems 15.6 and 15.14 we now find

$$A_k = \int_{-1}^{1} L_k(x)\,dx = \int_{-1}^{1}\frac{\pi(x)}{\pi'(x_k)(x - x_k)}\,dx = \int_{-1}^{1}\frac{P_n(x)}{P'_n(x_k)(x - x_k)}\,dx$$

leading at once to the result stated.

15.21. Prove that $(1-x^2)P'_n(x) + nxP_n(x) = nP_{n-1}(x)$, which is useful for simplifying the result of Problem 15.20.

We first notice that the combination $(1-x^2)P'_n + nxP_n$ is at most of degree $n+1$. However, with A representing the leading coefficient of $P_n(x)$, it is easy to see that x^{n+1} comes multiplied by $-nA + nA$ and so is not involved. Since P_n contains no term in x^{n-1}, our combination also has no term in x^n. Its degree is at most $n-1$ and by Problem 15.16 it can be expresed as

$$(1-x^2)P'_n(x) + nxP_n(x) = \sum_{i=0}^{n-1} c_i P_i(x)$$

Proceeding as in our development of the recursion formula, we now multiply by $P_k(x)$ and integrate. On the right only the kth term survives, because of the orthogonality, and we obtain

$$\frac{2}{2k+1}c_k = \int_{-1}^{1}(1-x^2)P'_n(x)P_k(x)\,dx + n\int_{-1}^{1}xP_n(x)P_k(x)\,dx$$

Integrating the first integral by parts, the integrated piece is zero because of the factor $(1-x^2)$. This leaves

$$\frac{2}{2k+1}c_k = -\int_{-1}^{1}P_n(x)\frac{d}{dx}[(1-x^2)P_k(x)]\,dx + n\int_{-1}^{1}xP_n(x)P_k(x)\,dx$$

For $k < n-1$ both integrands have $P_n(x)$ multiplied by a polynomial of degree $n-1$ or less. By Problem 15.9 all such c_k will be zero. For $k = n-1$ the last integral is covered by the Problem 15.17 bonus. In the first integral only the leading term of $P_{n-1}(x)$ contributes (again because of Problem 15.9) making this term

$$\int_{-1}^{1} P_n(x) \frac{d}{dx}\left\{x^2 \frac{(2n-2)!}{2^{n-1}[(n-1)!]^2} x^{n-1}\right\} dx$$

Using Problem 15.10, this now reduces to

$$\frac{(2n-2)!}{2^{n-1}[(n-1)!]^2}(n+1)\frac{2^{n+1}(n!)^2}{(2n+1)!} = \frac{2n(n+1)}{(2n+1)(2n-1)}$$

Substituting these various results, we find

$$c_{n-1} = \frac{2n-1}{2}\left[\frac{2n(n+1)}{(2n+1)(2n-1)} + \frac{2n^2}{(2n+1)(2n-1)}\right] = n$$

which completes the proof.

15.22. Apply Problem 15.21 to obtain $A_k = \dfrac{2(1-x_k^2)}{n^2[P_{n-1}(x_k)]^2}$.

Putting $x = x_k$, a zero of $P_n(x)$, we find $(1-x_k^2)P_n'(x_k) = nP_{n-1}(x_k)$. The derivative factor can now be replaced in our result of Problem 15.20, producing the required result.

15.23. The Gauss–Legendre integration formula can now be expressed as

$$\int_{-1}^{1} y(x)\, dx \simeq \sum_{i=1}^{n} A_i y(x_i)$$

where the arguments x_k are the zeros of $P_n(x)$ and the coefficients A_k are given in Problem 15.22. Tabulate these numbers for $n = 2, 4, 6, \ldots, 16$.

For $n = 2$ we solve $P_2(x) = \frac{1}{2}(3x^2 - 1) = 0$ to obtain $x_k = \pm\sqrt{\frac{1}{3}} = \pm.57735027$. The two coefficients prove to be the same. Problem 15.22 makes $A_k = 2(1-\frac{1}{3})/[4(\frac{1}{3})] = 1$.

For $n = 4$ we solve $P_4(x) = \frac{1}{8}(35x^4 - 30x^2 + 3) = 0$ to find $x_k^2 = (15 \pm 2\sqrt{30})/35$, leading to the four arguments $x_k = \pm[(15 \pm 2\sqrt{30})/35]^{1/2}$.

Computing these and inserting them into the formula of Problem 15.22 produces the x_k, A_k pairs given in Table 15.1. The results for larger integers n are found in the same way, the zeros of the high-degree polynomials being found by the familiar Newton method of successive approximations. (This method appears in a later chapter.)

15.24. Apply the two-point formula to $\int_0^{\pi/2} \sin t\, dt$.

The change of argument $t = \pi(x+1)/4$ converts this to our standard interval as

$$\int_{-1}^{1} \frac{\pi}{4} \sin \frac{\pi(x+1)}{4}\, dx$$

and the Gaussian arguments $x_k = \pm.57735027$ lead to $y(x_1) = .32589$, $y(x_2) = .94541$. The two-point formula now generates $(\pi/4)(.32589 + .94541) = .99848$ which is correct to almost three places. The *two*-point Gaussian formula has produced a better result than the trapezoidal rule with *seven* points (Problem 14.17). The error is two-tenths of 1 percent!

It is amusing to see what a *one*-point formula could have done. For $n = 1$ the Gauss–Legendre result is, as one may easily verify, $\int_{-1}^{1} y(x)\, dx \simeq 2y(0)$. For the sine function this becomes

$$\int_{-1}^{1} \frac{\pi}{4} \sin \frac{\pi(x+1)}{4}\, dx \simeq \frac{\pi}{4}\sqrt{2} \simeq 1.11$$

which is correct to within about 10 percent.

Table 15.1

n	x_k	A_k	n	x_k	A_k
2	±.57735027	1.00000000	14	±.98628381	.03511946
4	±.86113631	.34785485		±.92843488	.08015809
	±.33998104	.65214515		±.82720132	.12151857
6	±.93246951	.17132449		±.68729290	.15720317
	±.66120939	.36076157		±.51524864	.18553840
	±.23861919	.46791393		±.31911237	.20519846
8	±.96028986	.10122854		±.10805495	.21526385
	±.79666648	.22238103	16	±.98940093	.02715246
	±.52553241	.31370665		±.94457502	.06225352
	±.18343464	.36268378		±.86563120	.09515851
10	±.97390653	.06667134		±.75540441	.12462897
	±.86506337	.14945135		±.61787624	.14959599
	±.67940957	.21908636		±.45801678	.16915652
	±.43339539	.26926672		±.28160355	.18260342
	±.14887434	.29552422		±.09501251	.18945061
12	±.98156063	.04717534			
	±.90411725	.10693933			
	±.76990267	.16007833			
	±.58731795	.20316743			
	±.36783150	.23349254			
	±.12533341	.24914705			

15.25. Explain the accuracy of the extremely simple formulas used in Problem 15.24 by exhibiting the poynomials on which the formulas are based.

The $n = 1$ formula can be obtained by integrating the collocation polynomial of *degree zero*, $p(x) = y(x_1) = y(0)$. However, it can also be obtained, and this is the idea of the Gaussian method, from the osculating polynomial of degree $2n - 1 = 1$, which by Hermite's formula is $y(0) + xy'(0)$. Integrating this linear function between -1 and 1 produces the same $2y(0)$, the derivative term contributing zero. The zero-degree collocation polynomial produces the same integral as a first-degree polynomial, because the point of collocation was the Gaussian point (Fig. 15-1).

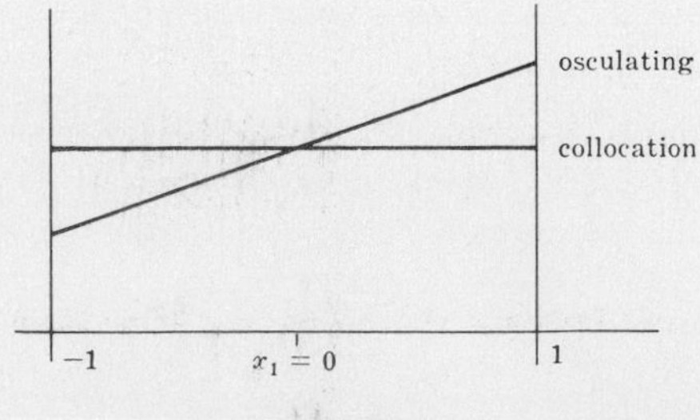

Fig. 15-1

Similarly, the $n = 2$ formula can be obtained by integrating the collocation polynomial of *degree one,* the points of collocation being the Gaussian points

$$\int_{-1}^{1} \left(\frac{x-r}{-2r} y_1 + \frac{x+r}{2r} y_2 \right) dx = y_1 + y_2$$

where $r = \sqrt{\frac{1}{3}}$. This same formula is obtained by integrating the osculating polynomial of degree three, since

$$\int_{-1}^{1} \left[\left(1 + \frac{x+r}{r}\right) \frac{3}{4}(x-r)^2 y_1 + \left(1 - \frac{x-r}{r}\right) \frac{3}{4}(x+r)^2 y_2 + \frac{3}{4}(x^2 - r^2)(x-r)y_1' + \frac{3}{4}(x^2-r^2)(x+r)y_2' \right] dx = y_1 + y_2$$

The polynomial of degree one performs so well because the points of collocation were the Gaussian points. (Fig. 15-2).

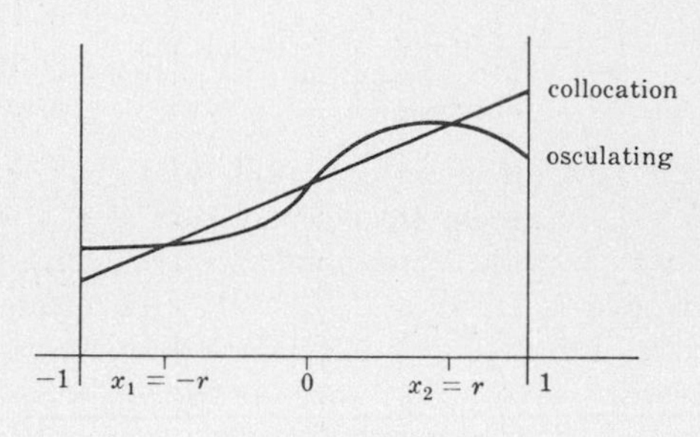

Fig. 15-2

15.26. Apply the Gaussian four-point formula to the integral of Problem 15.24.

Using the same change of argument, the four-point formula produces $\sum_{i=1}^{4} A_i y_i = 1.000000$, correct to six places. Comparing with the Simpson 32-point result of 1.0000003 and the Simpson 64-point result of .99999983, we find it superior to either.

15.27. Adapt the truncation error estimate of Problem 15.2 to the special case of Gauss–Legendre approximation.

Combining Problems 15.2, 15.11, and 15.14, we find the error to be

$$E = \frac{y^{(2n)}(\theta)}{(2n)!} \left[\frac{2^n (n!)^2}{(2n)!} \right]^2 \frac{2}{2n+1} = \frac{2^{2n+1}(n!)^4}{(2n+1)[(2n)!]^3} y^{(2n)}(\theta)$$

This is not an easy formula to apply if the derivatives of $y(x)$ are hard to compute. Some further idea of the accuracy of Gaussian formulas is, however, available by computing the coefficient of $y^{(2n)}$ for small n.

$$n = 2 \qquad E = .0074 y^{(4)}$$

$$n = 4 \qquad E = .0000003 y^{(8)}$$

$$n = 6 \qquad E = 1.5(10^{-12}) y^{(12)}$$

15.28. Apply the error estimates of Problem 15.27 to the integral of Problem 15.24 and compare with the actual errors.

After the change of argument which brings this integral to our standard form, we find

$$|y^{(4)}(x)| < \left(\frac{\pi}{4}\right)^5 \qquad |y^{(8)}(x)| < \left(\frac{\pi}{4}\right)^9$$

For $n=2$ this makes our error estimate $E=(.0074)(.298)=.00220$, while for $n=4$ we find $E=(.0000003)(.113)=.00000003$. The actual errors were .00152 and, to six places, zero. So our estimates are consistent with our results.

This example offers a favorable situation. The sine function is easy to integrate, even by approximate methods, because its derivatives are all bounded by the same constant, namely, 1. The powers of $\pi/4$ do enter with the change of argument, but they actually help in this case. The next example deals with a familiar function whose derivatives do not behave so favorably.

15.29. Apply the Gauss–Legendre formula to $\int_0^{\pi/2} \log(1+t)\,dt$.

The correct value of this integral is

$$\left(1+\frac{\pi}{2}\right)\left[\log\left(1+\frac{\pi}{2}\right)-1\right]+1=.856590$$

to six places. The change of argument $t=\pi(x+1)/4$ converts the integral to

$$\int_{-1}^{1} \frac{\pi}{4}\log\left[1+\frac{\pi(x+1)}{4}\right]dx$$

The fourth derivative of the new integrand is $(\pi/4)^5[-6/(1+t)^4]$. In the interval of integration this cannot exceed $6(\pi/4)^5$, so the truncation error cannot exceed $6(\pi/4)^5(.0074)$ if we use the two-point Gaussian formula. This is six times the corresponding estimate for the integral of the sine function. Similarly, the eighth derivative is $(\pi/4)^9[-7!/(1+t)^8]$. This means a truncation error of at most $(\pi/4)^9\cdot 7!\,(.0000003)$ which is 7! times the corresponding estimate for the integral of the sine function. While the successive derivatives of the sine function remain bounded by 1, those of the logarithm function increase as factorials. The difference has an obvious impact on the truncation errors of any of our formulas, perhaps especially on Gaussian formulas where especially high derivatives are involved. Even so, these formulas perform well. Using just two points we obtain .858, while four points manage .856592 which is off by just two units in the last place. The six-point Gaussian formula scores a bull's-eye to six places, even though its truncation error term involves $y^{(12)}(x)$, which is approximately of size 12!. For contrast, Simpson's rule requires 64 points to produce this same six-place result.

The function $\log(1+t)$ has a singularity at $t=-1$. This is not on the interval of integration, but it is close, and even a complex singularity nearby could produce the slow kind of convergence in evidence here.

15.30. How does the length of the interval of integration affect the Gaussian formulas?

For an integral over the interval $a \leqq t \leqq b$, the change of argument $t=a+\frac{b-a}{2}(x+1)$ produces the standard interval $-1 \leqq x \leqq 1$. It also makes

$$\int_a^b y(t)\,dt=\int_{-1}^{1}\frac{b-a}{2}\,y\left[a+\frac{b-a}{2}(x+1)\right]dx$$

The effect on truncation error is in the derivative factor, which is

$$\left(\frac{b-a}{2}\right)^{2n+1} y^{(2n)}(t)$$

In the examples just given $b-a$ was $\pi/2$ and this interval length actually helped to reduce error, but with a longer interval the potential of the powers of $b-a$ to magnify error is clear.

15.31. Apply the Gaussian method to $(2/\sqrt{\pi})\int_0^4 e^{-t^2}\,dt$

The higher derivatives of this error function are not easy to estimate realistically. Proceeding with computations, one finds the $n=4, 6, 8, 10$ formulas giving these results:

n	4	6	8	10
Approximation	.986	1.000258	1.000004	1.000000

For larger n the results agree with that for $n = 10$. This suggests accuracy to six places. We have already computed this integral by a patient application of Taylor's series (Problem 14.32) and found it to equal 1, correct to six places. For comparison, the Simpson formula requires 32 points to achieve six-place accuracy.

15.32. Apply the Gaussian method to $\int_0^4 \sqrt{1+\sqrt{t}}\,dt$.

The $n = 4, 8, 12, 16$ formulas give the results

n	4	8	12	16
Approximation	6.08045	6.07657	6.07610	6.07600

This suggests accuracy to four places. The exact integral can be found, by a change of argument, to be $\frac{8}{5}(2\sqrt{3}+\frac{1}{3})$, which is 6.07590 correct to five places. Observe that the accuracy obtained here is inferior to that of the previous problem. The explanation is that our square root integrand is not as smooth as the exponential function. Its higher derivatives grow very large, like factorials. Our other formulas also feel the influence of these large derivatives. Simpson's rule for instance produces these values:

No. of points	16	64	256	1024
Simpson values	6.062	6.07411	6.07567	6.07586

Even with a thousand points it has not managed the accuracy achieved in the previous problem with just 32 points.

15.33. Derive the Lanczos estimate for the truncation error of Gaussian formulas.

The relation $\int_{-1}^{1}[xy(x)]'\,dx = y(1)+y(-1)$ holds exactly. Let I be the approximate integral of $y(x)$ obtained by the Gaussian n-point formula, and I^* be the corresponding result for $[xy(x)]'$. Since $[xy(x)]' = y(x) + xy'(x)$,

$$I^* = I + \sum_{i=1}^{n} A_i x_i y'(x_i)$$

so that the error in I^* is

$$E^* = y(1) + y(-1) - I - \sum_{i=1}^{n} A_i x_i y'(x_i)$$

Calling the error in I itself E, we know that

$$E = C_n y^{(2n)}(\theta_1) \qquad E^* = C_n (xy)^{(2n+1)}(\theta_2)$$

for suitable θ_1 and θ_2 between -1 and 1. Suppose $\theta_1 = \theta_2 = 0$. On the one hand $(xy)^{(2n+1)}(0)/(2n)!$ is the coefficient of x^{2n} in the Taylor series expansion of $(xy)'$, while on the other hand

$$y(x) = \cdots + \frac{y^{(2n)}(0)x^{2n}}{(2n)!} + \cdots$$

leading directly to

$$[xy(x)]' = \cdots + \frac{(2n+1)y^{(2n)}(0)x^{2n}}{(2n)!} + \cdots$$

from which we deduce

$$(xy)^{(2n+1)}(0) = (2n+1)y^{(2n)}(0)$$

Thus $E^* = (2n+1)E$ approximately, making

$$E \simeq \frac{1}{2n+1}\left[y(1) + y(-1) - I - \sum_{i=1}^{n} A_i x_i y'(x_i)\right]$$

This involves applying the Gaussian formula to $xy'(x)$ as well as to $y(x)$ itself, but it avoids the often troublesome calculation of $y^{(2n)}(x)$. Putting $\theta_1 = \theta_2 = 0$ is the key move in deducing this formula. This has been found to be more reasonable for smooth integrands such as that of Problem 15.31, than for integrands with large derivatives, which seems reasonable since $y^{(2n)}(\theta_1)/y^{(2n)}(\theta_2)$ should be nearly 1 when $y^{(2n+1)}$ is small.

15.34. Apply the error estimate of the previous problem to the integral of Problem 15.31.

For $n = 8$ the Lanczos estimate is .000004 and is identical with the actual error. For $n = 10$ and above, the Lanczos estimate correctly predicts a six-place error of zero. If applied to the integral of Problem 15.32, however, in which the integrand is very unsmooth, the Lanczos estimate proves to be too conservative to be useful. The limits to the usefulness of this error formula are still to be determined.

OTHER GAUSSIAN FORMULAS

15.35. What are the Gauss–Laguerre formulas?

These formulas for approximate integration are of the form

$$\int_0^\infty e^{-x}y(x)\,dx \simeq \sum_{i=1}^n A_i y(x_i)$$

the arguments x_i being the zeros of the nth Laguerre polynomial

$$L_n(x) = e^x \frac{d^n}{dx^n}(e^{-x}x^n)$$

and the coefficients A_i being

$$A_i = \frac{1}{L_n'(x_i)} \int_0^\infty \frac{L_n(x)e^{-x}}{x - x_i}\,dx = \frac{(n!)^2}{x_i[L_n'(x_i)]^2}$$

The truncation error is

$$E = \frac{(n!)^2}{(2n)!}y^{(2n)}(\theta)$$

These results are found very much as the similar results for the Gauss–Legendre case. Here the weight function is $w(x) = e^{-x}$. The n-point formula is exact for polynomials of degree up to $2n - 1$. Arguments and coefficients are provided in Table 15.2.

15.36. Apply the Gauss–Laguerre one-point formula to the integration of e^{-x}.

Since $L_1(x) = 1 - x$, we have a zero at $x_1 = 1$. The coefficient is $A_1 = 1/[L_1'(1)]^2$ which is also 1. The one-point formula is therefore

$$\int_0^\infty e^{-x}y(x)\,dx \simeq y(1)$$

In this case $y(x) = 1$ and we obtain the exact integral, which is 1. This is no surprise, since with $n = 1$ we are guaranteed exact results for any polynomial of degree one or less. In fact with $y(x) = ax + b$ the formula produces

$$\int_0^\infty e^{-x}(ax + b)\,dx = y(1) = a + b$$

which is the correct value.

15.37. Apply the Gauss–Laguerre method to $\int_0^\infty e^{-x}\sin x\,dx$.

The exact value of this integral is easily found to be $\frac{1}{2}$. The smoothness of $\sin x$, by which is meant the boundedness of its derivatives, suggests that our formulas will perform well. The error estimate of $(n!)^2/(2n)!$, which replaces $y^{(2n)}$ by its maximum of 1, reduces to $\frac{1}{924}$ for $n = 6$ and suggests about

Table 15.2

n	x_k	A_k
2	.58578644	.85355339
	3.41421356	.14644661
4	.32254769	.60315410
	1.74576110	.35741869
	4.53662030	.03888791
	9.39507091	.00053929
6	.22284660	.45896467
	1.18893210	.41700083
	2.99273633	.11337338
	5.77514357	.01039920
	9.83746742	.00026102
	15.98287398	.00000090
8	.17027963	.36918859
	.90370178	.41878678
	2.25108663	.17579499
	4.26670017	.03334349
	7.04590540	.00279454
	10.75851601	.00009077
	15.74067864	.00000085
	22.86313174	.00000000
10	.13779347	.30844112
	.72945455	.40111993
	1.80834290	.21806829
	3.40143370	.06208746
	5.55249614	.00950152
	8.33015275	.00075301
	11.84378584	.00002826
	16.27925783	.00000042
	21.99658581	.00000000
	29.92069701	.00000000
12	.11572212	.26473137
	.61175748	.37775928
	1.51261027	.24408201
	2.83375134	.09044922
	4.59922764	.02010238
	6.84452545	.00266397
	9.62131684	.00020323
	13.00605499	.00000837
	17.11685519	.00000017
	22.15109038	.00000000
	28.48796725	.00000000
	37.09912104	.00000000
14	.09974751	.23181558
	.52685765	.35378469
	1.30062912	.25873461
	2.43080108	.11548289
	3.93210282	.03319209
	5.82553622	.00619287
	8.14024014	.00073989
	10.91649951	.00005491
	14.21080501	.00000241
	18.10489222	.00000006
	22.72338163	.00000000
	28.27298172	.00000000
	35.14944366	.00000000
	44.36608171	.00000000

three-place accuracy. Actually substituting into $\sum_{i=1}^{n} A_i \sin x_i$ brings the results

n	2	6	10	14
Σ	.43	.50005	.5000002	.50000000

so that our error formula is somewhat pessimistic.

15.38. Apply the Gauss–Laguerre method to $\int_1^\infty (e^{-t}/t)\, dt$.

The unsmoothness of $y(t) = 1/t$, meaning that its nth derivative

$$y^{(n)}(t) = (-1)^n n!\, t^{-(n+1)}$$

increases rapidly with n, does not suggest overconfidence in approximation formulas. Making the change of argument $t = x + 1$, this integral is converted into our standard interval as

$$\int_0^\infty e^{-x} \frac{1}{e(x+1)}\, dx$$

and the error formula becomes
$$E = \left[\frac{(n!)^2}{(2n)!}\right]\left[\frac{(2n)!}{e(\theta+1)^{2n+1}}\right]$$

which reduces to $(n!)^2/e(\theta+1)^{2n+1}$. If we replaced θ by 0 to obtain the maximum derivative this would surely be discouraging and yet no other choice nominates itself. Actual computations with the formula

$$\frac{1}{e}\sum_{i=1}^{n} \frac{A_i}{x_i+1}$$

bring these results:

n	2	6	10	14
Approximation	.21	.21918	.21937	.21938

Since the correct value to five places is .21938 we see that complete pessimism was unnecessary. The elusive argument θ appears to increase with n. A comparison of the actual and theoretical errors allows θ to be determined:

n	2	6	10
θ	1.75	3.91	5.95

In this example the function $y(x)$ has a singularity at $x = -1$. Even a complex singularity near the interval of integration can produce the slow convergence in evidence here. (Compare with Problem 15.29.) The convergence is more rapid if we move away from the singularity. For example, integration of the same function by the same method over the interval from 5 to ∞ brings these results:

n	2	6	10
Approximation	.001147	.0011482949	.0011482954

The last value is correct to almost ten places.

15.39. What are the Gauss–Hermite formulas?

These are of the form

$$\int_{-\infty}^{\infty} e^{-x^2} y(x)\, dx \simeq \sum_{i=1}^{n} A_i y(x_i)$$

the arguments x_i being the zeros of the nth Hermite polynomial

$$H_n(x) = (-1)^n e^{x^2} \frac{d^n}{dx^n}(e^{-x^2})$$

and the coefficients A_i being
$$A_i = \frac{2^{n+1} n!\sqrt{\pi}}{[H_n'(x_i)]^2}$$

The truncation error is
$$E = \frac{n!\sqrt{\pi}\, y^{(2n)}(\theta)}{2^n(2n)!}$$

These results are found very much as in the Gauss–Legendre case. Here the weight function is $w(x)=e^{-x^2}$. The n-point formula is exact for polynomials up to degree $2n-1$. Arguments and coefficients are provided in Table 15.3.

Table 15.3

n	x_k	A_k	n	x_k	A_k
2	± .70710678	.88622693	12	± .31424038	.57013524
4	± .52464762	.80491409		± .94778839	.26049231
	±1.65068012	.08131284		±1.59768264	.05160799
				±2.27950708	.00390539
6	± .43607741	.72462960		±3.02063703	.00008574
	±1.33584907	.15706732		±3.88972490	.00000027
	±2.35060497	.00453001			
			14	± .29174551	.53640591
8	± .38118699	.66114701		± .87871379	.27310561
	±1.15719371	.20780233		±1.47668273	.06850553
	±1.98165676	.01707798		±2.09518326	.00785005
	±2.93063742	.00019960		±2.74847072	.00035509
				±3.46265693	.00000472
10	± .34290133	.61086263		±4.30444857	.00000001
	±1.03661083	.24013861			
	±1.75668365	.03387439			
	±2.53273167	.00134365			
	±3.43615912	.00000764			

15.40. Apply the Gauss–Hermite two-point formula to the integral $\int_{-\infty}^{\infty} e^{-x^2}x^2\,dx$.

An exact result can be obtained, so we first compute

$$H_2(x)=e^{x^2}\frac{d^2}{dx^2}(e^{-x^2})=4x^2-2$$

The zeros of this polynomial are $x_k=\pm\sqrt{2}/2$. The coefficients A_i are easily found from the formula in Problem 15.39 to be $\sqrt{\pi}/2$. The two-point formula is therefore

$$\int_{-\infty}^{\infty} e^{-x^2}y(x)\,dx \simeq \frac{\sqrt{\pi}}{2}\left[y\left(\frac{\sqrt{2}}{2}\right)+y\left(-\frac{\sqrt{2}}{2}\right)\right]$$

With $y(x)=x^2$ this becomes $\int_{-\infty}^{\infty} e^{-x^2}x^2\,dx=\sqrt{\pi}/2$ which is the exact value of the integral.

15.41. Evaluate correct to six places $\int_{-\infty}^{\infty} e^{-x^2}\sin^2 x\,dx$.

The Gauss–Hermite formula produces these results:

n	2	4	6	8	10
Approximation	.748	.5655	.560255	.560202	.560202

This appears to suggest six-place accuracy and the result is actually correct to six places, the exact integral being $\sqrt{\pi}(1-e^{-1})/2$ which is to eight places .56020226.

15.42. Evaluate correct to three places $\int_{-\infty}^{\infty}(e^{-x^2}/\sqrt{1+x^2})\,dx$.

The square root factor is not so smooth as the sine function of the preceding problem so we should not expect quite so rapid convergence, and do not get it.

n	2	4	6	8	10	12
Approximation	.145	.151	.15202	.15228	.15236	.15239

The value .152 seems to be indicated.

15.43. What are the Gauss–Chebyshev formulas?

These are of Gaussian form with $w(x) = 1/\sqrt{1-x^2}$,

$$\int_{-1}^{1} \left[\frac{y(x)}{\sqrt{1-x^2}}\right] dx \simeq \frac{\pi}{n} \sum_{i=1}^{n} y(x_i)$$

the arguments x_i being the zeros of the nth Chebyshev polynomial

$$T_n(x) = \cos(n \arccos x)$$

Contrary to appearances this actually is a polynomial of degree n, and its zeros are

$$x_i = \cos\left[\frac{(2i-1)\pi}{2n}\right]$$

All coefficients A_i are simply π/n. The truncation error is

$$E = \frac{2\pi y^{(2n)}(\theta)}{2^{2n}(2n)!}$$

15.44. Apply the Gauss–Chebyshev formula for $n = 1$ to verify the familiar result

$$\int_{-1}^{1} \left(\frac{1}{\sqrt{1-x^2}}\right) dx = \pi$$

For $n = 1$ we find $T_n(x) = \cos(\arccos x) = x$. Since there is just one zero, our formula collapses to $\pi y(0)$. Since the Gaussian formula with $n = 1$ is exact for polynomials of degree one or less, the given integral is exactly $\pi \cdot y(0) = \pi$.

15.45. Apply the $n = 3$ formula to $\int_{-1}^{1} (x^4/\sqrt{1-x^2})\, dx$.

Directly from the definition we find $T_3(x) = 4x^3 - 3x$ so that $x_1 = 0$, $x_2 = \sqrt{3}/2$, $x_3 = -\sqrt{3}/2$. The Gauss–Chebyshev formula now yields $(\pi/3)(0 + \frac{9}{16} + \frac{9}{16}) = 3\pi/8$ which is also exact.

Supplementary Problems

15.46. Prove that $P_n'(x) = xP_{n-1}'(x) + nP_{n-1}(x)$, beginning as follows. From the definition of Legendre polynomials,

$$P_n'(x) = \frac{1}{2^n n!} \frac{d^n}{dx^n} [n(x^2-1)^{n-1}(2x)]$$

Apply the theorem on the nth derivative of a product to find

$$P_n'(x) = \frac{n}{2^n n!} \frac{d}{dx}\left[2x \frac{d^{n-1}}{dx^{n-1}}(x^2-1)^{n-1} + 2(n-1)\frac{d^{n-2}}{dx^{n-2}}(x^2-1)^{n-1}\right]$$

$$= \frac{d}{dx}[xP_{n-1}(x)] + (n-1)P_{n-1}(x)$$

15.47. Prove that $(1-x^2)P_n^{(2)}(x) - 2xP_n'(x) + n(n+1)P_n(x) = 0$, as follows. Let $z = (x^2-1)^n$. Then $z' = 2nx(x^2-1)^{n-1}$, making $(x^2-1)z' - 2nxz = 0$. Repeatedly differentiate this equation, obtaining

$$(x^2-1)z^{(2)} - (2n-2)xz' - 2nz = 0$$
$$(x^2-1)z^{(3)} - (2n-4)xz^{(2)} - [2n + (2n-2)]z' = 0$$
$$(x^2-1)z^{(4)} - (2n-6)xz^{(3)} - [2n + (2n-2) + (2n-4)]z^{(2)} = 0$$

and ultimately

$$(x^2-1)z^{(n+2)} - (2n-2n-2)xz^{(n+1)} - [2n + (2n-2) + (2n-4) + \cdots + (2n-2n)]z^{(n)} = 0$$

which simplifies to
$$(x^2-1)z^{(n+2)} + 2xz^{(n+1)} - n(n+1)z^{(n)} = 0$$

Since $P_n(x) = z^{(n)}/2^n n!$, the required result soon follows.

15.48. Differentiate the result of Problem 15.21 and compare with Problem 15.47 to prove

$$xP_n'(x) - P_{n-1}'(x) = nP_n(x)$$

15.49. Use Problem 15.21 to prove that for all n, $P_n(1) = 1$, $P_n(-1) = (-1)^n$.

15.50. Use Problem 15.46 to prove $P_n'(1) = \frac{1}{2}n(n+1)$, $P_n'(-1) = (-1)^{n+1}P_n'(1)$.

15.51. Use Problem 15.46 to show that

$$P_n^{(k)}(x) = xP_{n-1}^{(k)}(x) + (n+k-1)P_{n-1}^{(k-1)}(x)$$

Then apply the method of summing differences to verify

$$P_n^{(2)}(1) = \frac{(n+2)^{(4)}}{(2\cdot 4)} \qquad P_n^{(3)}(1) = \frac{(n+3)^{(6)}}{(2\cdot 4\cdot 6)}$$

and in general
$$P_n^{(k)}(1) = \frac{(n+k)^{(2k)}}{2^k k!} = \frac{(n+k)!}{(n-k)!\,2^k k!}$$

Since Legendre polynomials are either even or odd functions, also verify that

$$P_n^{(k)}(-1) = (-1)^{n+k}P_n^{(k)}(1)$$

15.52. Use Problems 15.46 and 15.48 to prove $P_{n+1}'(x) - P_{n-1}'(x) = (2n+1)P_n(x)$.

15.53. The leading coefficient in $P_n(x)$ is, as we know, $A_n = (2n)!/2^n(n!)^2$. Show that it can also be written as $A_n = 1\cdot\frac{3}{2}\cdot\frac{5}{3}\cdot\frac{7}{4}\cdots\frac{2n-1}{n} = \frac{1\cdot 3\cdot 5\cdots(2n-1)}{n!}$.

15.54. Compute the Gauss–Legendre arguments and coefficients for the case $n=3$, showing the arguments to be $x_k = 0, \pm\sqrt{\frac{3}{5}}$ and the coefficients to be $\frac{8}{9}$ for $x_k = 0$ and $\frac{5}{9}$ for the other arguments.

15.55. Verify these Gauss–Legendre arguments and coefficients for the case $n=5$:

x_k	A_k
0	.56888889
±.53846931	.47862867
±.90617985	.23692689

15.56. Apply the three-point Gaussian formula of Problem 15.54 to the integral of the sine function, $\int_0^{\pi/2} \sin t\, dt$. How does the result compare with that obtained by Simpson's rule using seven points (Problem 14.17)?

15.57. Apply the Gauss–Legendre two-point formula ($n = 2$) to $\int_{-1}^{1} \frac{1}{1+t^2}\,dt$ and compare with the exact value $\pi/2 \simeq 1.5708$.

15.58. Diagram the linear collocation and cubic osculating polynomials which lead to the $n = 2$ formula, using the function $y(t) = 1/(1+t^2)$ of Problem 15.57. (See Problem 15.25.)

15.59. How closely do our formulas verify $\int_0^1 x^x\,dx \simeq .7834$ to four places? Also apply some of our formulas for equally spaced arguments to this integral. Which algorithms work best? Which are easiest to apply "by hand"? Which are easiest to program for automatic computation?

15.60. As in Problem 15.59 apply various methods to $\int_0^{\pi/2} e^{\sin x}\,dx \simeq 3.1044$ and decide which algorithm is best for automatic computation.

15.61. Compute Laguerre polynomials through $n = 5$ from the definition given in Problem 15.35.

15.62. Find the zeros of $L_2(x)$ and verify the arguments and coefficients given in Table 15.2 for $n = 2$.

15.63. Use the method of Problem 15.9 to prove that $L_n(x)$ is orthogonal to any polynomial of degree less than n, in the sense that

$$\int_0^\infty e^{-x} L_n(x) p(x)\,dx = 0$$

where $p(x)$ is any such polynomial.

15.64. Prove that $\int_0^\infty e^{-x} L_n^2(x)\,dx = (n!)^2$ by the method of Problems 15.10 and 15.11.

15.65. Apply the Gauss–Laguerre two-point formula to obtain these exact results:

$$\int_0^\infty e^{-x} x^2\,dx = 2! \qquad \int_0^\infty e^{-x} x^3\,dx = 3!$$

15.66. Find the exact arguments and coefficients for three-point Gauss–Laguerre integration.

15.67. Use the formula of the previous problem to verify

$$\int_0^\infty e^{-x} x^4\,dx = 4! \qquad \int_0^\infty e^{-x} x^5\,dx = 5!$$

15.68. Apply the $n = 6$ and $n = 8$ formulas to the "smooth" integral $\int_0^\infty e^{-x} \cos x\,dx$.

15.69. Apply the $n = 6$ and $n = 8$ formulas to the "unsmooth" integral $\int_0^\infty e^{-x} \log(1+x)\,dx$.

15.70. Show that correct to four places $\int_0^\infty e^{-(x+1/x)}\,dx \simeq .2797$.

15.71. Compute Hermite polynomials through $n = 5$ from the definition given in Problem 15.39.

15.72. Show that the Gauss–Hermite one-point formula is $\int_{-\infty}^{\infty} e^{-x^2} y(x)\,dx \simeq \sqrt{\pi}\, y(0)$. This is exact for polynomials of degree one or less. Apply it to $y(x) = 1$.

15.73. Derive the exact formula for $n = 3$ Gauss–Hermite approximation. Apply it to the case $y(x) = x^4$ to obtain an exact result.

15.74. How closely do the four-point and eight-point formulas duplicate this result?

$$\int_{-\infty}^{\infty} e^{-x^2} \cos x\,dx = \sqrt{\pi}\, e^{-1/4} \simeq 1.3804$$

15.75. How closely do the four- and eight-point formulas duplicate this result?

$$\int_0^\infty e^{-x^2-1/x^2}\,dx = \frac{\sqrt{\pi}}{2e^2} \simeq .11994$$

15.76. Show that correct to three places $\int_{-\infty}^{\infty} [e^{-x^2}/(1+x^2)]\,dx \simeq 1.343$.

15.77. Evaluate correct to three places $\int_{-\infty}^{\infty} e^{-x^2}\sqrt{1+x^2}\,dx$.

15.78. Evaluate correct to three places $\int_{-\infty}^{\infty} e^{-x^2} \log(1+x^2)\,dx$.

15.79. Apply the Gauss–Chebyshev $n = 2$ formula to the exact verification of

$$\int_{-1}^{1} \frac{x^2}{\sqrt{1-x^2}}\,dx = \frac{\pi}{2}$$

15.80. Find the following integral correct to three places: $\int_{-1}^{1} [(\cos x)/\sqrt{1-x^2}]\,dx$.

15.81. Find the following integral correct to two places: $\int_{-1}^{1} (\sqrt{1+x^2}/\sqrt{1-x^2})\,dx$.

Chapter 16

Singular Integrals

It is unwise to apply the formulas of the preceding two chapters blindly. They are all based on the assumption that the function $y(x)$ can be conveniently approximated by a polynomial $p(x)$. If this is not true then the formulas may produce poor, if not completely deceptive, results. It would be comforting to be sure that the following application of Simpson's rule will never be made:

$$\int_1^2 \frac{dx}{x^2-2} \simeq \frac{1}{6}\left[-1+4(4)+\frac{1}{2}\right] = \frac{31}{12}$$

but less obvious singular points have probably been temporarily missed. Not quite so serious are the efforts to apply polynomial-based formulas to functions having singularities in their derivatives. Since polynomials breed endless generations of smooth derivatives, they are not ideally suited to such functions, and poor results are usually obtained.

PROCEDURES FOR SINGULAR INTEGRALS

A variety of procedures exist for dealing with singular integrals, whether for singular integrands or for an infinite range of integration. The following will be illustrated:

1. **Ignoring the singularity** may even be successful. Under certain circumstances it is enough to use more and more arguments x_i until a satisfactory result is obtained.
2. **Series expansions** of all or part of the integrand, followed by term-by-term integration, is a popular procedure provided convergence is adequately fast.
3. **Subtracting the singularity** amounts to splitting the integral into a singular piece which responds to the classic methods of analysis and a nonsingular piece to which our approximate integration formulas may be applied without anxiety.
4. **Change of argument** is one of the most powerful weapons of analysis. Here it may exchange a difficult singularity for a more cooperative one, or it may remove the singularity completely.
5. **Differentiation relative to a parameter** involves embedding the given integral in a family of integrals and then exposing some basic property of the family by differentiation.
6. **Gaussian methods** also deal with certain types of singularity, as reference to the previous chapter will show.
7. **Asymptotic series** are also relevant, but this procedure is treated in the following chapter.

Solved Problems

16.1. Compare the results of applying Simpson's rule to the integration of $\sqrt{x}$ near 0 and away from 0.

Take first the interval between 1 and 1.30 with $h = .05$, since we made this computation earlier (Problem 14.11). Simpson's rule gave a correct result to five places. Even the trapezoidal rule gave an error of only .00002. Applying Simpson's rule now to the interval between 0 and .30, which has the same length but includes a singular point of the derivative of $\sqrt{x}$, we obtain $\int_0^{0.3} \sqrt{x}\,dx \simeq .10864$. Since the correct figure is .10954, our result is not quite correct to three places. The error is more than a hundred times greater.

16.2. What is the effect of ignoring the singularity in the derivative of $\sqrt{x}$ and applying Simpson's rule with successively smaller intervals h?

Polya has proved (*Math. Z.*, 1933) that for functions of this type (continuous with singularities in derivatives) Simpson's rule and others of similar type should converge to the correct integral. Computations show these results:

$1/h$	8	32	128	512
$\int_0^1 \sqrt{x}\,dx$	.663	.6654	.66651	.666646

The convergence to $\frac{2}{3}$ is slow but does appear to be occurring.

16.3. Determine the effect of ignoring the singularity and applying Simpson's rule to the following integral: $\int_0^1 (1/\sqrt{x})\,dx = 2$.

Here the integrand itself has a discontinuity, and an infinite one, but Davis and Rabinowitz have proved (*SIAM Journal,* 1965) that convergence should occur. They also found Simpson's rule producing these results, which show that ignoring the singularity is sometimes successful:

$1/h$	64	128	256	512	1024	2048
Approx. integral	1.84	1.89	1.92	1.94	1.96	1.97

The convergence is again slow but does appear to be occurring. At current computing speeds slow convergence may not be enough to rule out a computing algorithm. There is, however, the usual question of how much roundoff error will affect a lengthy computation. For this same integral the trapezoidal rule with $h = \frac{1}{4096}$ managed 1.98, while application of the Gauss 48-point formula to quarters of the interval (192 points in all) produced 1.99.

16.4. Determine the result of ignoring the singularity and applying the Simpson and Gauss rules to the following integral: $\displaystyle\int_0^1 \frac{1}{x} \sin \frac{1}{x}\,dx \simeq .6347$.

Here the integrand has an infinite discontinuity and is also highly oscillatory. The combination can be expected to produce difficulty in numerical computation. Davis and Rabinowitz (see preceding problem) found Simpson's rule failing.

$1/h$	64	128	256	512	1024	2048
Approx. integral	2.31	1.69	$-.60$	1.21	.72	.32

and the Gauss 48-point formula doing no better. So the singularity cannot always be ignored.

16.5. Evaluate to three places the singular integral $\int_0^1 (e^x/\sqrt{x})\,dx$.

Direct use of the Taylor series leads to

$$\int_0^1 \left(\frac{e^x}{\sqrt{x}}\right) dx = \int_0^1 \left(\frac{1}{\sqrt{x}} + x^{1/2} + \frac{1}{2}x^{3/2} + \frac{1}{6}x^{5/2} + \cdots\right) dx$$

$$= 2 + \frac{2}{3} + \frac{1}{5} + \frac{1}{21} + \frac{1}{108} + \frac{1}{660} + \frac{1}{4680} + \frac{1}{37{,}800} + \cdots = 2.925$$

After the first few terms the series converges rapidly and higher accuracy is easily achieved if needed. Note that the singularity $1/\sqrt{x}$ has been handled as the first term of the series. (See also the next problem.)

16.6. Apply the method of "subtracting the singularity" to the integral of Problem 16.5.

Calling the integral I, we have

$$I = \int_0^1 \frac{1}{\sqrt{x}}\,dx + \int_0^1 \frac{e^x - 1}{\sqrt{x}}\,dx$$

The first integral is elementary and the second has no singularity. However, since $(e^x - 1)/\sqrt{x}$ behaves like $\sqrt{x}$ near zero, it does have a singularity in its first derivative. This is enough, as we saw in Problem 16.1, to make approximation integration inaccurate.

The subtraction idea can be extended to push the singularity into a higher derivative. For example, our integral can also be written as

$$I = \int_0^1 \frac{1+x}{\sqrt{x}}\,dx + \int_0^1 \frac{e^x - 1 - x}{\sqrt{x}}\,dx$$

Further terms of the series for the exponential function may be subtracted if needed. The first integral here is $\frac{8}{3}$, and the second could be handled by our formulas, though the series method still seems preferable in this case.

16.7. Evaluate the integral of Problem 16.5 by a change of argument.

The change of argument, or substitution, may be the most powerful device in integration. Here we let $t = \sqrt{x}$ and find $I = 2\int_0^1 e^{t^2}\,dt$ which has no singularity of any kind, even in its derivatives. This integral may be evaluated by any of our formulas or by a series development.

16.8. Evaluate correct to six decimal places $\int_0^1 (\cos x)(\log x)\,dx$.

Here a procedure like that of Problem 16.5 is adopted. Using the series for $\cos x$, the integral becomes

$$\int_0^1 \left(1 - \frac{x^2}{2!} + \frac{x^4}{4!} - \frac{x^6}{6!} + \cdots\right) \log x\,dx$$

Using the elementary integral

$$\int_0^1 x^i \log x\,dx = \frac{x^{i+1}}{i+1}\left(\log x - \frac{1}{i+1}\right)\Bigg|_0^1 = -\frac{1}{(i+1)^2}$$

the integral is replaced by the series

$$-1 + \frac{1}{3^2\,2!} - \frac{1}{5^2\,4!} + \frac{1}{7^2\,6!} - \frac{1}{9^2\,8!} + \cdots$$

which reduces to $-.946083$.

16.9. Evaluate $\displaystyle\int_1^\infty \frac{1}{t^2} \sin\frac{1}{t^2}\,dt$ by a change of variable which converts the infinite interval of integration into a finite interval.

Let $x = 1/t$. Then the integral becomes $\int_0^1 \sin(x^2)\,dx$ which can be computed by various approximate methods. Choosing a Taylor series expansion leads to

$$\int_0^1 \sin(x^2)\,dx = \frac{1}{3} - \frac{1}{42} + \frac{1}{1320} - \frac{1}{75{,}600} + \cdots$$

which is .310268 to six places, only four terms contributing.

16.10. Show that the change of variable used in Problem 16.9 converts $\int_1^\infty \frac{\sin t}{t}\,dt$ into a badly singular integral, so that reducing the interval of integration to finite length may not always be a useful step.

With $x = 1/t$ we obtain the integral $\int_0^1 \frac{1}{x}\sin\frac{1}{x}\,dx$ encountered in Problem 16.4, which oscillates badly near zero, making numerical integration nearly impossible. The integral of this problem may best be handled by asymptotic methods to be discussed in the next chapter.

16.11. Compute $\int_1^\infty \frac{1}{x^5}\sin \pi x\,dx$ by direct evaluation between the zeros of $\sin x$, thus developing part of an alternating series.

Applying the Gauss 8-point formula to each of the successive intervals $(1, 2)$, $(2, 3)$, and so on, these results are found:

Interval	Integral	Interval	Integral
(1, 2)	−.117242	(2, 3)	.007321
(3, 4)	−.001285	(4, 5)	.000357
(5, 6)	−.000130	(6, 7)	.000056
(7, 8)	−.000027	(8, 9)	.000014
(9, 10)	−.000008		

The total is −.11094, which is correct to five places.

This method of direct evaluation for an interval of finite length resembles in spirit the method of ignoring a singularity. The upper limit is actually replaced by a finite substitute, in this case ten, beyond which the contribution to the integral may be considered zero to the accuracy required.

16.12. Compute $\int_0^\infty e^{-x^2-1/x^2}\,dx$ by differentiation relative to a parameter.

This problem illustrates still another approach to the problem of integration. We begin by imbedding the problem in a family of similar problems. For t positive, let

$$F(t) = \int_0^\infty e^{-x^2-t^2/x^2}\,dx$$

Since the rapid convergence of this singular integral permits differentiation under the integral sign, we next find

$$F'(t) = -2t\int_0^\infty \frac{1}{x^2}e^{-x^2-t^2/x^2}\,dx$$

Now introduce the change of argument $y = t/x$, which allows the attractive simplification

$$F'(t) = -2\int_0^\infty e^{-y^2-t^2/y^2}\,dy = -2F(t)$$

Thus $F(t) = Ce^{-2t}$ and the constant C may be evaluated from the known result

$$F(0) = \int_0^\infty e^{-x^2}\,dx = \frac{\sqrt{\pi}}{2}$$

The result is

$$\int_0^\infty = e^{-x^2-t^2/x^2}\,dx = \frac{1}{2}\sqrt{\pi}\,e^{-2t}$$

For the special case $t = 1$, this produces .119938 correct to six digits.

Supplementary Problems

16.13. Compare the results of applying Simpson's rule with $h = \frac{1}{2}$ to $\int_0^1 x\,dx$ and $\int_0^1 x \log x\,dx$.

16.14. Use successively smaller h intervals for the second integral of Problem 16.13 and notice the convergence toward the exact value of $-\frac{1}{4}$.

16.15. Evaluate to three places by series development: $\int_0^1 (\sin x)/x^{3/2}\,dx$.

16.16. Apply the method of subtracting the singularity to the integral of Problem 16.15, obtaining an elementary integral and an integral which involves no singularity until the second derivative.

16.17. Ignore the singularity in the integral of Problem 16.15 and apply the Simpson and Gauss formulas, continually using more points. Do the results converge toward the value computed in Problem 16.15? (Define the integrand at zero as you wish.)

16.18. Evaluate $\int_0^1 e^{-x} \log x\,dx$ correct to three places by using the series for the exponential function.

16.19. Compute the integral of the preceding problem by ignoring the singularity and applying the Simpson and Gauss formulas. Do the results converge toward the value computed in Problem 16.18? (Define the integrand at zero as you wish.)

16.20. Use series to show that

$$-\int_0^1 \frac{\log x}{1-x}\,dx = -\frac{\pi^2}{6} \qquad \int_0^1 \frac{\log x}{1+x}\,dx = -\frac{\pi^2}{12} \qquad \int_0^1 \frac{\log x}{1-x^2}\,dx = -\frac{\pi^2}{8}$$

16.21. Verify that to four places $\int_0^\infty [e^{-x^2}/(1+x^2)]\,dx = .6716$.

16.22. Verify that to four places $\int_0^\infty e^{-x} \log x\,dx = -.5772$.

16.23. Verify that to four places $\int_0^\infty e^{-x-1/x}\,dx = .2797$.

16.24. Verify that to four places $\int_0^\infty e^{-x}\sqrt{x}\,dx = .8862$.

16.25. Verify that to four places $\int_0^1 (1/\sqrt{-\log x})\,dx = 1.772$.

16.26. Verify that to four places $\int_0^{\pi/2} (\sin x)(\log \sin x)\,dx = -.3069$.

Chapter 17

Sums and Series

REPRESENTATION OF NUMBERS AND FUNCTIONS AS SUMS

The representation of numbers and functions as finite or infinite sums has proved to be very useful in applied mathematics. Numerical analysis exploits such representations in many ways including the following:

1. The **telescoping method** makes it possible to replace long sums by short ones, with obvious advantage to the computer. The classic example is

$$\frac{1}{1\cdot 2}+\frac{1}{2\cdot 3}+\frac{1}{3\cdot 4}+\cdots+\frac{1}{n(n+1)}=\left(1-\frac{1}{2}\right)+\left(\frac{1}{2}-\frac{1}{3}\right)+\cdots+\left(\frac{1}{n}-\frac{1}{n+1}\right)=1-\frac{1}{n+1}$$

   in which the central idea of the method can be seen. Each term is replaced by a difference.

2. **Rapidly convergent infinite series** play one of the leading roles in numerical analysis. Typical examples are the series for the sine and cosine functions. Each such series amounts to a superb algorithm for generating approximations to the functions represented.

3. **Acceleration methods** have been developed for more slowly converging series. If too many terms must be used for the accuracy desired, then roundoffs and other troubles associated with long computations may prevent the attainment of this accuracy. Acceleration methods alter the course of the computation, or in other words, they change the algorithm, in order to make the overall job shorter.

   The **Euler transformation** is a frequently used acceleration method. This transformation was derived in an earlier chapter. It replaces a given series by another which often is more rapidly convergent.

   The **comparison method** is another acceleration device. Essentially the same as the method of subtracting singularities, it splits a series into a similar, but known, series and another which converges more rapidly than the original.

   Special methods may be devised to accelerate the series representations of certain functions. The logarithm and arctan functions will be used as illustrations.

4. The **Bernoulli polynomials** are given by

$$B_k(x)=\sum_{i=0}^{k}\binom{k}{i}B_{k-i}x^i$$

   with coefficients B_i determined by

$$B_0=1 \qquad \sum_{i=0}^{k-1}\binom{k}{i}B_i=0$$

   for $k=2$, 3, etc. Properties of Bernoulli polynomials include the following:

$$B_i'(x)=iB_{i-1}(x)$$

$$B_i(x+1)-B_i(x)=ix^{i-1}$$

$$\int_0^1 B_i(x)\,dx=0 \qquad \text{for } i>0$$

$$B_i(1)=B_i(0) \qquad \text{for } i>1$$

The **Bernoulli numbers** b_i are defined by

$$b_i = (-1)^{i+1} B_{2i}$$

for $i = 1, 2$, etc.

Sums of integer powers are related to the Bernoulli polynomials and numbers. Two such relationships are

$$\sum_{x=1}^{n} x^p = \frac{B_{p+1}(n+1) - B_{p+1}(0)}{p+1} \qquad \text{and} \qquad \sum_{k=1}^{\infty} \frac{1}{k^{2i}} = \frac{b_i (2\pi)^{2i}}{2(2i)!}$$

5. The **Euler–Maclaurin formula** may be derived carefully and an error estimate obtained through the use of Bernoulli polynomials. It may be used as an acceleration method. Euler's constant

$$C = \lim \left(1 + \frac{1}{2} + \frac{1}{3} + \cdots + \frac{1}{n} - \log n\right)$$

can be evaluated using the Euler–Maclaurin formula. Six terms are enough to produce almost ten-decimal-place accuracy.

6. **Wallis' product for π** is

$$\frac{\pi}{2} = \lim \frac{2 \cdot 2 \cdot 4 \cdot 4 \cdot 6 \cdot 6 \cdots 2k \cdot 2k}{1 \cdot 3 \cdot 3 \cdot 5 \cdot 5 \cdot 7 \cdots (2k-1)(2k+1)}$$

and is used to obtain **Stirling's series for large factorials,** which takes the form

$$\log \frac{n!\, e^n}{\sqrt{2\pi}\, n^{n+1/2}} \simeq \frac{b_1}{2n} - \frac{b_2}{3 \cdot 4n^3} + \frac{b_3}{5 \cdot 6n^5} - \cdots + \frac{(-1)^{k+1} b_k}{(2k)(2k-1)n^{2k-1}}$$

the b_i still being Bernoulli numbers. The simpler factorial approximation

$$n! \simeq \sqrt{2\pi}\, n^{n+1/2} e^{-n}$$

is the result of using just one term of the Stirling series.

7. **Asymptotic series** may be viewed as still another form of acceleration method. Though usually divergent, their partial sums have a property which makes them useful. The classic situation involves sums of the form

$$S_n(x) = \sum_{i=0}^{n} \frac{a_i}{x^i}$$

which diverge for all x as n tends to infinity, but such that

$$\lim x^n [f(x) - S_n(x)] = 0$$

for x tending to infinity. The error in using $S_n(x)$ as an approximation to $f(x)$ for large arguments x can then be estimated very easily, simply by looking at the first omitted term of the series. Stirling's series is a famous example of such an asymptotic series. This same general idea can also be extended to other types of sum.

Integration by parts converts many common integrals into asymptotic series. For large x this may be the best way for evaluating these integrals.

Solved Problems

THE TELESCOPING METHOD

17.1. Evaluate $\sum_{i=2}^{n} \log \frac{i-1}{i}$.

This is another telescoping sum. We easily find

$$\sum_{i=2}^{n} \log \frac{i-1}{i} = \sum_{i=2}^{n} [\log(i-1) - \log i] = -\log n$$

The telescoping method is of course the summation of differences as discussed in Chapter 5. The sum $\sum y_i$ can be easily evaluated if y_i can be expressed as a difference, for then $\sum_{i=a}^{b} y_i = \sum_{i=a}^{b} \Delta Y_i = Y_{b+1} - Y_a$.

17.2. Evaluate the power sum $\sum_{i=1}^{n} i^4$.

Since powers can be expressed in terms of factorial polynomials, which in turn can be expressed as differences (see Chapter 4), any such power sum can be telescoped. In the present example

$$\sum_{i=1}^{n} i^4 = \sum_{i=1}^{n} [i^{(1)} + 7i^{(2)} + 6i^{(3)} + i^{(4)}] = \sum_{i=1}^{n} \Delta\left[\frac{1}{2} i^{(2)} + \frac{7}{3} i^{(3)} + \frac{6}{4} i^{(4)} + \frac{1}{5} i^{(5)}\right]$$

$$= \frac{1}{2}(n+1)^{(2)} + \frac{7}{3}(n+1)^{(3)} + \frac{6}{4}(n+1)^{(4)} + \frac{1}{5}(n+1)^{(5)} = \frac{1}{30} n(n+1)(2n+1)(3n^2+3n-1)$$

Other power sums are treated in similar fashion.

17.3. Evaluate $\sum_{i=1}^{n} (i^2 + 3i + 2)$.

Since power sums may be evaluated by summing differences, sums of polynomial values are easy bonuses. For example,

$$\sum_{i=1}^{n} i^2 + 3\sum_{i=1}^{n} i + \sum_{i=1}^{n} 2 = \frac{n(n+1)(2n+1)}{6} + \frac{3n(n+1)}{2} + 2n$$

17.4. Evaluate $\sum_{i=1}^{n} \frac{1}{i(i+1)(i+2)}$.

This can also be written as a sum of differences. Recalling the factorial polynomials with negative exponent, of Chapter 4, we find $\frac{1}{2i(i+1)} - \frac{1}{2(i+1)(i+2)} = \frac{1}{i(i+1)(i+2)}$ and it follows that the given sum telescopes to $\frac{1}{4} - \frac{1}{2(n+1)(n+2)}$.

In this example the infinite series is convergent and $\sum_{i=1}^{\infty} \frac{1}{i(i+1)(i+2)} = \frac{1}{4}$.

17.5. Evaluate $\sum_{i=1}^{n} \frac{3}{i(i+3)}$.

Simple rational functions such as this (and in Problem 17.4) are easily summed. Here

$$\sum_{i=1}^{n} \frac{3}{i(i+3)} = \sum_{i=1}^{n} \left(\frac{1}{i} - \frac{1}{i+3}\right) = 1 + \frac{1}{2} + \frac{1}{3} - \frac{1}{n+1} - \frac{1}{n+2} - \frac{1}{n+3}$$

The infinite series converges to $\sum_{i=1}^{\infty} \frac{3}{i(i+3)} = \frac{11}{6}$.

RAPIDLY CONVERGENT SERIES

17.6. How many terms of the Taylor series for $\sin x$ in powers of x are needed to provide eight-place accuracy for all arguments between 0 and $\pi/2$?

Since the series $\sin x = \sum_{i=0}^{\infty} (-1)^i x^{2i+1}/(2i+1)!$ is alternating with steadily decreasing terms, the truncation error made by using only n terms will not exceed the $(n+1)$th term. This important property of such series makes truncation error estimation relatively easy. Here we find $(\pi/2)^{15}/15! \simeq 8 \cdot 10^{-10}$ so that seven terms of the sine series are adequate for eight-place accuracy over the entire interval.

This is an example of a rapidly convergent series. Since other arguments may be handled by the periodicity feature of this function, all arguments are covered. Notice, however, that a serious loss of significant digits can occur in argument reduction. For instance, with $x \simeq 31.4$ we find

$$\sin x \simeq \sin 31.4 = \sin(31.4 - 10\pi) \simeq \sin(31.4 - 31.416) = \sin(-.016) \simeq -.016$$

In the same way $\sin 31.3 \simeq -.116$ while $\sin 31.5 \simeq .084$. This means that although the input data 31.4 is known to three significant figures the output is not certain even to one significant figure. Essentially it is the number of digits to the right of the decimal point in the argument x which determines the accuracy obtainable in $\sin x$.

17.7. How many terms of the Taylor series for e^x in powers of x are needed to provide eight-place accuracy for all arguments between 0 and 1?

The series is the familiar $e^x = \sum_{i=0}^{\infty} x^i/i!$. Since this is not an alternating series, the truncation error may not be less than the first omitted term. Here we resort to a simple comparison test. Suppose we truncate the series after the x^n term. Then the error is

$$\sum_{i=n+1}^{\infty} \frac{x^i}{i!} = \frac{x^{n+1}}{(n+1)!}\left[1 + \frac{x}{n+2} + \frac{x^2}{(n+2)(n+3)} + \cdots\right]$$

and since $x < 1$ this error will not exceed

$$\frac{x^{n+1}}{(n+1)!}\left[1 + \frac{1}{n+2} + \frac{1}{(n+2)^2} + \cdots\right] = \frac{x^{n+1}}{(n+1)!}\frac{1}{1 - 1/(n+2)} = \frac{x^{n+1}}{(n+1)!}\frac{n+2}{n+1}$$

so that it barely exceeds the first omitted term. For $n = 11$ this error bound becomes about $2 \cdot 10^{-9}$ so that a polynomial of degree eleven is indicated. For example, at $x = 1$ the successive terms are as follows:

1.00000000	.50000000	.04166667	.00138889	.00002480	.00000028
1.00000000	.16666667	.00833333	.00019841	.00000276	.00000003

and their total is 2.71828184. This is wrong by one unit in the last place because of roundoff errors.

The error could also have been estimated using Lagrange's form (Problem 11.4), which gives

$$E = \frac{1}{(n+1)!} e^{\xi} x^{n+1} \qquad \text{with } 0 < \xi < x$$

17.8. Compute e^{-10} to six significant digits.

This problem illustrates an important difference. For six places we could proceed as in Problem 17.7, with $x = -10$. The series would however converge very slowly and there is trouble of another sort. In obtaining this small number as a difference of larger numbers we lose digits. Working to eight places we would obtain $e^{-10} \simeq .00004540$ which has only four significant digits. Such loss is frequent with alternating series. Occasionally double-precision arithmetic (working to twice as many places) overcomes the trouble. Here, however, we simply compute e^{10} and then take the reciprocal. The result is $e^{-10} \simeq .0000453999$ which is correct to the last digit.

17.9. In Problem 14.34 the integral $(2/\sqrt{\pi})\int_0^x e^{-t^2}\,dt$ was calculated by the Taylor series method for $x=1$. Suppose the series is used for larger x, but to avoid roundoff error growth no more than twenty terms are to be summed. How large can x be made, consistent with four-place accuracy?

The nth term of the integrated series is $2x^{2n-1}/\sqrt{\pi}(2n-1)(n-1)!$ apart from the sign. Since this series alternates, with steadily decreasing terms, the truncation error will not exceed the first omitted term.

Using 20 terms we require that $(2/\sqrt{\pi})x^{41}/41\cdot 20! < 5\cdot 10^{-5}$. This leads to $x<2.5$ approximately. For such arguments the series converges rapidly enough to meet our stipulations. For larger arguments it does not.

ACCELERATION METHODS

17.10. Not all series converge as rapidly as those of the previous problems. From the binomial series

$$\frac{1}{1+x^2} = 1 - x^2 + x^4 - x^6 + \cdots$$

one finds by integrating between 0 and x that

$$\arctan x = x - \frac{1}{3}x^3 + \frac{1}{5}x^5 - \frac{1}{7}x^7 + \cdots$$

At $x=1$ this gives the Leibnitz series

$$\frac{\pi}{4} = 1 - \frac{1}{3} + \frac{1}{5} - \frac{1}{7} + \cdots$$

How many terms of this series would be needed to yield four-place accuracy?

Since the series is alternating with steadily decreasing terms, the truncation error cannot exceed the first term omitted. If this term is to be .00005 or less, we must use terms out to about 1/20,000. This comes to 10,000 terms. In summing so large a number of terms we can expect roundoff errors to accumulate to 100 times the maximum individual roundoff. But the accumulation *could* grow to 10,000 times that maximum if we were unbelievably unlucky. At any rate this series does not lead to a pleasant algorithm for computing $\pi/4$.

17.11. Apply the Euler transformation of Chapter 11 to the series of the preceding problem to obtain four-place accuracy.

The best procedure is to sum the early terms and apply the transformation to the rest. For example, to five places,

$$1 - \frac{1}{3} + \frac{1}{5} - \cdots - \frac{1}{19} = .76046$$

The next few reciprocals and their differences are as follows:

.04762				
	−414			
.04348		66		
	−348		−14	
.04000		52		3
	−296		−11	
.03704		41		
	−255			
.03448				

The Euler transformation is

$$y_0 - y_1 + y_2 - y_3 + \cdots = \sum_{i=0}^{n} \frac{(-1)^i \Delta^i y_0}{2^{i+1}} = \frac{1}{2} y_0 - \frac{1}{4} \Delta y_0 + \frac{1}{8} \Delta^2 y_0 - \cdots$$

and applied to our table produces

$$.02381 + .00104 + .00008 + .00001 = .02494$$

Finally we have

$$\frac{\pi}{4} = 1 - \frac{1}{3} + \frac{1}{5} - \frac{1}{7} + \cdots = .76046 + .02494 = .78540$$

which is correct to five places. In all, 15 terms of the original series have seen action rather than 10,000. The Euler transformation often produces superb acceleration like this, but it can also fail.

17.12. Compute $\pi/4$ from the formula

$$\frac{\pi}{4} = 2 \arctan \frac{1}{5} + \arctan \frac{1}{7} + 2 \arctan \frac{1}{8}$$

working to eight digits.

This illustrates how special properties of the function involved may be used to bring accelerated convergence. The series

$$\arctan x = x - \frac{1}{3} x^3 + \frac{1}{5} x^5 - \frac{1}{7} x^7 + \cdots$$

converges quickly for the arguments now involved. We find using no more than five terms of the series:

$$2 \arctan \frac{1}{5} = .39479112 \qquad \arctan \frac{1}{7} = .14189705 \qquad 2 \arctan \frac{1}{8} = .24870998$$

with a total of .78539815. The last digit should be a 6.

17.13. How many terms of $\sum_{i=1}^{\infty} \frac{1}{i^2 + 1}$ would be needed to evaluate the series correct to three places?

Terms beginning with $i = 45$ are all smaller than .0005, so that none of these individually affects the third decimal place. Since all terms are positive, however, it is clear that collectively the terms from $i = 45$ onward will affect the third place, perhaps even the second. Stegun and Abramowitz (*Journal of SIAM*, 1956) showed that 5745 terms are actually required for three-place accuracy. This is a good example of a slowly convergent series of positive terms.

17.14. Evaluate the series of Problem 17.13 by the "comparison method," correct to three places. (This method is analogous to the evaluation of singular integrals by subtracting out the singularity.)

The comparison method involves introducing a known series of the same rate of convergence. For example,

$$\sum_{i=1}^{\infty} \frac{1}{i^2+1} = \sum_{i=1}^{\infty} \frac{1}{i^2} - \sum_{i=1}^{\infty} \frac{1}{i^2(i^2+1)}$$

We will prove later that the first series on the right is $\pi^2/6$. The second converges more rapidly than the others, and we find

$$\sum_{i=1}^{\infty} \frac{1}{i^2(i^2+1)} = \frac{1}{2} + \frac{1}{20} + \frac{1}{90} + \frac{1}{272} + \frac{1}{650} + \frac{1}{1332} + \frac{1}{2450} + \cdots \simeq .56798$$

with just ten terms being used. Subtracting from $\pi^2/6 \simeq 1.64493$ makes a final result of 1.07695, which can be rounded to 1.077.

17.15. Verify that the result obtained in Problem 17.14 is correct to at least three places.

The truncation error of our series computation is

$$E = \sum_{i=11}^{\infty} \frac{1}{i^2(i^2+1)} < \sum_{i=11}^{\infty} \frac{1}{i^4} = \sum_{i=1}^{\infty} \frac{1}{i^4} - \sum_{i=1}^{10} \frac{1}{i^4}$$

The first series on the right will later be proved to be $\pi^4/90$, and the second comes to at least 1.08200. This makes $E < 1.08234 - 1.08200 = .00034$. Roundoff errors cannot exceed $11 \cdot 5 \cdot 10^{-6}$ since 11 numbers of five-place accuracy have been summed. The combined error therefore does not exceed .0004, making our result correct to three places.

17.16. Apply the comparison method to $\sum_{i=1}^{\infty} \frac{1}{i^2(i^2+1)}$.

This series was summed directly in the preceding problem. To illustrate how the comparison method may be reapplied, however, notice that

$$\sum_{i=1}^{\infty} \frac{1}{i^2(i^2+1)} = \sum_{i=1}^{\infty} \frac{1}{i^4} - \sum_{i=1}^{\infty} \frac{1}{i^4(i^2+1)}$$

Direct evaluation of the last series brings $\frac{1}{2} + \frac{1}{80} + \frac{1}{810} + \frac{1}{4352} + \frac{1}{16{,}250} + \cdots$ which comes to .51403. Subtracting from $\pi^4/90$ we find

$$\sum_{i=1}^{\infty} \frac{1}{i^2(i^2+1)} \simeq 1.08234 - .51403 = .56831$$

which agrees nicely with the results of the previous two problems, in which this same sum was computed to be .56798 with an estimated error of .00034. The error estimate was almost perfect.

17.17. Evaluate $\sum_{i=1}^{\infty} \frac{1}{i^3}$ to four places.

The series converges a little too slowly for comfort. Applying the comparison method,

$$\sum_{i=1}^{\infty} \frac{1}{i^3} = 1 + \sum_{i=2}^{\infty} \frac{1}{(i-1)i(i+1)} - \sum_{i=2}^{\infty} \frac{1}{i^2(i^3-i)}$$

The first series on the right is telescoping and was found in Problem 17.4 to be exactly $\frac{1}{4}$. The last may be summed directly,

$$\frac{1}{24} + \frac{1}{216} + \frac{1}{960} + \frac{1}{3000} + \frac{1}{7560} + \frac{1}{16{,}464} + \cdots$$

and comes to .04787. Subtracting from 1.25, we have finally $\sum_{i=1}^{\infty} 1/i^3 = 1.20213$ which is correct to four places. See Problem 17.39 for a more accurate result.

THE BERNOULLI POLYNOMIALS

17.18. The Bernoulli polynomials $B_i(x)$ are defined by

$$e^{xt}\frac{t}{e^t-1}=\sum_{i=0}^{\infty}\frac{t^i}{i!}B_i(x)$$

Let $B_i(0)=B_i$ and develop a recursion for these B_i numbers.

Replacing x by 0, we have

$$t=(e^t-1)\sum_{i=0}^{\infty}\frac{t^iB_i}{i!}=\left(\sum_{j=1}^{\infty}\frac{t^j}{j!}\right)\left(\sum_{i=0}^{\infty}\frac{t^iB_i}{i!}\right)=\sum_{k=1}^{\infty}c_kt^k$$

with $c_k=\sum_{i=0}^{k-1}\frac{B_i}{i!\,(k-i)!}$. This makes $k!\,c_k=\sum_{i=0}^{k-1}\binom{k}{i}B_i$. Comparing the coefficients of t in the series equation above, we find that

$$B_0=1 \qquad \sum_{i=0}^{k-1}\binom{k}{i}B_i=0 \qquad \text{for } k=2,3,\ldots$$

Written out, this set of equations shows how the B_i may be determined one by one without difficulty:

$$B_0=1$$
$$B_0+2B_1=0$$
$$B_0+3B_1+3B_2=0$$
$$B_0+4B_1+6B_2+4B_3=0$$

etc. The first several B_i are therefore

$$B_0=1 \qquad B_1=-\frac{1}{2} \qquad B_2=\frac{1}{6} \qquad B_3=0 \qquad B_4=-\frac{1}{30} \qquad B_5=0 \qquad B_6=\frac{1}{42}$$

and so on. The set of equations used can also be described in the form

$$(B+1)^k-B^k=0 \qquad \text{for } k=2,3,\ldots$$

where it is understood that after applying the binomial theorem each "power" B^i is replaced by B_i.

17.19. Find an explicit formula for the Bernoulli polynomials.

From the defining equation and the special case $x=0$ treated above,

$$\left(\sum_{i=0}^{\infty}\frac{x^it^i}{i!}\right)\left(\sum_{j=0}^{\infty}\frac{B_jt^j}{j!}\right)=\sum_{k=0}^{\infty}\frac{t^k}{k!}B_k(x)$$

Comparing the coefficients of t^k on both sides makes $\frac{1}{k!}B_k(x)=\sum_{i=0}^{k}B_{k-i}\frac{1}{i!\,(k-i)!}x^i$ or

$$B_k(x)=\sum_{i=0}^{k}\binom{k}{i}B_{k-i}x^i$$

The first several Bernoulli polynomials are

$$B_0(x)=1 \qquad B_3(x)=x^3-\frac{3}{2}x^2+\frac{1}{2}x$$

$$B_1(x)=x-\frac{1}{2} \qquad B_4(x)=x^4-2x^3+x^2-\frac{1}{30}$$

$$B_2(x)=x^2-x+\frac{1}{6} \qquad B_5(x)=x^5-\frac{5}{2}x^4+\frac{5}{3}x^3-\frac{1}{6}x$$

etc. The formula can be summarized as $B_k(x) = (x + B)^k$ where once again it is to be understood that the binomial theorem is applied and then each "power" B^i is replaced by B_i.

17.20. Prove that $B_i'(x) = iB_{i-1}(x)$.

The defining equation can be written as

$$\frac{te^{xt}}{e^t - 1} = 1 + \sum_{i=1}^{\infty} \frac{t^i B_i(x)}{i!}$$

Differentiating relative to x and dividing through by t,

$$\frac{te^{xt}}{e^t - 1} = \sum_{i=1}^{\infty} \left[\frac{B_i'(x)}{i}\right] t^{i-1}(i-1)!$$

But the defining equations can also be written as

$$\frac{te^{xt}}{e^t - 1} = \sum_{i=1}^{\infty} [B_{i-1}(x)]\left[\frac{t^{i-1}}{(i-1)!}\right]$$

and comparing coefficients on the right, $B_i'(x) = iB_{i-1}(x)$ for $i = 1, 2, \ldots$. Notice also that the same result can be obtained instantly by formal differentiation of $B_i(x) = (x + B)^i$.

17.21. Prove $B_i(x + 1) - B_i(x) = ix^{i-1}$.

Proceeding formally (even though a rigorous proof would not be too difficult from $(B + 1)^k = B^k$, we find $\sum_{k=2}^{i} \binom{i}{k}(B + 1)^k x^{i-k} = \sum_{k=2}^{i} \binom{i}{k} B^k x^{i-k}$ or

$$(B + 1 + x)^i - i(B + 1)x^{i-1} = (B + x)^i - iBx^{i-1}$$

From the abbreviated formula for Bernoulli polynomials (Problem 17.19), this converts immediately to $B_i(x + 1) - B_i(x) = ix^{i-1}$.

17.22. Prove $B_i(1) = B_i(0)$ for $i > 1$.

This follows at once from the preceding problem with x replaced by zero.

17.23. Prove that $\int_0^1 B_i(x)\,dx = 0$ for $i = 1, 2, \ldots$.

By the previous problems

$$\int_0^1 B_i(x)\,dx = \frac{B_{i+1}(1) - B_{i+1}(0)}{i+1} = 0$$

17.24. The conditions of Problems 17.20 and 17.23 also determine the Bernoulli polynomials, given $B_0(x) = 1$. Determine $B_1(x)$ and $B_2(x)$ in this way.

From $B_1'(x) = B_0(x)$ it follows that $B_1(x) = x + C_1$ where C_1 is a constant. For the integral of $B_1(x)$ to be zero, C_1 must be $-\frac{1}{2}$. Then from $B_2'(x) = 2B_1(x) = 2x - 1$ it follows that $B_2(x) = x^2 - x + C_2$. For the integral of $B_2(x)$ to be zero, the constant C_2 must be $\frac{1}{6}$. In this way each $B_i(x)$ may be determined in its turn.

17.25. Prove $B_{2i-1} = 0$ for $i = 2, 3, \ldots$

Notice that

$$f(t) = \frac{t}{e^t - 1} + \frac{t}{2} = \frac{t}{2} \cdot \frac{e^t + 1}{e^t - 1} = B_0 + \sum_{i=2}^{\infty} \frac{B_i t^i}{i!}$$

is an even function, that is, $f(t)=f(-t)$. All odd powers of t must have zero coefficients, making B_i zero for odd i except $i=1$.

17.26. Define the Bernoulli numbers b_i.

These are defined as $b_i=(-1)^{i+1}B_{2i}$ for $i=1, 2, \ldots$. Thus

$$b_1=\frac{1}{6} \qquad b_4=\frac{1}{30} \qquad b_7=\frac{7}{6}$$

$$b_2=\frac{1}{30} \qquad b_5=\frac{5}{66} \qquad b_8=\frac{3617}{510}$$

$$b_3=\frac{1}{42} \qquad b_6=\frac{691}{2730} \qquad b_9=\frac{43,867}{798}$$

as is easily verified after computing the corresponding numbers B_i by the recursion formula of Problem 17.18.

17.27. Evaluate the sum of pth powers in terms of Bernoulli polynomials.

Since, by Problem 17.21, $\Delta B_i(x)=B_i(x+1)-B_i(x)=ix^{i-1}$, the Bernoulli polynomials provide "finite integrals" of the power functions. This makes it possible to telescope the power sum.

$$\sum_{x=0}^{n} x^p=\sum_{x=0}^{n}\frac{1}{p+1}\Delta B_{p+1}(x)=\frac{B_{p+1}(n+1)-B_{p+1}(0)}{p+1}$$

17.28. Evaluate the sums of the form $\sum_{k=1}^{\infty} 1/k^{2i}$ in terms of Bernoulli numbers.

It will be proved later (see chapter on trigonometric approximation) that the function

$$F_n(x)=B_n(x) \qquad 0 \leqq x<1$$

$$F_n(x \pm m)=F_n(x) \qquad \text{for } m \text{ an integer}$$

known as a Bernoulli function, having period 1, can be represented as

$$F_n(x)=(-1)^{n/2+1}\cdot n!\cdot\frac{2}{(2\pi)^n}\cdot\sum_{k=1}^{\infty}\frac{\cos 2\pi kx}{k^n}$$

for even n, and as

$$F_n(x)=(-1)^{(n+1)/2}\cdot n!\cdot\frac{2}{(2\pi)^n}\cdot\sum_{k=1}^{\infty}\frac{\sin 2\pi kx}{k^n}$$

when n is odd. For even n, say $n=2i$, we put $x=0$ and have

$$\sum_{k=1}^{\infty}\frac{1}{k^{2i}}=(-1)^{i+1}\frac{F_{2i}(0)(2\pi)^{2i}}{2(2i)!}$$

But $F_{2i}(0)=B_{2i}(0)=B_{2i}=(-1)^{i+1}b_i$ and so $\sum_{k=1}^{\infty} 1/k^{2i}=b_i(2\pi)^{2i}/2(2i)!$.

In particular, $\sum_{k=1}^{\infty} 1/k^2=\pi^2/6$, $\sum_{k=1}^{\infty} 1/k^4=\pi^4/90$, etc.

17.29. Show that all the Bernoulli numbers are positive and that they become arbitrarily large as i increases.

Noting that $1 < \sum_{k=1}^{\infty} 1/k^{2i} \leqq \sum_{k=1}^{\infty} 1/k^2 = \pi^2/6 < 2$, we see that

$$\frac{2(2i)!}{(2\pi)^{2i}} < b_i < \frac{4(2i)!}{(2\pi)^{2i}}$$

In particular all the b_i are positive and they grow limitlessly with increasing i.

17.30. Show that as i increases, $\lim \frac{(2\pi)^{2i}}{2(2i)!} b_i = 1$.

This also follows quickly from the series of Problem 17.28. All terms except the $k = 1$ term approach zero for increasing i, and because $1/x^p$ is a decreasing function of x,

$$\frac{1}{k^p} < \int_{k-1}^{k} \frac{1}{x^p}\,dx \qquad \text{so that, if } p > 1, \qquad \sum_{k=2}^{\infty} \frac{1}{k^p} < \int_1^{\infty} \frac{1}{x^p}\,dx = \frac{1}{p-1}$$

As p increases (in our case $p = 2i$) this entire series has limit zero, which establishes the required result. Since all terms of this series are positive, it also follows that $b_i > 2(2i)!/(2\pi)^{2i}$.

THE EULER–MACLAURIN FORMULA

17.31. Use the Bernoulli polynomials to derive the Euler–Maclaurin formula with an error estimate. (This formula was obtained in Chapter 11 by an operator computation, but without an error estimate.)

We begin with an integration by parts, using the facts that $B_1'(t) = B_0(t) = 1$ and $B_1(1) = -B_1(0) = \frac{1}{2}$.

$$\int_0^1 y(t)\,dt = \int_0^1 y(t)B_1'(t)\,dt = \frac{1}{2}(y_0 + y_1) - \int_0^1 y'(t)B_1(t)\,dt$$

Again integrate by parts using $B_2'(t) = 2B_1(t)$ from Problem 17.20 and $B_2(1) = B_2(0) = b_1$ to find

$$\int_0^1 y(t)\,dt = \frac{1}{2}(y_0 + y_1) - \frac{1}{2}b_1(y_1' - y_0') + \frac{1}{2}\int_0^1 y^{(2)}(t)B_2(t)\,dt$$

The next integration by parts brings

$$\frac{1}{2}\int_0^1 y^{(2)}(t)B_2(t)\,dt = \frac{1}{6}y^{(2)}(t)B_3(t)\Big|_0^1 - \frac{1}{6}\int_0^1 y^{(3)}(t)B_3(t)\,dt$$

But since $B_3(1) = B_3(0) = 0$, the integrated term vanishes and we proceed to

$$\frac{1}{2}\int_0^1 y^{(2)}(t)B_2(t)\,dt = -\frac{1}{24}y^{(3)}(t)B_4(t)\Big|_0^1 + \frac{1}{24}\int_0^1 y^{(4)}(t)B_4(t)\,dt$$

$$= \frac{1}{24}b_2(y_1^{(3)} - y_0^{(3)}) + \frac{1}{24}\int_0^1 y^{(4)}(t)B_4(t)\,dt$$

since $B_4(1) = B_4(0) = B_4 = -b_2$. Continuing in this way, we develop the result

$$\int_0^1 y(t)\,dt = \frac{1}{2}(y_0 + y_1) + \sum_{i=1}^{k} \frac{(-1)^i b_i}{(2i)!}(y_1^{(2i-1)} - y_0^{(2i-1)}) + R_k$$

where $$R_k = \frac{1}{(2k)!}\int_0^1 y^{(2k)}(t)B_{2k}(t)\,dt$$

Integrating R_k by parts the integrated part again vanishes, leaving

$$R_k = \frac{-1}{(2k+1)!}\int_0^1 y^{(2k+1)}(t)B_{2k+1}(t)\,dt$$

Corresponding results hold for the intervals between other consecutive integers. Summing, we find substantial telescoping and obtain

$$\sum_{i=0}^{n} y_i = \int_0^n y(t)\,dt + \frac{1}{2}(y_0 + y_n) - \sum_{i=1}^{k} \frac{(-1)^i b_i}{(2i)!}\left(y_n^{(2i-1)} - y_0^{(2i-1)}\right)$$

with an error of

$$E_k = \frac{-1}{(2k+1)!}\int_0^n y^{(2k+1)}(t) F_{2k+1}(t)\,dt$$

where $F_{2k}(t)$ is the Bernoulli function of Problem 17.28, the periodic extension of the Bernoulli polynomial $B_{2k}(t)$. The same argument may be used between integer arguments a and b rather than 0 and n. We may also allow b to become infinite, provided that the series and the integral we encounter are convergent. In this case we assume that $y(t)$ and its derivatives all become zero at infinity, so that the formula becomes

$$\sum_{i=a}^{\infty} y_i = \int_a^{\infty} y(t)\,dt + \frac{1}{2}y_a + \sum_{i=1}^{k} \frac{(-1)^i b_i}{(2i)!} y_a^{(2i-1)}$$

17.32. Evaluate the power sum $\sum_{i=0}^{n} i^4$ by use of the Euler–Maclaurin formula.

In this case the function $y(t) = t^4$, so that with $k = 2$ the series of the preceding problem terminates. Moreover, the error E_k becomes zero since $y^{(5)}(t)$ is zero. The result is

$$\sum_{i=0}^{n} i^4 = \frac{1}{5}n^5 + \frac{1}{2}n^4 + \frac{1}{12}(4n^3) - \frac{1}{720}(24n) = \frac{1}{30}n(n+1)(2n+1)(3n^2+3n-1)$$

as in Problem 17.2. This is an example in which increasing k in the Euler–Maclaurin formula leads to a finite sum. (The method of Problem 17.27 could also have been applied to this sum.)

17.33. Compute Euler's constant $C = \lim\left(1 + \frac{1}{2} + \frac{1}{3} + \cdots + \frac{1}{n} - \log n\right)$ assuming convergence. (See also Problem 17.77.)

Using Problem 17.1, this can be rewritten as $C = 1 + \sum_{i=2}^{\infty}\left(\frac{1}{i} + \log\frac{i-1}{i}\right)$.

The Euler–Maclaurin formula may now be applied with $y(t) = 1/t - \log t + \log(t-1)$. Actually it is more convenient to sum the first few terms directly and then apply the Euler–Maclaurin formula to the rest of the series. To eight places,

$$1 + \sum_{i=2}^{9}\left(\frac{1}{i} + \log\frac{i-1}{i}\right) = .63174368$$

Using 10 and ∞ as limits, we first compute

$$\int_{10}^{\infty}\left[\frac{1}{t} - \log t + \log(t-1)\right]dt = (1-t)\log\frac{t}{t-1}\Big|_{10}^{\infty}$$

$$= -1 + 9\log 10 - 9\log 9 \simeq -.05175536$$

the first term coming from the upper limit by evaluation of the "indeterminate form." Next

$$\frac{1}{2}y_{10} = -.00268026 \qquad -\frac{1}{12}y'_{10} = -.00009259 \qquad \frac{1}{720}y_{10}^{(3)} = .00000020$$

all values at infinity being zero. Summing the five terms just computed, we have $C \simeq .57721567$. Carrying ten places and computing only one more term would lead to the better approximation $C \simeq .5772156650$ which is itself one unit too large in the tenth place.

In this example the accuracy obtainable by the Euler–Maclaurin formula is limited. After a point, using more terms (increasing k) leads to poorer approximations to Euler's constant rather than better. In other words, we have used a few terms of a divergent series to obtain our results. To see this we need only note that the ith term of the series is $\frac{(-1)^{i+1}b_i}{(2i)(2i-1)}\left(\frac{2i+9}{10^{2i}}-\frac{1}{9^{2i-1}}\right)$ and that by Problem 17.29 the b_i exceed $2(2i)!/(2\pi)^{2i}$ which guarantees the unlimited growth of this term. Divergence is more typical than convergence for the Euler–Maclaurin series.

17.34. A truck can travel a distance of one "leg" on the maximum load of fuel it is capable of carrying. Show that if an unlimited supply of fuel is available at the edge of a desert, then the truck can cross the desert no matter what its width. Estimate how much fuel would be needed to cross a desert 10 "legs" wide.

On just one load of fuel the truck could cross a desert one leg wide. With two loads available this strategy could be followed: Loading up, the truck is driven out into the desert to a distance of one-third leg. One-third load of fuel is left in a cache and the truck returns to the fuel depot just as its fuel vanishes. On the second load it drives out to the cache, which is then used to fill up. With a full load the truck can then be driven one more leg, thereby cross a desert of width $(1+\frac{1}{3})$ legs, as shown in Fig. 17-1. With three loads of fuel available at the depot two trips can be made to establish a cache of $\frac{6}{5}$ loads at a distance of $\frac{1}{5}$ leg out into the desert. The third load then brings the truck to the cache with $(\frac{4}{5}+\frac{6}{5})$ loads available. Repeating the previous strategy then allows a journey of $1+\frac{1}{3}+\frac{1}{5}$ legs, as shown in Fig. 17-2.

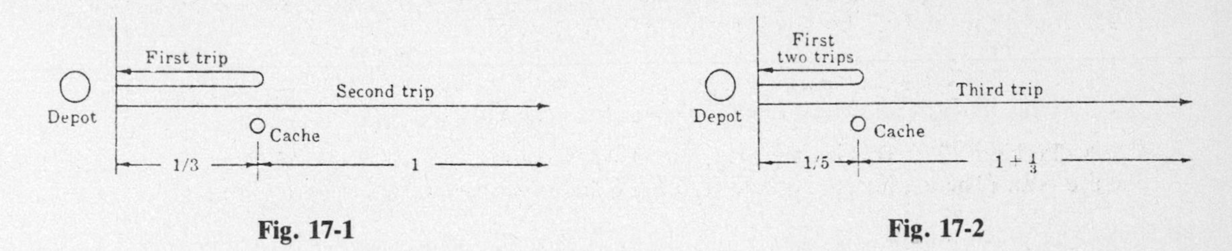

Fig. 17-1 **Fig. 17-2**

A similar strategy allows a desert of width $\left(1+\frac{1}{3}+\frac{1}{5}+\cdots+\frac{1}{2n-1}\right)$ to be crossed using n loads of fuel. Since this sum grows arbitrarily large with increasing n, a desert of any width can be crossed if sufficient fuel is available at the depot.

To estimate how much fuel is needed to cross a desert ten legs wide, we write

$$1+\frac{1}{3}+\cdots+\frac{1}{2n-1}=\left(1+\frac{1}{2}+\frac{1}{3}+\cdots+\frac{1}{2n}\right)-\frac{1}{2}\left(1+\frac{1}{2}+\frac{1}{3}+\cdots+\frac{1}{n}\right)$$

and apply the approximation of Problem 17.33:

$$1+\frac{1}{3}+\cdots+\frac{1}{2n-1}\simeq\log(2n)+C-\frac{1}{2}(\log n+C)$$

$$=\frac{1}{2}\log n+\log 2+\frac{1}{2}C\simeq\frac{1}{2}\log n+.98$$

This reaches ten for n equal to almost 100 million loads of fuel.

WALLIS' INFINITE PRODUCT

17.35. Obtain Wallis' product for π.

Repeated applications of the recursion formula

$$\int_0^{\pi/2}\sin^n x\,dx=\frac{n-1}{n}\int_0^{\pi/2}\sin^{n-2}x\,dx\qquad\text{for } n>1$$

available in integral tables, easily brings the results

$$\int_0^{\pi/2} \sin^{2k} x\,dx = \frac{2k-1}{2k} \cdot \frac{2k-3}{2k-2} \cdots \frac{1}{2} \cdot \int_0^{\pi/2} dx$$

$$\int_0^{\pi/2} \sin^{2k+1} x\,dx = \frac{2k}{2k+1} \cdot \frac{2k-2}{2k-1} \cdots \frac{2}{3} \cdot \int_0^{\pi/2} \sin x\,dx$$

Evaluating the remaining integrals and dividing one result by the other,

$$\frac{\pi}{2} = \frac{2 \cdot 2 \cdot 4 \cdot 4 \cdot 6 \cdot 6 \cdots 2k \cdot 2k}{1 \cdot 3 \cdot 3 \cdot 5 \cdot 5 \cdot 7 \cdots (2k-1)(2k+1)} \cdot \frac{\int_0^{\pi/2} \sin^{2k} x\,dx}{\int_0^{\pi/2} \sin^{2k+1} x\,dx}$$

The quotient of the two integrals converges to 1 as k increases. This can be proved as follows. Since $0 < \sin x < 1$,

$$0 < \int_0^{\pi/2} \sin^{2k+1} x\,dx \leqq \int_0^{\pi/2} \sin^{2k} x\,dx \leqq \int_0^{\pi/2} \sin^{2k-1} x\,dx$$

Dividing by the first integral and using the original recursion formula,

$$1 \leqq \frac{\int_0^{\pi/2} \sin^{2k} x\,dx}{\int_0^{\pi/2} \sin^{2k+1} x\,dx} \leqq \frac{2k+1}{2k}$$

so that the quotient does have limit 1. Thus

$$\frac{\pi}{2} = \lim \frac{2 \cdot 2 \cdot 4 \cdot 4 \cdot 6 \cdot 6 \cdots 2k \cdot 2k}{1 \cdot 3 \cdot 3 \cdot 5 \cdot 5 \cdot 7 \cdots (2k-1)(2k+1)}$$

which is Wallis' infinite product.

17.36. Obtain Wallis' infinite product for $\sqrt{\pi}$.

Since $\lim 2k/(2k+1) = 1$, the result of the previous problem can be written as

$$\frac{\pi}{2} = \lim \frac{2^2 \cdot 4^2 \cdots (2k-2)^2}{3^2 \cdot 5^2 \cdots (2k-1)^2} 2k$$

Taking the square root and then filling in missing integers, we find

$$\sqrt{\frac{\pi}{2}} = \lim \frac{2 \cdot 4 \cdots (2k-2)}{3 \cdot 5 \cdots (2k-1)} \sqrt{2k} = \lim \frac{2^{2k}(k!)^2}{(2k)!\,\sqrt{2k}}$$

from which Wallis' product follows at once in the form

$$\sqrt{\pi} = \lim \frac{2^{2k}(k!)^2}{(2k)!\,\sqrt{k}}$$

This will be needed in the next problem.

STIRLING'S SERIES FOR LARGE FACTORIALS

17.37. Derive Stirling's series for large factorials.

In the Euler–Maclaurin formula let $y(t) = \log t$ and use the limits 1 and n. Then

$$\log 1 + \log 2 + \cdots + \log n = n \log n - n + \frac{1}{2} \log n + \sum_{i=1}^{k} \frac{(-1)^i b_i}{(2i)(2i-1)} \left(1 - \frac{1}{n^{2i-1}}\right) - \int_1^n \frac{F_{2k+1}(t)}{(2k+1)t^{2k+1}}\,dt$$

This can be rearranged into

$$\log n! = \left(n + \frac{1}{2}\right)\log n - n + c - \sum_{i=1}^{k} \frac{(-1)^i b_i}{(2i)(2i-1)n^{2i-1}} + \int_n^\infty \frac{F_{2k+1}(t)}{(2k+1)t^{2k+1}}\,dt$$

where
$$c = \sum_{i=1}^{k} \frac{(-1)^i b_i}{(2i)(2i-1)} - \int_1^\infty \frac{F_{2k+1}(t)}{(2k+1)t^{2k+1}}\,dt$$

To evaluate c let $n \to \infty$ in the previous equation. The finite sum has limit zero. The integral, since F_{2k+1} is periodic and hence bounded, behaves as $1/n^{2k}$ and so also has limit zero. Thus

$$c = \lim \log \frac{n!\, e^n}{n^{n+1/2}} = \lim \log \alpha_n$$

A simple artifice now evaluates this limit. Since $\alpha_n^2 = \dfrac{(n!)^2 e^{2n}}{n^{2n+1}}$, $\alpha_{2n} = \dfrac{(2n)!\, e^{2n}}{(2n)^{2n+1/2}}$ we find

$$\lim \alpha_n = \lim \frac{\alpha_n^2}{\alpha_{2n}} = \lim \left[\sqrt{2}\,\frac{(n!)^2 2^{2n}}{\sqrt{n}\,(2n)!}\right] = \sqrt{2\pi}$$

by Wallis' product for $\sqrt{\pi}$. Thus $c = \log\sqrt{2\pi}$. Our result can now be written as the Stirling series

$$\log \frac{n!\, e^n}{\sqrt{2\pi}\, n^{n+1/2}} \sim \frac{b_1}{2n} - \frac{b_2}{3\cdot 4n^3} + \frac{b_3}{5\cdot 6n^5} - \cdots + \frac{(-1)^{k+1} b_k}{(2k)(2k-1)n^{2k-1}}$$

the error being $E_n = \displaystyle\int_n^\infty \frac{F_{2k+1}(t)}{(2k+1)t^{2k+1}}\,dt$. For large n this means that the logarithm is near zero, making $n! \simeq \sqrt{2\pi}\, n^{n+1/2} e^{-n}$.

17.38. Approximate 20! by Stirling's series.

For $n = 20$ the series itself becomes $\dfrac{1}{240} - \dfrac{1}{2{,}880{,}000} + \cdots \simeq .00417$ to five places, only *one* term being used. We now have

$$\log 20! \simeq .00417 - 20 + \log\sqrt{2\pi} + 20.5 \log 20 \simeq 42.33558$$

$$20! \simeq 2.43281 \cdot 10^{18}$$

This is correct to almost five digits. More terms of the Stirling series could be used for even greater accuracy, but it is important to realize that this series is not convergent. As k is increased beyond a certain point, for fixed n, the terms increase and the error E grows larger. This follows from the fact (see Problem 17.29) that $b_k > 2(2k)!/(2\pi)^{2k}$. As will be proved shortly, the Stirling series is an example of an *asymptotic series*.

17.39. Compute $\sum_{i=1}^{\infty} 1/i^3$ to seven places.

Sum the first nine terms directly to find $\sum_{i=1}^{9} 1/i^3 = 1.19653199$. With $f(t) = 1/t^3$ the Euler–Maclaurin formula now involves

$$\int_{10}^\infty \frac{dx}{x^3} = .005 \qquad \frac{1}{2} f(10) = .0005 \qquad -\frac{1}{12} f'(10) = .000025 \qquad \frac{1}{720} f^{(3)}(10) = .00000008$$

and the total is 1.2020569. This improves the result of Problem 17.17.

ASYMPTOTIC SERIES

17.40. Define an asymptotic series.

Let $S_n(x) = \sum_{i=0}^{n} a_i x^i$. If for $x \to 0$, $\lim\,[f(x) - S_n(x)]/x^n = 0$ for any fixed positive integer n, then

$f(x)$ is said to be asymptotic to $\sum_{i=0}^{\infty} a_i x^i$ at zero. This is represented by the symbol

$$f(x) \simeq \sum_{i=0}^{\infty} a_i x^i$$

With x replaced by $x - x_0$ the same definition applies, the series being asymptotic to $f(x)$ at x_0.

Perhaps the most useful case of all is the asymptotic expansion at infinity. If for $x \to \infty$,

$$\lim x^n [f(x) - S_n(x)] = 0$$

where now $S_n(x) = \sum_{i=0}^{n} a_i / x^i$, then $f(x)$ has an asymptotic series at infinity, and we write

$$f(x) \simeq \sum_{i=0}^{\infty} \frac{a_i}{x^i}$$

The idea can be further generalized. If, for example,

$$\frac{f(x) - g(x)}{h(x)} \simeq \sum_{i=0}^{\infty} \frac{a_i}{x^i}$$

then we also say that $f(x)$ has the following asymptotic representation:

$$f(x) \simeq g(x) + h(x) \sum_{i=0}^{\infty} \frac{a_i}{x^i}$$

Note that none of these series is assumed to converge.

17.41. Obtain an asymptotic series for $\int_x^\infty (e^{-t}/t)\, dt$.

Successive integrations by parts bring

$$f(x) = \int_x^\infty \frac{e^{-t}}{t}\, dt = \frac{e^{-x}}{x} - \int_x^\infty \frac{e^{-t}}{t^2}\, dt = \frac{e^{-x}}{x} - \frac{e^{-x}}{x^2} + 2! \int_x^\infty \frac{e^{-t}}{t^3}\, dt$$

and so on. Ultimately one finds

$$f(x) = \int_x^\infty \frac{e^{-t}}{t}\, dt = e^{-x}\left[\frac{1}{x} - \frac{1}{x^2} + \frac{2!}{x^3} - \frac{3!}{x^4} + \cdots + (-1)^{n+1} \frac{(n-1)!}{x^n}\right] + R_n$$

where $R_n = (-1)^n n! \int_x^\infty \frac{e^{-t}}{t^{n+1}}\, dt$. Since $|R_n| < n!\, e^{-x}/x^{n+1}$, we have

$$\left| x^n \left[e^x f(x) - \sum_{i=1}^{n} \frac{(-1)^{i+1}(i-1)!}{x^i} \right] \right| < \frac{n!}{x}$$

so that as $x \to \infty$ this does have limit 0. This makes $e^x f(x)$ asymptotic to the series and by our generalized definition

$$f(x) \simeq e^{-x}\left(\frac{1}{x} - \frac{1}{x^2} + \frac{2!}{x^3} - \frac{3!}{x^4} + \cdots\right)$$

Notice that the series diverges for every value of x.

17.42. Show that the truncation error involved in using the series of the preceding problem does not exceed the first omitted term.

The truncation error is precisely R_n. The first omitted term is $(-1)^{n+2} e^{-x} n!/x^{n+1}$ which is identical with the estimate of R_n occurring in Problem 17.41.

17.43. Use the asymptotic series of Problem 17.41 to compute $f(5)$.

We find

$$e^5 f(5) \simeq .2 - .04 + .016 - .0096 + .00746 - .00746 + \cdots$$

after which terms increase. Since the error does not exceed the first term we omit, only four terms need be used, with the result

$$f(5) \simeq e^{-5}(.166) \simeq .00112$$

with the last digit doubtful. The point is, the series cannot produce $f(5)$ more accurately than this. For larger x arguments the accuracy attainable improves substantially but is still limited.

17.44. Use the series of Problem 17.41 to compute $f(10)$.

We find, carrying six places,

$$e^{10} f(10) \simeq .1 - .01 + .002 - .0006 + .00024 - .000120 + .000072$$
$$- .000050 + .000040 - .000036 + .000036 - \cdots$$

after which the terms increase. Summing the first nine terms, we have

$$f(10) \simeq e^{-10}(.091582) \simeq .0000041579$$

with the last digit doubtful. In the previous problem two-place accuracy was attainable. Here we have managed four places. The essential idea of asymptotic series is that for increasing x arguments the error tends to zero.

17.45. Prove that the Stirling series is asymptotic.

With n playing the role of x and the logarithm the role of $f(x)$ (see Problem 17.37), we must show that

$$\lim n^{2k-1} E_n = \lim n^{2k-1} \int_n^\infty \frac{F_{2k+1}(t)}{(2k+1)t^{2k+1}}\, dt = 0$$

Since $F_{2k+1}(t)$ repeats, with period 1, the behavior of $B_{2k+1}(t)$ in the interval $(0, 1)$ it is bounded, say $|F| < M$. Then

$$|n^{2k-1} E_n| < \frac{n^{2k-1} M}{2k(2k+1)n^{2k}}$$

and with increasing n this becomes arbitrarily small.

17.46. Find an asymptotic series for $\int_x^\infty e^{-t^2/2}\, dt$.

The method of successive integrations by parts is again successful. First

$$\int_x^\infty e^{-t^2/2}\, dt = \int_x^\infty -\frac{1}{t}(-te^{-t^2/2})\, dt = \frac{1}{x} e^{-x^2/2} - \int_x^\infty \frac{1}{t^2} e^{-t^2/2}\, dt$$

and continuing in this way we find

$$\int_x^\infty e^{-t^2/2}\, dt = e^{-x^2/2}\left[\frac{1}{x} - \frac{1}{x^3} + \frac{1 \cdot 3}{x^5} - \cdots + (-1)^{n-1} \frac{1 \cdot 3 \cdots (2n-3)}{x^{2n-1}}\right] + R_n$$

where $R_n = 1 \cdot 3 \cdot 5 \cdots (2n-1) \int_x^\infty e^{-t^2/2} \frac{1}{t^{2n}}\, dt$. The remainder can be rewritten as

$$R_n = \frac{1 \cdot 3 \cdot 5 \cdots (2n-1)}{x^{2n+1}} e^{-x^2/2} - R_{n+1}$$

Since both remainders are positive, it follows that

$$R_n < \frac{1 \cdot 3 \cdot 5 \cdots (2n-1)}{x^{2n+1}} e^{-x^2/2}$$

This achieves a double purpose. It shows that the truncation error does not exceed the first omitted term. And since it also makes $\lim e^{x^2/2}x^{2n-1}R_n = 0$, it proves the series asymptotic.

$$\int_x^\infty e^{-t^2/2}\,dt \simeq e^{-x^2/2}\left(\frac{1}{x} - \frac{1}{x^3} + \frac{1\cdot 3}{x^5} - \frac{1\cdot 3\cdot 5}{x^7} + \cdots\right)$$

17.47. Compute $\sqrt{2/\pi}\int_4^\infty e^{-t^2/2}\,dt$ by the series of Problem 17.46.

With $x = 4$ we find

$$\sqrt{\frac{2}{\pi}}e^{-8}[.25 - .015625 + .002930 - .000916 + .000401 - .000226$$
$$+ .000155 - .000126 + .000118 - .000125 + \cdots]$$

to the point where terms begin to increase. The result of stopping before the smallest term is

$$\sqrt{\frac{2}{\pi}}\int_4^\infty e^{-t^2/2}\,dt \simeq .0000633266$$

with the 2 digit in doubt. This agrees nicely with our results of Problem 14.32. Independent computations which confirm one another are very reassuring. Note the difference in methods in these two problems, and the simplicity of the present computation.

17.48. Find an asymptotic series for the sine integral.

Once again integration by parts proves useful. First

$$\text{Si}(x) = \int_x^\infty \frac{\sin t}{t}\,dt = \frac{\cos x}{x} - \int_x^\infty \frac{\cos t}{t^2}\,dt$$

after which similar steps generate the series

$$\int_x^\infty \frac{\sin t}{t}\,dt \simeq \frac{\cos x}{x} + \frac{\sin x}{x^2} - \frac{2!\cos x}{x^3} - \frac{3!\sin x}{x^4} + \cdots$$

which can be proved asymptotic as in previous problems.

17.49. Compute Si (10).

Putting $x = 10$ in the previous problem,

$$\text{Si}(10) \simeq -.083908 - .005440 + .001678 + .000326 - .000201$$
$$- .000065 + .000060 + .000027 - .000034 - .000019$$

after which both the cosine and sine terms start to grow larger. The total of these ten terms rounds to $-.0876$, which is correct to four places.

Supplementary Problems

17.50. Express as a sum of differences and so evaluate $\sum_{i=1}^n (i^2 - 3i + 2)$.

17.51. Express as a sum of differences and so evaluate $\sum_{i=1}^n i^5$

17.52. Express as a sum of differences and so evaluate $\sum_{i=1}^n \frac{1}{i(i+2)}$.

17.53. Evaluate the sum in Problem 17.51 by the Euler–Maclaurin formula.

17.54. Evaluate the sum in Problem 17.50 by the Euler–Maclaurin formula.

17.55. How many terms of the cosine series are needed to provide eight-place accuracy for arguments from 0 to $\pi/2$?

17.56. Show that

$$y_0 - y_1 + y_2 - \cdots = \frac{1}{1+E} y_0 = \frac{1}{D}\left(\frac{D}{e^D - 1} - \frac{2D}{e^{2D} - 1}\right) y_0$$
$$= \left(\frac{1}{2} - B_2 \frac{4-1}{2!} D + B_4 \frac{16-1}{4!} D^3 - \frac{64-1}{6!} D^5 + \cdots\right) y_0$$

where the B_i are Bernoulli numbers. Apply this to the Leibnitz series for $\pi/4$ to obtain the six-place result .785398.

17.57. Apply the Euler transformation to evaluate $1 - \frac{1}{\sqrt{2}} + \frac{1}{\sqrt{3}} - \frac{1}{\sqrt{4}} + \cdots$ to four places.

17.58. Use the Euler transformation to evaluate $1 - \frac{1}{9} + \frac{1}{25} - \frac{1}{49} + \cdots$ to eight places, confirming the result .91596559.

17.59. Use the Euler transformation to show that $1 - \frac{1}{\log 2} + \frac{1}{\log 3} - \frac{1}{\log 4} + \cdots$ to four places equals .0757.

17.60. Apply the Euler transformation to $\log 2 = 1 - \frac{1}{2} + \frac{1}{3} - \frac{1}{4} + \frac{1}{5} - \cdots$.

17.61. For how large an argument x will twenty terms of the series

$$\log(1+x) = x - \frac{1}{2}x^2 + \frac{1}{3}x^3 - \frac{1}{4}x^4 + \cdots$$

produce four-place accuracy?

17.62. How many terms of the cosine series $\cos x = 1 - \frac{1}{2}x^2 + \frac{1}{4!}x^4 - \cdots$ are needed to guarantee eight-place accuracy for the interval from 0 to $\pi/2$?

17.63. For how large an argument x will twenty terms of the series

$$\arctan x = x - \frac{1}{3}x^3 + \frac{1}{5}x^5 - \frac{1}{7}x^7 + \cdots$$

produce six-place accuracy?

17.64. For the series $\sinh x = x + \frac{x^3}{3!} + \frac{x^5}{5!} + \frac{x^7}{7!} + \cdots$ estimate the truncation error in terms of the first term omitted. (See Problem 17.7 for a possible method.) For how large an argument x will twenty terms be enough for eight-place accuracy?

17.65. Apply the comparison method of Problem 17.14 to compute $\sum_{i=1}^{\infty} 1/(i^2+i+1)$ to three places. [Use $\sum_{i=1}^{\infty} 1/(i+1)i = 1$ as the comparison series.]

17.66. Compute $\sum_{i=1}^{\infty} 1/(i^3+1)$ to three places by the comparison method using the result of Problem 17.17.

17.67. Compute $\sum_{i=1}^{\infty} 1/(i^2+2i+2)$ to three places by the comparison method.

17.68. Compute $\sum_{i=1}^{\infty} i^2/(i^4+1)$ to three places by the comparison method.

17.69. Determine the first ten b_i numbers from the recursion of Problem 17.18.

17.70. Write out $B_6(x)$ through $B_{10}(x)$ from the formula of Problem 17.19.

17.71. Prove $\int_x^{x+1} B_i(x)\,dx = x^i$.

17.72. Determine $B_3(x)$ and $B_4(x)$ as in Problem 17.24.

17.73. What polynomials are determined by the conditions

$$Q_i'(x) = iQ_{i-1}(x) \qquad Q_i(0) = 0$$

starting with $Q_0(x) = 1$?

17.74. Use Problem 17.28 to evaluate $\sum_{k=1}^{\infty} 1/k^p$ for $p = 6$, 8, and 10, verifying the results $\pi^6/945$, $\pi^8/9450$, and $\pi^{10}/93{,}555$.

17.75. Use the Euler–Maclaurin formula to prove $\sum_{i=0}^{n} i^3 = n^2(n+1)^2/4$.

17.76. Use the Euler–Maclaurin formula to evaluate $\sum_{i=1}^{n} (i^2+3i+2)$. Compare with Problem 17.3.

17.77. Use the Euler–Maclaurin formula to show that

$$S_n = \sum_{i=1}^{n} \frac{1}{i} - \log n = C + \frac{1}{2n} + \int_n^{\infty} \frac{F_1(t)}{t^2}\,dt$$

where C is Euler's constant and $F_1(t)$ is the periodic extension of $B_1(t)$. This proves the convergence of S_n and also allows estimation of the difference between S_n and C for large n.

17.78. By applying the Euler–Maclaurin formula, show that

$$C = \frac{1}{2}\log 2 + \frac{1}{4} + \sum_{i=1}^{k} \frac{(-1)^{i+1} b_i}{(2i)(2i-1)} \left(\frac{2i+1}{2^{2i}} - 1\right) + \text{error term}$$

and use this to evaluate Euler's constant C. Show that as k increases, the sum on the right becomes a divergent series. At what point do the terms of this series begin to grow larger?

17.79. Referring to Problem 17.34, show that a desert of width five legs requires more than 3000 loads of fuel.

17.80. Compute $\sum_{k=1}^{\infty} 1/k^{5/2}$ to six places.

17.81. Compute $\sum_{k=1}^{\infty} 1/(2k-1)^2$ to three places.

17.82. Evaluate $\frac{1}{1} - \frac{1}{4} + \frac{1}{9} - \frac{1}{16} + \frac{1}{25} - \cdots$ exactly.

17.83. Evaluate the sum of Problem 17.81 exactly.

17.84. Show that the Euler transformation converts $\sum_{k=0}^{\infty} (-\frac{1}{2})^k$ into a more rapidly convergent series.

17.85. Show that the Euler transformation converts $\sum_{k=0}^{\infty} (-\frac{1}{3})^k$ into a more slowly convergent series.

17.86. How accurately does the Stirling series produce 2! and at what point do the terms of the series start to increase?

17.87. Derive the asymptotic series

$$\int_x^\infty \sin t^2\,dt \simeq \cos x^2\left(\frac{1}{2x}-\frac{3}{2^3x^5}+\frac{3\cdot 5\cdot 7}{2^5x^9}-\cdots\right)+\sin x^2\left(\frac{1}{2^2x^3}-\frac{3\cdot 5}{2^4x^7}+\frac{3\cdot 5\cdot 7\cdot 9}{2^6x^{11}}-\cdots\right)$$

and use it when $x=10$, obtaining as much accuracy as you can.

Chapter 18

Difference Equations

DEFINITIONS

The term difference equation might be expected to refer to an equation involving differences. However, an example such as

$$\Delta^2 y_k + 2\Delta y_k + y_k = 0$$

which quickly collapses to $y_{k+2} = 0$, shows that combinations of differences are not always convenient, may even obscure information. As a result, difference equations are usually written directly in terms of the y_k values. As an example take

$$y_{k+1} = a_k y_k + b_k$$

where a_k and b_k are given functions of the integer argument k. This could be rewritten as $\Delta y_k = (a_k - 1)y_k + b_k$ but this is not normally found to be useful. In summary, a difference equation is a relation between the values y_k of a function defined on a discrete set of arguments x_k. Assuming the arguments equally spaced, the usual change of argument $x_k = x_0 + kh$ leaves us with an integer argument k.

A **solution of a difference equation** will be a sequence of y_k values for which the equation is true, for some set of consecutive integers k. The nature of a difference equation allows solution sequences to be computed recursively. In the above example, for instance, y_{k+1} may be computed very simply if y_k is known. One known value thus triggers the computation of the entire sequence.

The **order of a difference equation** is the difference between the largest and smallest arguments k appearing in it. The last example above has order one.

ANALOGY TO DIFFERENTIAL EQUATIONS

A strong analogy exists between the theory of difference equations and the theory of differential equations. For example, a first-order equation normally has exactly one solution satisfying the initial condition $y_0 = A$. And a second-order equation normally has exactly one solution satisfying two initial conditions $y_0 = A$, $y_1 = B$. Several further aspects of this analogy will be emphasized, such as the following:

1. **Procedures for finding solutions** are similar in the two subjects. First-order linear equations are solved in terms of sums, as the corresponding differential equations are solved in terms of integrals. For example, the equation $y_{k+1} = xy_k + c_{k+1}$ with $y_0 = c_0$ has the polynomial solution

$$y_n = c_0 x^n + c_1 x^{n-1} + \cdots + c_n$$

   Computation of this polynomial recursively, from the difference equation itself, is known as Horner's method for evaluating the polynomial. It is more economical than the standard evaluation by powers.

2. The **digamma function** is defined as

$$\psi(x) = \sum_{i=1}^{\infty} \frac{x}{i(i+x)} - C$$

   where C is Euler's constant. It is one summation form of the solution of the first-order difference equation

$$\Delta\psi(x) = \frac{1}{x+1}$$

This also gives it the character of a finite integral of $1/(x+1)$. For integer arguments n, it follows that

$$\psi(n) = \sum_{k=1}^{n} \frac{1}{k} - C$$

This function plays a role in difference calculus somewhat analogous to that of the logarithm function in differential calculus. Compare, for instance, these two formulas:

$$\sum_{k=1}^{\infty} \frac{1}{(k+a)(k+b)} = \frac{\psi(b)-\psi(a)}{b-a} \qquad \int_1^{\infty} \frac{dx}{(x+a)(x+b)} = \frac{\log(b+1)-\log(a+1)}{b-a}$$

Various sums may be expressed in terms of the digamma function and its derivatives. The above is one example. Another is

$$\sum_{k=1}^{\infty} \frac{2k+1}{k(k+1)^2} = \psi(1) - \psi(0) - \psi'(1)$$

which also proves to be $\pi^2/6$.

The **gamma function** is related to the digamma function by

$$\frac{\Gamma'(x+1)}{\Gamma(x+1)} = \psi(x)$$

3. The **linear homogeneous second-order equation**

$$y_{k+2} + a_1 y_{k+1} + a_2 y_k = 0$$

has the solution family

$$y_k = c_1 u_k + c_2 v_k$$

where u_k and v_k are themselves solutions and c_1, c_2 are arbitrary constants. As in the theory of differential equations, this is called the **principle of superposition**. Any solution of the equation can be expressed as such a superposition of u_k and v_k, by proper choice of c_1 and c_2, provided the Wronskian determinant

$$w_k = \begin{vmatrix} u_k & v_k \\ u_{k-1} & v_{k-1} \end{vmatrix}$$

is not zero.

4. The **case of constant coefficients**, where a_1 and a_2 are constants, allows easy determination of the solutions u_k and v_k. With r_1 and r_2 the roots of the characteristic equation

$$r^2 + a_1 r + a_2 = 0$$

these solutions are

$$u_k = r_1^k \qquad v_k = r_2^k \qquad \text{when } a_1^2 > 4a_2$$

$$u_k = r^k \qquad v_k = kr^k \qquad \text{when } a_1^2 = 4a_2,\ r_1 = r_2 = r$$

$$u_k = R^k \sin k\theta \qquad v_k = R^k \cos k\theta \qquad \text{when } a_1^2 < 4a_2,\ r_1, r_2 = R(\cos\theta \pm i \sin\theta)$$

The analogy with differential equations is apparent. The Wronskian determinants of these u_k, v_k pairs are not zero, and so by superposition we may obtain all possible solutions of the difference equation.

The **Fibonacci numbers** are solution values of

$$y_{k+2} = y_{k+1} + y_k$$

and by case 1 above may be represented by real power functions. They have some applications in information theory.

5. The **nonhomogeneous equation**

$$y_{k+2} + a_1 y_{k+1} + a_2 y_k = b_k$$

has the solution family

$$y_k = c_1 u_k + c_2 v_k + Y_k$$

where u_k, v_k are as above and Y_k is one solution of the given equation. This is also analogous to a result of differential equations. For certain elementary functions b_k it is possible to deduce the corresponding solution Y_k very simply.

IMPORTANCE OF DIFFERENCE EQUATIONS

Our interest in difference equations is twofold. First, they do occur in applications. And second, numerous methods for the approximate solution of differential equations involve replacing them by difference equations as substitutes.

Solved Problems

FIRST-ORDER EQUATIONS

18.1. Solve the first-order equation $y_{k+1} = ky_k + k^2$ recursively, given the initial condition $y_0 = 1$.

This problem illustrates the appeal of difference equations in computation. Successive y_k values are found simply by doing the indicated additions and multiplications,

$$y_1 = 0 \qquad y_2 = 1 \qquad y_3 = 6 \qquad y_4 = 27 \qquad y_5 = 124$$

and so on. Initial-value problems of difference equations may always be solved in this simple recursive fashion. Often, however, one wishes to know the character of the solution function, making an analytic representation of the solution desirable. Only in certain cases have such representations been found.

18.2. Given the functions a_k and b_k, what is the character of the solution of the linear first-order equation $y_{k+1} = a_k y_k + b_k$ with initial condition $y_0 = A$?

Proceeding as in the previous problem, we find

$$y_1 = a_0 A + b_0$$
$$y_2 = a_1 y_1 + b_1 = a_0 a_1 A + a_1 b_0 + b_1$$
$$y_3 = a_2 y_2 + b_2 = a_0 a_1 a_2 A + a_1 a_2 b_0 + a_2 b_1 + b_2$$

etc. With p_n denoting the product $p_n = a_0 a_1 \cdots a_{n-1}$, the indicated result appears to be

$$y_n = p_n \left(A + \frac{b_0}{p_1} + \frac{b_1}{p_2} + \cdots + \frac{b_{n-1}}{p_n} \right)$$

This could be verified formally by substitution. As in the case of linear first-order differential equations, this result is only partially satisfactory. With differential equations the solution can be expressed in terms of an integral. Here we have a sum. In certain cases, however, further progress is possible. It is important to notice that there is exactly one solution which satisfies the difference equation and assumes the prescribed initial value $y_0 = A$.

18.3. What is the character of the solution function in the special case $a_k = r$, $b_k = 0$?

Here the result of Problem 18.2 simplifies to the power function $y_n = Ar^n$. Such power functions play an important role in the solution of other equations also.

18.4. What is the character of the solution function when $a_k = r$ and $b_k = 1$, with $y_0 = A = 1$?

Now the result of Problem 18.2 simplifies to

$$y_n = r^n + r^{n-1} + \cdots + 1 = \frac{r^{n+1} - 1}{r - 1}$$

18.5. What is the character of the solution function of $y_{k+1} = xy_k + c_{k+1}$ with $y_0 = A = c_0$?

This problem serves as a good illustration of how simple functions are sometimes best evaluated by difference equation procedures. Here the result of Problem 18.2 becomes

$$y_n = c_0x^n + c_1x^{n-1} + \cdots + c_n$$

The solution takes the form of a polynomial. Horner's method for evaluating this polynomial at argument x involves computing $y_1, y_2, \ldots, y_n$ successively. This amounts to n multiplications and n additions, and is equivalent to rearranging the polynomial into

$$y_n = c_n + x(c_{n-1} + \cdots + x(c_3 + x(c_2 + x(c_1 + xc_0))))$$

It is more efficient than building up the powers of x one by one and then evaluating by the standard polynomial form.

18.6. What is the character of the solution of $y_{k+1} = \dfrac{k+1}{x} y_k + 1$ with initial value $y_0 = 1$?

Here the p_n of Problem 18.2 becomes $p_n = n!/x^n$, while all $b_k = 1$. The solution is therefore expressible as

$$\frac{y_n}{p_n} = \frac{x^n y_n}{n!} = 1 + x + \frac{1}{2}x^2 + \cdots + \frac{1}{n!}x^n$$

so that for increasing n, $\lim x^n y_n / n! = e^x$.

18.7. What is the character of the solution of $y_{k+1} = [1 - x^2/(k+1)^2]y_k$ with $y_0 = 1$?

Here all the b_k of Problem 18.2 are zero and $A = 1$, making

$$y_n = p_n = (1 - x^2)\left(1 - \frac{x^2}{2^2}\right)\left(1 - \frac{x^2}{3^2}\right) \cdots \left(1 - \frac{x^2}{n^2}\right)$$

This product vanishes for $x = \pm 1, \pm 2, \ldots, \pm n$. For increasing n we encounter the infinite product

$$\lim y_n = \prod_{k=0}^{\infty} \left[1 - \frac{x^2}{(k+1)^2}\right]$$

which can be shown to represent $(\sin \pi x)/\pi x$.

THE DIGAMMA FUNCTION

18.8. The method of summing by "telescoping" depends upon being able to express a sum as a sum of differences,

$$\sum_{k=0}^{n} b_k = \sum_{k=0}^{n} \Delta y_k = y_{n+1} - y_0$$

That is, it requires solving the first-order difference equation

$$\Delta y_k = y_{k+1} - y_k = b_k$$

Apply this method when $b_k = 1/(k+1)$, solving the difference equation and evaluating the sum.

Start by defining the *digamma function* as $\psi(x) = \sum_{i=1}^{\infty} \dfrac{x}{i(i+x)} - C$ where C is Euler's constant.

Directly we find for any $x \neq -i$,

$$\Delta\psi(x) = \psi(x+1) - \psi(x) = \sum_{i=1}^{\infty}\left[\frac{x+1}{i(i+x+1)} - \frac{x}{i(i+x)}\right]$$
$$= \sum_{i=1}^{\infty}\left(\frac{1}{i+x} - \frac{1}{i+x+1}\right) = \frac{1}{x+1}$$

When x takes integer values, say $x = k$, this provides a new form for the sum of integer reciprocals, since

$$\sum_{k=0}^{n-1}\frac{1}{k+1} = \sum_{k=0}^{n-1}\Delta\psi(k) = \psi(n) - \psi(0) = \psi(n) + C$$

We may also rewrite this as

$$\psi(n) = \sum_{k=1}^{n}\frac{1}{k} - C$$

so that the digamma function for integer arguments is a familiar quantity. Its behavior is shown in Fig. 18-1, and the logarithmic character for large positive x is no surprise when one recalls the definition of Euler's constant. In a sense $\psi(x)$ generalizes from $\psi(n)$ much as the gamma function generalizes factorials.

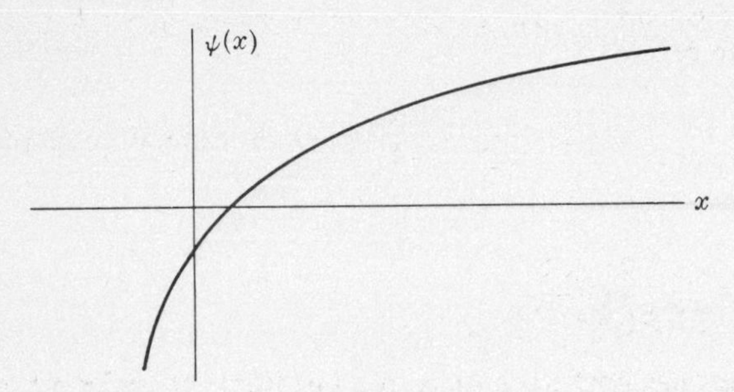

Fig. 18-1

18.9. Evaluate the sum $\sum_{k=1}^{n} 1/(k+t)$ for arbitrary t.

From Problem 18.8, for any x, $\psi(x+1) - \psi(x) = 1/(x+1)$. Replace x by $k+t-1$ to obtain

$$\psi(k+t) - \psi(k+t-1) = \frac{1}{k+t}$$

Now we have the ingredients of a telescoping sum and find

$$\sum_{k=1}^{n}\frac{1}{k+t} = \sum_{k=1}^{n}[\psi(k+t) - \psi(k+t-1)] = \psi(n+t) - \psi(t)$$

18.10. Evaluate the series $\sum_{k=1}^{\infty} 1/(k+a)(k+b)$ in terms of the digamma function.

Using partial fractions, we find

$$s_n = \sum_{k=1}^{n}\frac{1}{(k+a)(k+b)} = \frac{1}{b-a}\sum_{k=1}^{n}\left(\frac{1}{k+a} - \frac{1}{k+b}\right)$$

Now applying the previous problem, this becomes

$$s_n = \frac{1}{b-a}[\psi(n+a) - \psi(a) - \psi(n+b) + \psi(b)]$$

From the series definition in Problem 18.8 it follows after a brief calculation that

$$\psi(n+a)-\psi(n+b)=(a-b)\sum_{i=1}^{\infty}\frac{1}{(i+n+a)(i+n+b)}$$

so that for $n\to\infty$ this difference has limit zero. Finally,

$$\sum_{k=1}^{\infty}\frac{1}{(k+a)(k+b)}=\lim s_n=\frac{\psi(b)-\psi(a)}{b-a}$$

18.11. Find formulas for $\psi'(x)$, $\psi^{(2)}(x)$, etc., in series form.

Differentiating the series of Problem 18.8 produces $\psi'(x)=\sum_{k=1}^{\infty}1/(k+x)^2$. Since this converges uniformly in x on any interval not including a negative integer, the computation is valid. Repeating,

$$\psi^{(2)}(x)=\sum_{k=1}^{\infty}\frac{-2!}{(k+x)^3}\qquad \psi^{(3)}(x)=\sum_{k=1}^{\infty}\frac{3!}{(k+x)^4}\qquad \text{etc.}$$

In particular, for integer arguments, Problem 17.28 makes $\psi'(0)=\sum_{k=1}^{\infty}1/k^2=\pi^2/6$ after which we lose one term at a time to obtain

$$\psi'(1)=\frac{\pi^2}{6}-1\qquad \psi'(2)=\frac{\pi^2}{6}-1-\frac{1}{4}\qquad \text{and in general}\qquad \psi'(n)=\frac{\pi^2}{6}-1-\frac{1}{4}-\cdots-\frac{1}{n^2}$$

18.12. Evaluate the series $\sum_{k=1}^{\infty}\frac{2k+1}{k(k+1)^2}$.

This further illustrates how sums and series involving rational terms in k may be evaluated in terms of the digamma function. Again introducing partial fractions,

$$\sum_{k=1}^{\infty}\frac{2k+1}{k(k+1)^2}=\sum_{k=1}^{\infty}\left[\frac{1}{k}-\frac{1}{k+1}+\frac{1}{(k+1)^2}\right]$$

The first two terms cannot be handled separately since the series would diverge. They can, however, be handled together as in Problem 18.10. The result is

$$\sum_{k=1}^{\infty}\left[\frac{1}{k(k+1)}+\frac{1}{(k+1)^2}\right]=\psi(1)-\psi(0)+\psi'(1)=\frac{\pi^2}{6}$$

Other sums of rational terms may be treated in similar fashion.

18.13. Evaluate the series $\sum_{k=1}^{\infty}\frac{1}{1^2+2^2+\cdots+k^2}$.

Summing the squares as in Problem 5.2 we may replace this by

$$\sum_{k=1}^{\infty}\frac{6}{k(k+1)(2k+1)}=\sum_{k=1}^{\infty}\left(\frac{6}{k}+\frac{6}{k+1}-\frac{24}{2k+1}\right)$$

Since no one of these three series is individually convergent, we do not treat each separately. Extending the device used in the problem just solved we may, however, rewrite the combination as

$$\sum_{k=1}^{\infty}\left[\left(\frac{6}{k}-\frac{6}{k}\right)+\left(\frac{6}{k+1}-\frac{6}{k}\right)-\left(\frac{24}{2k+1}-\frac{24}{2k}\right)\right]=\sum_{k=1}^{\infty}\left[\frac{-6}{k(k+1)}+\frac{6}{k(k+\frac{1}{2})}\right]$$

$$=-6[\psi(1)-\psi(0)]+12\left[\psi\left(\frac{1}{2}\right)-\psi(0)\right]$$

where Problem 18.10 has been used twice in the last step. Finally,

$$\sum_{k=1}^{\infty} \frac{1}{1^2 + 2^2 + \cdots + k^2} = 12\psi\left(\frac{1}{2}\right) - 6 + 12C$$

18.14. Show that $\mathscr{S}(x) = \Gamma'(x+1)/\Gamma(x+1)$ also has the property $\Delta\mathscr{S}(x) = 1/(x+1)$, where $\Gamma(x)$ is the gamma function.

The gamma function is defined for positive x by

$$\Gamma(x) = \int_0^\infty e^{-t} t^{x-1}\, dt$$

Integration by parts exposes the familiar feature

$$\Gamma(x+1) = x\Gamma(x)$$

and then differentiation brings $\Gamma'(x+1) = x\Gamma'(x) + \Gamma(x)$, or

$$\frac{\Gamma'(x+1)}{\Gamma(x+1)} - \frac{\Gamma'(x)}{\Gamma(x)} = \frac{1}{x}$$

from which the required result follows upon replacing x by $x + 1$.

Since $\psi(x+1) - \psi(x) = 1/(x+1)$, we find that

$$\frac{\Gamma'(x+1)}{\Gamma(x+1)} - \psi(x) = A$$

where A is a constant, and where x is restricted to a discrete set with unit spacing. The same result can be proved for all x except negative integers, the constant A being zero.

LINEAR SECOND-ORDER EQUATION, HOMOGENEOUS CASE

18.15. The difference equation $y_{k+2} + a_1 y_{k+1} + a_2 y_k = 0$ in which a_1 and a_2 may depend upon k is called linear and homogeneous. Prove that if u_k and v_k are solutions, then so are $c_1 u_k + c_2 v_k$ for arbitrary constants c_1 and c_2. (It is this feature that identifies a linear homogeneous equation. The equation is homogeneous because $y_k \equiv 0$ is a solution.)

Since $u_{k+2} + a_1 u_{k+1} + a_2 u_k = 0$ and $v_{k+2} + a_1 v_{k+1} + a_2 v_k = 0$, it follows at once by multiplying the first equation by c_1, the second equation by c_2, and adding that

$$c_1 u_{k+2} + c_2 v_{k+2} + a_1(c_1 u_{k+1} + c_2 v_{k+1}) + a_2(c_1 u_k + c_2 v_k) = 0$$

which was to be proved.

18.16. Show that for a_1 and a_2 constant, two real solutions can be found in terms of elementary functions.

First suppose $a_1^2 > 4a_2$. Then we may take

$$u_k = r_1^k \qquad v_k = r_2^k$$

where r_1 and r_2 are the distinct real roots of the quadratic equation $r^2 + a_1 r + a_2 = 0$. To prove this we verify directly that

$$u_{k+2} + a_1 u_{k+1} + a_2 u_k = r^k(r^2 + a_1 r + a_2) = 0$$

where r is either root. The quadratic equation involved here is known as the characteristic equation.

Next suppose $a_1^2 = 4a_2$. Then the characteristic equation has only one root, say r, and can be rewritten as

$$r^2 + a_1 r + a_2 = \left(r + \frac{1}{2}a_1\right)^2 = 0$$

Two real solutions are now available in

$$u_k = r^k \qquad v_k = kr^k$$

The solution u_k may be verified exactly as above. As for v_k,

$$(k+2)r^{k+2} + a_1(k+1)r^{k+1} + a_2kr^k = r^k[k(r^2 + a_1r + a_2) + (2r + a_1)r] = 0$$

since both parentheses are zero.

Finally suppose $a_1^2 < 4a_2$. Then the characteristic equation has complex conjugate roots $Re^{\pm i\theta}$. Substituting, we find

$$\begin{aligned} R^2e^{\pm i2\theta} + a_1Re^{\pm i\theta} + a_2 &= R^2(\cos 2\theta \pm i \sin 2\theta) + a_1R(\cos\theta \pm i\sin\theta) + a_2 \\ &= (R^2\cos 2\theta + a_1R\cos\theta + a_2) \pm i(R^2\sin 2\theta + a_1R\sin\theta) = 0 \end{aligned}$$

This requires that both parentheses vanish:

$$R^2\cos 2\theta + a_1R\cos\theta + a_2 = 0 \qquad R^2\sin 2\theta + a_1R\sin\theta = 0$$

We now verify that two real solutions of the difference equation are

$$u_k = R^k \sin k\theta \qquad v_k = R^k \cos k\theta$$

For example,

$$\begin{aligned} u_{k+2} + a_1u_{k+1} + a_2u_k &= R^{k+2}\sin(k+2)\theta + a_1R^{k+1}\sin(k+1)\theta + a_2R^k\sin k\theta \\ &= R^k(\sin k\theta)(R^2\cos 2\theta + a_1R\cos\theta + a_2) + R^k(\cos k\theta)(R^2\sin 2\theta + a_1R\sin\theta) = 0 \end{aligned}$$

since both parentheses vanish. The proof for v_k is almost identical.

It now follows that for a_1 and a_2 constant, the equation $y_{k+2} + a_1y_{k+1} + a_2y_k = 0$ always has a family of elementary solutions $y_k = c_1u_k + c_2v_k$.

18.17. Solve the difference equation $y_{k+2} - 2Ay_{k+1} + y_k = 0$ in terms of power functions, assuming $A > 1$.

Let $y_k = r^k$ and substitute to find that $r^2 - 2Ar + 1 = 0$ is necessary.

This leads to $r = A \pm \sqrt{A^2 - 1} = r_1, r_2$ and $y_k = c_1r_1^k + c_2r_2^k = c_1u_k + c_2v_k$.

One of these power functions grows arbitrarily large with k, and the other tends to zero, since $r_1 > 1$ but $0 < r_2 < 1$. [The fact that $r_2 = A - \sqrt{A^2 - 1} < 1$ follows from $(A-1)^2 = A^2 + 1 - 2A < A^2 - 1$ after taking square roots and transposing terms.]

18.18. Solve the equation $y_{k+2} - 2y_{k+1} + y_k = 0$.

Here we have $a_1^2 = 4a_2 = 4$. The only root of $r^2 - 2r + 1 = 0$ is $r = 1$. This means that $u_k = 1$, $v_k = k$ are solutions and that $y_k = c_1 + c_2k$ is a family of solutions. This is hardly surprising in view of the fact that this difference equation may be written as $\Delta^2 y_k = 0$.

18.19. Solve $y_{k+2} - 2Ay_{k+1} + y_k = 0$ where $A < 1$.

Now $a_1^2 < 4a_2$. The roots of the characteristic equation become

$$Re^{\pm i\theta} = A \pm i\sqrt{1 - A^2} = \cos\theta \pm i\sin\theta$$

where $A = \cos\theta$ and $R = 1$. Thus $u_k = \sin k\theta$, $v_k = \cos k\theta$ and the family of solutions

$$y_k = c_1 \sin k\theta + c_2 \cos k\theta$$

is available.

The v_k functions, when expressed as polynomials in A, are known as Chebyshev polynomials. For example,

$$v_0 = 1 \qquad v_1 = A \qquad v_2 = 2A^2 - 1 \qquad \cdots$$

The difference equation of this problem is the recursion for the Chebyshev polynomials.

18.20. Show that if two solutions of $y_{k+2}+a_1y_{k+1}+a_2y_k=0$ agree in value at two consecutive integers k, then they must agree for all integers k. (Assume $a_2 \neq 0$.)

Let u_k and v_k be solutions which agree in value at k equal to m and $m+1$. Then their difference $d_k=u_k-v_k$ is a solution (by Problem 18.15) for which $d_m=d_{m+1}=0$. But then

$$d_{m+2}+a_1d_{m+1}+a_2d_m=0 \qquad d_{m+1}+a_1d_m+a_2d_{m-1}=0$$

from which it follows that $d_{m+2}=0$ and $d_{m-1}=0$. In the same way we may prove d_k to be zero for $k>m+2$ and for $k<m-1$, taking each integer in its turn. Thus d_k is identically zero and $u_k \equiv v_k$. (The assumption $a_2 \neq 0$ merely guarantees that we do have a second-order difference equation.)

18.21. Show that any solution of $y_{k+2}+a_1y_{k+1}+a_2y_k=0$ may be expressed as a combination of two particular solutions u_k and v_k,

$$y_k=c_1u_k+c_2v_k$$

provided that the Wronskian determinant

$$w_k=\begin{vmatrix} u_k & v_k \\ u_{k-1} & v_{k-1}\end{vmatrix} \neq 0$$

We know that $c_1u_k+c_2v_k$ is a solution. By the previous problem it will be identical with the solution y_k if it agrees with y_k for two consecutive integer values of k. In order to obtain such agreement we choose $k=0$ and $k=1$ (any other consecutive integers would do) and determine the coefficients c_1 and c_2 by the equations

$$c_1u_0+c_2v_0=y_0 \qquad c_1u_1+c_2v_1=y_1$$

The unique solution is $c_1=(y_1v_0-y_0v_1)/w_1$, $c_2(y_0u_1-y_1u_0)/w_1$ since $w_1 \neq 0$.

18.22. Show that if the Wronskian determinant is zero for one value of k, it must be identically zero, assuming u_k, v_k to be solutions of the equation of Problem 18.20. Apply this to the particular case of Problem 18.16 to prove $w_k \neq 0$.

We compute the difference

$$\begin{aligned}\Delta w_k &= (u_{k+1}v_k-v_{k+1}u_k)-(u_kv_{k-1}-v_ku_{k-1}) \\ &= v_k(-a_1u_k-a_2u_{k-1})-u_k(-a_1v_k-a_2v_{k-1})-u_kv_{k-1}+v_ku_{k-1} \\ &= (a_2-1)w_k=w_{k+1}-w_k\end{aligned}$$

from which it soon follows that $w_k=a_2^kw_0$. Since $a_2 \neq 0$, the only way for w_k to be zero is to have $w_0=0$. But then w_k is identically zero.

When w_k is identically zero, it follows that u_k/v_k is the same as u_{k-1}/v_{k-1} for all k, that is, $u_k/v_k=$ constant. Since this is definitely not true for the u_k, v_k of Problem 18.16, w_k cannot be zero there.

18.23. Solve by direct computation the second-order initial-value problem

$$y_{k+2}=y_{k+1}+y_k \qquad y_0=0 \qquad y_1=1$$

Taking $k=0, 1, 2, \ldots$ we easily find the successive y_k values 1, 2, 3, 5, 8, 13, 21, 34, 55, 89, 144, . . . which are known as Fibonacci numbers. The computation clearly shows a growing solution but does not bring out its exact character.

18.24. Determine the character of the solution of the previous problem.

Following the historical path mapped in Problems 18.15, 18.16, etc., we consider the characteristic equation $r^2-r-1=0$.

Since $a_1^2 > 4a_2$, there are two real roots, namely $r_1, r_2 = (1 \pm \sqrt{5})/2$. All solutions can therefore be expressed in the form

$$y_k = c_1 u_k + c_2 v_k = c_1\left(\frac{1+\sqrt{5}}{2}\right)^k + c_2\left(\frac{1-\sqrt{5}}{2}\right)^k$$

To satisfy the initial conditions, we need $c_1 + c_2 = 0$ and $c_1\left(\frac{1+\sqrt{5}}{2}\right) + c_2\left(\frac{1-\sqrt{5}}{2}\right) = 1$. This makes $c_1 = -c_2 = \frac{1}{\sqrt{5}}$ and $y_k = \frac{1}{\sqrt{5}}\left[\left(\frac{1+\sqrt{5}}{2}\right)^k - \left(\frac{1-\sqrt{5}}{2}\right)^k\right]$.

18.25. Show that for the Fibonacci numbers, $\lim (y_{k+1}/y_k) = (1+\sqrt{5})/2$.

For such results it is convenient to know the character of the solution function. Using the previous problem we find, after a brief calculation,

$$\frac{y_{k+1}}{y_k} = \frac{1+\sqrt{5}}{2} \cdot \frac{1 - [(1-\sqrt{5})/(1+\sqrt{5})]^{k+1}}{1 - [(1-\sqrt{5})/(1+\sqrt{5})]^k}$$

and $(1-\sqrt{5})/(1+\sqrt{5})$ has absolute value less than 1, so that the required result follows.

18.26. The Fibonacci numbers occur in certain problems involving the transfer of information along a communications channel. The capacity C of a channel is defined as $C = \lim (\log y_k)/k$, the logarithm being to base 2. Evaluate this limit.

Again the analytic character of the solution y_k is needed. But it is available, and we find

$$\begin{aligned} \log y_k &= \log \frac{1}{\sqrt{5}} + \log\left[\left(\frac{1+\sqrt{5}}{2}\right)^k - \left(\frac{1-\sqrt{5}}{2}\right)^k\right] \\ &= \log\frac{1}{\sqrt{5}} + \log\left(\frac{1+\sqrt{5}}{2}\right)^k + \log\left[1 - \left(\frac{1-\sqrt{5}}{1+\sqrt{5}}\right)^k\right] \end{aligned}$$

making
$$C = \lim \left\{\frac{\log(1/\sqrt{5})}{k} + \log\frac{1+\sqrt{5}}{2} + \frac{1}{k}\log\left[1 - \left(\frac{1-\sqrt{5}}{1+\sqrt{5}}\right)^k\right]\right\} = \log\frac{1+\sqrt{5}}{2}$$

THE NONHOMOGENEOUS CASE

18.27. The equation $y_{k+2} + a_1 y_{k+1} + a_2 y_k = b_k$ is linear and nonhomogeneous. Show that if u_k and v_k are solutions of the associated homogeneous equation (with b_k replaced by 0) with nonvanishing Wronskian and if Y_k is one *particular* solution of the equation as it stands, then every solution can be expressed as $y_k = c_1 u_k + c_2 v_k + Y_k$ where c_1 and c_2 are suitable constants.

With y_k denoting any solution of the nonhomogeneous equation, and Y_k the particular solution,

$$y_{k+2} + a_1 y_{k+1} + a_2 y_k = b_k$$

$$Y_{k+2} + a_1 Y_{k+1} + a_2 Y_k = b_k$$

and subtracting,
$$d_{k+2} + a_1 d_{k+1} + a_2 d_k = 0$$

where $d_k = y_k - Y_k$. But this makes d_k a solution of the homogeneous equation, so that $d_k = c_1 u_k + c_2 v_k$. Finally, $y_k = c_1 u_k + c_2 v_k + Y_k$ which is the required result.

18.28. By the previous problem, to find all solutions of a nonhomogeneous equation we may find just one such particular solution and attach it to the solution of the associated homogeneous problem. Follow this procedure for $y_{k+2} - y_{k+1} - y_k = Ax^k$.

When the term b_k is a power function, a solution can usually be found which is itself a power function. Here we try to determine the constant C so that $Y_k = Cx^k$.

Substitution leads to $Cx^k(x^2 - x - 1) = Ax^k$, making $C = A/(x^2 - x - 1)$. All solutions are therefore expressible as

$$y_k = c_1\left(\frac{1+\sqrt{5}}{2}\right)^k + c_2\left(\frac{1-\sqrt{5}}{2}\right)^k + \frac{Ax^k}{x^2 - x - 1}$$

Should $x^2 - x - 1 = 0$, this effort fails.

18.29. For the preceding problem, how can a particular solution Y_k be found in the case where $x^2 - x - 1 = 0$?

Try to determine C so that $Y_k = Ckx^k$.

Substitution leads to $Cx^k[(k+2)x^2 - (k+1)x - k] = Ax^k$ from which $C = A/(2x^2 - x)$. This makes $Y_k = Akx^k/(2x^2 - x)$.

18.30. For what sort of b_k term may an elementary solution Y_k be found?

Whenever b_k is a power function or a sine or cosine function, the solution Y_k has similar character. Table 18.1 makes this somewhat more precise. If the Y_k suggested in Table 18.1 includes a solution of the associated homogeneous equation, then this Y_k should be multiplied by k until no such solutions are included. Further examples of the effectiveness of this procedure will be given.

Table 18.1

b_k	Y_k
Ax^k	Cx^k
k^n	$C_0 + C_1k + C_2k^2 + \cdots + C_nk^n$
$\sin Ak$ or $\cos Ak$	$C_1 \sin Ak + C_2 \cos Ak$
k^nx^k	$x^k(C_0 + C_1k + C_2k^2 + \cdots + C_nk^n)$
$x^k \sin Ak$ or $x^k \cos Ak$	$x^k(C_1 \sin Ak + C_2 \cos Ak)$

Supplementary Problems

18.31. Given $y_{k+1} = ry_k + k$ and $y_0 = A$, compute $y_1, \ldots, y_4$ directly. Then discover the character of the solution function.

18.32. Given $y_{k+1} = -y_k + 4$ and $y_0 = 1$, compute $y_1, \ldots, y_4$ directly. What is the character of the solution function? Can you discover the solution character for arbitrary y_0?

18.33. If a debt is amortized by regular payments of size R, and is subject to interest rate i, the unpaid balance is P_k where $P_{k+1} = (1+i)P_k - R$. The initial debt being $P_0 = A$, show that $P_k = A(1+i)^k - R\dfrac{(1+i)^k - 1}{i}$. Also show that to reduce P_k to zero in exactly n payments ($P_n = 0$) we must take $R = Ai/[1 - (1+i)^{-n}]$.

18.34. Show that the difference equation $y_{k+1} = (k+1)y_k + (k+1)!$ with initial condition $y_0 = 2$ has the solution $y_k = k!\,(k+2)$.

18.35. Solve $y_{k+1} = ky_k + 2^k k!$ with $y_0 = 0$.

18.36. Apply Horner's method of Problem 18.5 to evaluate $p(x) = 1 + x + x^2 + \cdots + x^6$ at $x = \frac{1}{2}$.

18.37. Adapt Horner's method to $p(x) = x - x^3/3! + x^5/5! - x^7/7! + x^9/9!$.

18.38. Show that for $k > 0$, $(k+1)y_{k+1} + ky_k = 2k - 3$ has the solution $y_k = 1 - 2/k$.

18.39. Show that the nonlinear equation $y_{k+1} = y_k/(1 + y_k)$ has the solutions $y_k = C/(1 + Ck)$.

18.40. Solve the equation $\Delta y_k = (1/k - 1)y_k$ with initial condition $y_1 = 1$.

18.41. Compute $\psi^{(3)}(0)$, $\psi^{(3)}(1)$, and $\psi^{(3)}(2)$ from the results in Problem 18.11. What general result is indicated for integer arguments?

18.42. Evaluate $\sum_{k=1}^{\infty} 1/k(k+2)$ in terms of the ψ function.

18.43. Evaluate $\sum_{k=1}^{\infty} 1/k^2(k+2)^2$, using Problem 18.41.

18.44. Compute $\psi(\frac{1}{2})$ to three places from the series definition, using an acceleration device. Then compute $\psi(\frac{3}{2})$ and $\psi(-\frac{1}{2})$ from $\Delta\psi(x) = 1/(x+1)$.

18.45. What is the behavior of $\psi(x)$ as x approaches -1 from above?

18.46. Evaluate $\sum_{x=1}^{\infty} 1/P_3(x)$ where $P_3(x)$ is the Legendre polynomial of degree three.

18.47. Evaluate $\sum_{x=1}^{\infty} 1/T_3(x)$ where $T_3(x) = 4x^3 - 3x$ and is the Chebyshev polynomial of degree three.

18.48. Evaluate $\sum_{x=1}^{\infty} 1/P_4(x)$ where $P_4(x)$ is the Legendre polynomial of degree four.

18.49. Given $y_{k+2} + 3y_{k+1} + 2y_k = 0$ with initial conditions $y_0 = 2$, $y_1 = 1$, compute $y_2, \ldots, y_{10}$ directly.

18.50. Solve the preceding problem by the method of Problem 18.16.

18.51. Show that the solutions of $y_{k+2} - 4y_{k+1} + 4y_k = 0$ are $y_k = 2^k(c_1 + c_2k)$, where c_1, c_2 are arbitrary constants.

18.52. Find the solution family of $y_{k+2} - y_k = 0$. Also find the solution satisfying the initial conditions $y_0 = 0$, $y_1 = 1$.

18.53. Solve $y_{k+2} - 7y_{k+1} + 12y_k = \cos k$ with $y_0 = 0$, $y_1 = 0$.

18.54. Solve $4y_{k+2} + 4y_{k+1} + y_k = k^2$ with $y_0 = 0$, $y_1 = 0$.

18.55. Show that the solutions of $y_{k+2} - 2y_{k+1} + 2y_k = 0$ are

$$y_k = c_1(\sqrt{2})^k \sin\frac{\pi k}{4} + c_2(\sqrt{2})^k \cos\frac{\pi k}{4}$$

18.56. Solve $2y_{k+2} - 5y_{k+1} + 2y_k = 0$ with initial conditions $y_0 = 0$, $y_1 = 1$.

18.57. Solve $y_{k+2} + 6y_{k+1} + 25y_k = 2^k$ with $y_0 = 0$, $y_1 = 0$.

18.58. Solve $y_{k+2} - 4y_{k+1} + 4y_k = \sin k + 2^k$ with initial conditions $y_0 = y_1 = 0$.

18.59. For what values of a are the solutions of $y_{k+2} - 2y_{k+1} + (1-a)y_k = 0$ oscillatory in character?

18.60. Solve $y_{k+2} - 2y_{k+1} - 3y_k = P_2(k)$ where $P_2(k)$ is the second-degree Legendre polynomial and $y_0 = y_1 = 0$.

18.61. What is the character of the solutions of $y_{k+2} - 2ay_{k+1} + ay_k = 0$ for $0 < a < 1$? For $a = 1$? For $a > 1$?

18.62. Show that the nonlinear equation $Q_{k+1} = a - b/Q_k$ can be converted to the linear equation $y_{k+2} - ay_{k+1} + by_k = 0$ by the change of argument $Q_k = y_{k+1}/y_k$.

18.63. Show that for N even there is no solution of $y_{k+2} - y_k = 0$ satisfying the boundary conditions $y_0 = 0$, $y_N = 1$.

18.64. Show that there are infinitely many solutions of the equation of the preceding problem satisfying $y_0 = y_N \equiv 0$.

18.65. Show that there is exactly one solution of $y_{k+2} - y_k = 0$ satisfying the boundary conditions $y_0 = 0$, $y_N = 1$ if N is odd. Find this solution. Also show that there is exactly one solution satisfying $y_0 = y_N = 0$, namely $y_k = 0$.

Chapter 19

Differential Equations

THE CLASSICAL PROBLEM

Solving differential equations is one of the major problems of numerical analysis. This is because such a wide variety of applications lead to differential equations, and so few can be solved analytically. The *classical initial value problem* is to find a function $y(x)$ which satisfies the first-order differential equation $y' = f(x, y)$ and takes the initial value $y(x_0) = y_0$. A broad variety of methods have been devised for the approximate solution of this problem, most of which have then been generalized for treating higher-order problems as well. The present chapter is focused on solution methods for this one problem.

1. The **method of isoclines** is presented first. Based upon the geometrical interpretation of $y'(x)$ as the slope of the solution curve, it gives a qualitative view of the entire solution family. The function $f(x, y)$ defines the prescribed slope at each point. This "direction field" determines the character of the solution curves.
2. The **historical method of Euler** involves computing a discrete set of y_k values, for arguments x_k, using the difference equation
$$y_{k+1} = y_k + hf(x_k, y_k)$$
where $h = x_{k+1} - x_k$. This is an obvious and no too accurate approximation of $y' = f(x, y)$.
3. **More efficient algorithms** for computing solutions are then developed. Polynomial approximation is the basis of the most popular algorithms. Except for certain series methods, what is actually computed is a sequence of values y_k corresponding to a discrete set of arguments x_k, as in the Euler method. Most methods are equivalent to the replacement of the given differential equation by a difference equation. The particular difference equation obtained depends upon the choice of polynomial approximation.
4. The **Taylor series** is heavily used. If $f(x, y)$ is an analytic function the successive derivatives of $y(x)$ may be obtained and the series for $y(x)$ written out in standard Taylor format. Sometimes a single series will serve for all arguments of interest. In other problems a single series may converge too slowly to produce the required accuracy for all arguments of interest and several Taylor series with different points of expansion may be used. The eventual truncation of any such series means that the solution is being approximated by a Taylor polynomial.
5. **Runge–Kutta methods** were developed to avoid the computation of high-order derivatives which the Taylor method may involve. In place of these derivatives extra values of the given function $f(x, y)$ are used, in a way which duplicates the accuracy of a Taylor polynomial. The most common formulas are
$$k_1 = hf(x, y)$$
$$k_2 = hf\left(x + \frac{1}{2}h,\ y + \frac{1}{2}k_1\right)$$
$$k_3 = hf\left(x + \frac{1}{2}h,\ y + \frac{1}{2}k_2\right)$$
$$k_4 = hf(x + h,\ y + k_3)$$
$$y(x + h) \simeq y(x) + \frac{1}{6}(k_1 + 2k_2 + 2k_3 + k_4)$$
but there are numerous variations.

6. **Predictor–corrector methods** involve the use of one formula to make a prediction of the next y_k value, followed by the application of a more accurate corrector formula which then provides successive improvements. Though slightly complex, such methods have the advantage that from successive approximations to each y_k value an estimate of the error may be made. A simple predictor–corrector pair is

$$y_{k+1} \simeq y_k + hy'_k$$

$$y_{k+1} \simeq y_k + \frac{1}{2}h(y'_k + y'_{k+1})$$

the predictor being Euler's formula and the corrector being known as the modified Euler formula. Since $y'_k = f(x_k, y_k)$ and $y'_{k+1} = f(x_{k+1}, y_{k+1})$ the predictor first estimates y_{k+1}. This estimate then leads to a y'_{k+1} value and then to a corrected y_{k+1}. Further corrections of y'_{k+1} and y_{k+1} successively can be made until a satisfactory result is achieved.

7. The **Milne method** uses the predictor–corrector pair

$$y_{k+1} \simeq y_{k-3} + \frac{4h}{3}(2y'_{k-2} - y'_{k-1} + 2y'_k)$$

$$y_{k+1} \simeq y_{k-1} + \frac{h}{3}(y'_{k+1} + 4y'_k + y'_{k-1})$$

in which Simpson's rule is easily recognized. It requires four previous values $(y_k, y_{k-1}, y_{k-2}, y_{k-3})$ to prime it. These must be obtained by a different method, often the Taylor series.

8. The **Adams method** uses the predictor–corrector pair

$$y_{k+1} \simeq y_k + \frac{h}{24}(55y'_k - 59y'_{k-1} + 37y'_{k-2} - 9y'_{k-3})$$

$$y_{k+1} \simeq y_k + \frac{h}{24}(9y'_{k+1} + 19y'_k - 5y'_{k-1} + y'_{k-2})$$

and like the Milne method requires four previous values.

ERROR

Truncation error is made when a partial sum is used to approximate the value of an infinite series and this is perhaps the original use of the term, which is now used more loosely. When a differential equation is replaced by a difference equation, a local truncation error is made with each forward step from k to $k+1$. These local errors then blend together in some obscure way to produce the cumulative or global truncation error. It is rarely possible to follow error development through a differential equations algorithm with any realism but some rough estimates are possible.

A **convergent method** is one which, when continually refined (more and more terms of a series being used, or smaller and smaller intervals between successive arguments), yields a sequence of approximate solutions converging to the exact solution. The Taylor, Runge–Kutta, and some predictor–corrector methods will be proved convergent under appropriate circumstances. Convergence proofs deal with truncation error only, ignoring the problem of roundoffs.

Roundoff error is, needless to say, present in all these methods, sometimes in an important way. It is more elusive than truncation error and very limited success has rewarded the efforts made to analyze it.

The **relative error** of an approximation, the ratio of error to exact solution, is usually of greater interest than the error itself, since if the solution grows larger, then a larger error can probably be tolerated. Even more important, if the exact solution diminishes, then error must do the same or it will overwhelm the solution and computed results will be meaningless. The simple problem $y' = Ay$,

$y(0) = 1$, for which the exact solution is $y = e^{Ax}$, often serves as a test case for tracing relative error behavior in our various methods. One hopes that information obtained in this way will have some relevance for the use of the same methods on the general equation $y' = f(x, y)$. This may seem optimistic, but the study of error has its limitations.

A **stable method** is one for which relative error remains bounded, hopefully by its initial value. This is a strong requirement and one that may be hard to verify. Also, a method may be stable for some equations and unstable for others. Only partial results can be offered, in particular for the equation $y' = Ay$.

Error monitoring refers to a step-by-step effort to measure local truncation error and to use this information for determining whether or not the current step size is suitable. With predictor–corrector methods, a practical error estimate can be made using the predicted and corrected values. With Runge–Kutta methods, a parallel computation using double step size leads to an error estimate much as in adaptive integration. Here, as there, the objective is to attain a final result of specified accuracy with minimum effort.

Solved Problems

THE METHOD OF ISOCLINES

19.1. Use the *method of isoclines* to determine the qualitative behavior of the solutions of $y'(x) = xy^{1/3}$.

This equation can of course be solved by elementary methods but we shall use it as a test case for various approximation methods. The method of isoclines is based on the family of curves $y'(x) =$ constant which are not themselves solutions but are helpful in determining the character of solutions. In this example the isoclines are the family $xy^{1/3} = M$ where M is the constant value of $y'(x)$. Some of these curves are sketched (dashed) in Fig. 19-1, with M values indicated. Where a solution of the differential equation crosses one of these isoclines it must have for its slope the M number of that isocline. A few solution curves are also included (solid) in Fig. 19-1. Others can be sketched in, at least roughly.

Accuracy is not the goal of the isocline method but rather the general character of the solution family. For example, there is symmetry about both axes. One solution through $(0, 0)$ and those above it have a U shape. Solutions below this are more unusual. Along $y = 0$ different solutions can come together. A solution can even include a piece of the x axis. One such solution might enter $(0, 0)$ on a descending arc, follow the x axis to $(2, 0)$ and then start upwards again as shown in Fig. 19-2. The possible combinations of line and arc are countless. Information of this sort is often a useful guide when efforts to compute accurate solutions are made.

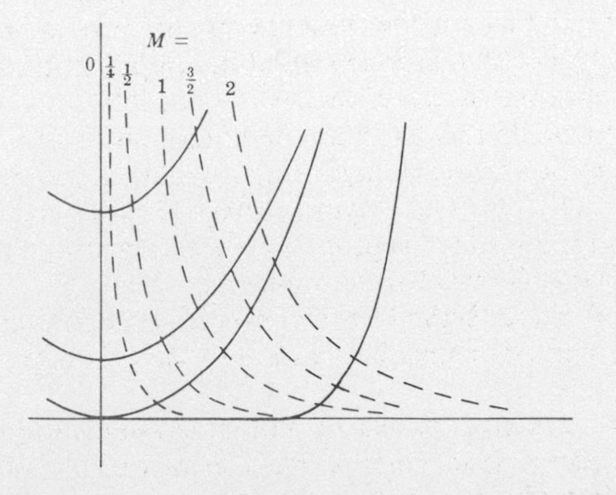

Fig. 19-1

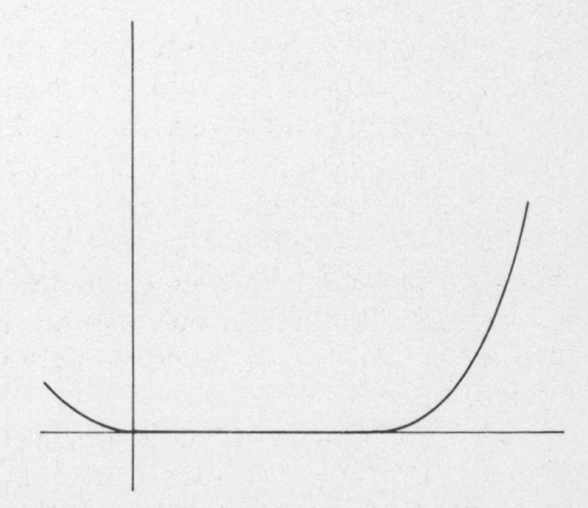

Fig. 19-2

THE EULER METHOD

19.2. Illustrate the simplest Euler method for computing a solution of

$$y' = f(x, y) = xy^{1/3} \qquad y(1) = 1$$

This is perhaps the original device for converting the method of isoclines into a computational scheme. It uses the formula

$$y_{k+1} - y_k = \int_{x_k}^{x_{k+1}} y'\,dx \simeq hy'_k$$

which amounts to considering y' constant between x_k and x_{k+1}. It also amounts to the linear part of a Taylor series, so that if y_k and y'_k were known exactly the error in y_{k+1} would be $\frac{1}{2}h^2y^{(2)}(\xi)$. This is called the *local truncation error,* since it is made *in this step* from x_k to x_{k+1}. Since it is fairly large, it follows that rather small increments h would be needed for high accuracy.

The formula is seldom used in practice but serves to indicate the nature of the task ahead and some of the difficulties to be faced. With x_0, $y_0 = 1$ three applications of this Euler formula, using $h = .01$, bring

$$y_1 \simeq 1 + (.01)(1) = 1.0100$$
$$y_2 \simeq 1.0100 + (.01)(1.01)(1.0033) \simeq 1.0201$$
$$y_3 \simeq 1.0201 + (.01)(1.02)(1.0067) \simeq 1.0304$$

Near $x = 1$ we have $y^{(2)} = y^{1/3} + \frac{1}{3}xy^{-2/3}(xy^{1/3}) \simeq \frac{4}{3}$, which makes the truncation error in each step about .00007. After three such errors, the fourth decimal place is already open to suspicion. A smaller increment h is necessary if we hope for greater accuracy. The accumulation of truncation error is further illustrated in Fig. 19-3 where the computed points have been joined to suggest a solution curve. Our approximation amounts to following successively the tangent lines to various solutions of the equation. As a result the approximation tends to follow the convex side of the solution curve. Notice also that Euler's formula is a nonlinear difference equation of order one: $y_{k+1} = y_k + hx_k y_k^{1/3}$.

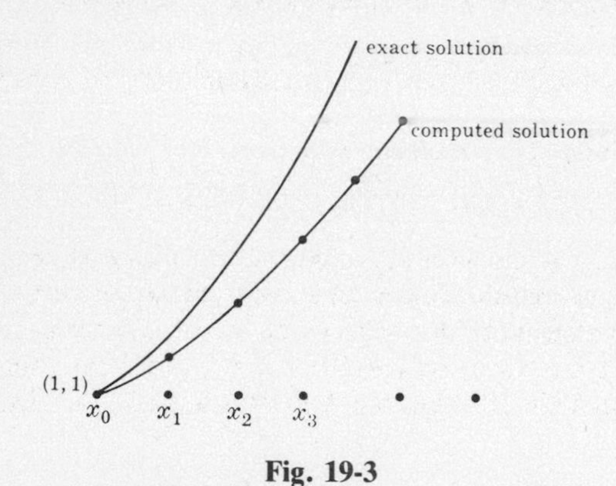

Fig. 19-3

19.3. Illustrate the concept of convergence by comparing the results of applying Euler's method with $h = .10$, $.05$, and $.01$ with the correct solution $y = [(x^2 + 2)/3]^{3/2}$.

Convergence refers to the improvement of approximations as the interval h tends to zero. A method which does not converge is of doubtful value as an approximation scheme. Convergence for the various schemes to be introduced will be proved later, but as circumstantial evidence the data of Table 19.1, obtained by Euler's method, are suggestive. Only values for integer x arguments are included, all others being suppressed for brevity.

Notice that across each row there is a reassuring trend toward the exact value. Using smaller intervals means more computing. The value 25.96 in the bottom row, for instance, was obtained in 50 steps whereas the value 26.89 required 500 steps. The extra labor has brought an improvement, which seems only fair. As h tends to zero the computation grows even longer and we hope that the results

Table 19.1

x	$h=.10$	$h=.05$	$h=.01$	Exact
1	1.00	1.00	1.00	1.00
2	2.72	2.78	2.82	2.83
3	6.71	6.87	6.99	7.02
4	14.08	14.39	14.63	14.70
5	25.96	26.48	26.89	27.00

approach the exact values as limits. This is the convergence concept. Needless to say, roundoff errors will limit the accuracy attainable but they are not a part of the convergence issue.

THE TAYLOR METHOD

19.4. Apply the local Taylor series method to obtain a solution of $y' = xy^{1/3}$, $y(1) = 1$ correct to three places for arguments up to $x = 5$.

Generally speaking the method involves using $p(x+h)$ in place of $y(x+h)$, where $p(x)$ is the Taylor polynomial for argument x. We may write directly

$$y(x+h) \simeq y(x) + hy'(x) + \frac{1}{2}h^2y^{(2)}(x) + \frac{1}{6}h^3y^{(3)}(x) + \frac{1}{24}h^4y^{(4)}(x)$$

accepting a local truncation error of amount $E = h^5y^{(5)}(\xi)/120$.

The higher derivatives of $y(x)$ are computed from the differential equation:

$$y^{(2)}(x) = \frac{1}{3}x^2y^{-1/3} + y^{1/3} \qquad y^{(3)}(x) = -\frac{1}{9}x^3y^{-1} + xy^{-1/3} \qquad y^{(4)}(x) = \frac{1}{9}x^4y^{-5/3} - \frac{2}{3}x^2y^{-1} + y^{-1/3}$$

The initial condition $y(1) = 1$ has been prescribed, so with $x = 1$ and $h = .1$ we find

$$y(1+.1) \simeq 1 + .1 + \frac{2}{3}(.1)^2 + \frac{4}{27}(.1)^3 + \frac{1}{54}(.1)^4 \simeq 1.10682$$

Next apply the Taylor formula at $x = 1.1$ and find

$$y(1.1+.1) \simeq 1.22788 \qquad y(1.1-.1) \simeq 1.00000$$

The second of these serves as an accuracy check since it reproduces our first result to five-place accuracy. (This is the same procedure used in Chapter 14 for the error function integral.) Continuing in this way, the results presented in Table 19.2 are obtained. The exact solution is again included for comparison. Though $h = .1$ was used, only values for $x = 1(.5)5$ are listed. Notice that the errors are much smaller than those made in the Euler method with $h = .01$. The Taylor method is a more rapidly convergent algorithm.

Table 19.2

x	Taylor result	Exact result	Error
1.0	1.00000	1.00000	
1.5	1.68618	1.68617	−1
2.0	2.82846	2.82843	−3
2.5	4.56042	4.56036	−6
3.0	7.02123	7.02113	−10
3.5	10.35252	10.35238	−14
4.0	14.69710	14.69694	−16
4.5	20.19842	20.19822	−20
5.0	27.00022	27.00000	−22

19.5. Apply the Taylor method to $y' = -xy^2$ to obtain the solution satisfying $y(0) = 2$.

The procedure of the preceding problem could be applied. Instead, however, an alternative will be illustrated, essentially a method of undetermined coefficients. Assuming convergence at the outset, we write the Taylor series $y(x) = \sum_{i=0}^{\infty} a_i x^i$. Then

$$y^2(x) = \left(\sum_{i=0}^{\infty} a_i x^i\right)\left(\sum_{j=0}^{\infty} a_j x^j\right) = \sum_{k=0}^{\infty}\left(\sum_{i=0}^{k} a_i a_{k-i}\right)x^k \qquad y'(x) = \sum_{i=1}^{\infty} i a_i x^{i-1}$$

Substituting into the differential equation and making minor changes in the indices of summation,

$$\sum_{j=0}^{\infty}(j+1)a_{j+1}x^j = -\sum_{j=1}^{\infty}\left(\sum_{i=0}^{j-1} a_i a_{j-1-i}\right)x^j$$

Comparing coefficients of x^j makes $a_1 = 0$ and

$$(j+1)a_{j+1} = -\sum_{i=0}^{j-1} a_i a_{j-1-i} \qquad \text{for } j = 1, 2, \ldots$$

The initial condition forces $a_0 = 2$ and then we find recursively

$$a_2 = -\frac{1}{2}a_0^2 = -2 \qquad a_6 = -\frac{1}{6}(2a_0a_4 + 2a_1a_3 + a_2^2) = -2$$

$$a_3 = -\frac{1}{3}(2a_0a_1) = 0 \qquad a_7 = -\frac{1}{7}(2a_0a_5 + 2a_1a_4 + 2a_2a_3) = 0$$

$$a_4 = -\frac{1}{4}(2a_0a_2 + a_1^2) = 2 \qquad a_8 = -\frac{1}{8}(2a_0a_6 + 2a_1a_5 + 2a_2a_4 + a_3^2) = 2$$

$$a_5 = -\frac{1}{5}(2a_0a_3 + 2a_1a_2) = 0$$

and so on. The recursion can be programmed so that coefficients could be computed automatically as far as desired. The indicated series is

$$y(x) = 2(1 - x^2 + x^4 - x^6 + x^8 - \cdots)$$

Since the exact solution is easily found to be $y(x) = 2/(1+x^2)$, the series obtained is no surprise.

This method sees frequent application. The principal assumption involved is that the solution does actually have a series representation. In this case the series converges only for $-1 < x < 1$. For $-\frac{1}{2} < x < \frac{1}{2}$ only six terms are needed to give three-place accuracy. In the previous problem a new Taylor polynomial was used for each value computed. Here just one such polynomial is enough. The issue is one of range and accuracy required. To proceed up to $x = 5$, for example, the earlier method can be used. In further contrast we may also note that in Problem 19.4 polynomials of fixed degree are used and the convergence issue does not arise explicitly. Here in Problem 19.5 we introduce the entire series into the differential equation, assuming $y(x)$ analytic in the interval of interest.

RUNGE–KUTTA METHODS

19.6. Find coefficients a, b, c, d, m, n, and p in order that the Runge–Kutta formulas

$$k_1 = hf(x, y)$$
$$k_2 = hf(x + mh, y + mk_1)$$
$$k_3 = hf(x + nh, y + nk_2)$$
$$k_4 = hf(x + ph, y + pk_3)$$
$$y(x+h) - y(x) \simeq ak_1 + bk_2 + ck_3 + dk_4$$

duplicate the Taylor series through the term in h^4. Note that the last formula, though not a polynomial approximation, is then near the Taylor polynomial of degree four.

We begin by expressing the Taylor series in a form which facilitates comparisons. Let

$$F_1 = f_x + ff_y \qquad F_2 = f_{xx} + 2ff_{xy} + f^2 f_{yy} \qquad F_3 = f_{xxx} + 3ff_{xxy} + 3f^2 f_{xyy} + f^3 f_{yyy}$$

Then differentiating the equation $y' = f(x, y)$, we find

$$y^{(2)} = f_x + f_y y' = f_x + f_y f = F_1$$

$$y^{(3)} = f_{xx} + 2ff_{xy} + f^2 f_{yy} + f_y(f_x + ff_y) = F_2 + f_y F_1$$

$$y^{(4)} = f_{xxx} + 3ff_{xxy} + 3f^2 f_{xyy} + f^3 f_{yyy} + f_y(f_{xx} + 2ff_{xy} + f^2 f_{yy}) + 3(f_x + ff_y)(f_{xy} + ff_{yy}) + f_y^2(f_x + ff_y)$$

$$= F_3 + f_y F_2 + 3F_1(f_{xy} + ff_{yy}) + f_y^2 F_1$$

which allows the Taylor series to be written as

$$y(x+h) - y(x) = hf + \frac{1}{2}h^2 F_1 + \frac{1}{6}h^3(F_2 + f_y F_1) + \frac{1}{24}h^4[F_3 + f_y F_2 + 3(f_{xy} + ff_{yy})F_1 + f_y^2 F_1] + \cdots$$

Turning now to the various k values, similar computations produce

$$k_1 = hf$$

$$k_2 = h\left[f + mhF_1 + \frac{1}{2}m^2h^2F_2 + \frac{1}{6}m^3h^3F_3 + \cdots\right]$$

$$k_3 = h\left[f + nhF_1 + \frac{1}{2}h^2(n^2F_2 + 2mnf_yF_1) + \frac{1}{6}h^3(n^3F_3 + 3m^2nf_yF_2 + 6mn^2(f_{xy} + ff_{yy})F_1) + \cdots\right]$$

$$k_4 = h\left[f + phF_1 + \frac{1}{2}h^2(p^2F_2 + 2npf_yF_1) + \frac{1}{6}h^3(p^3F_3 + 3n^2pf_yF_2 + 6np^2(f_{xy} + ff_{yy})F_1 + 6mnpf_y^2F_1) + \cdots\right]$$

Combining these as suggested by the final Runge–Kutta formula,

$$y(x+h) - y(x) = (a+b+c+d)hf + (bm + cn + dp)h^2F_1 + \frac{1}{2}(bm^2 + cn^2 + dp^2)h^3F_2$$

$$+ \frac{1}{6}(bm^3 + cn^3 + dp^3)h^4F_3 + (cmn + dnp)h^3 f_y F_1 + \frac{1}{2}(cm^2n + dn^2p)h^4 f_y F_2$$

$$+ (cmn^2 + dnp^2)h^4(f_{xy} + ff_{yy})F_1 + dmnph^4 f_y^2 F_1 + \cdots$$

Comparison with the Taylor series now suggests the eight conditions

$$a + b + c + d = 1 \qquad cmn + dnp = \frac{1}{6}$$

$$bm + cn + dp = \frac{1}{2} \qquad cmn^2 + dnp^2 = \frac{1}{8}$$

$$bm^2 + cn^2 + dp^2 = \frac{1}{3} \qquad cm^2n + dn^2p = \frac{1}{12}$$

$$bm^3 + cn^3 + dp^3 = \frac{1}{4} \qquad dmnp = \frac{1}{24}$$

These eight equations in seven unknowns are actually somewhat redundant. The classical solution set is

$$m = n = \frac{1}{2} \qquad p = 1 \qquad a = d = \frac{1}{6} \qquad b = c = \frac{1}{3}$$

leading to the Runge–Kutta formulas

$$k_1 = hf(x, y) \qquad k_2 = hf\left(x + \frac{1}{2}h, y + \frac{1}{2}k_1\right) \qquad k_3 = hf\left(x + \frac{1}{2}h, y + \frac{1}{2}k_2\right)$$

$$k_4 = hf(x + h, y + k_3) \qquad y(x+h) \simeq y(x) + \frac{1}{6}(k_1 + 2k_2 + 2k_3 + k_4)$$

It is of some interest to notice that for $f(x, y)$ independent of y this reduces to Simpson's rule applied to $y'(x) = f(x)$.

19.7. What is the advantage of Runge–Kutta formulas over the Taylor method?

Though approximately the same as the Taylor polynomial of degree four, these formulas do not require prior calculation of the higher derivatives of $y(x)$, as the Taylor method does. Since the differential equations arising in applications are often complicated, the calculation of derivatives can be onerous. The Runge–Kutta formulas involve computation of $f(x, y)$ at various positions instead and this function occurs in the given equation. The method is very extensively used.

19.8. Apply the Runge–Kutta formula to $y' = f(x, y) = xy^{1/3}$, $y(1) = 1$.

With $x_0 = 1$ and $h = .1$ we find

$$k_1 = (.1)f(1, 1) = .1 \qquad k_3 = (.1)f(1.05, 1.05336) \simeq .10684$$
$$k_2 = (.1)f(1.05, 1.05) \simeq .10672 \qquad k_4 = (.1)f(1.1, 1.10684) \simeq .11378$$

from which we compute

$$y_1 = 1 + \frac{1}{6}(.1 + .21344 + .21368 + .11378) \simeq 1.10682$$

This completes one step and we begin another with x_1 and y_1 in place of x_0 and y_0, and continue in this way. Since the method duplicates the Taylor series through h^4, it is natural to expect results similar to those found by the Taylor method. Table 19.3 makes a few comparisons and we do find differences in the last two places. These are partly explained by the fact that the local truncation errors of the two methods are not identical. Both are of the form Ch^5, but the factor C is not the same. Also, roundoff errors usually differ even between algorithms which are algebraically identical, which these are not. Here the advantage is clearly with the Runge–Kutta formulas.

Table 19.3

x	Taylor	Runge–Kutta	Exact
1	1.00000	1.00000	1.00000
2	2.82846	2.82843	2.82843
3	7.02123	7.02113	7.02113
4	14.69710	14.69693	14.69694
5	27.00022	26.99998	27.00000

19.9. Illustrate variations of the Runge–Kutta formulas.

It is not hard to verify that

$$y(x + h) = y(x) + hf\left(x + \frac{1}{2}h, y + \frac{1}{2}hf(x, h)\right)$$

in which y denotes $y(x)$, duplicates the Taylor series through terms of second degree. (See Problem 19.63). It is, therefore, known as a Runge–Kutta method of order two. Similarly,

$$k_1 = hf(x, y)$$
$$k_2 = hf\left(x + \frac{1}{2}h, y + \frac{1}{2}k_1\right)$$
$$k_3 = hf(x + h, y - k_1 + 2k_2)$$
$$y(x + h) = y(x) + \frac{1}{6}(k_1 + 4k_2 + k_3)$$

has order three. Other methods of order two and three also exist. The set

$$k_1 = hf(x, y)$$
$$k_2 = hf\left(x + \frac{1}{2}h, y + \frac{1}{2}k_1\right)$$
$$k_3 = hf\left(x + \frac{1}{2}h, y + \frac{1}{4}k_1 + \frac{1}{4}k_2\right)$$
$$k_4 = hf(x + h, y - k_2 + 2k_3)$$
$$y(x + h) = y(x) + \frac{1}{6}(k_1 + 4k_3 + k_4)$$

is an alternate method of order four, while the more exotic

$$k_1 = f(x, y)$$
$$k_2 = hf\left(x + \frac{1}{2}h, y + \frac{1}{2}k_1\right)$$
$$k_3 = hf\left(x + \frac{1}{2}h, y + \frac{1}{4}k_1 + \frac{1}{4}k_2\right)$$
$$k_4 = hf(x + h, y - k_2 + 2k_3)$$
$$k_5 = hf\left(x + \frac{2}{3}h, y + \frac{7}{27}k_1 + \frac{10}{27}k_2 + \frac{1}{27}k_4\right)$$
$$k_6 = hf\left(x + \frac{1}{5}h, y + \frac{28}{625}k_1 - \frac{1}{5}k_2 + \frac{546}{625}k_3 + \frac{54}{625}k_4 - \frac{378}{625}k_5\right)$$
$$y(x + h) = y(x) + \frac{1}{24}k_1 + \frac{5}{48}k_4 + \frac{27}{56}k_5 + \frac{125}{336}k_6$$

has order five. The higher the order, the greater is the diversity of possible methods, and the lower the truncation error. A method of order n duplicates the Taylor series through terms of nth degree, and so has truncation error

$$T = cy^{(n+1)}h^{n+1}$$

which means that for a smooth function $y(x)$ the computation can proceed with a relatively large h and progress more rapidly. The development of high-order methods involves some strenuous algebra, and it has been feasible only with the aid of computer programs for doing the manipulations.

CONVERGENCE OF THE TAYLOR METHOD

19.10. The equation $y' = y$ with $y(0) = 1$ has the exact solution $y(x) = e^x$. Show that the approximate values y_k obtained by the Taylor method converge to this exact solution for h tending to zero and p fixed. (The more familiar convergence concept keeps h fixed and lets p tend to infinity.)

The Taylor method involves approximating each correct value y_{k+1} by

$$Y_{k+1} = Y_k + hY'_k + \frac{1}{2}h^2Y_k^{(2)} + \cdots + \frac{1}{p!}h^pY_k^{(p)}$$

For the present problem all the derivatives are the same, making

$$Y_{k+1} = \left(1 + h + \frac{1}{2}h^2 + \cdots + \frac{1}{p!}h^p\right)Y_k = rY_k$$

When $p = 1$ this reduces to the Euler method. In any case it is a difference equation of order one. Its solution with $Y_0 = 1$ is

$$Y_k = r^k = \left(1 + h + \frac{1}{2}h^2 + \cdots + \frac{1}{p!}h^p\right)^k$$

But by Taylor's polynomial formula,

$$e^h = 1 + h + \frac{1}{2}h^2 + \cdots + \frac{1}{p!}h^p + \frac{h^{p+1}}{(p+1)!}e^{\xi h}$$

with ξ between 0 and 1. Now recalling the identity

$$a^k - r^k = (a-r)(a^{k-1} + a^{k-2}r + \cdots + ar^{k-2} + r^{k-1})$$

we find for the case $a > r > 0$,
$$a^k - r^k < (a-r)ka^{k-1}$$

Choosing $a = e^h$ and r as above, this last inequality becomes

$$0 < e^{kh} - Y_k < \frac{h^{p+1}}{(p+1)!}e^{\xi h}ke^{(k-1)h} < \frac{kh^{p+1}}{(p+1)!}e^{kh}$$

the last step being a consequence of $0 < \xi < 1$. The question of convergence concerns the behavior of values computed for a fixed argument x as h tends to zero. Accordingly we put $x_k = kh$ and rewrite our last result as

$$0 < e^{x_k} - Y_k < \frac{h^p}{(p+1)!}x_k e^{x_k}$$

Now choose a sequence of step sizes h, in such a way that x_k recurs endlessly in the finite argument set of each computation. (The simplest way is to continually halve h.) By the above inequality the sequence of Y_k values obtained at the fixed x_k argument converges to the exact e^{x_k} as h^p. The practical implication is, of course, that the smaller h is chosen the closer the computed result draws to the exact solution. Naturally roundoff errors, which have not been considered in this problem, will limit the accuracy attainable.

19.11. How does the error of the Taylor approximation, as developed in the previous problem, behave for a fixed step size as k increases, in other words as the computation is continued to larger and larger amounts?

Note that this is not a convergence question, since h is fixed. It is a question of how the error, due to truncation of the Taylor series at the term h^p, accumulates as the computation continues. By the last inequality we see that the error contains the true solution as a factor. Actually it is the relative error which may be more significant, since it is related to the number of significant digits in our computed values. We find

$$\text{Relative error} = \left|\frac{e^{x_k} - Y_k}{e^{x_k}}\right| < \frac{h^p}{(p+1)!}x_k$$

which, for fixed h, grows linearly with x_k.

19.12. Prove the convergence of the Taylor method for the general first-order equation $y' = f(x, y)$ with initial condition $y(x_0) = y_0$ under appropriate assumptions on $f(x, y)$.

This generalizes the result of Problem 19.10. Continuing to use capital Y for the approximate solution, the Taylor method makes

$$Y_{k+1} = Y_k + hY'_k + \frac{1}{2}h^2Y_k^{(2)} + \cdots + \frac{1}{p!}h^pY_k^{(p)}$$

where all entries $Y_k^{(i)}$ are computed from the differential equation. For example,

$$Y'_k = f(x_k, Y_k) \qquad Y_k^{(2)} = f_x(x_k, Y_k) + f_y(x_k, Y_k)f(x_k, Y_k) = f'(x_k, Y_k)$$

and suppressing arguments for brevity,

$$Y_k^{(3)} = f_{xx} + 2f_{xy}f + f_{yy}f^2 + (f_x + f_yf)f_y = f''(x_k, Y_k)$$

it being understood that f and its derivatives are evaluated at x_k, Y_k and that Y_k denotes the computed value at arguments x_k. The other $Y_k^{(i)}$ are obtained from similar, but more involved, formulas. If we use $y(x)$ to represent the exact solution of the differential problem, then Taylor's formula offers a similar expression for $y(x_{k+1})$,

$$y(x_{k+1}) = y(x_k) + hy'(x_k) + \frac{1}{2}h^2 y^{(2)}(x_k) + \cdots + \frac{1}{p!}h^p y^{(p)}(x_k) + \frac{h^{p+1}}{(p+1)!}y^{(p+1)}(\xi)$$

provided the exact solution actually has such derivatives. As usual ξ is between x_k and x_{k+1}. In view of $y'(x) = f(x, y(x))$, we have

$$y'(x_k) = f(x_k, y(x_k))$$

and differentiating,

$$y^{(2)}(x_k) = f_x(x_k, y(x_k)) + f_y(x_k, y(x_k))f(x_k, y(x_k)) = f'(x_k, y(x_k))$$

In the same way

$$y^{(3)}(x_k) = f''(x_k, y(x_k))$$

and so on. Subtraction now brings

$$y(x_{k+1}) - Y_{k+1} = y(x_k) - Y_k + h[y'(x_k) - Y_k'] + \frac{1}{2}h^2[y^{(2)}(x_k) - Y_k^{(2)}]$$

$$+ \cdots + \frac{1}{p!}h^p[y^{(p)}(x_k) - Y_k^{(p)}] + \frac{h^{p+1}}{(p+1)!}y^{(p+1)}(\xi)$$

Now notice that if $f(x, y)$ satisfies a Lipschitz condition,

$$|y'(x_k) - Y_k'| = |f(x_k, y(x_k)) - f(x_k, Y_k)| \leqq L\,|y(x_k) - Y_k|$$

We will further assume that $f(x, y)$ is such that

$$|y^{(i)}(x_k) - Y_k^{(i)}| = |f^{(i-1)}(x_k, y(x_k)) - f^{(i-1)}(x_k, Y_k)| \leqq L\,|y(x_k) - Y_k|$$

This can be proved to be true, for instance, for $i = 1, \ldots, p$ if $f(x, y)$ has continuous derivatives through order $p + 1$. This same condition also guarantees that the exact solution $y(x)$ has continuous derivatives through order $p + 1$, a fact assumed above. Under these assumptions on $f(x, y)$ we now let $d_k = y(x_k) - Y_k$ and have

$$|d_{k+1}| \leqq |d_k|\left(1 + hL + \frac{1}{2}h^2L + \cdots + \frac{1}{p!}h^pL\right) + \frac{h^{p+1}}{(p+1)!}B$$

where B is a bound on $|y^{p+1}(x)|$. For brevity, this can be rewritten as

$$|d_{k+1}| \leqq (1 + \alpha)\,|d_k| + \beta$$

where

$$\alpha = L\left(h + \frac{1}{2}h^2 + \cdots + \frac{1}{p!}h^p\right) \qquad \beta = \frac{h^{p+1}}{(p+1)!}B$$

We now prove that

$$|d_k| \leqq \beta\frac{e^{k\alpha} - 1}{\alpha}$$

The numbers α and β are positive. Since the exact and approximate solutions both satisfy the initial condition, $d_0 = 0$ and the last inequality holds for $k = 0$. To prove it by induction we assume it for some nonnegative integer k and find

$$|d_{k+1}| \leqq (1 + \alpha)\beta\frac{e^{k\alpha} - 1}{\alpha} + \beta = \frac{(1 + \alpha)e^{k\alpha} - 1}{\alpha}\beta < \frac{e^{(k+1)\alpha} - 1}{\alpha}\beta$$

the last step following since $1 + \alpha < e^{\alpha}$. The induction is therefore valid and the inequality holds for nonnegative integers k. Since $\alpha = Lh + \epsilon h < Mh$ where ϵ tends to zero with h, we can replace L by the slightly larger M and obtain

$$|y(x_k) - Y_k| \leqq \frac{h^pB}{(p+1)!} \cdot \frac{e^{M(x_k - x_0)} - 1}{M}$$

with the usual change of argument $x_k = x_0 + kh$, so that convergence is again like h^p.

19.13. What does the result of Problem 19.12 tell about the error for fixed h as the computation continues to larger arguments x_k?

The result is adequate for proving convergence, but since the exact solution is unknown it does not lead at once to an estimate of the relative error. Further error analysis and an extrapolation to the limit process have been explored.

19.14. Are Runge–Kutta methods also convergent?

Since these methods duplicate the Taylor series up to a point (in our example up to the term in h^4), the proof of convergence is similar to that just offered for the Taylor method itself. The details are more complicated and will be omitted.

THE PREDICTOR–CORRECTOR METHOD

19.15. Derive the modified Euler formula $y_{k+1} \simeq y_k + \frac{1}{2}h(y'_k + y'_{k+1})$ and its local truncation error.

The formula can be produced by applying the trapezoidal rule to the integration of y' as follows:

$$y_{k+1} - y_k = \int_{x_k}^{x_{k+1}} y'\,dx \simeq \frac{1}{2}h(y'_k + y'_{k+1})$$

By Problem 14.66, the error in this application of the trapezoidal rule to y' will be $-h^3y^{(3)}(\xi)/12$, and this is the local truncation error. (Recall that local truncation error refers to error introduced by the approximation made in the step from x_k to x_{k+1}, that is, in the integration process. Effectively we pretend that y_k and earlier values are known correctly.) Comparing our present result with that for the simpler Euler method, we of course find the present error substantially smaller. This may be viewed as the natural reward for using the trapezoidal rule rather than a still more primitive integation rule. It is also interesting to note that instead of treating y' as constant between x_k and x_{k+1}, so that $y(x)$ is supposed linear, we now treat y' as linear in this interval, so that $y(x)$ is supposed quadratic.

19.16. Apply the modified Euler formula to the problem $y' = xy^{1/3}$, $y(1) = 1$.

Though this method is seldom used for serious computing, it serves to illustrate the nature of the predictor–corrector method. Assuming y_k and y'_k already in hand, the two equations

$$y_{k+1} \simeq y_k + \frac{1}{2}h(y'_k + y'_{k+1}) \qquad y'_{k+1} = f(x_{k+1}, y_{k+1})$$

are used to determine y_{k+1} and y'_{k+1}. An iterative algorithm much like those to be presented in Chapter 25 for determining roots of equations will be used. Applied successively, beginning with $k = 0$, this algorithm generates sequences of values y_k and y'_k. It is also interesting to recall a remark made in the solution of the previous problem, that we are treating $y(x)$ as though it were quadratic between the x_k values. Our overall approximation to $y(x)$ may thus be viewed as a chain of parabolic segments. Both $y(x)$ and $y'(x)$ will be continuous, while $y''(x)$ will have jumps at the "corner points" (x_k, y_k).

To trigger each forward step of our computation, the simpler Euler formula will be used as a *predictor*. It provides a first estimate of y_{k+1}. Here, with $x_0 = 1$ and $h = .05$ it offers

$$y(1.05) \simeq 1 + (.05)(1) = 1.05$$

The differential equation then presents us with

$$y'(1.05) \simeq (1.05)(1.016) \simeq 1.0661$$

Now the modified Euler formula serves as a corrector, yielding

$$y(1.05) \simeq 1 + (.025)(1 + 1.0661) \simeq 1.05165$$

With this new value the differential equation corrects $y'(1.05)$ to 1.0678, after which the corrector is reapplied and produces

$$y(1.05) \simeq 1 + (.025)(1 + 1.0678) \simeq 1.0517$$

Another cycle reproduces these four-place values, so we stop. This iterative use of the corrector formula, together with the differential equation, is the core of the predictor–corrector method. One iterates until convergence occurs, assuming it will. (See Problem 19.29 for a proof.) It is then time for the next step forward, again beginning with a single application of the predictor formula. Since more powerful predictor–corrector formulas are now to be obtained, we shall not continue the present computation further. Notice, however, that the one result we have is only two units too small in the last place, verifying that our corrector formula is more accurate than the simpler Euler predictor, which was barely yielding four-place accuracy with $h = .01$. More powerful predictor–corrector combinations will now be developed.

19.17. Derive the "predictor" formula $y_{k+1} \simeq y_{k-3} + \frac{4}{3}h(2y'_{k-2} - y'_{k-1} + 2y'_k)$.

Earlier (Chapter 14) we integrated a collocation polynomial over the entire interval of collocation (Cotes formulas) and also over just a part of that interval (formulas with end corrections). The second procedure leads to more accurate, if more troublesome, results. Now we integrate a collocation polynomial over more than its interval of collocation. Not too surprisingly, the resulting formula will have somewhat diminished accuracy but it has an important role to play nevertheless. The polynomial

$$p_k = y'_0 + k\frac{y'_1 - y'_{-1}}{2} + k^2\frac{y'_1 - 2y'_0 + y'_{-1}}{2}$$

satisfies $p_k = y'_k$ for $k = -1, 0, 1$. It is a collocation polynomial for $y'(x)$ in the form of Stirling's formula of degree two, a parabola. Integrating from $k = -2$ to $k = 2$, we obtain

$$\int_{-2}^{2} p_k\, dk = 4y'_0 + \frac{8}{3}(y'_1 - 2y'_0 + y'_{-1}) = \frac{4}{3}(2y'_1 - y'_0 + 2y'_{-1})$$

With the usual change of argument $x = x_0 + kh$ this becomes

$$\int_{x_{-2}}^{x_2} p(x)\, dx = \frac{4}{3}h(2y'_1 - y'_0 + 2y'_{-1})$$

Since we are thinking of $p(x)$ as an approximation to $y'(x)$,

$$\int_{x_{-2}}^{x_2} y'(x)\, dx = y_2 - y_{-2} \simeq \frac{4}{3}h(2y'_1 - y'_0 + 2y'_{-1})$$

Since the same argument applies on other intervals, the indices may all be increased by $k - 1$ to obtain the required predictor formula. It is so called because it allows the y_2 to be predicted from data for smaller arguments.

19.18. What is the local truncation error of this predictor?

It may be estimated by the Taylor series method. Using zero as a temporary reference point,

$$y_k = y_0 + (kh)y'_0 + \frac{1}{2}(kh)^2y_0^{(2)} + \frac{1}{6}(kh)^3y_0^{(3)} + \frac{1}{24}(kh)^4y_0^{(4)} + \frac{1}{120}(kh)^5y_0^{(5)} + \cdots$$

it follows that

$$y_2 - y_{-2} = 4hy'_0 + \frac{8}{3}h^3y_0^{(3)} + \frac{8}{15}h^5y_0^{(5)} + \cdots$$

Differentiation also brings

$$y'_k = y'_0 + (kh)y_0^{(2)} + \frac{1}{2}(kh)^2y_0^{(3)} + \frac{1}{6}(kh)^3y_0^{(4)} + \frac{1}{24}(kh)^4y_0^{(5)} + \cdots$$

from which we find

$$2y'_1 - y'_0 + 2y'_{-1} = 3y'_0 + 2h^2y_0^{(3)} + \frac{1}{6}h^4y_0^{(5)} + \cdots$$

The local truncation error is therefore

$$(y_2 - y_{-2}) - \frac{4}{3}h(2y'_1 - y'_0 + 2y'_{-1}) = \frac{14}{45}h^5y_0^{(5)} + \cdots$$

of which the first term will be used as an estimate. For our shifted interval this becomes

$$E_p \simeq \frac{14}{45} h^5 y_{k-1}^{(5)}$$

19.19. Compare the predictor error with that of the "corrector" formula

$$y_{k+1} \simeq y_{k-1} + \frac{1}{3} h(y'_{k-1} + 4y'_k + y'_{k+1})$$

This corrector is actually Simpson's rule applied to $y'(x)$. The local truncation error is therefore

$$E_c = \int_{x_{k-1}}^{x_{k+1}} y'(x)\, dx - \frac{1}{3} h(y'_{k-1} + 4y'_k + y'_{k+1}) \simeq -\frac{1}{90} h^5 y_k^{(5)}(\xi)$$

by Problem 14.65. Thus $E_p \simeq -28E_c$ where the difference in the arguments of $y^{(5)}$ has been ignored.

19.20. Show that the error of the corrector formula of Problem 19.19 can be estimated in terms of the difference between predictor and corrector values.

Considering just the local truncation errors made in the step from x_k to x_{k+1}, we have

$$y_{k+1} = P + E_p = C + E_c$$

with P and C denoting the predictor and corrector values. Then

$$P - C = E_c - E_p = 29E_c$$

and

$$E_c = \frac{P - C}{29}$$

more or less. It is not uncommon to apply this estimate as a further correction, yielding

$$y_{k+1} = C + \frac{P - C}{29}$$

and this formula does have truncation error of order h^6. Under some conditions, however, the use of such "mop-up" terms can make a computation unstable.

19.21. The Milne method uses the formula

$$y_{k+1} \simeq y_{k-3} + \frac{4}{3} h(2y'_{k-2} - y'_{k-1} - 2y'_k)$$

as a predictor, together with

$$y_{k+1} \simeq y_{k-1} + \frac{1}{3} h(y'_{k+1} + 4y'_k + y'_{k-1})$$

as a corrector. Apply this method using $h = .2$ to the problem $y' = -xy^2$, $y(0) = 2$.

The predictor requires four previous values, which it blends into y_{k+1}. The initial value $y(0) = 2$ is one of these. The others must be obtained. Since the entire computation will be based on these starting values, it is worth an extra effort to get them reasonably accurate. The Taylor method or Runge–Kutta method may be used to obtain

$$y(.2) = y_1 \simeq 1.92308 \qquad y(.4) = y_2 \simeq 1.72414 \qquad y(.6) = y_3 \simeq 1.47059$$

correct to five places. The differential equation then yields

$$y'(0) = y'_0 = 0 \qquad y'(.2) = y'_1 \simeq -.73964 \qquad y'(.4) = y'_2 \simeq -1.18906 \qquad y'(.6) = y'_3 \simeq -1.29758$$

correct to five places. The Milne predictor then manages

$$y_4 \simeq y_0 + \frac{4}{3}(.2)(2y'_3 - y'_2 + 2y'_1) \simeq 1.23056$$

In the differential equation we now find our first estimate of y'_4,

$$y'_4 \simeq -(.8)(1.23056)^2 \simeq -1.21142$$

The Milne corrector then provides the new approximation,

$$y_4 \simeq y_2 + \frac{1}{3}(.2)(-1.21142 + 4y'_3 + y'_2) \simeq 1.21808$$

Recomputing y' from the differential equation brings the new estimate $y'_4 \simeq -1.18698$. Reapplying the corrector, we next have

$$y_4 \simeq y_2 + \frac{1}{3}(.2)(-1.18698 + 4y'_3 + y'_2) \simeq 1.21971$$

Once again applying the differential equation, we find

$$y'_4 \simeq -1.19015$$

and returning to the corrector,

$$y_4 \simeq y_2 + \frac{1}{3}(.2)(-1.19015 + 4y'_3 + y'_2) \simeq 1.21950$$

The next two rounds produce

$$y'_4 \simeq -1.18974 \qquad y_4 \simeq 1.21953 \qquad y'_4 \simeq -1.18980 \qquad y_4 \simeq 1.21953$$

and since our last two estimates of y_4 agree, we can stop. The iterative use of the corrector formula and differential equation has proved to be a convergent process, and the resulting y_4 value is actually correct to four places. In this case four applications of the corrector have brought convergence. If h is chosen too large in a process of this sort, an excessive number of iterative cycles may be needed for convergence or the algorithm may not converge at all. Large differences between predictor and corrector outputs suggest reduction of the interval. On the other hand, insignificant differences between predictor and corrector outputs suggest increasing h and perhaps speeding up the computation. The computation of y_5 and y'_5 may now be made in the same way. Results up to $x = 10$ are provided in Table 19.4. Though $h = .2$ was used, only values for integer arguments are printed in the interest of brevity. The exact values are included for comparison.

Table 19.4

x	y (correct)	y (predictor)	Error	y (corrector)	Error
0	2.00000				
1	1.00000	1.00037	−37	1.00012	−12
2	.40000	.39970	30	.39996	4
3	.20000	.20027	−27	.20011	−11
4	.11765	.11737	28	.11750	15
5	.07692	.07727	−35	.07712	−20
6	.05405	.05364	41	.05381	14
7	.04000	.04048	−48	.04030	−30
8	.03077	.03022	55	.03041	36
9	.02439	.02500	−61	.02481	−42
10	.01980	.01911	69	.01931	49

19.22. Discuss the error of the previous computation.

Since the exact solution is known for this test case, it is easy to see some things which would usually be quite obscure. The fifth derivative of $y(x) = 2/(1+x^2)$ has the general behavior shown in Fig. 19.4.

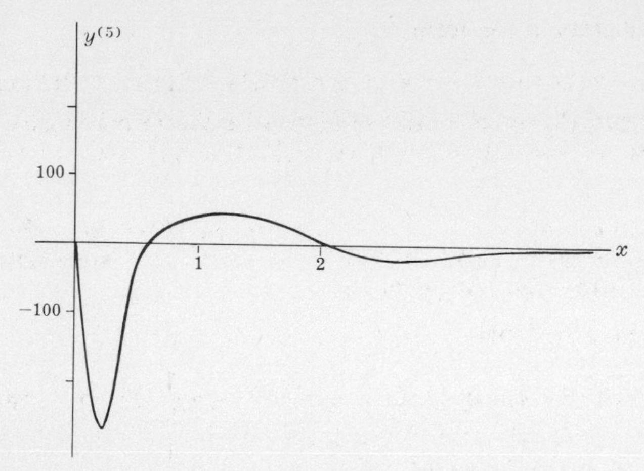

Fig. 19-4

The large fluctuations between 0 and 1 would usually make it difficult to use our truncation error formulas. For example, the local error of the predictor is $14h^5y^{(5)}/45$ and in our first step (to $x = .8$) we actually find the predictor in error by $-.011$. This corresponds to $y^{(5)} \simeq -100$. The local corrector error is $-h^5y^{(5)}/90$ and in the same first step the error was actually $-.00002$. This corresponds to $y^{(5)} \simeq 6$. This change of sign in $y^{(5)}$ annuls the anticipated change in sign of error between the predictor and corrector results. It also means that an attempt to use the extrapolation to the limit idea would lead to worse results rather than better, in this case. The oscillating sign of the error as the computation continues will be discussed later.

19.23. Derive the Adams predictor formula

$$\begin{aligned} y_{k+1} &= y_k + h\left(y'_k + \frac{1}{2}\nabla y'_k + \frac{5}{12}\nabla^2 y'_k + \frac{3}{8}\nabla^3 y'_k\right) \\ &= y_k + \frac{1}{24}h(55y'_k - 59y'_{k-1} + 37y'_{k-2} - 9y'_{k-3}) \end{aligned}$$

As in Problem 19.17, we obtain this predictor by integrating a collocation polynomial beyond the interval of collocation. The Newton backward formula of degree three, applied to $y'(x)$, is

$$p_k = y'_0 + k\,\nabla y'_0 + \frac{1}{2}k(k+1)\,\nabla^2 y'_0 + \frac{1}{6}k(k+1)(k+2)\,\nabla^3 y'_0$$

where as usual $x_k = x_0 + kh$. Integrating from $k = 0$ to $k = 1$ (though the points of collocation are $k = 0, -1, -2, -3$), we obtain

$$\int_0^1 p_k\,dk = y'_0 + \frac{1}{2}\nabla y'_0 + \frac{5}{12}\nabla^2 y'_0 + \frac{3}{8}\nabla^3 y'_0$$

In terms of the argument x and using $p(x) \simeq y'(x)$, this becomes

$$\int_{x_0}^{x_1} y'(x)\,dx = y_1 - y_0 \simeq h\left(y'_0 + \frac{1}{2}\nabla y'_0 + \frac{5}{12}\nabla^2 y'_0 + \frac{3}{8}\nabla^3 y'_0\right)$$

Since the same reasoning may be applied between x_k and x_{k+1}, we may raise all indices by k to obtain the first result required. The second then follows by writing out the differences in terms of the y values.

19.24. What is the local truncation error of the Adams predictor?

The usual Taylor series approach leads to $E = 251h^5y^{(5)}/720$.

19.25. Derive other predictors of the form

$$y_{k+1} = a_0 y_k + a_1 y_{k-1} + a_2 y_{k-2} + h(b_0 y'_k + b_1 y'_{k-1} + b_2 y'_{k-2} + b_3 y'_{k-3})$$

Varying the approach, we shall make this formula exact for polynomials through degree four. The convenient choices are $y(x) = 1$, $(x - x_k)$, $(x - x_k)^2$, $(x - x_k)^3$, and $(x - x_k)^4$. This leads to the five conditions

$$\begin{aligned} 1 &= a_0 + a_1 + a_2 & 1 &= -a_1 - 8a_2 + 3b_1 + 12b_2 + 27b_3 \\ 1 &= -a_1 - 2a_2 + b_0 + b_1 + b_2 + b_3 & 1 &= a_1 + 16a_2 - 4b_1 - 32b_2 - 108b_3 \\ 1 &= a_1 + 4a_2 - 2b_1 - 4b_2 - 6b_3 \end{aligned}$$

which may be solved in the form

$$\begin{aligned} a_0 &= 1 - a_1 - a_2 & b_2 &= \frac{1}{24}(37 - 5a_1 + 8a_2) \\ b_0 &= \frac{1}{24}(55 + 9a_1 + 8a_2) & b_3 &= \frac{1}{24}(-9 + a_1) \\ b_1 &= \frac{1}{24}(-59 + 19a_1 + 32a_2) \end{aligned}$$

with a_1 and a_2 arbitrary. The choice $a_1 = a_2 = 0$ leads us back to the previous problem. Two other simple and popular choices are $a_1 = \frac{1}{2}$, $a_2 = 0$ which leads to

$$y_{k+1} = \frac{1}{2}(y_k + y_{k-1}) + \frac{1}{48}h(119y'_k - 99y'_{k-1} + 69y'_{k-2} - 17y'_{k-3})$$

with local truncation error $161h^5y^{(5)}/480$ and $a_1 = \frac{2}{3}$, $a_2 = \frac{1}{3}$ which leads to

$$y_{k+1} = \frac{1}{3}(2y_{k-1} + y_{k-2}) + \frac{1}{72}h(191y'_k - 107y'_{k-1} + 109y'_{k-2} - 25y'_{k-3})$$

with local truncation error $707h^5y^{(5)}/2160$.

Clearly, one could use these two free parameters to further reduce truncation error, even to order h^7, but another factor to be considered shortly suggests that truncation error is not our only problem. It is also clear that other types of predictor, perhaps using a y_{k-3} term, are possible, but we shall limit ourselves to the abundance we already have.

19.26. Illustrate the possibilities for other corrector formulas.

The possibilities are endless, but suppose we seek a corrector of the form

$$y_{k+1} \simeq a_0 y_k + a_1 y_{k-1} + a_2 y_{k-2} + h(cy'_{k+1} + b_0 y'_k + b_1 y'_{k-1} + b_2 y'_{k-2})$$

for which the local truncation error is of the order h^5. Asking that the corrector be exact for $y(x) = 1$, $(x - x_k), \ldots, (x - x_k)^4$ leads to the five conditions

$$\begin{aligned} a_0 + a_1 + a_2 &= 1 & 13a_1 + 32a_2 - 24b_1 &= 5 \\ a_1 + 24c &= 9 & a_1 - 8a_2 + 24b_2 &= 1 \\ 13a_1 + 8a_2 - 24b_0 &= -19 \end{aligned}$$

involving seven unknown constants. It would be possible to make this corrector exact for even more powers of x, thus lowering the local truncation error still further. However, the two degrees of freedom will be used to bring other desirable features instead to the resulting algorithm. With $a_0 = 0$ and $a_1 = 1$

the remaining constants prove to be those of the Milne corrector:

$$a_2 = 0 \qquad c = \frac{1}{3} \qquad b_0 = \frac{4}{3} \qquad b_1 = \frac{1}{3} \qquad b_2 = 0$$

Another choice, which matches to some extent the Adams predictor, involves making $a_1 = a_2 = 0$, which produces the formula

$$y_{k+1} \simeq y_k + \frac{1}{24}h(9y'_{k+1} + 19y'_k - 5y'_{k-1} + y'_{k-2})$$

If $a_1 = \frac{2}{3}$, $a_2 = \frac{1}{3}$, then we have a formula which resembles another predictor just illustrated:

$$y_{k+1} \simeq \frac{1}{3}(2y_{k-1} + y_{k-2}) + \frac{1}{72}h(25y'_{k+1} + 91y'_k + 43y'_{k-1} + 9y'_{k-2})$$

Still another formula has $a_0 = a_1 = \frac{1}{2}$, making

$$y_{k+1} \simeq \frac{1}{2}(y_k + y_{k-1}) + \frac{1}{48}h(17y'_{k+1} + 51y'_k + 3y'_{k-1} + y'_{k-2})$$

The various choices differ somewhat in their truncation errors.

19.27. Compare the local truncation errors of the predictor and corrector formulas just illustrated.

The Taylor series method can be applied as usual to produce the following error estimates:

Predictor: $y_{k+1} = y_k + \frac{1}{24}h(55y'_k - 59y'_{k-1} + 37y'_{k-2} - 9y'_{k-3}) + \frac{251h^5y^{(5)}}{720}$

Corrector: $y_{k+1} = y_k + \frac{1}{24}h(9y'_{k+1} + 19y'_k - 5y'_{k-1} + y'_{k-2}) - \frac{19h^5y^{(5)}}{720}$

Predictor: $y_{k+1} = \frac{1}{2}(y_k + y_{k-1}) + \frac{1}{48}h(119y'_k - 99y'_{k-1} + 69y'_{k-2} - 17y'_{k-3}) + \frac{161h^5y^{(5)}}{480}$

Corrector: $y_{k+1} = \frac{1}{2}(y_k + y_{k-1}) + \frac{1}{48}h(17y'_{k+1} + 51y'_k + 3y'_{k-1} + y'_{k-2}) - \frac{9h^5y^{(5)}}{480}$

Predictor: $y_{k+1} = \frac{1}{3}(2y_{k-1} + y_{k-2}) + \frac{1}{72}h(191y'_k - 107y'_{k-1} + 109y'_{k-2} - 25y'_{k-3}) + \frac{707h^5y^{(5)}}{2160}$

Corrector: $y_{k+1} = \frac{1}{3}(2y_{k-1} + y_{k-2}) + \frac{1}{72}h(25y'_{k+1} + 91y'_k + 43y'_{k-1} + 9y'_{k-2}) - \frac{43h^5y^{(5)}}{2160}$

In each case the corrector error is considerably less than that of its predictor mate. It is also of opposite sign, which can be helpful information in a computation. The lower corrector error can be explained by its pedigree. It uses information concerning y'_{k+1} while the predictor must take the leap forward from y_k. This also explains why the burden of the computation falls on the corrector, the predictor being used only as a primer.

For each pair of formulas a mop-up term may be deduced. Take the Adams predictor and the corrector below it, the first pair above. Proceeding in the usual way, considering local truncation errors only and remaining aware that results so obtained must be viewed with some skepticism, we find

$$I = P + E_1 = C + E_2$$

where I is the exact value. Since $19E_1 \simeq -251E_2$, we have $E_2 \simeq \frac{19}{270}(P - C)$. This is the mop-up term and $I \simeq C + \frac{19}{270}(P - C)$ is the corresponding extrapolation to the limit. Once again it must be remembered that $y^{(5)}$ does not really mean the same thing in both formulas, so that there is still a possibility of sizable error in this extrapolation.

19.28. Apply the Adams method to $y' = -xy^2$ with $y(0) = 2$, using $h = .2$.

The method is now familiar, each step involving a prediction and then an iterative use of the corrector formula. The Adams method uses the first pair of formulas of Problem 19.27 and leads to the results in Table 19.5.

Table 19.5

x	y (correct)	y (predicted)	Error	y (corrected)	Error
0	2.000000				
1	1.000000	1.000798	−789	1.000133	−133
2	.400000	.400203	−203	.400158	−158
3	.200000	.200140	−140	.200028	−28
4	.117647	.117679	−32	.117653	−6
5	.076923	.076933	−10	.076925	−2
6	.054054	.054058	−4	.054055	−1
7	.040000	.040002	−2	.040000	
8	.030769	.030770	−1	.030769	
9	.024390	.024391	−1	.024390	
10	.019802	.019802		.019802	

The error behavior suggests that $h = .2$ is adequate for six-place accuracy for large x, but that a smaller h (say .1) might be wise at the start. The diminishing error is related to the fact (see Problem 19.36) that for this method the "relative error" remains bounded.

19.29. Prove that, for h sufficiently small, iterative use of a corrector formula does produce a convergent sequence, and that the limit of this sequence is the unique value Y_{k+1} satisfying the corrector formula.

We are seeking a number Y_{k+1} with the property

$$Y_{k+1} = hcf(x_{k+1}, Y_{k+1}) + \cdots$$

the dots indicating terms containing only previously computed results, and so independent of Y_{k+1}. Assume as usual that $f(x, y)$ satisfies a Lipschitz condition on y in some region R. Now define a sequence

$$Y^{(0)}, Y^{(1)}, Y^{(2)}$$

subscripts $k+1$ being suppressed for simplicity, by the iteration

$$Y^{(i)} = hcf(x_{k+1}, Y^{(i-1)}) + \cdots$$

and assume all points $(x_{k+1}, Y^{(i)})$ are in R. Subtracting, we find

$$Y^{(i+1)} - Y^{(i)} = hc[f(x_{k+1}, Y^{(i)}) - f(x_{k+1}, Y^{(i-1)})]$$

Repeated use of the Lipschitz condition then brings

$$|Y^{(i+1)} - Y^{(i)}| \leqq hcK\,|Y^{(i)} - Y^{(i-1)}| \leqq \cdots \leqq (hcK)^i\,|Y^{(1)} - Y^{(0)}|$$

Now choose h small enough to make $|hcK| = r < 1$, and consider the sum

$$Y^{(n)} - Y^{(0)} = (Y^{(1)} - Y^{(0)}) + \cdots + (Y^{(n)} - Y^{(n-1)})$$

For n tending to infinity the series produced on the right is dominated (apart from a factor) by the geometric series $1 + r + r^2 + \cdots$ and so converges. This proves that $Y^{(n)}$ has a limit. Call this limit Y_{k+1}.

Now, because of the Lipschitz condition,

$$|f(x_{k+1}, Y^{(n)}) - f(x_{k+1}, Y_{k+1})| \leqq K\,|Y^{(n)} - Y_{k+1}|$$

and it follows that $\lim f(x_{k+1}, Y^{(n)}) = f(x_{k+1}, Y_{k+1})$. We may thus let n tend to infinity in the iteration

$$Y^{(n)} = hcf(x_{k+1}, Y^{(n-1)}) + \cdots$$

and obtain at once, as required,

$$Y_{k+1} = hcf(x_{k+1}, Y_{k+1}) + \cdots$$

To prove uniqueness, suppose Z_{k+1} were another value satisfying the corrector formula at x_{k+1}. Then much as before,

$$|Y_{k+1} - Z_{k+1}| \leqq hcK\,|Y_{k+1} - Z_{k+1}| \leqq \cdots \leqq (hcK)^{(i)}\,|Y_{k+1} - Z_{k+1}|$$

for arbitrary i. Since $|hcK| = r < 1$, this forces $Y_{k+1} = Z_{k+1}$. Notice that this uniqueness result proves the correct Y_{k+1} to be independent of $Y^{(0)}$, that is, independent of the choice of predictor formula, at least for small h. The choice of predictor is therefore quite free. It seems reasonable to use a predictor of comparable accuracy, from the local truncation error point of view, with a given corrector. This leads to an attractive "mop-up" argument as well. The pairings in Problem 19.27 keep these factors, and some simple esthetic factors, in mind.

CONVERGENCE OF PREDICTOR-CORRECTOR METHODS

19.30. Show that the modified Euler method is convergent.

In this method the simple Euler formula is used to make a first prediction of each y_{k+1} value, but then the actual approximation is found by the modified formula

$$Y_{k+1} = Y_k + \frac{1}{2}h(Y'_{k+1} + Y'_k)$$

The exact solution satisfies a similar relation with a truncation error term. Calling the exact solution $y(x)$ as before, we have

$$y(x_{k+1}) = y(x_k) + \frac{1}{2}h[y'(x_{k+1}) + y'(x_k)] - \frac{1}{12}h^3y^{(3)}(\xi)$$

the truncation error term having been evaluated in Problem 19.15. Subtracting and using d_k for $y(x_k) - Y_k$, we have

$$|d_{k+1}| \leqq |d_k| + \frac{1}{2}hL(|d_{k+1}| + |d_k|) + \frac{1}{12}h^3B$$

provided we assume the Lipschitz condition, which makes

$$|y'(x_k) - Y'_k| = |f(x_k, y(x_k)) - f(x_k, Y_k)| \leqq L\,|d_k|$$

with a similar result at argument $k+1$. The number B is a bound for $|y^{(3)}(x)|$, which we also assume to exist. Our inequality can also be written as

$$\left(1 - \frac{1}{2}hL\right)|d_{k+1}| \leqq \left(1 + \frac{1}{2}hL\right)|d_k| + \frac{1}{12}h^3B$$

Suppose no initial error ($d_0 = 0$) and consider also the solution of

$$\left(1 - \frac{1}{2}hL\right)D_{k+1} = \left(1 + \frac{1}{2}hL\right)D_k + \frac{1}{12}h^3B$$

with initial value $D_0 = 0$. For purposes of induction we assume $|d_k| \leqq D_k$ and find as a consequence

$$\left(1 - \frac{1}{2}hL\right)|d_{k+1}| \leqq \left(1 - \frac{1}{2}hL\right)D_{k+1}$$

so that $|d_{k+1}| \leqq D_{k+1}$. Since $d_0 = D_0$ the induction is complete and guarantees $|d_k| \leqq D_k$ for positive integers k. To find D_k we solve the difference equation and find the solution family

$$D_k = C\left(\frac{1 + \frac{1}{2}hL}{1 - \frac{1}{2}hL}\right)^k - \frac{h^2B}{12L}$$

with C an arbitrary constant. To satisfy the initial condition $D_0 = 0$, we must have $C = (h^2B/12L)$ so that

$$|y(x_k) - Y_k| \leqq \frac{h^2B}{12L}\left[\left(\frac{1+\frac{1}{2}hL}{1-\frac{1}{2}hL}\right)^k - 1\right]$$

To prove convergence at a fixed argument $x_k = x_0 + kh$ we must investigate the second factor, since as h tends to zero k will increase indefinitely. But since

$$\left(\frac{1+\frac{1}{2}hL}{1-\frac{1}{2}hL}\right)^k = \left[\frac{1+L(x_k-x_0)/2k}{1-L(x_k-x_0)/2k}\right]^k \to \frac{e^{L(x_k-x_0)/2}}{e^{-L(x_k-x_0)/2}} = e^{L(x_k-x_0)}$$

we have

$$y(x_k) - Y_k = 0(h^2)$$

Thus as h tends to zero, $\lim Y_k = y(x_k)$, which is the meaning of convergence. Our result also provides a measure of the way truncation errors propagate through the computation.

19.31. Prove the convergence of Milne's method.

The Milne corrector formula is essentially Simpson's rule and provides the approximate values

$$Y_{k+1} = Y_{k-1} + \frac{1}{3}h(Y'_{k+1} + 4Y'_k + Y'_{k-1})$$

The exact solution $y(x)$ satisfies a similar relation, but with a truncation error term

$$y_{k+1} = y_{k-1} + \frac{1}{3}h(y'_{k+1} + 4y'_k + y'_{k-1}) - \frac{1}{90}h^5y^{(5)}(\xi)$$

with ξ between x_{k-1} and x_{k+1}. Subtracting and using $d_k = y(x_k) - Y_k$,

$$|d_{k+1}| \leqq |d_{k-1}| + \frac{1}{3}hL(|d_{k+1}| + 4|d_k| + |d_{k-1}|) + \frac{1}{90}h^5B$$

with the Lipschitz condition again involved and B a bound on $y^{(5)}(x)$. Rewriting the inequality as

$$\left(1 - \frac{1}{3}hL\right)|d_{k+1}| \leqq \frac{4}{3}hL\,|d_k| + \left(1 + \frac{1}{3}hL\right)|d_{k-1}| + \frac{1}{90}h^5B$$

we compare it with the difference equation

$$\left(1 - \frac{1}{3}hL\right)D_{k+1} = \frac{4}{3}hLD_k + \left(1 + \frac{1}{3}hL\right)D_{k-1} + \frac{1}{90}h^5B$$

Suppose initial errors of d_0 and d_1. We will seek a solution D_k such that $d_0 \leqq D_0$ and $d_1 \leqq D_1$. Such a solution will dominate $|d_k|$, that is, it will have the property $|d_k| \leqq D_k$ for nonnegative integers k. This can be proved by induction much as in the previous problem, for if we assume $|d_{k-1}| \leqq D_{k-1}$ and $|d_k| \leqq D_k$ we at once find that $|d_{k+1}| \leqq D_{k+1}$ also, and the induction is already complete. To find the required solution the characteristic equation

$$\left(1 - \frac{1}{3}hL\right)r^2 - \frac{4}{3}hLr - \left(1 + \frac{1}{3}hL\right) = 0$$

may be solved. It is easy to discover that one root is slightly greater than 1, say r_1, and another in the vicinity of -1, say r_2. More specifically,

$$r_1 = 1 + hL + 0(h^2) \qquad r_2 = -1 + \frac{1}{3}hL + 0(h^2)$$

The associated homogeneous equation is solved by a combination of the kth powers of these roots. The nonhomogeneous equation itself has the constant solution $-h^4B/180L$. And so we have

$$D_k = c_1r_1^k + c_2r_2^k - \frac{h^4B}{180L}$$

Let E be the greater of the two numbers d_0 and d_1. Then

$$D_k = \left(E + \frac{h^4B}{180L}\right)r_1^k - \frac{h^4B}{180L}$$

will be a solution with the required initial features. It has $D_0 = E$, and since $1 < r_1$ it grows steadily larger. Thus

$$|d_k| \leqq \left(E + \frac{h^4B}{180L}\right)r_1^k - \frac{h^4B}{180L}$$

If we make no initial error, then $d_0 = 0$. If also as h is made smaller we improve our value Y_1 (which must be obtained by some other method such as the Taylor series) so that $d_1 = 0(h)$, then we have $E = 0(h)$ and as h tends to zero so does d_k. This proves the convergence of the Milne method.

19.32. Generalizing the previous problems, prove the convergence of methods based on the corrector formula

$$Y_{k+1} = a_0Y_k + a_1Y_{k-1} + a_2Y_{k-2} + h(cY'_{k+1} + b_0Y'_k + b_1Y'_{k-1} + b_2Y'_{k-2})$$

We have chosen the available coefficients to make the truncation error of order h^5. Assuming this to be the case, the difference $d_k = y(x_k) - Y_k$ is found by the same procedure just employed for the Milne corrector to satisfy

$$(1 - |c|\, hL)\, |d_{k+1}| \leqq \sum_{i=0}^{2} (|a_i| + hL\, |b_i|)\, |d_{k-i}| + T$$

where T is the truncation error term. This corrector requires three starting values, perhaps found by the Taylor series. Call the maximum error of these values E, so that $|d_k| \leqq E$ for $k = 0, 1, 2$. Consider also the difference equation

$$(1 - |c|\, hL)D_{k+1} = \sum_{i=0}^{2} (|a_i| + hL\, |b_i|)D_{k-i} + T$$

We will seek a solution satisfying $E \leqq D_k$ for $k = 0, 1, 2$. Such a solution will dominate $|d_k|$. For, assuming $|d_{k-i}| \leqq D_{k-i}$ for $i = 0, 1, 2$ we at once have $|d_{k+1}| \leqq D_{k+1}$. This completes an induction and proves $|d_k| \leqq D_k$ for nonnegative integers k. To find the required solution we note that the characteristic equation

$$(1 - |c|\, hL)r^3 - \sum_{i=0}^{2} (|a_i| + hL\, |b_i|)r^{2-i} = 0$$

has a real root greater than one. This follows since at $r = 1$ the left side becomes

$$A = 1 - |c|\, hL - \sum_{i=0}^{2} (|a_i| + hL\, |b_i|)$$

which is surely negative since $a_0 + a_1 + a_2 = 1$, while for large r the left side is surely positive if we choose h small enough to keep $1 - |c|\, hL$ positive. Call the root in question r_1. Then a solution with the required features is

$$D_k = \left(E - \frac{T}{A}\right)r_1^k + \frac{T}{A}$$

since at $k = 0$ this becomes E and as k increases it grows still larger. Thus

$$|y(x_k) - Y_k| \leqq \left(E - \frac{T}{A}\right)r_1^k + \frac{T}{A}$$

As h tends to zero the truncation error T tends to zero. If we also arrange that the initial errors tend to zero, then $\lim y(x_k) = Y_k$ and convergence is proved.

ERROR AND STABILITY

19.33. What is meant by a stable method for solving differential equations?

The idea of stability has been described in many ways. Very loosely, a computation is stable if it doesn't "blow up," but this would hardly be appropriate as a formal definition. In the introduction to this chapter stability was defined as boundedness of the relative error and without question this would be a desirable feature for an algorithm. Gradual deterioration of the relative error means gradual loss of significant digits, which is hardly something to look forward to. The trouble is, over the long run relative error often does deteriorate. An easy example may be useful to gain insight. Consider the modified Euler method.

$$y_{k+1} = y_k + \frac{1}{2}h(y'_{k+1} + y'_k)$$

Apply it to the trivial problem $\quad y' = Ay \qquad y(0) = 1$

for which the exact solution is $y = e^{Ax}$. The Euler formula becomes

$$\left(1 - \frac{1}{2}Ah\right)y_{k+1} = \left(1 + \frac{1}{2}Ah\right)y_k$$

which is a difference equation of order one with solution

$$y_k = r^k = \left(\frac{1 + \frac{1}{2}Ah}{1 - \frac{1}{2}Ah}\right)^k$$

For small h this is close to $\quad \left(\frac{e^{(1/2)Ah}}{e^{-(1/2)Ah}}\right)^k = e^{Akh} = e^{Ax}$

giving us an intuitive proof of convergence. But our goal here lies in another direction. The exact solution satisfies

$$\left(1 - \frac{1}{2}Ah\right)y(x_{k+1}) = \left(1 + \frac{1}{2}Ah\right)y(x_k) + T$$

where T is the truncation error $-h^3A^3y(\xi)/12$. Subtracting, and using $d_k = y(x_k) - y_k$, we find the similar equation

$$\left(1 - \frac{1}{2}Ah\right)d_{k+1} = \left(1 + \frac{1}{2}Ah\right)d_k - \frac{1}{12}h^3A^3y(\xi)$$

for the error d_k. Now divide by $(1 - \frac{1}{2}Ah)y_{k+1}$ and assume Ah small to obtain

$$R_{k+1} = R_k - \frac{1}{12}h^3A^3$$

for the relative error $R_k = y_k/y(x_k)$. Solving

$$R_k = R_0 - \frac{1}{12}kh^3A^3 = R_0 - \frac{1}{12}x_kh^2A^3$$

suggesting that the relative error grows like x_k, or linearly, as the computation proceeds. This may be far from a blow-up, but neither is it a case of relative error remaining bounded.

Taking another view, we will watch the progress of a single error as it penetrates through the solution process, say an initial error d_0. Assuming no other errors committed, we omit T and have

$$d_k = d_0\left(\frac{1 + \frac{1}{2}Ah}{1 - \frac{1}{2}Ah}\right)^k \simeq d_0e^{Akh}$$

which makes the relative error $R_k = d_k/e^{Akh} \simeq d_0$. So the long-range effect of any single error is an imitation of the behavior of the solution itself. If A is positive, the error and the solution grow in the same proportion, while if A is negative, they decay in the same proportion. In both cases the relative error holds firm. That this view is slightly optimistic is suggested by the linear growth predicted above, but at least no blow-up is forecast. By some definitions this is enough to consider the Euler algorithm stable. This informal, relaxed usage of the term can be convenient.

There remains the question of how small Ah should be to justify the approximations made in these arguments. Since the true solution is monotone, it seems advisable to keep the value of $(1 + \frac{1}{2}Ah)/(1 - \frac{1}{2}Ah)$

positive. This is true only for Ah between -2 and 2. Prudence suggests keeping one's distance from both of these extremes.

19.34. Analyze error behavior in the Milne corrector formula.

$$y_{k+1} = y_{k-1} + \frac{h}{3}(y'_{k+1} + 4y'_k + y'_{k-1})$$

Again choosing the special equation $y' = Ay$, the error d_k is easily found to satisfy the difference equation of order two

$$\left(1 - \frac{1}{3}Ah\right)d_{k+1} = \frac{4}{3}Ahd_k + \left(1 + \frac{1}{3}Ah\right)d_{k-1} + T$$

for which the characteristic equation is (see Chapter 18)

$$\left(1 - \frac{1}{3}Ah\right)r^2 - \frac{4}{3}Ahr - \left(1 + \frac{1}{3}Ah\right) = 0$$

The roots are

$$r_1 = 1 + Ah + 0(h^2) \qquad r_2 = -1 + \frac{1}{3}Ah + 0(h^2)$$

which makes

$$d_k \simeq c_1(1 + Ah)^k + c_2\left(-1 + \frac{1}{3}Ah\right)^k$$
$$\simeq c_1 e^{Ahk} + (d_0 - c_1)(-1)^k e^{-Ahk/3}$$

Now it is possible to see the long-range effect of the initial error d_0. If A is positive, then d_k behaves very much like the correct solution e^{Ahk}, since the second term tends to zero. In fact, the relative error can be estimated as

$$\frac{d_k}{e^{Ahk}} = c_1 + (d_0 - c_1)(-1)^k e^{-4Ahk/3}$$

which approaches a constant. If A is negative, however, the second term does not disappear. Indeed it soon becomes the dominant term. The relative error becomes an unbounded oscillation and the computation degenerates into nonsense beyond a certain point.

The Milne method is said to be stable for A positive and unstable for A negative. In this second case the computed "solution" truly blows up.

19.35. Do the computations made earlier confirm these theoretical predictions?

Referring once again to Table 19.4 the following relative errors may be computed. Though the equation $y' = -xy^2$ is not linear its solution is decreasing, as that of the linear equation does for negative A. The oscillation in the above data is apparent. The substantial growth of relative error is also apparent.

x_k	1	2	3	4	5	6	7	8	9	10
d_k/y_k	−.0001	.0001	−.0005	.0013	−.0026	.0026	−.0075	.0117	−.0172	.0247

19.36. Analyze error behavior for the Adams corrector

$$Y_{k+1} = Y_k + \frac{1}{24}h(9Y'_{k+1} + 19Y'_k - 5Y'_{k-1} + Y'_{k-2})$$

The usual process in this case leads to

$$\left(1-\frac{9}{24}Ah\right)d_{k+1}=\left(1+\frac{19}{24}Ah\right)d_k-\frac{5}{24}Ahd_{k-1}+\frac{1}{24}Ahd_{k-2}+T$$

Ignoring T we attempt to discover how a solitary error would propagate, in particular what its influence on relative error would be over the long run. The first step is once again to consider the roots of the characteristic equation.

$$\left(1-\frac{9}{24}Ah\right)r^3-\left(1+\frac{19}{24}Ah\right)r^2+\frac{5}{24}Ahr-\frac{1}{24}Ah=0$$

This has one root near 1, which may be verified to be $r_1 \simeq 1+Ah$. If this root is removed, the quadratic factor

$$(24-9Ah)r^2-4Ahr+Ah=0$$

remains. If Ah were zero this quadratic would have a double root at zero. For Ah nonzero but small the roots, call them r_2 and r_3, will still be near zero. Actually for small positive Ah they are complex with moduli $|r| \simeq \sqrt{Ah/24}$, while for small negative Ah they are real and approximately $\pm\sqrt{-6Ah}/12$. Either way we have

$$|r_2|,\ |r_3| < 1+Ah \simeq e^{Ah}$$

for small Ah. The solution of the difference equation can now be written as

$$d_k \simeq c_1(1+Ah)^k + 0(|Ah|^{k/2}) \simeq c_1e^{Akh}+0(e^{Akh})$$

The constant c_1 depends upon the solitary error which has been assumed. Dividing by the exact solution, we find that relative error remains bounded. The Adams corrector is therefore stable for both positive and negative A. A single error will not ruin the computation.

19.37. Do the computations made earlier confirm these theoretical predictions?

Referring once again to Table 19.5, the following relative errors may be computed:

x_k	1	2	3	4	5	6	7 to 10
d_k/y_k	−.00013	−.00040	−.00014	−.00005	−.00003	−.00002	zero

As predicted the errors are diminishing, even the relative error. Once again results obtained for a linear problem prove to be informative about the behavior of computations for a nonlinear problem.

19.38. What are parasitic solutions and what is their connection with the idea of computational stability which underlies the preceding problems?

The methods in question involve substituting a difference equation for the differential equation, and for the case $y'=Ay$ it is a difference equation which is linear with constant coefficients. Its solution is, therefore, a combination of terms of the form r_i^k with the r_i the roots of the characteristic equation. One of these roots will be $r_1=1+Ah$, apart from terms of higher degree in h, and r_1^k will then be close to $e^{Ahk}=e^{Ax}$ when h is small. This is the solution we want, the one that converges to the differential solution. Other components, corresponding to the other r_i, are called parasitic solutions. They are the price paid for the lower truncation error that methods such as Milne and Adams bring.

If the parasitic terms are dominated by the r_1 term, then their contribution will be negligible and the relative error will remain acceptable. If, on the other hand, a parasitic solution becomes dominant, it will ruin the computation. In Problem 19.33, for the modified Euler method, the relevant difference equation had only the root

$$r_1=\frac{1+Ah/2}{1-Ah/2}=1+Ah+0(h^2)$$

There were no parasitic solutions. In Problem 19.34, the Milne method offered us

$$r_1 = 1 + Ah \qquad r_2 = -1 + \frac{1}{3}Ah$$

up to the terms in h^2. For $A > 0$ it is r_1 that dominates, but for $A < 0$ it is r_2 that takes over and the desired solution is buried. In Problem 19.36, apart from the usual $r_1 = 1 + Ah$, we found two parasitic solution terms, both of size about Ah. Both are dominated by the r_1 term, whether A is positive or negative. The Adams method means stable computing in either case.

We are drawn to the conclusion that to avoid a computational blow-up any parasitic term should be dominated by the principal term, that is, we want

$$|r_i| \leqq r_1$$

for $i \neq 1$. Any method for which these conditions are violated is called unstable. In fact, it is best if the inequalities are satisfied by a wide margin.

19.39. Apply the second-order Runge–Kutta method

$$y_{k+1} = y_k + hf\left(x_k + \frac{1}{2}h,\ y_k + \frac{1}{2}hf(x_k, y_k)\right)$$

to $y' = Ay$. What does this reveal about the stability of this formula?

Substituting Ay for $f(x, y)$ brings

$$y_{k+1} = \left(1 + Ah + \frac{1}{2}A^2h^2\right)y_k$$

making

$$y_k = \left(1 + Ah + \frac{1}{2}A^2h^2\right)^k$$

which is close to the true solution $y_k = e^{kh} = e^{x_k}$ if Ah is small. But how small should Ah be? Figure 19-5 provides a view of the quadratic $r = 1 + Ah + \frac{1}{2}A^2h^2$. When A is positive, r will be greater than one, so both r^k and e^{kh} will be increasing. The qualitative behavior of r^k is, therefore, correct. But when A is negative, we want a decreasing solution, and this will occur only if Ah is between -2 and 0. Below this interval the approximate solution r^k will be increasing and will bear no resemblance whatsoever to e^{kh}. Here there are no parasitic solutions, since Runge–Kutta methods do not reach back beyond y_k to do their work. The blow-up of relative error has a different origin, in the nature of the root r_1 itself.

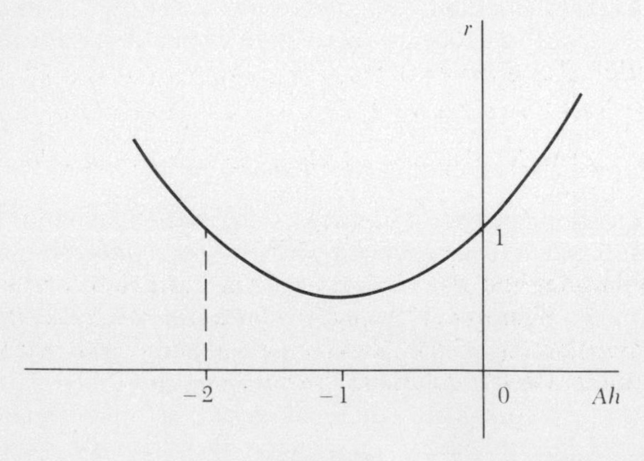

Fig. 19-5

19.40. Apply the fourth-order Runge–Kutta formulas of Problem 19.12 to $y' = Ay$. For what range of Ah values is it stable?

With a little care we find

$$y_{k+1}=\left(1+Ah+\frac{1}{2}A^2h^2+\frac{1}{6}A^3h^3+\frac{1}{24}A^4h^4\right)y_k$$

in which the approximation to e^{Ah} is prominent. Denoting it by r, our approximate solution is again $y_k=r^k$. A plot of r against Ah appears as Fig. 19-6 and, as with the second-order method, suggests that for positive A the true and the approximate solutions will have the same character, both increasing steadily. But for negative A, just as in the preceding problem, there is a lower bound below which the r^k values will not follow the decreasing trend of the true solution. Here that bound is near -2.78. For Ah smaller than this, we find r greater than one and an exploding computation.

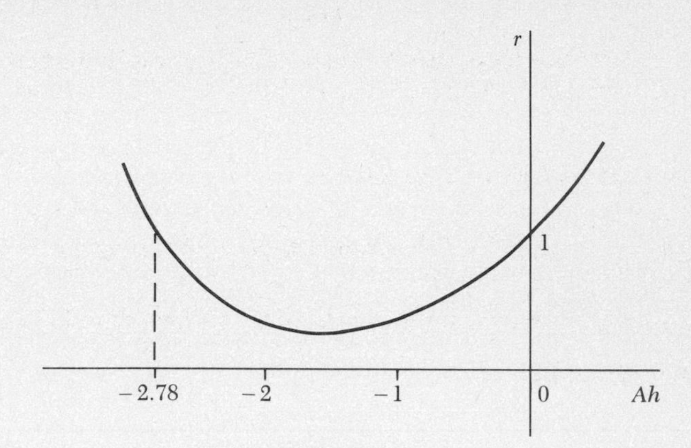

Fig. 19-6

19.41. How can an analysis based upon the equation $y'=Ay$ tell us anything useful about the general problem $y'=f(x,y)$?

There are certainly no guarantees, but the general equation is too difficult for such analysis so the issue is really one of doing what is possible. One link that can be established between the two problems is the identification of our constant A with the partial derivative f_y, evaluated originally in the vicinity of the initial point (x_0, y_0), and later at other regions of the plane to which the solution has penetrated. If f_y changes sign along the way, we would expect the stability of Milne's method to react quickly and that of Runge–Kutta methods to show some sensitivity as well.

19.42. Apply the fourth-order Runge–Kutta method to the nonlinear equation $y'=-100xy^2$ with $y(0)=2$. The exact solution is $y=2/(1+100x^2)$. Test the stability for different step sizes.

Since $f_y=-200xy=-400x/(1+100x^2)$, which is zero initially but climbs quickly to -20 at $x=.1$, we recall the stability condition

$$-2.78\leqq Ah=-20h$$

and decide to test h values around .14. With $h=.10$ the computed solution decays nicely to .0197 at $x=1$ and to .0050 at $x=2$. With $h=.12$, a similar descent is observed. But with $h=.13$, three steps bring us to the very unsatisfactory -29.11, followed by overflow. This definite blow-up speaks well for efforts to transfer our linear stability criteria to the nonlinear scene.

19.43. What can be done to control roundoff error?

In a long solution process, roundoff can become a serious factor. If double precision arithmetic is available, it should probably be used, in spite of the additional expense. It may be the only recourse. There is an intermediate step which may be helpful if the use of higher precision throughout the entire computation is deemed too time consuming. To illustrate, many of our formulas for solving differential equations amount to

$$y_{k+1} = y_k + h\,\Delta y_k$$

with the term Δy_k small compared with y_k itself. To perform the addition on the right, this small correction term has to be shifted (to line up the binary points) and this is where many roundoffs occur. To avoid them, the y_k are stored in double precision and this addition is done in double precision. The job of computing Δy_k, usually the heaviest work, is still done in single precision because this term is expected to be small anyway. In this way double precision is used only where it is needed most.

ADAPTIVE METHODS, VARIABLE STEP SIZE

19.44. How can the idea of adaptive integration, introduced in Problem 14.27, be extended to treat differential equations?

Suppose the goal is to solve $y' = f(x, y)$ approximately from an initial point $x = a$ to a terminal point $x = b$, arriving with an error no greater than e. Assume that the error will accumulate linearly, so that over a step of length h we can allow an error of size $eh/(b-a)$. This is precisely the idea of adaptive integration used earlier. Let T be an estimate of the truncation error made in taking the step of length h. Then if T does not exceed $eh/(b-a)$, this step is accepted and we move on to the next. Otherwise, the step size h is reduced (to $.5h$ or a suitable alternative) and the step repeated. With a convergent method the requirements will eventually be met, provided the step size h does not become so small that roundoff becomes the dominant error source.

If the Milne predictor–corrector method is being used, then Problem 19.20 provides the needed truncation error estimate $(P-C)/29$ and the condition for acceptance is

$$|P - C| \leqq \frac{29eh}{b-a}$$

which is easily computed from ingredients already in hand. If the Adams method is being used, then Problem 19.27 leads to the similar condition of acceptance

$$|P - C| \leqq \frac{\frac{270}{19}eh}{b-a}$$

In either case, rejection will require reactivating the supplementary start-up procedure.

19.45. To make Runge–Kutta methods adaptive, a practical way of estimating local truncation error is needed. Develop such an estimate, one that does not involve the higher derivatives of $y(x)$.

The now familiar idea of comparing errors for step sizes h and $2h$ will be used. Take the classical fourth-order method and make a step of size $2h$ from the current position x_k. The local error is about

$$T_{2h} = C(2h)^5 = 32Ch^5$$

Now cover the same interval in two steps of size h. The combined error is about

$$2T_h = 2Ch^5$$

leading to these two estimates of the true value y_{k+2}:

$$y_{k+2} = A_{2h} + 32Ch^5 = A_h + 2Ch^5$$

The subscripts $2h$ and h indicate the step sizes used in getting the two approximations. Subtraction now yields the value of C and the error estimate

$$T_h = Ch^5 = \frac{A_h - A_{2h}}{30}$$

which may be doubled for the full forward run. This estimate assumes that Ch^5 is an appropriate error measure and that C (with the higher derivatives imbedded) does not change much over the interval.

19.46. Use the error estimate of the preceding problem to make the Runge–Kutta method adaptive.

For the interval (a, b) let the allowable error be e. For this to be distributed proportionately, we ask that between x_k and x_{k+2} the local error not exceed $2eh/(b-a)$. If $2T_h$ as just estimated does not exceed this, that is, if

$$|A_h - A_{2h}| \leqq \frac{30eh}{b-a}$$

the value A_h can be accepted at x_{k+2} and one moves on. Otherwise a smaller step size h^* is needed such that the new truncation error T_{h^*} will be suitable. Returning to basics, we assume

$$T_h = Ch^5 \qquad T_{h^*} = Ch^{*5} = \frac{T_h h^{*5}}{h^5}$$

with the latter not to exceed $h^*e/(b-a)$ in magnitude. Putting the pieces together, the new step size is determined.

$$h^* = \left[\frac{eh^5}{(b-a)T_h}\right]^{1/4}$$

In view of the various assumptions made in deriving this formula, it is suggested that it not be pushed to the limit. An insurance factor of .8 is usually introduced. Moreover, if h is already quite small, and T_h small with it, the computation of h^* may even cause an overflow. The formula should be used with discretion.

19.47. Which methods are better for adaptive computing, the predictor–corrector pairs or Runge–Kutta?

Predictor–corrector methods have the advantage that ingredients for estimating local error are already in hand when needed. With Runge–Kutta a separate application of the formulas must be made, as just outlined. This almost doubles the number of times that $f(x, y)$ has to be evaluated, and since this is where the major computing effort is involved, running time may be almost doubled. On the other hand, and as said before, whenever the step size is changed it will be necessary to assist a predictor–corrector method in making a restart. This means extra programming, and if frequent changes are anticipated, it may be just as well to use Runge–Kutta throughout.

19.48. Try varying the step in the classical Runge–Kutta method as it solves the problem

$$y' = -xy^2 \qquad y(0) = 2$$

for which we have the exact solution $y = 2/(1+x^2)$.

The solution starts with a relatively sharp downward turn, then gradually levels off and becomes rather flat. So we anticipate the need for a small step size at the start and a gradual relaxation as things move along. It is interesting to watch these expectations develop in a run to $x = 27$.

x	.15	1	2	3	4	9	12	17	27
h	.07	.05	.1	.2	.3	.9	1.4	2.7	4.3

19.49. What are variable order methods?

Varying the order of the formulas used in integrating a differential equation is another way of trying to achieve a given level of accuracy with a minimum of computing. Starting with a low-order formula to make the process self-starting, and a small step size to keep it accurate, both are adjusted as computation proceeds. The idea is to find an optimal order and step size for the current step. A variety of professional programs are available for doing this, all somewhat complex, but the underlying strategy is similar to that in Problems 19.44 to 19.46.

STIFF EQUATIONS

19.50. What is a stiff differential equation?

The term is usually associated with a system of equations, but can be illustrated in principle at a simpler level. Take the equation

$$y' = -100y + 99e^{-x}$$

which has the solution

$$y = e^{-x} - e^{-100x}$$

satisfying the initial condition $y(0) = 0$. Both terms of this solution tend to zero, but the point is, the second decays much faster than the first. At $x = .1$, this term is already zero to four decimal places. It is truly a transient term compared with the first, which could almost be called the "steady state." Systems in which different components operate on quite different time scales are called stiff systems and offer more than normal resistance to numerical solution.

19.51. In view of the rapid decay of the above transient term, one might expect a step size of $h = .1$ to generate values of the remaining term e^{-x}. What does the classic Runge–Kutta method actually produce?

Much as in Problem 19.42, we have $f_y = -100$ and associate this with the A of our stability criterion, which becomes

$$-2.78 \leqq Ah = -100h$$

and suggests that we keep the step size h less than .0278. This is something of a surprise because it seems to imply that the transient term, negligible in size after $x = .1$, can still influence the computation in an important, underground way. Putting theory to the test, a run was made with $h = .03$. The predicted blow-up did occur, values of y quickly descending to the vicinity of -10^{14}. But using $h = .025$ led to a successful run, producing .04980 at $x = 3$. This is just one unit high in the fifth place.

19.52. Develop the Gear formula

$$\nabla y_{n+1} + \frac{1}{2}\nabla^2 y_{n+1} + \frac{1}{3}\nabla^3 y_{n+1} = hy'_{n+1}$$

where ∇ is the backward difference operator. Show that it is equivalent to

$$y_{n+1} = \frac{18}{11}y_n - \frac{9}{11}y_{n-1} + \frac{2}{11}y_{n-2} + \frac{6h}{11}y'_{n+1}$$

where

$$y'_{n+1} = f(x_{n+1}, y_{n+1})$$

Starting with the Newton backward formula

$$p_k = y_{n+1} + k\,\nabla y_{n+1} + \frac{k(k+1)}{2}\nabla^2 y_{n+1} + \frac{k(k+1)(k+2)}{6}\nabla^3 y_{n+1}$$

(see Problem 7.9) in which $x - x_{n+1} = kh$ and p_k is a polynomial of degree three in k collocating with y at $k = 0, -1, -2, -3$, we differentiate and set $k = 0$

$$\left.\frac{dp}{dx}\right|_{k=0} = \left.\frac{dp}{dk}\frac{1}{h}\right|_{k=0} = \frac{1}{h}\left(\nabla y_{n+1} + \frac{1}{2}\nabla^2 y_{n+1} + \frac{1}{3}\nabla^3 y_{n+1}\right)$$

Adopting this as an approximation to y'_{n+1}, we already have the first Gear formula. The second follows easily by replacing the backward differences with their equivalents in terms of the y_i.

These formulas can also be found by the method of undetermined coefficients, requiring exactness for polynomials of degree up to three. Corresponding formulas of higher order are available by extension. For example, if the Newton formula is extended back to $k = -4$, by introducing the fourth difference term, then $\frac{1}{4}\nabla^4 y_{n+1}$ is added to the left side above.

19.53. Why are the formulas of Gear preferred for solving stiff equations?

They prove to be stable for considerably larger values of h than our other formulas. Take once again the equation of Problem 19.50. We have found the Runge–Kutta method unstable for $h = .03$. In contrast, the Gear formula now reduces to

$$y_{n+1} = \frac{18y_n - 9y_{n-1} + 2y_{n-2} + 594he^{-(x_n+h)}}{11 + 600h}$$

upon inserting y' from the equation and then solving for y_{n+1}. With $h = .1$, this generated (using three correct starting values)

x	2	4	6
y	.135336	.018316	.002479

the first of which is one unit high in the final place. Even $h = .5$ can be considered a modest success.

x	2	4	6
y	.1350	.01833	.002480

The larger h brings more truncation error but there is no cause to complain about the stability.

19.54. The Gear formulas are usually nonlinear in y_{n+1}. Develop the Newton iteration as it applies to the extraction of this unknown.

In the above example $f(x, y)$ was linear in y, permitting a direct solution for y_{n+1}. Generally, however, we must view the Gear formula as

$$F(y) = y - \frac{6h}{11} f(x_{n+1}, y) - S = 0$$

where y_{n+1} has been abbreviated to y and S stands for the sum of three terms not involving y_{n+1}. Newton's iteration is then

$$y^{(k+1)} = y^{(k)} - \frac{F(y^{(k)})}{F'(y^{(k)})}$$

where $$F'(y) = 1 - \frac{6h}{11} f_y(x_{n+1}, y)$$

Supplementary Problems

19.55. By considering the direction field of the equation $y' = x^2 - y^2$, deduce the qualitative behavior of its solutions. Where will the solutions have maxima and minima? Where will they have zero curvature? Show that for large positive x we must have $y(x) < x$.

19.56. For the equation of the preceding problem try to estimate graphically where the solution through $(-1, 1)$ will be for $x = 0$.

19.57. By considering the direction field of the equation $y' = -2xy$, deduce the qualitative behavior of its solutions.

19.58. Apply the simple Euler method to $y' = -xy^2$, $y(0) = 2$, computing up to $x = 1$ with a few h intervals such as .5, .2, .1, .01. Do the results appear to converge toward the exact value $y(1) = 1$?

19.59. Apply the "midpoint formula" $y_{k+1} \simeq y_{k-1} + 2hf(x_k, y_k)$ to $y' = -xy^2$, $y(0) = 2$, using $h = .1$ and verifying the result $y(1) \simeq .9962$.

19.60. Apply the modified Euler method to $y' = -xy^2$, $y(0) = 2$ and compare the predictions of $y(1)$ obtained in the last three problems. Which of these very simple methods is performing best for the same h interval? Can you explain why?

19.61. Apply the local Taylor series method to the solution of $y' = -xy^2$, $y(0) = 2$, using $h = .2$. Compare your results with those in the solved problems.

19.62. Apply a Runge–Kutta method to the above problem and again compare your results.

19.63. Verify the first statement in Problem 19.9.

19.64. Apply the Milne predictor–corrector method to $y' = xy^{1/3}$, $y(1) = 1$, using $h = .1$. Compare results with those in the solved problems.

19.65. Apply the Adams predictor–corrector method to the above problem and again compare results.

19.66. Apply two or three other predictor–corrector combinations to Problem 19.64. Are there any substantial differences in the results?

19.67. Apply various methods to $y' = x^2 - y^2$, $y(-1) = 1$. What is $y(0)$ and how close was your estimate made in Problem 19.56?

19.68. Apply various methods to $y' = -2xy$, $y(0) = 1$. How do the results compare with the exact solution $y = e^{-x^2}$?

19.69. Show that Milne's method applied to $y' = y$ with $y(0) = 1$, using $h = .3$ and carrying four decimal places, leads to the following relative errors:

x	1.5	3.0	4.5	6.0
Rel. error	.00016	.00013	.00019	.00026

This means that the computation has steadily produced almost four significant digits.

19.70. Show that Milne's method applied to $y' = -y$ with $y(0) = 1$, using $h = .3$ and carrying five decimal places, leads to the following relative errors:

x	1.5	3.0	4.5	6.0
Rel. error	0	$-.0006$	.0027	$-.0248$

Though four almost correct decimal places are produced, the relative error has begun its growing oscillation.

19.71. Prove the instability of the midpoint method,

$$Y_{k+1} = Y_{k-1} + 2hf(x_k, Y_k)$$

Show that this formula has a lower truncation error than the Euler method, the exact solution satisfying

$$y_{k+1} = y_{k-1} + 2hf(x_k, y_k) + \frac{1}{3}h^3y^{(3)}(\xi)$$

For the special case $f(x, y) = Ay$, show that

$$d_{k+1} = d_{k-1} + 2hAd_k$$

ignoring the truncation error term in order to focus once again on the long-range effect of a single error d_0. Solve this difference equation by proving the roots of $r^2 - 2hAr - 1 = 0$ to be

$$r = hA \pm \sqrt{h^2A^2 + 1} = hA \pm 1 + 0(h^2)$$

For small hA these are near e^{hA} and $-e^{-hA}$ and the solution is

$$d_k = c_1(1 + Ah)^k + c_2(-1)^k(1 - Ah)^k \simeq c_1e^{Ahk} + c_2(-1)^ke^{-Ahk}$$

Setting $k = 0$, show that $d_0 = c_1 + c_2$. Dividng by y_k, the relative error becomes

$$r_k \simeq c_1 + (d_0 - c_1)(-1)^ke^{-2Ahk}$$

Show that for positive A this remains bounded, but that for negative A it grows without bound as k increases. The method is therefore unstable in this case.

19.72. The results in Table 19.6 were obtained by applying the midpoint method to the equation $y' = -xy^2$ with $y(0) = 2$. The interval $h = .1$ was used but only values for $x = .5(.5)5$ are printed. This equation is not linear, but calculate the relative error of each value and discover the rapidly increasing oscillation forecast by the analysis of the previous linear problem.

Table 19.6

x_k	Computed y_k	Exact y_k	x_k	Computed y_k	Exact y_k
.5	1.5958	1.6000	3.0	.1799	.2000
1.0	.9962	1.0000	3.5	.1850	.1509
1.5	.6167	.6154	4.0	.0566	.1176
2.0	.3950	.4000	4.5	.1689	.0941
2.5	.2865	.2759	5.0	−.0713	.0769

19.73. Analyze relative error for the other corrector formulas listed in Problem 19.27.

19.74. Show that the formula

$$y_{k+1} \simeq y_k + \frac{1}{2}h(y'_{k+1} + y'_k) + \frac{1}{12}h^2(-y''_{k+1} + y''_k)$$

has truncation error $h^5y^{(5)}(\xi)/720$, while the similar predictor

$$y_{k+1} \simeq y_k + \frac{1}{2}h(-y'_k + 3y'_{k-1}) + \frac{1}{12}h^2(17y''_k + 7y''_{k-1})$$

has truncation error $31h^5y^{(5)}(\xi)/6!$. These formulas use values of the second derivative to reduce truncation error.

19.75. Apply the formulas of the preceding problem to $y' = -xy^2$, $y(0) = 2$, using $h = .2$. One extra starting value is required and may be taken from an earlier solution of this same equation, say the Taylor series.

19.76. As a test case compute $y(\pi/2)$, given $y' = \sqrt{1 - y^2}$, $y(0) = 0$, using any of our approximation methods.

19.77. Use any of our approximation methods to find $y(2)$, given $y' = x - y$, $y(0) = 2$.

19.78. Solve by any of our approximation methods $y' = \dfrac{y(1 - x^2y^4)}{x(1 + x^2y^4)}$, $y(1) = 1$ up to $x = 2$.

19.79. Solve by any of our approximation methods $y' = -\dfrac{2xy + e^y}{x^2 + xe^y}$, $y(1) = 0$ up to $x = 2$.

19.80. Solve by any of our approximation methods $y' = -\dfrac{2x + y}{2y - x}$, $y(1) = 0$ up to $x = 2$.

19.81. An object falling toward the earth progresses, under the Newtonian theory with only the gravitational attraction of the earth considered, according to the equation (also see Problem 20.16)

$$\frac{dy}{dt} = -\sqrt{2gR^2}\sqrt{\frac{H - y}{Hy}}$$

where y = distance from the earth's center, $g = 32$, $R = 4000(5280)$, and H = initial distance from the earth's center. The exact solution of this equation can be shown to be

$$t = \frac{H^{3/2}}{8y}\left[\sqrt{\frac{y}{H} - \left(\frac{y}{H}\right)^2} + \frac{1}{2}\arccos\left(\frac{2y}{H} - 1\right)\right]$$

the initial speed being zero. But apply one of our approximation methods to the differential equation itself with initial condition $y(0) = H = 237{,}000(5280)$. At what time do you find that $y = R$? This result may be interpreted as the time required for the moon to fall to earth if it were stopped in its course and the earth remained stationary.

19.82. A raindrop of mass m has speed v after falling for time t. Suppose the equation of motion to be

$$\frac{dv}{dt} = 32 - \frac{cv^2}{m}$$

where c is a measure of air resistance. It can then be proved that the speed approaches a limiting value. Confirm this result by directly applying one of our approximate methods to the differential equation itself for the case $c/m = 2$. Use any initial speed.

19.83. A shot is fired upwards against air resistance of cv^2. Assume the equation of motion to be

$$\frac{dv}{dt} = -32 - \frac{cv^2}{m}$$

If $c/m = 2$ and $v(0) = 1$, apply one of our methods to find the time required for the shot to reach maximum height.

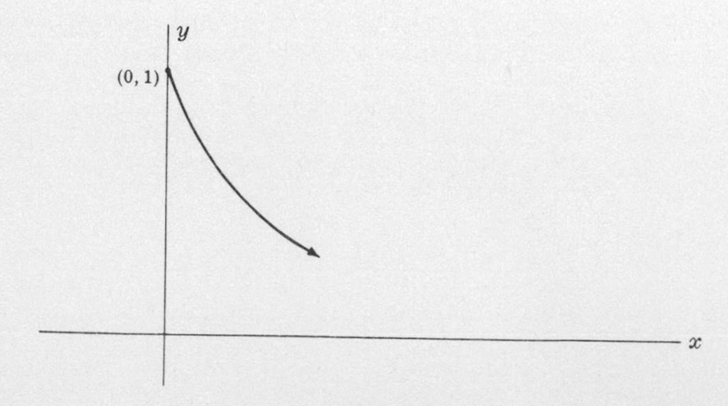

Fig. 19-7

19.84. One end of a rope of length L is carried along a straight line. The path of a weight attached to the other end is determined by (see Fig. 19-7)

$$y' = -\frac{y}{\sqrt{L^2 - y^2}}$$

The exact solution may be found. However, use one of our approximation methods to compute the path of the weight, starting from $(0, L)$. Take $L = 1$.

Chapter 20

Differential Problems of Higher Order

THE BASIC PROBLEM

A system of first-order differential equations such as

$$y_i' = f_i(x, y_1, \ldots, y_n) \qquad i = 1, \ldots, n$$

for determining the n functions $y_i(x)$, with given initial conditions $y_i(x_0) = a_i$, is the basic problem to be considered in this chapter. It arises in a wide variety of applications. That it is a direct generalization of the initial value problem treated in Chapter 19 is made especially plain by writing it in the vector form

$$Y'(x) = F(x, Y) \qquad Y(x_0) = A$$

where Y, F, and A have components y_i, f_i, and a_i, respectively.

An equation of higher order can be replaced by such a system of first-order equations and this is the standard method of treatment. As the simplest example, the second-order equation

$$y'' = f(x, y, y')$$

becomes the system

$$y' = p \qquad p' = f(x, y, p)$$

for the two functions y and p. The accompanying initial conditions $y(x_0) = a$, $y'(x_0) = b$ are replaced by $y(x_0) = a$ and $p(x_0) = b$. The basic problem above is then in hand. With a third-order equation, the definitions $y' = p$ and $y'' = q$ quickly lead to a system of three first-order equations, and so on. Systems of higher-order equations are handled by treating each as just described. The option is thus available to reduce any higher-order problem to a system of first-order equations.

SOLUTION METHODS

The methods of the preceding chapter are easily extended to systems of first-order equations. Taylor series are frequently appropriate, their application being quite direct.

Runge–Kutta methods also apply, each equation of the system being treated almost exactly as in Chapter 19. The same is true of predictor–corrector methods. Examples of such extensions will be provided in the solved problems.

Solved Problems

20.1. Illustrate the Taylor series procedure for simultaneous equations by solving the system

$$x' = -x - y$$
$$y' = x - y$$

for the two functions $x(t)$ and $y(t)$ satisfying initial conditions $x(0) = 1$, $y(0) = 0$.

We substitute directly into the two series

$$x(t) = x(0) + tx'(0) + \frac{1}{2}t^2x''(0) + \cdots$$

$$y(t) = y(0) + ty'(0) + \frac{1}{2}t^2y''(0) + \cdots$$

obtaining the needed ingredients from the given system. First $x'(0) = -1$ and $y'(0) = 1$. Then from $x'' = -x' - y'$ and $y'' = x' - y'$ come $x''(0) = 0$, $y''(0) = -2$. Higher derivatives follow in the same way. The series begin as follows:

$$x(t) = 1 - t + \frac{1}{3}t^3 - \frac{1}{6}t^4 + \cdots$$

$$y(t) = t - t^2 + \frac{1}{3}t^3 + \cdots$$

The given system is not only linear but also has constant coefficients. Writing it in the form

$$X'(t) = AX(t)$$

with

$$X = \begin{pmatrix} x \\ y \end{pmatrix} \qquad \text{and} \qquad A = \begin{bmatrix} -1 & -1 \\ 1 & -1 \end{bmatrix}$$

the exact solution can be found by trying

$$X = e^{\lambda t}\begin{pmatrix} a \\ b \end{pmatrix}$$

Substituting into the system leads to an eigenvalue problem for the matrix A. For the present A we have

$$\begin{aligned} (-1-\lambda)a - \qquad\qquad b &= 0 \\ a + (-1-\lambda)b &= 0 \end{aligned}$$

yielding $\lambda = -1 \pm i$ and after a slight effort

$$x(t) = e^{-t}\cos t \qquad y(t) = e^{-t}\sin t$$

The Taylor series begun above is, of course, the series for these functions.

The process as illustrated is easily extended to larger systems of equations.

20.2. Write out the Runge–Kutta formulas for two simultaneous first-order equations using the classical fourth-order set.

Let the given equations be

$$y' = f_1(x, y, p) \qquad p' = f_2(x, y, p)$$

with initial conditions $y(x_0) = y_0$, $p(x_0) = p_0$. The formulas

$$\begin{aligned} k_1 &= hf_1(x_n, y_n, p_n) & k_3 &= hf_1(x_n + \tfrac{1}{2}h, y_n + \tfrac{1}{2}k_2, p_n + \tfrac{1}{2}l_2) \\ l_1 &= hf_2(x_n, y_n, p_n) & l_3 &= hf_2(x_n + \tfrac{1}{2}h, y_n + \tfrac{1}{2}k_2, p_n + \tfrac{1}{2}l_2) \\ k_2 &= hf_1(x_n + \tfrac{1}{2}h, y_n + \tfrac{1}{2}k_1, p_n + \tfrac{1}{2}l_1) & k_4 &= hf_1(x_n + h, y_n + k_3, p_n + l_3) \\ l_2 &= hf_2(x_n + \tfrac{1}{2}h, y_n + \tfrac{1}{2}k_1, p_n + \tfrac{1}{2}l_1) & l_4 &= hf_2(x_n + h, y_n + k_3, p_n + l_3) \end{aligned}$$

$$y_{n+1} = y_n + \tfrac{1}{6}(k_1 + 2k_2 + 2k_3 + k_4)$$

$$p_{n+1} = p_n + \tfrac{1}{6}(l_1 + 2l_2 + 2l_3 + l_4)$$

may be shown to duplicate the Taylor series for both functions up through terms of order four. The details are identical with those for a single equation and will be omitted. For more than two simultaneous equations, say n, the extension of the Runge–Kutta method parallels the above, with n sets of formulas instead of two. For an example of such formulas in use see Problem 20.7.

20.3. Write out the Adams type predictor–corrector formula for the simultaneous equations of the preceding problem.

Assume that four starting values of each function are available, say y_0, y_1, y_2, y_3 and p_0, p_1, p_2, p_3.

Then the predictor formulas

$$y_{k+1} \simeq y_k + \frac{1}{24}h(55y'_k - 59y'_{k-1} + 37y'_{k-2} - 9y'_{k-3})$$

$$p_{k+1} \simeq p_k + \frac{1}{24}h(55p'_k - 59p'_{k-1} + 37p'_{k-2} - 9p'_{k-3})$$

may be applied with $\quad y'_k = f_1(x_k, y_k, p_k) \qquad p'_k = f_2(x_k, y_k, p_k)$

The results may be used to prime the corrector formulas

$$y_{k+1} \simeq y_k + \frac{1}{24}h(9y'_{k+1} + 19y'_k - 5y'_{k-1} + y'_{k-2})$$

$$p_{k+1} \simeq p_k + \frac{1}{24}h(9p'_{k+1} + 19p'_k - 5p'_{k-1} + p'_{k-2})$$

which are then iterated until consecutive outputs agree to a specified tolerance. The process hardly differs from that for a single equation. Extension to more equations or to other predictor–corrector combinations is similar.

HIGHER-ORDER EQUATIONS AS SYSTEMS

20.4. Show that a second-order differential equation may be replaced by a system of two first-order equations.

Let the second-order equation be $y'' = f(x, y, y')$. Then introducing $p = y'$ we have at once $y' = p$, $p' = f(x, y, p)$. As a result of this standard procedure a second-order equation may be treated by system methods if this seems desirable.

20.5. Show that the general nth-order equation

$$y^{(n)} = f(x, y, y', y^{(2)}, \dots, y^{(n-1)})$$

may also be replaced by a system of first-order equations.

For convenience we assign $y(x)$ the alias $y_1(x)$ and introduce the additional functions $y_2(x), \dots, y_n(x)$ by

$$y'_1 = y_2 \qquad y'_2 = y_3 \qquad \cdots \qquad y'_{n-1} = y_n$$

Then the original nth-order equation becomes

$$y'_n = f(x, y_1, y_2, \dots, y_n)$$

These n equations are of first order and may be solved by system methods.

20.6. Replace the following equations for the motion of a particle in three dimensions:

$$x'' = f_1(t, x, y, z, x', y', z') \qquad y'' = f_2(t, x, y, z, x', y', z') \qquad z'' = f_3(t, x, y, z, x', y', z')$$

by an equivalent system of first-order equations.

Let $x' = u$, $y' = v$, $z' = w$ be the velocity components. Then

$$u' = f_1(t, x, y, z, u, v, w) \qquad v' = f_2(t, x, y, z, u, v, w) \qquad w' = f_3(t, x, y, z, u, v, w)$$

These six equations are the required first-order system. Other systems of higher-order equations may be treated in the same way.

20.7. Compute the solution of van der Pol's equation

$$y'' - (.1)(1 - y^2)y' + y = 0$$

with initial values $y(0)=1$, $y'(0)=0$ up to the third zero of $y(t)$. Use the Runge–Kutta formulas for two first-order equations.

An equivalent first-order system is

$$y' = p = f_1(t, y, p)$$
$$p' = -y + (.1)(1-y^2)p = f_2(t, y, p)$$

The Runge–Kutta formulas for this system are

$$k_1 = hp_n \qquad l_1 = h[-y_n + (.1)(1-y_n^2)p_n]$$
$$k_2 = h\left(p_n + \frac{1}{2}l_1\right) \qquad l_2 = h\left\{-\left(y_n + \frac{1}{2}k_1\right) + (.1)\left[1 - \left(y_n + \frac{1}{2}k_1\right)^2\right]\left(p_n + \frac{1}{2}l_1\right)\right\}$$
$$k_3 = h\left(p_n + \frac{1}{2}l_2\right) \qquad l_3 = h\left\{-\left(y_n + \frac{1}{2}k_2\right) + (.1)\left[1 - \left(y_n + \frac{1}{2}k_2\right)^2\right]\left[p_n + \frac{1}{2}l_2\right)\right\}$$
$$k_4 = h(p_n + l_3) \qquad l_4 = h\{-(y_n + k_3) + (.1)[1 - (y_n + k_3)^2](p_n + l_3)\}$$

and
$$y_{n+1} = y_n + \frac{1}{6}(k_1 + 2k_2 + 2k_3 + k_4) \qquad p_{n+1} = p_n + \frac{1}{6}(l_1 + 2l_2 + 2l_3 + l_4)$$

Choosing $h = .2$, computations produce the following results to three places:

$$k_1 = (.2)(0) = 0 \qquad l_1 = (.2)[-1 + (.1)(1-1)(0)] = -.2$$
$$k_2 = (.2)(-.1) = -.02 \qquad l_2 = (.2)[-1 + (.1)(1-1)(-.1)] = -.2$$
$$k_3 \simeq (.2)(-.1) = -.02 \qquad l_3 = (.2)[-.99 + (.1)(.02)(-.1)] \simeq -.198$$
$$k_4 \simeq (.2)(-.198) \simeq -.04 \qquad l_4 = (.2)[-(.98) + (.1)(.04)(-.198)] \simeq -.196$$

These values now combine into

$$y_1 \simeq 1 + \frac{1}{6}(-.04 - .04 - .04) = .98$$

$$p_1 \simeq 0 + \frac{1}{6}(-.2 - .4 - .396 - .196) \simeq -.199$$

The second step now follows with $n = 1$ and the computation is continued in this way. Results up to $t = 6.4$ when the curve has crossed below the y axis again are illustrated in Fig. 20-1, in which y and p values serve as coordinates. This "phase plane" is often used in the study of oscillatory systems. Here the oscillation (shown solid) is growing and will approach the periodic oscillation (shown dashed) as x tends to infinity. This is proved in the theory of nonlinear oscillations.

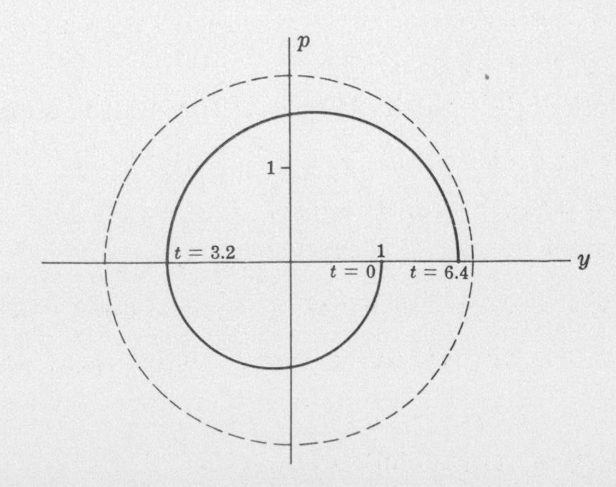

Fig. 20-1

HIGHER-ORDER EQUATIONS SOLVED BY SERIES

20.8. Obtain a series solution of the linear equation $y'' + (1+x^2)y = e^x$ in the neighborhood of $x=0$.

Let the series be $y(x) = \sum_{i=0}^{\infty} a_i x^i$ and substitute to obtain

$$\sum_{i=2}^{\infty} a_i i(i-1)x^{i-2} + (1+x^2)\sum_{i=0}^{\infty} a_i x^i = \sum_{i=0}^{\infty} \frac{x^i}{i!}$$

which can be converted by changes of indices to

$$(a_0 + 2a_2) + (a_1 + 6a_3)x + \sum_{k=2}^{\infty} [(k+2)(k+1)a_{k+2} + a_k + a_{k-2}]x^k = \sum_{k=0}^{\infty} \frac{x^k}{k!}$$

Comparing coefficients of the powers of x brings $a_2 = (1-a_0)/2$, $a_3 = (1-a_1)/6$, and then the recursion

$$(k+2)(k+1)a_{k+2} = -a_k - a_{k-2} + \frac{1}{k!}$$

which yields successively $a_4 = -a_0/24$, $a_5 = -a_1/24$, $a_6 = (13a_0 - 11)/720$, and so on. The numbers a_0 and a_1 would be determined by initial conditions.

A similar series could be developed near any other argument x, since the ingredients of our differential equation are analytic functions. Such series may be adequate for computation of the solution over the interval required, or if not, serve to generate starting values for other methods.

20.9. Obtain a series solution of the nonlinear equation $y'' = 1 + y^2$ in the neighborhood of $x=0$, with $y(0) = y'(0) = 0$.

The method of the preceding problem could be used, but the alternative of computing the higher derivatives directly will be illustrated once again. We easily compute

$$y^{(3)} = 2yy' \qquad y^{(4)} = 2y(1+y^2) + 2(y')^2 \qquad y^{(5)} = 10y^2y' + 6y' \qquad y^{(6)} = 20y(y')^2 + (1+y^2)(10y^2+6)$$

and so on. With the initial conditions given these are all zero except for $y^{(6)}$, and by Taylor's theorem $y = \frac{1}{2}x^2 + \frac{1}{120}x^6 + \cdots$.

20.10. Apply the Gear method of Problem 19.52 to the stiff system

$$y' = p$$
$$p' = -100y - 101p$$

with initial conditions $y(0) = 1$ and $p(0) = -1$. This system is equivalent to the second-order equation

$$y'' + 101y' + 100y = 0$$

with $y = 1$ and $y' = -1$ initially. The exact solution is $y(x) = e^{-x}$.

Runge–Kutta methods could handle this system, but the classic fourth-order set would require a step size less than .0278 for a stable computation. Writing out the Gear formula for both y and p we have

$$y_{n+1} = \frac{1}{11}(18y_n - 9y_{n-1} + 2y_{n-2}) + \frac{6h}{11}p_{n+1}$$

$$p_{n+1} = \frac{1}{11}(18p_n - 9p_{n-1} + 2p_{n-2}) + \frac{6h}{11}(-100y_{n+1} - 101p_{n+1})$$

which can be rewritten as a linear system for y_{n+1} and p_{n+1}:

$$y_{n+1} - \frac{6h}{11} p_{n+1} = \frac{1}{11}(18y_n - 9y_{n-1} + 2y_{n-2})$$

$$\frac{600h}{11} y_{n+1} + \left(1 + \frac{606h}{11}\right) p_{n+1} = \frac{1}{11}(18p_n - 9p_{n-1} + 2p_{n-2})$$

Since the system is linear, there is no need to use the Newton iteration for its solution. Results for two choices of step size h appear below, both much larger than what is needed for Runge–Kutta. The true values are also listed for comparison.

x	$y = e^{-x}$	$h = .1$	$h = .2$
2	.1353	.1354	.1359
4	.01832	.01833	.0185
6	.002479	.002483	.00251
8	.0003355	.0003362	.000342
10	.0000454	.0000455	.0000465

20.11. A dog, out in a field, sees his master walking along the road and runs toward him. Assuming that the dog always aims directly at his master, and that the road is straight, the equation governing the dog's path is (see Fig. 20-2)

$$xy'' = c\sqrt{1 + (y')^2}$$

with c the ratio of the man's speed to the dog's. A well-known line of attack leads to the exact solution

$$y = \frac{1}{2}\left(\frac{x^{1+c}}{1+c} - \frac{x^{1-c}}{1-c}\right) + \frac{c}{1-c^2}$$

for c less than one. As x approaches zero, the dog catches his master at position $y = c/(1-c^2)$. Solve this problem by an approximate method for the case $c = \frac{1}{2}$. The chase should end at $y = \frac{2}{3}$.

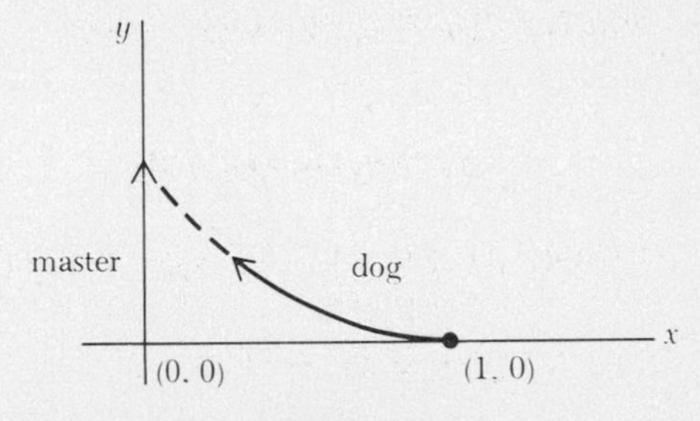

Fig. 20-2

The second-order equation is first replaced by the system

$$y' = p$$

$$p' = \frac{c\sqrt{1+p^2}}{x}$$

and the initial conditions by $y(1) = 0$, $p(1) = 0$. The Runge–Kutta formulas of Problem 20.2 can again be used, this time with a negative h. The only difficulty here is that as x nears zero the slope p grows very large. An adaptive method, with h decreasing in size, seems to be indicated. A primitive strategy was attempted, with $h = -.1$ down to $x = .1$, then $h = -.01$ down to $x = .01$, and so on. The results appear as Table 20.1. The last two x entries appear to contain roundoff error. Values of p are not listed but rose to nearly 1000 in size.

Table 20.1

x	y
.1	.3608
.01	.5669
.001	.6350
.0001	.6567
.00001	.6636
.0000006	.6659
−.0000003	.6668

20.12. The equations

$$r'' = \frac{9}{r^3} - \frac{2}{r^2} \qquad \theta' = \frac{3}{r^2}$$

in which primes refer to differentiation relative to time t, describe the Newtonian orbit of a particle in an inverse square gravitational field, after suitable choices of some physical constants. If $t = 0$ at the position of minimum r (Fig. 20-3) and

$$r(0) = 3 \qquad \theta(0) = 0 \qquad r'(0) = 0$$

then the orbit proves to be the ellipse $r = 9/(2 + \cos\theta)$. Use one of our approximation methods and compare with this exact result.

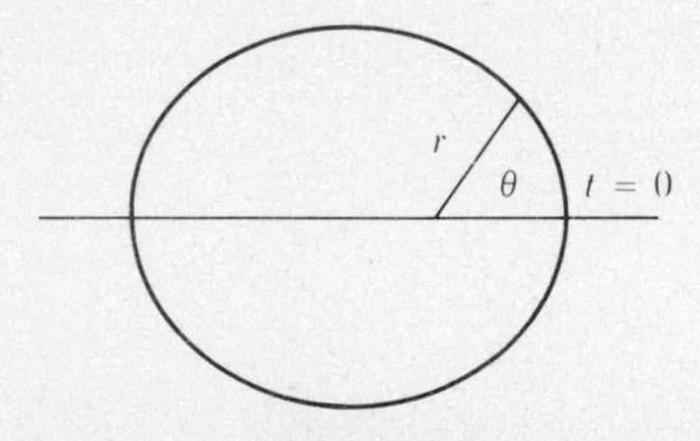

Fig. 20-3

The application is quite straightforward. The familiar reduction to a first-order system comes first,

$$r' = p \qquad p' = \frac{9}{r^3} - \frac{2}{r^2} \qquad \theta' = \frac{3}{r^2}$$

followed by the programming of three sets of Runge–Kutta formulas, still following the model of Problem 20.2. Integration continued until the angle θ exceeded 2π. A selected fragment of the output is provided as Table 20.2 (step size $h = .1$ was used) and it clearly has the desired orbital quality. As a further check, theory offers the period $T = 12\pi\sqrt{3}$, or about 65.3, and this fits in very nicely.

Table 20.2

t	r	θ	p
0	3.00	.00	.00
6	4.37	1.51	.33
7	4.71	1.66	.33
32	9.00	3.12	.01
33	9.00	3.15	−.004
59	4.47	4.73	−.33
65	3.00	6.18	−.03
66	3.03	6.52	.08

Supplementary Problems

20.13. The equations

$$x'(t) = -\frac{2x}{\sqrt{x^2+y^2}} \qquad y'(t) = 1 - \frac{2y}{\sqrt{x^2+y^2}}$$

describe the path of a duck attempting to swim across a river by aiming steadily at the target position T. The speed of the river is 1 and the duck's speed is 2. The duck starts at S, so that $x(0) = 1$ and $y(0) = 0$. (See Fig. 20-4.) Apply the Runge–Kutta formulas for two simultaneous equations to compute the duck's path. Compare with the exact trajectory $y = \frac{1}{2}(x^{1/2} - x^{3/2})$. How long does it take the duck to reach the target?

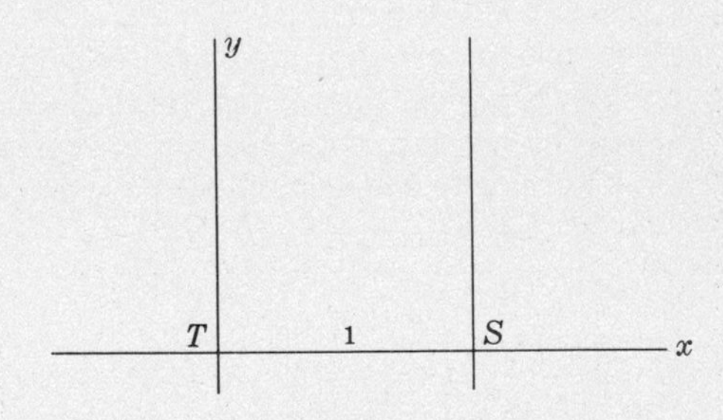

Fig. 20-4

20.14. Solve the preceding problem by the Adams predictor–corrector method.

20.15. Apply the Milne method to Problem 20.13.

20.16. The classical inverse square law for an object falling toward an attracting gravitational mass (say the earth) is

$$y''(t) = -\frac{gR^2}{y^2}$$

where g is a constant and R is the earth's radius. This has the well-known and somewhat surprising solution

$$t = \frac{H^{3/2}}{8y}\left[\sqrt{\frac{y}{H} - \left(\frac{y}{H}\right)^2} + \frac{1}{2}\arccos\left(\frac{2y}{H} - 1\right)\right]$$

where H is the initial altitude and the initial speed is zero. Introducing the equivalent system

$$y' = p \qquad p' = -\frac{gR^2}{y^2}$$

apply the Runge–Kutta formulas to compute the velocity $p(t)$ and position $y(t)$. When does the falling object reach the earth's surface? Compare with the exact result. (If miles and seconds are used as units, then $g = \frac{32}{5280}$, $R = 4000$, and take H to be 200,000 which is the moon's distance from earth. The problem illustrates some of the difficulties of computing space trajectories.)

20.17. Apply the Adams method to Problem 20.16.

20.18. Show that the solution of $yy'' + 3(y')^2 = 0$ with $y(0) = 1$ and $y'(0) = \frac{1}{4}$ can be expressed as

$$y(x) = 1 + \frac{x}{4} - \frac{3x^2}{32} + \frac{7x^3}{128} - \frac{77x^4}{2048} + \cdots$$

20.19. Show that $x^2y'' - 2x^2y' + (\frac{1}{4} + x^2)y = 0$ has a solution of the form

$$y(x) = \sqrt{x}(a_0 + a_1x + a_2x^2 + \cdots)$$

and determine the coefficients if the condition $\lim \frac{y(x)}{\sqrt{x}} = 1$ is required for x approaching zero.

20.20. Apply the Runge–Kutta formulas to

$$y' = -12y + 9z \qquad z' = 11y - 10z$$

which have the exact solution

$$y = 9e^{-x} + 5e^{-21x} \qquad z = 11e^{-x} - 5e^{-21x}$$

using $y(1) \simeq 9e^{-1}$, $z(1) \simeq 11e^{-1}$ as initial conditions. Work to three or four decimal places with $h = .2$ and carry the computation at least to $x = 3$. Notice that $11y/9z$, which should remain close to one, begins to oscillate badly. Explain this by comparing the fourth-degree Taylor approximation to e^{-21x} (which the Runge–Kutta method essentially uses) with the exact exponential.

Chapter 21

Least-Squares Polynomial Approximation

THE LEAST-SQUARES PRINCIPLE

The basic idea of choosing a polynomial approximation $p(x)$ to a given function $y(x)$ in a way which minimizes the squares of the errors (in some sense) was developed first by Gauss. There are several variations, depending on the set of arguments involved and the error measure to be used.

First of all, *when the data are discrete* we may minimize the sum

$$S = \sum_{i=0}^{N} (y_i - a_0 - a_1 x_i - \cdots - a_m x_i^m)^2$$

for given data x_i, y_i, and $m < N$. The condition $m < N$ makes it unlikely that the polynomial

$$p(x) = a_0 + a_1 x + a_2 x^2 + \cdots + a_m x^m$$

can collocate at all N data points. So S probably cannot be made zero. The idea of Gauss is to make S as small as we can. Standard techniques of calculus then lead to the *normal equations,* which determine the coefficients a_j. These equations are

$$\begin{aligned} s_0 a_0 + s_1 a_1 + \cdots + s_m a_m &= t_0 \\ s_1 a_0 + s_2 a_1 + \cdots + s_{m+1} a_m &= t_1 \\ \cdots\cdots\cdots\cdots\cdots\cdots\cdots\cdots \\ s_m a_0 + s_{m+1} a_1 + \cdots + s_{2m} a_m &= t_m \end{aligned}$$

where $s_k = \sum_{i=0}^{N} x_i^k$, $t_k = \sum_{i=0}^{N} y_i x_i^k$. This system of linear equations does determine the a_i uniquely, and the resulting a_j do actually produce the minimum possible value of S. For the case of a linear polynomial

$$p(x) = Mx + B$$

the normal equations are easily solved and yield

$$M = \frac{s_0 t_1 - s_1 t_0}{s_0 s_2 - s_1^2} \qquad B = \frac{s_2 t_0 - s_1 t_1}{s_0 s_2 - s_1^2}$$

In order to provide a unifying treatment of the various least-squares methods to be presented, including this first method just described, a general problem of minimization in a vector space is considered. The solution is easily found by an algebraic argument, using the idea of *orthogonal projection.* Naturally the general problem reproduces our $p(x)$ and normal equations. It will be reinterpreted to solve other variations of the least-squares principle as we proceed. In most cases a duplicate argument for the special case in hand will also be provided.

Except for very low degree polynomials, the above system of normal equations proves to be *ill-conditioned.* This means that, although it does define the coefficients a_j uniquely, in practice it may prove to be impossible to extricate these a_j. Standard methods for solving linear systems (to be presented in Chapter 26) may either produce no solution at all, or else badly magnify data errors. As a result, *orthogonal polynomials* are introduced. (This amounts to choosing an orthogonal basis for the abstract vector space.) For the case of discrete data these are polynomials $P_{m,N}(t)$ of degree $m = 0, 1, 2, \ldots$ with the property

$$\sum_{t=0}^{N} P_{m,N}(t) P_{n,N}(t) = 0$$

This is the **orthogonality property**. The explicit representation

$$P_{m,N}(t)=\sum_{i=0}^{m}(-1)^i\binom{m}{i}\binom{m+i}{i}\frac{t^{(i)}}{N^{(i)}}$$

will be obtained, in which binomial coefficients and factorial polynomials are prominent.

An alternate form of our least-squares polynomial now becomes convenient, namely

$$p(t)=\sum_{k=0}^{m}a_kP_{k,N}(t)$$

with new coefficients a_k. The equations determining these a_k prove to be extremely easy to solve. In fact,

$$a_k=\frac{\sum_{t=0}^{N}y_tP_{k,N}(t)}{\sum_{t=0}^{N}P_{k,N}^2(t)}$$

These a_k do minimize the error sum S, the minimum being

$$S_{\min}=\sum_{t=0}^{N}y_t^2-\sum_{k=0}^{m}W_ka_k^2$$

where W_k is the denominator sum in the expression for a_k.

APPLICATIONS

There are two major applications of least-squares polynomials for discrete data.

1. **Data smoothing.** By accepting the polynomial

$$p(x)=a_0+a_1x+\cdots+a_mx^m$$

in place of the given $y(x)$, we obtain a smooth line, parabola, or other curve in place of the original, probably irregular, data function. What degree $p(x)$ should have depends on the circumstances. Frequently a five-point least-squares parabola is used, corresponding to points (x_i, y_i) with $i=k-2, k-1, \ldots, k+2$. It leads to the smoothing formula

$$y(x_k)\simeq p(x_k)=y_k-\frac{3}{35}\delta^4y_k$$

This formula blends together the five values $y_{k-2}, \ldots, y_{k+2}$ to provide a new estimate to the unknown exact value $y(x_k)$. Near the ends of a finite data supply, minor modifications are required.

The **root-mean-square error** of a set of approximations A_i to corresponding true values T_i is defined as

$$\text{RMS error}=\left[\sum_{i=0}^{N}\frac{(T_i-A_i)^2}{N}\right]^{1/2}$$

In various test cases, where the T_i are known, we shall use this error measure to estimate the effectiveness of least-squares smoothing.

2. **Approximate differentiation.** As we saw earlier, fitting a collocation polynomial to irregular data leads to very poor estimates of derivatives. Even small errors in the data are magnified to troublesome size. But a least-squares polynomial does not collocate. It passes between the data values and provides smoothing. This smoother function usually brings better estimates of derivatives, namely, the values of $p'(x)$. The five-point parabola just mentioned

leads to the formula

$$y'(x_k) \simeq p'(x_k) = \frac{1}{10h}(-2y_{k-2} - y_{k-1} + y_{k+1} + 2y_{k+2})$$

Near the ends of a finite data supply this also requires modification. The formula usually produces results much superior to those obtained by differentiating collocation polynomials. However, reapplying it to the $p'(x_k)$ values in an effort to estimate $y''(x_k)$ again leads to questionable accuracy.

CONTINUOUS DATA

For continuous data $y(x)$ we may minimize the integral

$$I = \int_{-1}^{1} [y(x) - a_0 P_0(x) - \cdots - a_m P_m(x)]^2 \, dx$$

the $P_j(x)$ being Legendre polynomials. [We must assume $y(x)$ integrable.] This means that we have chosen to represent our least-squares polynomial $p(x)$ from the start in terms of orthogonal polynomials, in the form

$$p(x) = a_0 P_0(x) + \cdots + a_m P_m(x)$$

The coefficients prove to be

$$a_k = \frac{2k+1}{2} \int_{-1}^{1} y(x) P_k(x) \, dx$$

For convenience in using the Legendre polynomials, the interval over which the data $y(x)$ are given is first normalized to $(-1, 1)$. Occasionally it is more convenient to use the interval $(0, 1)$. In this case the Legendre polynomials must also be subjected to a change of argument. The new polynomials are called **shifted Legendre polynomials.**

Some type of discretization is usually necessary when $y(x)$ is of complicated structure. Either the integrals which give the coefficients must be computed by approximation methods, or the continuous argument set must be discretized at the outset and a sum minimized rather than an integral. Plainly there are several alternate approaches and the computer must decide which to use for a particular problem.

Smoothing and approximate differentiation of the given continuous data function $y(x)$ are again the foremost applications of our least-squares polynomial $p(x)$. We simply accept $p(x)$ and $p'(x)$ as substitutes for the more irregular $y(x)$ and $y'(x)$.

A generalization of the least-squares principle involves minimizing the integral

$$I = \int_a^b w(x)[y(x) - a_0 Q_0(x) - \cdots - a_m Q_m(x)]^2 \, dx$$

where $w(x)$ is a nonnegative weight function. The $Q_k(x)$ are orthogonal polynomials in the generalized sense

$$\int_a^b w(x) Q_j(x) Q_k(x) \, dx = 0$$

for $j \neq k$. The details parallel those for the case $w(x) = 1$ already mentioned, the coefficients a_k being given by

$$a_k = \frac{\int_a^b w(x) y(x) Q_k(x) \, dx}{\int_a^b w(x) Q_k^2(x) \, dx}$$

The minimum value of I can be expressed as

$$I_{\min} = \int_a^b w(x)y^2(x)\,dx - \sum_{k=0}^{m} W_k a_k^2$$

where W_k is the denominator integral in the expression for a_k. This leads to **Bessel's inequality**

$$\sum_{k=0}^{m} W_k a_k^2 \leqq \int_a^b w(x)y^2(x)\,dx$$

and to the fact that for m tending to infinity the series $\sum_{k=0}^{\infty} W_k a_k^2$ is convergent. If the orthogonal family involved has a property known as **completeness** and if $y(x)$ is sufficiently smooth, then the series actually converges to the integral which appears in $I_{\min}$. This means that the error of approximation tends to zero as the degree of $p(x)$ is increased.

CHEBYSHEV POLYNOMIALS

Approximation using Chebyshev polynomials is the important special case $w(x) = 1/\sqrt{1-x^2}$ of the generalized least-squares method, the interval of integration being normalized to $(-1, 1)$. In this case the orthogonal polynomials $Q_k(x)$ are the Chebyshev polynomials

$$T_k(x) = \cos(k \arccos x)$$

The first few prove to be $\quad T_0(x) = 1 \quad T_1(x) = x \quad T_2(x) = 2x^2 - 1 \quad T_3(x) = 4x^3 - 3x$

Properties of the Chebyshev polynomials include

$$T_{n+1}(x) = 2xT_n(x) - T_{n-1}(x)$$

$$\int_{-1}^{1} \frac{T_m(x)T_n(x)}{\sqrt{1-x^2}}\,dx = \begin{cases} 0 & \text{if } m \neq n \\ \pi/2 & \text{if } m = n \neq 0 \\ \pi & \text{if } m = n = 0 \end{cases}$$

$$T_n(x) = 0 \qquad \text{for } x = \cos[(2i+1)\pi/2n],\ i = 0, 1, \ldots, n-1$$

$$T_n(x) = (-1)^i \qquad \text{for } x = \cos(i\pi/n),\ i = 0, 1, \ldots, n$$

An especially attractive property is the *equal-error property,* which refers to the oscillation of the Chebyshev polynomials between extreme values of ± 1, reaching these extremes at $n+1$ arguments inside the interval $(-1, 1)$. As a consequence of this property the error $y(x) - p(x)$ is frequently found to oscillate between maxima and minima of approximately $\pm E$. Such an almost-equal-error is desirable since it implies that our approximation has almost uniform accuracy across the entire interval. For an *exact* equal-error property see the next chapter.

The powers of x may be expressed in terms of Chebyshev polynomials by simple manipulations. For example,

$$1 = T_0 \qquad x = T_1 \qquad x^2 = \frac{1}{2}(T_0 + T_2) \qquad x^3 = \frac{1}{4}(3T_1 + T_3)$$

This has suggested a process known as *economization of polynomials,* by which each power of x in a polynomial is replaced by the corresponding combination of Chebyshev polynomials. It is often found that a number of the higher-degree Chebyshev polynomials may then be dropped, the terms retained then constituting a least-squares approximation to the original polynomial, of sufficient accuracy for many purposes. The result obtained will have the almost-equal-error property. This process of economization may be used as an approximate substitute for direct evaluation of the coefficient integrals of an approximation by Chebyshev polynomials. The unpleasant weight factor $w(x)$ makes these integrals formidable for most $y(x)$.

Another variation of the least-squares principle is to minimize the sum

$$\sum_{i=0}^{N-1} [y(x_i) - a_0 T_0(x_i) - \cdots - a_m T_m(x_i)]^2$$

the arguments being $x_i = \cos[(2i+1)\pi/2N]$. These arguments may be recognized as the zeros of $T_N(x)$. The coefficients are easily determined using a second orthogonality property of the Chebyshev polynomials,

$$\sum_{i=0}^{N-1} T_m(x_i)T_n(x_i) = \begin{cases} 0 & \text{if } m \neq n \\ N/2 & \text{if } m = n \neq 0 \\ N & \text{if } m = n = 0 \end{cases}$$

and prove to be
$$a_0 = \frac{1}{N}\sum_{i=0}^{N-1} y(x_i) \qquad a_k = \frac{2}{N}\sum_{i=0}^{N-1} y(x_i)T_k(x_i)$$

The approximating polynomial is then, of course,

$$p(x) = a_0 T_0(x) + \cdots + a_m T_m(x)$$

This polynomial also has an almost-equal-error.

THE L_2 NORM

The underlying theme of this chapter is to minimize the norm

$$\|y - p\|_2$$

where y represents the given data and p the approximating polynomial.

Solved Problems

DISCRETE DATA, THE LEAST-SQUARES LINE

21.1. Find the straight line $p(x) = Mx + B$ for which $\sum_{i=0}^{N} (y_i - Mx_i - B)^2$ is a minimum, the data (x_i, y_i) being given.

Calling the sum S, we follow a standard minimum-finding course and set derivatives to zero.

$$\frac{\partial S}{\partial B} = -2\sum_{i=0}^{N} 1 \cdot (y_i - Mx_i - B) = 0 \qquad \frac{\partial S}{\partial M} = -2\sum_{i=0}^{N} x_i \cdot (y_i - Mx_i - B) = 0$$

Rewriting we have

$$(N+1)B + \left(\sum x_i\right)M = \sum y_i \qquad \left(\sum x_i\right)B + \left(\sum x_i^2\right)M = \sum x_i y_i$$

which are the "normal equations." Introducing the symbols

$$s_0 = N+1 \qquad s_1 = \sum x_i \qquad s_2 = \sum x_i^2 \qquad t_0 = \sum y_i \qquad t_1 = \sum x_i y_i$$

these equations may be solved in the form

$$M = \frac{s_0 t_1 - s_1 t_0}{s_0 s_2 - s_1^2} \qquad B = \frac{s_2 t_0 - s_1 t_1}{s_0 s_2 - s_1^2}$$

To show that $s_0 s_2 - s_1^2$ is not zero, we may first notice that squaring and adding terms such as $(x_0 - x_1)^2$ leads to

$$0 < \sum_{i<j} (x_i - x_j)^2 = N \cdot \sum x_i^2 - 2\sum_{i<j} x_i x_j$$

But also
$$(\sum x_i)^2 = \sum x_i^2 + 2\sum_{i<j} x_i x_j$$
so that $s_0 s_2 - s_1^2$ becomes
$$(N+1)\sum x_i^2 - (\sum x_i)^2 = N \cdot \sum x_i^2 - 2\sum_{i<j} x_i x_j > 0$$
Here we have assumed that the x_i are not all the same, which is surely reasonable. This last inequality also helps to prove that the M and B chosen actually produce a minimum. Calculating second derivatives, we find
$$\frac{\partial^2 S}{\partial B^2} = 2s_0 \qquad \frac{\partial^2 S}{\partial M^2} = 2s_2 \qquad \frac{\partial^2 S}{\partial B\,\partial M} = 2s_1$$
Since the first two are positive and since
$$(2s_1)^2 - 2(N+1)(2s_2) = 4(s_1^2 - s_0 s_2) < 0$$
the second derivative test for a minimum of a function of two arguments B and M is satisfied. The fact that the first derivatives can vanish together only once shows that our minimum is an absolute minimum.

21.2. The average scores reported by golfers of various handicaps on a difficult par-three hole are as follows:

Handicap	6	8	10	12	14	16	18	20	22	24
Average	3.8	3.7	4.0	3.9	4.3	4.2	4.2	4.4	4.5	4.5

Find the least-squares linear function for this data by the formulas of Problem 21.1.

Let h represent handicap and $x = (h-6)/2$. Then the x_i are the integers $0, \ldots, 9$. Let y represent average score. Then $s_0 = 10$, $s_1 = 45$, $s_2 = 285$, $t_0 = 41.5$, $t_1 = 194.1$ and so
$$M = \frac{(10)(194.1) - (45)(41.5)}{(10)(285) - (45)^2} \simeq .089 \qquad B = \frac{(285)(41.5) - (45)(194.1)}{(10)(285) - (45)^2} \simeq 3.76$$
This makes $y \simeq p(x)$ where $p(x) = .09x + 3.76 \simeq .045h + 3.49$.

21.3. Use the least-squares line of the previous problem to smooth the reported data.

The effort to smooth data proceeds on the assumption that the reported data contain inaccuracies of a size to warrant correction. In this case the data seem to fall roughly along a straight line, but there are large fluctuations, due perhaps to the natural fluctuations in a golfer's game. (See Fig. 21-1 below.) The

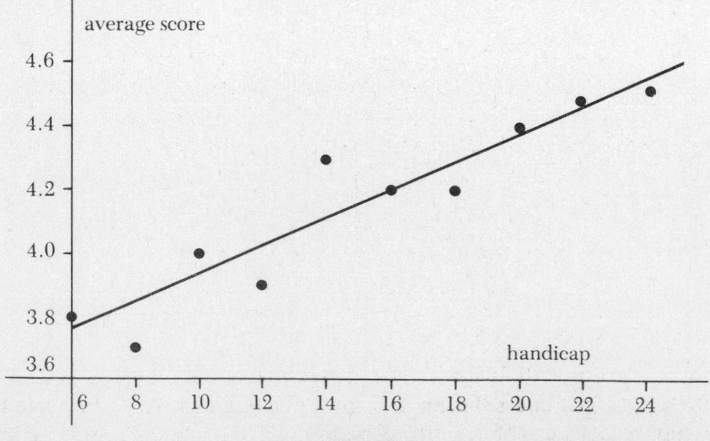

Fig. 21-1

least-squares line may be assumed to be a better representation of the true relationship between the handicap and the average scores than the original data are. It yields the following smoothed values:

Handicap	6	8	10	12	14	16	18	20	22	24
Smoothed y	3.76	3.85	3.94	4.03	4.12	4.21	4.30	4.39	4.48	4.57

21.4. Estimate the rate at which the average score increases per unit handicap.

From the least-squares line of Problem 21.2 we obtain the estimate .045 stroke per unit handicap.

21.5. Obtain a formula of the type $P(x) = Ae^{Mx}$ from the following data:

x_i	1	2	3	4
P_i	7	11	17	27

Let $y = \log P$, $B = \log A$. Then taking logarithms, $\log P = \log A + Mx$ which is equivalent to $y(x) = Mx + B$.

We now decide to make this the least-squares line for the (x_i, y_i) data points.

x_i	1	2	3	4
y_i	1.95	2.40	2.83	3.30

Since $s_0 = 4$, $s_1 = 10$, $s_2 = 30$, $t_0 = 10.48$, $t_1 = 28.44$, the formulas of Problem 21.1 make $M \simeq .45$ and $B \simeq 1.5$. The resulting formula is $P = 4.48e^{.45x}$.

It should be noted that in this procedure we do not minimize $\sum [P(x_i) - P_i]^2$, but instead choose the simpler task of minimizing $\sum [y(x_i) - y_i]^2$. This is a very common decision in such problems.

DISCRETE DATA, THE LEAST-SQUARES POLYNOMIAL

21.6. Generalizing Problem 21.1, find the polynomial $p(x) = a_0 + a_1x + \cdots + a_mx^m$ for which $S = \sum_{i=0}^{N} (y_i - a_0 - a_1x_i - \cdots - a_mx_i^m)^2$ is a minimum, the data (x_i, y_i) being given, and $m < N$.

We proceed as in the simpler case of the straight line. Setting the derivatives relative to $a_0, a_1, \ldots, a_m$ to zero produces the $m+1$ equations

$$\frac{\partial S}{\partial a_k} = -2\sum_{i=0}^{N} x_i^k(y_i - a_0 - a_1x_i - \cdots - a_mx_i^m) = 0$$

where $k = 0, \ldots, m$. Introducing the symbols $s_k = \sum_{i=0}^{N} x_i^k$, $t_k = \sum_{i=0}^{N} y_ix_i^k$, these equations may be rewritten as

$$\begin{aligned}
s_0a_0 + s_1a_1 &+ \cdots + s_ma_m = t_0 \\
s_1a_0 + s_2a_1 &+ \cdots + s_{m-1}a_m = t_1 \\
&\cdots\cdots\cdots\cdots \\
s_ma_0 + s_{m+1}a_1 &+ \cdots + s_{2m}a_m = t_m
\end{aligned}$$

and are called normal equations. Solving for the coefficients a_i, we obtain the least-squares polynomial. We will show that there is just one solution and that it does minimize S. For smaller integers m, these normal equations may be solved without difficulty. For larger m the system is badly ill-conditioned and an alternative procedure will be suggested.

21.7. Show how the least-squares idea, as just presented in Problem 21.6 and earlier in Problem 21.1, may be generalized to arbitrary vector spaces. What is the relationship with orthogonal projection?

This more general approach will also serve as a model for other variations of the least-squares idea to be presented later in this chapter and focuses attention on the common features which all these variations share. First recall that in Euclidean plane geometry, given a point y and a line S, the point on S closest to y is the unique point p such that $\overline{py}$ is orthogonal to S, p being the *orthogonal projection* point of y onto S. Similarly in Euclidean solid geometry, given a point y and a plane S, the point on S closest to y is the unique point p such that $\overline{py}$ is orthogonal to all vectors in S. Again p is the orthogonal projection of y. This idea is now extended to a more general vector space.

We are given a vector y in a vector space E and are to find a vector p in a given subspace S such that

$$\|y-p\|<\|y-q\|$$

where q is any other vector in S and the **norm** of a vector v is

$$\|v\|=\sqrt{(v, v)}$$

parentheses denoting the scalar product associated with the vector space. We begin by showing that there is a unique vector p for which $y-p$ is orthogonal to every vector in S. This p is called the *orthogonal projection* of y.

Let $e_0, \ldots, e_m$ be an orthogonal basis for S and consider the vector

$$p=(y, e_0)e_0+(y, e_1)e_1+\cdots+(y, e_m)e_m$$

Direct calculation shows that $(p, e_k)=(y, e_k)$ and therefore $(p-y, e_k)=0$ for $k=0, \ldots, m$. It then follows that $(p-y, q)=0$ for any q in S, simply by expressing q in terms of the orthogonal basis. If another vector p' also had this property $(p'-y, q)=0$, then it would follow that for any q in S $(p-p', q)=0$. Since $p-p'$ is itself in S, this forces $(p-p', p-p')=0$ which by required properties of any scalar product implies $p=p'$. The orthogonal projection p is thus unique.

But now, if q is a vector other than p in S,

$$\begin{aligned}\|y-q\|^2&=\|(y-p)+(p-q)\|^2\\&=\|y-p\|^2+\|p-q\|^2+2(y-p, p-q)\end{aligned}$$

Since the last term is zero, $p-q$ being in S, we deduce that $\|y-p\|<\|y-q\|$ as required.

21.8. If $u_0, u_1, \ldots, u_m$ is an arbitrary basis for S, determine the vector p of the preceding problem in terms of the u_k.

We must have $(y-p, u_k)=0$ or $(p, u_k)=(y, u_k)$ for $k=0, \ldots, m$. Since p has the unique representation $p=a_0u_0+a_1u_1+\cdots+a_mu_m$, substitution leads directly to

$$(u_0, u_k)a_0+(u_1, u_k)a_1+\cdots+(u_m, u_k)a_m=(y, u_k)$$

for $k=0, \ldots, m$. These are the *normal equations* for the given problem and are to be solved for the coefficients $a_0, \ldots, a_m$. A unique solution is guaranteed by the previous problem. Note that in the special case where the $u_0, u_1, \ldots, u_m$ are orthonormal, these normal equations reduce to $a_i=(y, u_i)$ as in the proof given in Problem 21.7.

Note also the following important corollary. If y itself is represented in terms of an orthogonal basis in E which includes $u_0, \ldots, u_m$, say

$$y=a_0u_0+a_1u_1+\cdots+a_mu_m+a_{m+1}u_{m+1}+\cdots$$

then the orthogonal projection p, which is the least-squares approximation, is available by simple *truncation* of the representation after the a_mu_m term:

$$p=a_0u_0+a_1u_1+\cdots+a_mu_m$$

21.9. How is the specific case treated in Problem 21.6 related to the generalization given in Problems 21.7 and 21.8?

The following identifications must be made:

E: The space of discrete real-valued functions on the set of arguments $x_0, \ldots, x_N$

S: The subset of E involving polynomials of degree m or less

y: The data function having values $y_0, \ldots, y_N$

(v_1, v_2): The scalar product $\sum_{i=0}^{N} v_1(x_i)v_2(x_i)$

$\|v\|^2$: The norm $\sum_{i=0}^{N} [v(x_i)]^2$

u_k: The function with values x_i^k

p: The polynomial with values $p_i = a_0 + a_1x_i + \cdots + a_mx_i^m$

$\|y-p\|^2$: The sum $S = \sum_{i=0}^{N} (y_i - p_i)^2$

(y, u_k): $t_k = \sum_{i=0}^{N} y_ix_i^k$

(u_j, u_k): $s_{j+k} = \sum_{i=0}^{N} x_i^{j+k}$

With these identifications we also learn that the polynomial p of Problem 21.6 is unique and actually does provide the minimum sum. The general result of Problems 21.7 and 21.8 establishes this.

21.10. Determine the least-squares quadratic function for the data of Problem 21.2.

The sums s_0, s_1, s_2, t_0, and t_1 have already been computed. We also need $s_3 = 2025$, $s_4 = 15{,}333$, and $t_2 = 1292.9$ which allow the normal equations to be written

$$10a_0 + 45a_1 + 285a_2 = 41.5 \qquad 45a_0 + 285a_1 + 2025a_2 = 194.1 \qquad 285a_0 + 2025a_1 + 15{,}333a_2 = 1248$$

After some labor these yield $a_0 = 3.73$, $a_1 = .11$, and $a_2 = -.0023$ so that our quadratic function is $p(x) = 3.73 + .11x - .0023x^2$.

21.11. Apply the quadratic function of the preceding problem to smooth the reported data.

Assuming that the data should have been values of our quadratic function, we obtain these values:

Handicap	6	8	10	12	14	16	18	20	22	24
Smoothed y	3.73	3.84	3.94	4.04	4.13	4.22	4.31	4.39	4.46	4.53

These hardly differ from the predictions of the straight-line hypothesis, and the parabola corresponding to our quadratic function would not differ noticeably from the straight line of Fig. 21-1. The fact that a_2 is so small already shows that the quadratic hypothesis may be unnecessary in the golfing problem.

SMOOTHING AND DIFFERENTIATION

21.12. Derive the formula for a least-squares parabola for five points (x_i, y_i) where $i = k-2, k-1, k, k+1, k+2$.

Let the parabola be $p(t) = a_0 + a_1t + a_2t^2$ where $t = (x - x_k)/h$, the arguments x_i being assumed equally spaced at interval h. The five points involved now have arguments $t = -2, -1, 0, 1, 2$. For this symmetric arrangement the normal equations simplify to

$$5a_0 \qquad + 10a_2 = \sum y_i$$

$$10a_1 \qquad = \sum t_iy_i$$

$$10a_0 \qquad + 34a_2 = \sum t_i^2y_i$$

and are easily solved. We find first

$$70a_0 = 34 \sum y_i - 10 \sum t_i^2 y_i$$

$$= -6y_{k-2} + 24y_{k-1} + 34y_k + 24y_{k+1} - 6y_{k+2}$$

$$= 70y_k - 6(y_{k-2} - 4y_{k-1} + 6y_k - 4y_{k+1} + y_{k+2})$$

from which
$$a_0 = y_k - \frac{3}{35}\,\delta^4 y_k$$

Substituting back we also obtain

$$a_2 = \frac{1}{14}(2y_{k-2} - y_{k-1} - 2y_k - y_{k+1} + 2y_{k+2})$$

And directly from the middle equation

$$a_1 = \frac{1}{10}(-2y_{k-2} - y_{k-1} + y_{k+1} + 2y_{k+2})$$

21.13. With $y(x_k)$ representing the exact value of which y_k is an approximation, derive the smoothing formula $y(x_k) \simeq y_k - \frac{3}{35}\,\delta^4 y_k$.

The least-squares parabola for the five points (x_{k-2}, y_{k-2}) to (x_{k+2}, y_{k+2}) is

$$p(x) = a_0 + a_1 t + a_2 t^2$$

At the center argument $t = 0$ this becomes $p(x_k) = a_0 = y_k - \frac{3}{35}\,\delta^4 y_k$ by Problem 21.12. Using this formula amounts to accepting the value of p on the parabola as better than the data value y_k.

21.14. The square roots of the integers from 1 to 10 were rounded to two decimal places and a random error of $-.05, -.04, \ldots, .05$ added to each (determined by drawing cards from a pack of 11 cards so labeled). The results form the top row of Table 21.1. Smooth these values using the formula of the preceding problem.

Table 21.1

x_k	**1**		**2**		**3**		**4**		**5**		**6**		**7**		**8**		**9**		**10**
y_k	1.04		1.37		1.70		2.00		2.26		2.42		2.70		2.78		3.00		3.14
δy		33		33		30		26		16		28		8		22		14	
$\delta^2 y$			0		−3		−4		−10		12		−20		14		−8		
$\delta^3 y$				−3		−1		−6		22		−32		34		−22			
$\delta^4 y$					2		−5		28		−54		66		−56				
$\frac{3}{35}\delta^4 y$					0		0		2		−5		6		−5				
$p(x_k)$					1.70		2.00		2.24		2.47		2.64		2.83				

Differences through the fourth also appear in Table 21.1, as well as $\frac{3}{35}\delta^4 y$. Finally the bottom row contains the smoothed values.

21.15. The smoothing formula of Problem 21.13 requires two data values on each side of x_k for producing the smoothed value $p(x_k)$. It cannot therefore be applied to the two first and last entries of a data table. Derive the formulas

$$y(x_0) \simeq y_0 + \frac{1}{5}\Delta^3 y_0 + \frac{3}{35}\Delta^4 y_0 \qquad y(x_{N-1}) \simeq y_{N-1} + \frac{2}{5}\nabla^3 y_N - \frac{1}{7}\nabla^4 y_N$$

$$y(x_1) \simeq y_1 - \frac{2}{5}\Delta^3 y_0 - \frac{1}{7}\Delta^4 y_0 \qquad y(x_N) \simeq y_N - \frac{1}{5}\nabla^3 y_N + \frac{3}{35}\nabla^4 y_N$$

for smoothing end values.

If we let $t = (x - x_2)/h$, then the quadratic function of Problem 21.12 is the least-squares quadratic for the first five points. We shall use the values of this function at x_0 and x_1 as smoothed values of y. First

$$p(x_0) = a_0 - 2a_1 + 4a_2$$

and inserting our expressions for the a_i, with k replaced by 2,

$$p(x_0) = \frac{1}{70}(62y_0 + 18y_1 - 6y_2 - 10y_3 + 6y_4)$$

$$= y_0 + \frac{1}{70}[(-14y_0 + 42y_1 - 42y_2 + 14y_3) + (6y_0 - 24y_1 + 36y_2 - 24y_3 + 6y_4)]$$

which reduce to the above formula for $y(x_0)$. For $p(x_1)$ we have

$$p(x_1) = a_0 - a_1 + a_2$$

and insertion of our expressions for the a_i again leads to the required formula. At the other end of our data supply the change of argument $t = (x - x_{N-2})/h$ applies, the details being similar.

21.16. Apply the formulas of the preceding problem to complete the smoothing of the y values in Table 21.1.

We find these changes to two places

$$y(x_0) \simeq 1.04 + \frac{1}{5}(-.03) + \frac{3}{35}(.02) \simeq 1.03 \qquad y(x_{N-1}) \simeq 3.00 + \frac{2}{5}(-.22) - \frac{1}{7}(-.56) \simeq 2.99$$

$$y(x_1) \simeq 1.37 - \frac{2}{5}(-.03) - \frac{1}{7}(.02) \simeq 1.38 \qquad y(x_N) \simeq 3.14 - \frac{1}{5}(-.22) + \frac{3}{35}(-.56) \simeq 3.14$$

21.17. Compute the RMS error of both the original data and the smoothed values.

The root-mean-square error of a set of approximations A_i corresponding to exact values T_i is defined by

$$\text{RMS error} = \left[\sum_{i=0}^{N} \frac{(T_i - A_i)^2}{N}\right]^{1/2}$$

In this example we have the following values:

T_i	1.00	1.41	1.73	2.00	2.24	2.45	2.65	2.83	3.00	3.16
y_i	1.04	1.37	1.70	2.00	2.26	2.42	2.70	2.78	3.00	3.14
$p(x_i)$	1.03	1.38	1.70	2.00	2.24	2.47	2.64	2.83	2.99	3.14

The exact roots are given by two places. By the above formula,

$$\text{RMS error of } y_i \simeq \left(\frac{.0108}{10}\right)^{1/2} \simeq .033$$

$$\text{RMS error of } p(x_i) \simeq \left(\frac{.0037}{10}\right)^{1/2} \simeq .019$$

so that the error is less by nearly half. The improvement over the center portion is greater. If the two values at each end are ignored we find RMS errors of .035 and .015, respectively, for a reduction of more than half. The formula of Problem 21.13 appears more effective than those of Problem 21.15.

21.18. Use the five-point parabola to obtain the formula

$$y'(x_k) \simeq \frac{1}{10h}(-2y_{k-2} - y_{k-1} + y_{k+1} + 2y_{k+2})$$

for approximate differentiation.

With the symbols of Problem 21.13 we shall use $y'(x_k)$, which is the derivative of our five-point parabola, as an approximation to the exact derivative at x_k. This again amounts to assuming that our data values y_i are approximate values of an exact but unknown function, but that the five-point parabola will be a better approximation, especially in the vicinity of the center point. On the parabola

$$p = a_0 + a_1 t + a_2 t^2$$

and according to plan, we calculate $p'(t)$ at $t = 0$ to be a_1. To convert this to a derivative relative to x involves merely division by h, and so, recovering the value a_1 found in Problem 21.12 and taking $p'(x)$ as an approximation to $y'(x)$, we come to the required formula.

21.19. Apply the preceding formula to estimate $y'(x)$ from the y_k values given in Table 21.1.

At $x_2 = 3$ we find

$$y'(3) \simeq \frac{1}{10}(-2.08 - 1.37 + 2.00 + 4.52) = .307$$

and at $x_3 = 4$, $$y'(4) \simeq \frac{1}{10}(-2.74 - 1.70 + 2.26 + 4.84) = .266$$

The other entries in the top row shown are found in the same way. The second row was computed using the approximation

$$y'(x_k) \simeq \frac{1}{12h}(y_{k-2} - 8y_{k-1} + 8y_{k+1} - y_{k+2})$$

found earlier from Stirling's five-point collocation polynomial. Notice the superiority of the present formula. Errors in data were found earlier to be considerably magnified by approximate differentiation formulas. Preliminary smoothing can lead to better results, by reducing such data errors.

$y'(x)$ by least squares	.31	.27	.24	.20	.18	.17
$y'(x)$ by collocation	.31	.29	.20	.23	.18	.14
Correct $y'(x)$	.29	.25	.22	.20	.19	.18

21.20. The formula of Problem 21.18 does not apply near the ends of the data supply. Use a four-point parabola at each end to obtain the formulas

$$y'(x_0) \simeq \frac{1}{20h}(-21y_0 + 13y_1 + 17y_2 - 9y_3)$$

$$y'(x_1) \simeq \frac{1}{20h}(-11y_0 + 3y_1 + 7y_2 + y_3)$$

$$y'(x_{N-1}) \simeq \frac{1}{20h}(11y_N - 3y_{N-1} - 7y_{N-2} - y_{N-3})$$

$$y'(x_N) \simeq \frac{1}{20h}(21y_N - 13y_{N-1} - 17y_{N-2} + 9y_{N-3})$$

Four points will be used rather than five, with the thought that a fifth point may be rather far from the position x_0 or x_N where a derivative is required. Depending on the size of h, the smoothness of the data, and perhaps other factors, one could use formulas based on five points or more. Proceeding to the four-point parabola we let $t = (x - x_1)/h$ so that the first four points have arguments $t = -1, 0, 1, 2$. The normal equations become

$$4a_0 + 2a_1 + 6a_2 = y_0 + y_1 + y_2 + y_3 \qquad 2a_0 + 6a_1 + 8a_2 = -y_0 + y_2 + 2y_3$$

$$6a_0 + 8a_1 + 18a_2 = y_0 + y_2 + 4y_3$$

and may be solved for

$$20a_0 = 3y_0 + 11y_1 + 9y_2 - 3y_3 \qquad 20a_1 = -11y_0 + 3y_1 + 7y_2 + y_3 \qquad 4a_2 = y_0 - y_1 - y_2 + y_3$$

With these and $y'(x_0) = (a_1 - 2a_2)/h$, $y'(x_1) = a_1/h$ the required results follow. Details at the other end of the data supply are almost identical.

21.21. Apply the formulas of the preceding problem to the data of Table 21.1.

We find

$$y'(1) \simeq \frac{1}{20}[-21(1.04) + 13(1.37) + 17(1.70) - 9(2.00)] \simeq .35$$

$$y'(2) \simeq \frac{1}{20}[-11(1.04) + 3(1.37) + 7(1.70) + 2.00] \simeq .33$$

Similarly $y'(9) \simeq .16$ and $y'(10) \simeq .19$. The correct values are .50, .35, .17, and .16. The poor results obtained at the endpoints are further evidence of the difficulties of numerical differentiation. Newton's original formula

$$y'(x_0) \simeq \Delta y_0 - \frac{1}{2}\Delta^2 y_0 + \frac{1}{3}\Delta^3 y_0 - \frac{1}{4}\Delta^4 y_0 + \cdots$$

produces from this data the value .32, which is worse than our .35. At the other extreme the corresponding backward difference formula manages .25 which is much worse than our .19.

21.22. Apply the formulas for approximate derivatives a second time to estimate $y''(x)$, using the data of Table 21.1.

We have already obtained estimates of the first derivative, of roughly two-place accuracy. They are as follows:

x	1	2	3	4	5	6	7	8	9	10
$y'(x)$	.35	.33	.31	.27	.24	.20	.18	.17	.16	.19

Now applying the same formulas to the $y'(x)$ values will produce estimates of $y''(x)$. For example, at $x = 5$,

$$y''(5) \simeq \frac{1}{10}[-2(.31) - (.27) + (.20) + 2(.18)] \simeq -.033$$

which is half again as large as the correct $-.022$. Complete results from our formulas and correct values are as follows:

$-y''$ (computed)	.011	.021	.028	.033	.033	.026	.019	.004	.012	-0.32
$-y''$ (correct)	.250	.088	.048	.031	.022	.017	.013	.011	.009	.008

Near the center we have an occasional ray of hope but at the ends the disaster is evident.

21.23. The least-squares parabola for seven points leads to the smoothing formula

$$y(x_k) \simeq y_k - \frac{3}{7}\delta^4 y_k - \frac{2}{21}\delta^6 y_k$$

(The derivation is requested as a supplementary problem.) Apply this to the data of Table 21.1. Does it yield better values than the five-point smoothing formula?

A row of sixth differences may be added to Table 21.1:

$$40 \qquad -115 \qquad 202 \qquad -242$$

Then the formula yields
$$y(4) \simeq 2.00 - \frac{3}{7}(-.05) - \frac{2}{21}(.40) \simeq 1.98$$

$$y(5) \simeq 2.26 - \frac{3}{7}(.28) - \frac{2}{21}(-1.15) \simeq 2.25$$

and similarly $y(6) \simeq 2.46$, $y(7) \simeq 2.65$. These are a slight improvement over the results from the five-point formula, except for $y(4)$ which is slightly worse.

ORTHOGONAL POLYNOMIALS, DISCRETE CASE

21.24. For large N and m the set of normal equations may be badly ill-conditioned. To see this show that for equally spaced x_i from 0 to 1 the matrix of coefficients is approximately

$$\begin{bmatrix} 1 & \frac{1}{2} & \frac{1}{3} & \cdots & \frac{1}{m+1} \\ \frac{1}{2} & \frac{1}{3} & \frac{1}{4} & \cdots & \frac{1}{m+2} \\ \cdots & \cdots & \cdots & \cdots & \cdots \\ \frac{1}{m+1} & \frac{1}{m+2} & \frac{1}{m+3} & \cdots & \frac{1}{2m+1} \end{bmatrix}$$

if a factor of N is deleted from each term. This matrix is the Hilbert matrix of order $m+1$.

For large N the area under $y(x) = x^k$ between 0 and 1 will be approximately the sum of N rectangular areas. (See Fig. 21-2.) Since the exact area is given by an integral, we have

$$\frac{1}{N}\sum_{i=0}^{N} x_i^k \simeq \int_0^1 x^k\,dx = \frac{1}{k+1}$$

Thus $s_k \simeq N/(k+1)$, and deleting the N we have at once the Hilbert matrix. This matrix will later be shown to be extremely troublesome for large N.

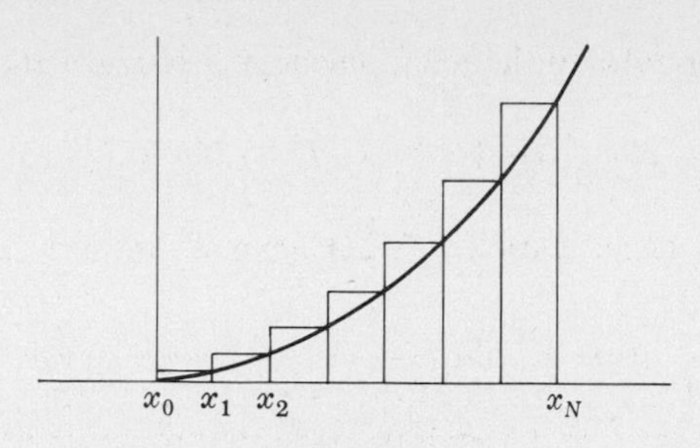

Fig. 21-2

21.25. How can the Hilbert matrices be avoided?

The preceding problem shows that the normal equations which arise with the basis $1, x, \ldots, x^m$ and equally spaced arguments involve an approximately Hilbert matrix, which is troublesome. It is computationally more efficient to find an orthogonal basis so that the corresponding normal equations become trivial. A convenient orthogonal basis is constructed in the next problem. It is interesting to note that in developing this basis we will deal directly with the Hilbert matrix itself, not with approximations to it, and that the system of equations encountered will be solved exactly, thus avoiding the pitfalls of computing with ill-conditioned systems. (See also Chapter 26.)

21.26. Construct a set of polynomials $P_{m,N}(t)$ of degrees $m = 0, 1, 2, \ldots$ such that

$$\sum_{t=0}^{N} P_{m,N}(t)P_{n,N}(t) = 0 \qquad \text{for } m > n$$

Such polynomials are called orthogonal over the set of arguments t.

Let the polynomial be

$$P_{m,N}(t) = 1 + c_1 t + c_2 t^{(2)} + \cdots + c_m t^{(m)}$$

where $t^{(i)}$ is the factorial $t(t-1)\cdots(t-i+1)$. We first make the polynomial orthogonal to $(t+s)^{(s)}$ for $s = 0, 1, \ldots, m-1$, which means that we require

$$\sum_{t=0}^{N} (t+s)^{(s)} P_{m,N}(t) = 0$$

Since
$$(t+s)^{(s)} P_{m,N}(t) = (t+s)^{(s)} + c_1(t+s)^{(s+1)} + \cdots + c_m(t+s)^{(s+m)}$$

summing over the arguments t and using Problem 4.10 brings

$$\sum_{t=0}^{N} (t+s)^{(s)} P_{m,N}(t) = \frac{(N+s+1)^{(s+1)}}{s+1} + c_1 \frac{(N+s+1)^{(s+2)}}{s+2} + \cdots + c_m \frac{(N+s+1)^{(s+m+1)}}{s+m+1}$$

which is to be zero. Removing the factor $(N+s+1)^{(s+1)}$, the sum becomes

$$\frac{1}{s+1} + \frac{Nc_1}{s+2} + \frac{N^{(2)}c_2}{s+3} + \cdots + \frac{N^{(m)}c_m}{s+m+1} = 0$$

and setting $N^{(i)}c_i = a_i$ this simplifies to

$$\frac{1}{s+1} + \frac{a_1}{s+2} + \frac{a_2}{s+3} + \cdots + \frac{a_m}{s+m} = 0$$

for $s = 0, 1, \ldots, m-1$. The Hilbert matrix again appears in this set of equations, but solving the system exactly will still lead us to a useful algorithm. If the last sum were merged into a single quotient it would take the form $Q(s)/(s+m+1)^{(m+1)}$ with $Q(s)$ a polynomial of degree at most m. Since $Q(s)$ must be zero at the m arguments $s = 0, 1, \ldots, m-1$, we must have $Q(s) = Cs^{(m)}$ where C is independent of s.

To determine C we multiply both the sum and the equivalent quotient by $(s+1)$ and have

$$1+(s+1)\left(\frac{a_1}{s+2}+\cdots+\frac{a_m}{s+m+1}\right)=\frac{Cs^{(m)}}{(s+2)\cdots(s+m+1)}$$

which must be true for all s except zeros of denominators. Setting $s=-1$, we see that $C=m!/[(-1)(-2)\cdots(-m)]=(-1)^m$. We now have

$$\frac{1}{s+1}+\frac{a_1}{s+2}+\cdots+\frac{a_m}{s+m+1}=\frac{(-1)^m s^{(m)}}{(s+m+1)^{(m+1)}}$$

The device which produced C now produces the a_i. Multiply by $(s+m+1)^{(m+1)}$ and then set $s=-i-1$ to find for $i=1,\ldots,m$

$$(-1)^i i!\,(m-i)!\,a_i=(-1)^m(-i-1)^{(m)}=(m+i)^{(m)}$$

and then solve for

$$a_i=(-1)^i\frac{(m+i)^{(m)}}{(m-i)!\,i!}=(-1)^i\binom{m}{i}\binom{m+i}{i}$$

Recalling that $a_i=c_iN^{(i)}$, the required polynomials may be written as

$$P_{m,N}(t)=\sum_{i=0}^{m}(-1)^i\binom{m}{i}\binom{m+i}{i}\frac{t^{(i)}}{N^{(i)}}$$

What we have proved is that each $P_{m,N}(t)$ is orthogonal to the functions

$$1 \qquad t+1 \qquad (t+2)(t+1) \qquad \cdots \qquad (t+m-1)^{(m-1)}$$

but in Problem 4.18, we saw that the powers $1, t, t^2, \ldots, t^{m-1}$ may be expressed as combinations of these, so that $P_{m,N}(t)$ is orthogonal to each of these powers as well. Finally, since $P_{n,N}(t)$ is a combination of these powers we find $P_{m,N}(t)$ and $P_{n,N}(t)$ to be themselves orthogonal. The first five of these polynomials are

$$P_{0,N}=1$$

$$P_{1,N}=1-\frac{2t}{N}$$

$$P_{2,N}=1-\frac{6t}{N}+\frac{6t(t-1)}{N(N-1)}$$

$$P_{3,N}=1-\frac{12t}{N}+\frac{30t(t-1)}{N(N-1)}-\frac{20t(t-1)(t-2)}{N(N-1)(N-2)}$$

$$P_{4,N}=1-\frac{20t}{N}+\frac{90t(t-1)}{N(N-1)}-\frac{140t(t-1)(t-2)}{N(N-1)(N-2)}+\frac{70t(t-1)(t-2)(t-3)}{N(N-1)(N-2)(N-3)}$$

21.27. Determine the coefficients a_k so that

$$p(x)=a_0P_{0,N}(t)+a_1P_{1,N}(t)+\cdots+a_mP_{m,N}(t)$$

[with $t=(x-x_0)/h$] will be the least-squares polynomial of degree m for the data (x_t, y_t), $t=0, 1, \ldots, N$.

We are to minimize

$$S=\sum_{t=0}^{N}[y_t-a_0P_{0,N}(t)-\cdots-a_mP_{m,N}(t)]^2$$

Setting derivatives relative to the a_k equal to zero, we have

$$\frac{\partial S}{\partial a_k}=-2\sum_{t=0}^{N}[y_t-a_0P_{0,N}(t)-\cdots-a_mP_{m,N}(t)]P_{k,N}(t)=0$$

for $k = 0, 1, \ldots, m$. But by the orthogonality property most terms here are zero, only two contributing.

$$\sum_{t=0}^{N} [y_t - a_k P_{k,N}(t)] P_{k,N}(t) = 0$$

Solving for a_k, we find

$$a_k = \frac{\sum_{t=0}^{N} y_t P_{k,N}(t)}{\sum_{t=0}^{N} P_{k,N}^2(t)}$$

This is one advantage of the orthogonal functions. The coefficients a_k are uncoupled, each appearing in a single normal equation. Substituting the a_k into $p(x)$, we have the least-squares polynomial.

The same result follows directly from the general theorem of Problems 21.7 and 21.8. Identifying E, S, y, (v_1, v_2), and $\|v\|$ exactly as before, we now take $u_k = P_{k,N}(t)$ so that the orthogonal projection is still $p = a_0 u_0 + \cdots + a_m u_m$. The kth normal equation is $(u_k, u_k) a_k = (y, u_k)$ and leads to the expression for a_k already found. Our general theory now also guarantees that we have actually minimized S, and that our $p(x)$ is the unique solution. An argument using second derivatives could also establish this but is now not necessary.

21.28. Show that the minimum value of S takes the form $\sum_{t=0}^{N} y_t^2 - \sum_{k=0}^{m} W_k a_k^2$ where $W_k = \sum_{t=0}^{N} P_{k,N}^2(t)$.

Expansion of the sum brings

$$S = \sum_{t=0}^{N} y_t^2 - 2 \sum_{t=0}^{N} y_t \sum_{k=0}^{m} a_k P_{k,N}(t) + \sum_{t=0}^{N} \sum_{j,k=0}^{m} a_j a_k P_{j,N}(t) P_{k,N}(t)$$

The second term on the right equals $-2 \sum_{k=0}^{m} a_k (W_k a_k) = -2 \sum_{k=0}^{m} W_k a_k^2$. The last term vanishes by the orthogonality except when $j = k$, in which case it becomes $\sum_{k=0}^{m} W_k a_k^2$. Putting the pieces back together,

$$S_{\min} - \sum_{t=0}^{N} y_t^2 - \sum_{k=0}^{m} W_k a_k^2$$

Notice what happens to the minimum of S as the degree m of the approximating polynomial is increased. Since S is nonnegative, the first sum in $S_{\min}$ clearly dominates the second. But the second increases with m, steadily diminishing the error. When $m = n$ we know by our earlier work that a collocation polynomial exists, equal to y_t at each argument $t = 0, 1, \ldots, N$. This reduces S to zero.

21.29. Apply the orthogonal functions algorithm to find a least-squares polynomial of degree three for the following data:

x_i	0	1	2	3	4	5	6	7	8	9	10
y_i	1.22	1.41	1.38	1.42	1.48	1.58	1.84	1.79	2.03	2.04	2.17

x_i	11	12	13	14	15	16	17	18	19	20
y_i	2.36	2.30	2.57	2.52	2.85	2.93	3.03	3.07	3.31	3.48

The coefficients a_j are computed directly by the formula of the preceding problem. For hand computing, tables of the W_k and $P_{k,N}(t)$ exist and should be used. Although we have "inside information" that degree three is called for, it is instructive to go slightly further. Up through $m = 5$ we find $a_0 = 2.2276$, $a_1 = -1.1099$, $a_2 = .1133$, $a_3 = .0119$, $a_4 = .0283$, $a_5 = -.0038$; and with $x = t$,

$$p(x) = 2.2276 - 1.1099 P_{1,20} + .1133 P_{2,20} + .0119 P_{3,20} + .0283 P_{4,20} - .0038 P_{5,20}$$

By the nature of orthogonal function expansions we obtain least-squares approximations of various degrees by truncation of this result. The values of such polynomials from degree one to degree five are given in Table 21.2 below, along with the original data. The final column lists the values of $y(x) = (x + 50)^3/10^5$ from which the data were obtained by adding random errors of size up to .10. Our goal has been to recover this cubic, eliminating as much error as we can by least-squares smoothing. Without prior knowledge that a cubic polynomial was our target, there would be some difficulty in choosing our approximation. Fortunately the results do not disagree violently after the linear approximation. A computation of the RMS error shows that the quadratic has, in this case, outperformed the cubic approximation.

Degree	1	2	3	4	5	Raw data
RMS	.060	.014	.016	.023	.023	.069

Table 21.2

x	**Given data**	**1**	**2**	**3**	**4**	**5**	**Correct results**
0	1.22	1.12	1.231	1.243	1.27	1.27	1.250
1	1.41	1.23	1.308	1.313	1.31	1.31	1.327
2	1.38	1.34	1.389	1.388	1.37	1.38	1.406
3	1.42	1.45	1.473	1.469	1.45	1.45	1.489
4	1.48	1.56	1.561	1.554	1.54	1.54	1.575
5	1.58	1.67	1.652	1.645	1.63	1.63	1.663
6	1.84	1.78	1.747	1.740	1.74	1.73	1.756
7	1.79	1.89	1.845	1.839	1.84	1.84	1.852
8	2.03	2.01	1.947	1.943	1.95	1.95	1.951
9	2.04	2.12	2.053	2.051	2.07	2.07	2.054
10	2.17	2.23	2.162	2.162	2.18	2.18	2.160
11	2.36	2.34	2.275	2.277	2.29	2.29	2.270
12	2.30	2.45	2.391	2.395	2.41	2.41	2.383
13	2.57	2.56	2.511	2.517	2.52	2.52	2.500
14	2.52	2.67	2.635	2.642	2.64	2.64	2.621
15	2.85	2.78	2.762	2.769	2.76	2.76	2.746
16	2.93	2.89	2.892	2.899	2.88	2.88	2.875
17	3.03	3.00	3.027	3.031	3.01	3.01	3.008
18	3.07	3.12	3.164	3.165	3.15	3.15	3.144
19	3.31	3.23	3.306	3.301	3.30	3.30	3.285
20	3.48	3.34	3.451	3.439	3.47	3.47	3.430

CONTINUOUS DATA, THE LEAST-SQUARES POLYNOMIAL

21.30. Determine the coefficients a_i so that

$$I = \int_{-1}^{1} [y(x) - a_0P_0(x) - a_1P_1(x) - \cdots - a_mP_m(x)]^2 \, dx$$

will be a minimum, the function $P_k(x)$ being the kth Legendre polynomial.

Here it is not a sum of squares which is to be minimized but an integral, and the data are no longer discrete values y_i but a function $y(x)$ of the continuous argument x. The use of the Legendre polynomials is very convenient. As in the previous section it will reduce the normal equations, which determine the a_k, to a very simple set. And since any polynomial can be expressed as a combination of Legendre polynomials, we are actually solving the problem of least-squares polynomial approximation for continuous data. Setting the usual derivatives to zero, we have

$$\frac{\partial I}{\partial a_k} = -2\int_{-1}^{1}[y(x) - a_0P_0(x) - \cdots - a_mP_m(x)]P_k(x)\,dx = 0$$

for $k = 0, 1, \ldots, m$. By the orthogonality of these polynomials, these equations simplify at once to

$$\int_{-1}^{1}[y(x) - a_kP_k(x)]P_k(x)\,dx = 0$$

Each equation involves only one of the a_k so that

$$a_k = \frac{\int_{-1}^{1} y(x)P_k(x)\,dx}{\int_{-1}^{1} P_k^2(x)\,dx} = \frac{2k+1}{2}\int_{-1}^{1} y(x)P_k(x)\,dx$$

Here again it is true that our problem is a special case of Problems 21.7 and 21.8, with these identifications:

E:	The space of real-valued functions on $-1 \leqq x \leqq 1$
S:	Polynomials of degree m or less
y:	The data function $y(x)$
(v_1, v_2):	The scalar product $\int_{-1}^{1} v_1(x)v_2(x)\,dx$
$\|v\|$:	The norm $\int_{-1}^{1} [v(x)]^2\,dx$
u_k:	$P_k(x)$
p:	$a_kP_0(x) + \cdots + a_mP_m(x)$
a_k:	$(y, u_k)/(u_k, u_k)$

These problems therefore guarantee that our solution $p(x)$ is unique and does minimize the integral I.

21.31. Find the least-squares approximation to $y(t) = t^2$ on the interval $(0, 1)$ by a straight line.

Here we are approximating a parabolic arc by a line segment. First let $t = (x+1)/2$ to obtain the interval $(-1, 1)$ in the argument x. This makes $y = (x+1)^2/4$. Since $P_0(x) = 1$ and $P_1(x) = x$, the coefficients a_0 and a_1 are

$$a_0 = \frac{1}{2}\int_{-1}^{1}\frac{1}{4}(x+1)^2\,dx = \frac{1}{3} \qquad a_1 = \frac{3}{2}\int_{-1}^{1}\frac{1}{4}(x+1)^2x\,dx = \frac{1}{2}$$

and the least-squares line is $y = \frac{1}{3}P_0(x) + \frac{1}{2}P_1(x) = \frac{1}{3} + \frac{1}{2}x = t - \frac{1}{6}$.

Both the parabolic arc and the line are shown in Fig. 21-3. The difference between y values on the line and the parabola is $t^2 - t + \frac{1}{6}$, and this takes extreme values at $t = 0, \frac{1}{2}$, and 1 of amounts $\frac{1}{6}$, $-\frac{1}{12}$, and $\frac{1}{6}$. The error made in substituting the line for the parabola is therefore slightly greater at the ends than at the center of the interval. This error can be expressed as

$$\frac{1}{4}(x+1)^2 - \frac{1}{3}P_0(x) - \frac{1}{2}P_1(x) = \frac{1}{6}P_2(x)$$

and the shape of $P_2(x)$ corroborates this error behavior.

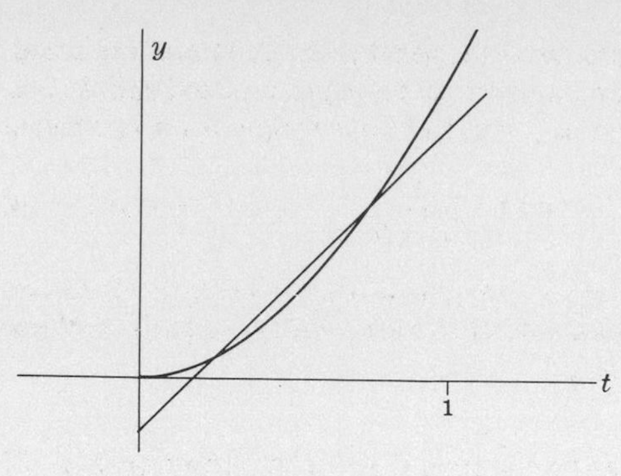

Fig. 21-3

21.32. Find the least-squares approximation to $y(t) = \sin t$ on the interval $(0, \pi)$ by a parabola.

Let $t = \pi(x+1)/2$ to obtain the interval $(-1, 1)$ in the argument x. Then $y = \sin[\pi(x+1)/2]$. The coefficients are

$$a_0 = \frac{1}{2}\int_{-1}^{1} \sin\left[\frac{\pi(x+1)}{2}\right] dx = \frac{2}{\pi}$$

$$a_1 = \frac{3}{2}\int_{-1}^{1} \sin\left[\frac{\pi(x+1)}{2}\right] x\, dx = 0$$

$$a_2 = \frac{5}{2}\int_{-1}^{1} \sin\left[\frac{\pi(x+1)}{2}\right] \frac{1}{2}(3x^2-1)\, dx = \frac{10}{\pi}\left(1 - \frac{12}{\pi^2}\right)$$

so that the parabola is

$$y = \frac{2}{\pi} + \frac{10}{\pi}\left(1 - \frac{12}{\pi^2}\right)\frac{1}{2}(3x^2-1) = \frac{2}{\pi} + \frac{10}{\pi}\left(1 - \frac{12}{\pi^2}\right)\left[\frac{6}{\pi^2}\left(t - \frac{\pi}{2}\right)^2 - \frac{1}{2}\right]$$

The parabola and sine curve are shown in Fig. 21-4, with slight distortions to better emphasize the over and under nature of the approximation.

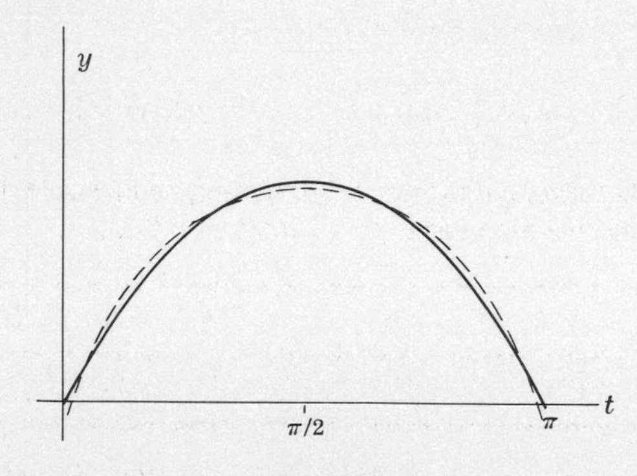

Fig. 21-4

21.33. What are the "shifted Legendre polynomials"?

These result from a change of argument which converts the interval $(-1, 1)$ into $(0, 1)$. Let $t = (1-x)/2$ to effect this change. The familiar Legendre polynomials in the argument x then become

$$P_0 = 1 \qquad P_2 = \frac{1}{2}(3x^2 - 1) = 1 - 6t + 6t^2$$

$$P_1 = x = 1 - 2t \qquad P_3 = \frac{1}{2}(5x^3 - 3x) = 1 - 12t + 30t^2 - 20t^3$$

and so on. These polynomials are orthogonal over $(0, 1)$ and we could have used them as the basis of our least-squares analysis of continuous data in place of the standard Legendre polynomials. With this change of argument the integrals involved in our formulas for coefficients become

$$\int_0^1 [P_n(t)]^2\, dt = \frac{1}{2n+1} \qquad a_k = (2k+1)\int_0^1 y(t)P_k(t)\, dt$$

The argument change $t = (x+1)/2$ might also have been used, altering the sign of each odd-degree polynomial, but the device used leads to a close analogy with the orthogonal polynomials for the discrete case developed in Problem 21.26.

21.34. Suppose that an experiment produces the curve shown in Fig. 21-5. It is known or suspected that the curve should be a straight line. Show that the least-squares line is approximately given by $y = .21t + .11$, which is shown dashed in the diagram.

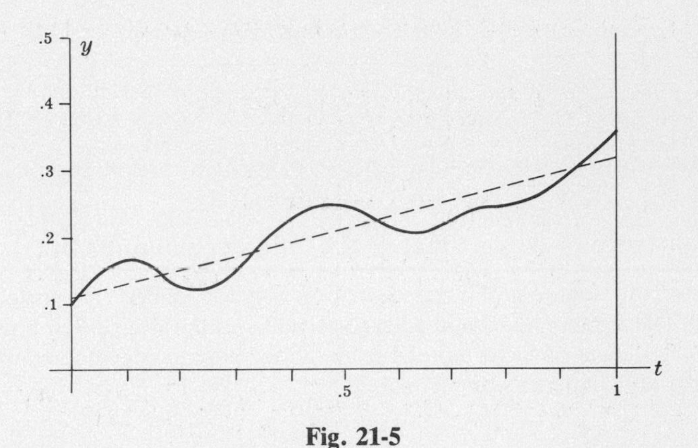

Fig. 21-5

Instead of reducing the interval to $(-1, 1)$ we work directly with the argument t and the shifted Legendre polynomials. Two coefficients are needed,

$$a_0 = \int_0^1 y(t)\, dt \qquad a_1 = 3\int_0^1 y(t)(1-2t)\, dt$$

Since $y(t)$ is not available in analytic form, these integrals must be evaluated by approximate methods. Reading from the diagram, we may estimate y values as follows:

t	0	.1	.2	.3	.4	.5	.6	.7	.8	.9	1.0
y	.10	.17	.13	.15	.23	.25	.21	.22	.25	.29	.36

Applying Simpson's rule now makes $a_0 \simeq .214$ and $a_1 \simeq -.105$. The resulting line is

$$y = .214 - .105(1-2t) = .21t + .11$$

and this appears in Fig. 21-5. An alternative treatment of this problem could involve applying the methods for discrete data to the y values read from the diagram.

CONTINUOUS DATA, A GENERALIZED TREATMENT

21.35. Develop the least-squares polynomial in terms of a set of orthogonal polynomials on the interval (a, b) with nonnegative weight function $w(x)$.

The details are very similar to those of earlier derivations. We are to minimize

$$I = \int_a^b w(x)[y(x) - a_0 Q_0(x) - \cdots - a_m Q_m(x)]^2 \, dx$$

by choice of the coefficients a_k, where the functions $Q_k(x)$ satisfy the orthogonality condition

$$\int_a^b w(x) Q_j(x) Q_k(x) \, dx = 0$$

for $j \neq k$. Without stopping for the duplicate argument involving derivatives, we appeal at once to Problems 21.7 and 21.8, with the scalar product

$$(v_1, v_2) = \int_a^b w(x) v_1(x) v_2(x) \, dx$$

and other obvious identifications, and find

$$a_k = \frac{\int_a^b w(x) y(x) Q_k(x) \, dx}{\int_a^b w(x) Q_k^2(x) \, dx}$$

With these a_k the least-squares polynomial is $p(x) = a_0 Q_0(x) + \cdots + a_m Q_m(x)$.

21.36. What is the importance of the fact that a_k does not depend upon m?

This means that the degree of the approximation polynomial does not have to be chosen at the start of a computation. The a_k may be computed successively and the decision of how many terms to use can be based on the magnitudes of the computed a_k. In nonorthogonal developments a change of degree will usually require that all coefficients be recomputed.

21.37. Show that the minimum value of I can be expressed in the form

$$\int_a^b w(x) y^2(x) \, dx - \sum_{k=0}^m W_k a_k^2 \qquad \text{where} \qquad W_k = \int_a^b w(x) Q_k^2(x) \, dx$$

Explicitly writing out the integral makes

$$I = \int_a^b w(x) y^2(x) \, dx - 2 \sum_{k=0}^m \int_a^b w(x) y(x) a_k Q_k(x) \, dx + \sum_{j,k=0}^m \int_a^b w(x) a_j a_k Q_j(x) Q_k(x) \, dx$$

The second term on the right equals $-2 \sum_{k=0}^m a_k (W_k a_k) = -2 \sum_{k=0}^m W_k a_k^2$. The last term vanishes by the orthogonality except when $j = k$, in which case it becomes $\sum_{k=0}^m W_k a_k^2$. Putting the pieces back together, $I_{\min} = \int_a^b w(x) y^2(x) \, dx - \sum_{k=0}^m W_k a_k^2$.

21.38. Prove Bessel's inequality, $\sum_{k=0}^m W_k a_k^2 \leqq \int_a^b w(x) y^2(x) \, dx$.

Assuming $w(x) \geqq 0$, it follows that $I \geqq 0$ so that Bessel's inequality is an immediate consequence of the preceding problem.

21.39. Prove the series $\sum_{k=0}^\infty W_k a_k^2$ to be convergent.

It is a series of positive terms with partial sums bounded above by the integral in Bessel's inequality.

This guarantees convergence. Of course, it is assumed all along that the integrals appearing in our analysis exist, in other words that we are dealing with functions which are integrable on the interval (a, b).

21.40. Is it true that as m tends to infinity the value of $I_{\min}$ tends to zero?

With the families of orthogonal functions ordinarily used, the answer is yes. The process is called *convergence in the mean* and the set of orthogonal functions is called *complete*. The details of proof are more extensive than will be attempted here.

APPROXIMATION WITH CHEBYSHEV POLYNOMIALS

21.41. The Chebyshev polynomials are defined for $-1 \leqq x \leqq 1$ by $T_n(x) = \cos(n \arccos x)$. Find the first few such polynomials directly from this definition.

For $n = 0$ and 1 we have at once $T_0(x) = 1$, $T_1(x) = x$. Let $A = \arccos x$. Then

$$T_2(x) = \cos 2A = 2\cos^2 A - 1 = 2x^2 - 1$$

$$T_3(x) = \cos 3A = 4\cos^3 A - 3\cos A = 4x^3 - 3x, \qquad \text{etc.}$$

21.42. Prove the recursion relation $T_{n+1}(x) = 2xT_n(x) - T_{n-1}(x)$.

The trigonometric relationship $\cos(n+1)A + \cos(n-1)A = 2\cos A \cos nA$ translates directly into $T_{n+1}(x) + T_{n-1}(x) = 2xT_n(x)$.

21.43. Use the recursion to produce the next few Chebyshev polynomials

Beginning with $n = 3$,

$$T_4(x) = 2x(4x^3 - 3x) - (2x^2 - 1) = 8x^4 - 8x^2 + 1$$

$$T_5(x) = 2x(8x^4 + 1) - (4x^3 - 3x) = 16x^5 - 20x^3 + 5x$$

$$T_6(x) = 2x(16x^5 - 20x^3 + 5x) - (8x^4 - 8x^2 + 1) = 32x^6 - 48x^4 + 18x^2 - 1$$

$$T_7(x) = 2x(32x^6 - 48x^4 + 18x^2 - 1) - (16x^5 - 20x^3 + 5x) = 64x^7 - 112x^5 + 56x^3 - 7x \qquad \text{etc.}$$

21.44. Prove the orthogonality property

$$\int_{-1}^{1} \frac{T_m(x)T_n(x)}{\sqrt{1-x^2}}\,dx = \begin{cases} 0 & m \neq n \\ \pi/2 & m = n \neq 0 \\ \pi & m = n = 0 \end{cases}$$

Let $x = \cos A$ as before. The above integral becomes

$$\int_0^{\pi} (\cos mA)(\cos nA)\,dA = \left[\frac{\sin(m+n)A}{2(m+n)} + \frac{\sin(m-n)A}{2(m-n)}\right]_0^{\pi} = 0$$

for $m \neq n$. If $m = n = 0$, the result π is immediate. If $m = n \neq 0$, the integral is

$$\int_0^{\pi} \cos^2 nA\,dA = \left[\frac{1}{2}\left(\frac{\sin nA \cos nA}{n} + A\right)\right]_0^{\pi} = \frac{\pi}{2}$$

21.45. Express the powers of x in terms of Chebyshev polynomials.

We find

$$1 = T_0 \qquad x = T_1 \qquad x^2 = \frac{1}{2}(T_0 + T_2) \qquad x^3 = \frac{1}{4}(3T_1 + T_3)$$

$$x^4=\frac{1}{8}(3T_0+4T_2+T_4) \qquad x^5=\frac{1}{16}(10T_1+5T_3+T_5)$$

$$x^6=\frac{1}{32}(10T_0+15T_2+6T_4+T_6) \qquad x^7=\frac{1}{64}(35T_1+21T_3+7T_5+T_7)$$

and so on. Clearly the process may be continued to any power.

21.46. Find the least-squares polynomial which minimizes the integral

$$\int_{-1}^{1}\frac{1}{\sqrt{1-x^2}}[y(x)-a_0T_0(x)-\cdots-a_mT_m(x)]^2\,dx$$

By results of the previous section the coefficients a_k are

$$a=\frac{\int_{-1}^{1}w(x)y(x)T_k(x)\,dx}{\int_{-1}^{1}w(x)T_k^2(x)\,dx}=\frac{2}{\pi}\int_{-1}^{1}\frac{y(x)T_k(x)}{\sqrt{1-x^2}}dx$$

except for a_0 which is $a_0=\frac{1}{\pi}\int_{-1}^{1}\frac{y(x)}{\sqrt{1-x^2}}dx$. The least-squares polynomial is $a_0T_0(x)+\cdots+a_mT_m(x)$.

21.47. Show that $T_n(x)$ has n zeros inside the interval $(-1,1)$ and none outside. What is the "equal-ripple" property?

Since $T_n(x)=\cos n\theta$, with $x=\cos\theta$ and $-1\leqq x\leqq 1$, we may require $0\leqq\theta\leqq\pi$ without loss. Actually this makes the relationship between θ and x more precise. Clearly $T_n(x)$ is zero for $\theta=(2i+1)\pi/2n$, or

$$x_i=\cos\frac{(2i+1)\pi}{2n}, \qquad i=0,1,\ldots,n-1$$

These are n distinct arguments between -1 and 1. Since $T_n(x)$ has only n zeros, there can be none outside the interval. Being equal to a cosine in the interval $(-1,1)$, the polynomial $T_n(x)$ cannot exceed one in magnitude there. It reaches this maximum size at $n+1$ arguments, including the endpoints.

$$T_n(x)=(-1)^i \quad \text{at} \quad x=\cos\frac{i\pi}{n} \qquad i=0,1,\ldots,n$$

This oscillation between extreme values of equal magnitude is known as the **equal-ripple property**. This property is illustrated in Fig. 21-6 which shows $T_2(x)$, $T_3(x)$, $T_4(x)$, and $T_5(x)$.

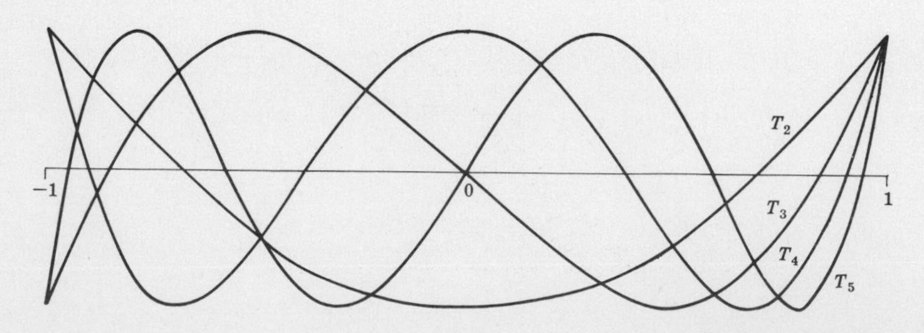

Fig. 21-6

21.48. In what way does the equal-ripple property make the least-squares approximation

$$y(x) \simeq a_0 T_0(x) + \cdots + a_m T_m(x)$$

superior to similar approximations using other polynomials in place of the $T_k(x)$?

Suppose we assume that, for the $y(x)$ concerned, the series obtained by letting m tend to infinity converges to $y(x)$ and also that it converges quickly enough so that

$$y(x) - a_0 T_0(x) - \cdots - a_m T_m(x) \simeq a_{m+1} T_{m+1}(x)$$

In other words, the error made in truncating the series is essentially the first omitted term. Since $T_{m+1}(x)$ has the equal-ripple property, the error of our approximation will fluctuate between a_{m+1} and $-a_{m+1}$ across the entire interval $(-1, 1)$. The error will not be essentially greater over one part of the interval compared with another. This error uniformity may be viewed as a reward for accepting the unpleasant weighting factor $1/\sqrt{1-x^2}$ in the integrals.

21.49. Find the least-squares line for $y(t) = t^2$ over the interval $(0, 1)$ using the weight function $1/\sqrt{1-x^2}$.

The change of argument $t = (x+1)/2$ converts the interval to $(-1, 1)$ in the argument x, and makes $y = \frac{1}{4}(x^2 + 2x + 1)$. If we note first the elementary result

$$\int_{-1}^{1} \frac{x^p}{\sqrt{1-x^2}}\,dx = \int_0^{\pi} (\cos A)^p\,dA = \begin{cases} \pi & p = 0 \\ 0 & p = 1 \\ \pi/2 & p = 2 \\ 0 & p = 3 \end{cases}$$

then the coefficient a_0 becomes (see Problem 21.46) $a_0 = \frac{1}{4}(\frac{1}{2} + 0 + 1) = \frac{3}{8}$; and since $y(x)T_1(x)$ is $\frac{1}{4}(x^3 + 2x^2 + x)$, we have $a_1 = \frac{1}{4}(0 + 2 + 0) = \frac{1}{2}$. The least-squares polynomial is, therefore,

$$\frac{3}{8} T_0(x) + \frac{1}{2} T_1(x) = \frac{3}{8} + \frac{1}{2}x$$

There is a second and much briefer path to this result. Using the results in Problem 21.45,

$$y(x) = \frac{1}{4}\left(\frac{1}{2}T_0 + \frac{1}{2}T_2 + 2T_1 + T_0\right) = \frac{3}{8}T_0 + \frac{1}{2}T_1 + \frac{1}{8}T_2$$

Truncating this after the linear terms, we have at once the result just found. Moreover we see that the error is, in the case of this quadratic $y(x)$, precisely the equal ripple function $T_2(x)/8$. This is, of course, a consequence of the series of Chebyshev polynomials terminating with this term. For most functions the error will only be approximately the first omitted term, and therefore only approximately an equal-ripple error. Comparing the extreme errors here $(\frac{1}{8}, -\frac{1}{8}, \frac{1}{8})$ with those in Problem 21.31 which were $(\frac{1}{6}, -\frac{1}{12}, \frac{1}{6})$, we see that the present approximation sacrifices some accuracy in the center for improved accuracy at the extremes plus the equal ripple feature. Both lines are shown in Fig. 21-7.

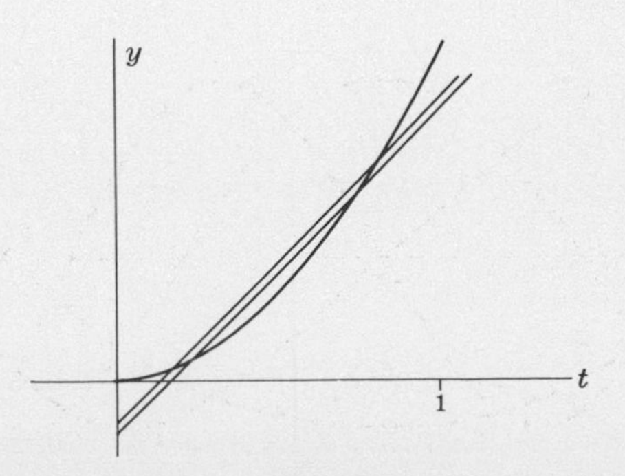

Fig. 21-7

21.50. Find a cubic approximation in terms of Chebyshev polynomials for $y(x) = \sin x$.

The integrals which must be computed to obtain the coefficients of the least-squares polynomial with weight function $w(x) = 1/\sqrt{1-x^2}$ are too complicated in this case. Instead we will illustrate the process of *economization of polynomials*. Beginning with

$$\sin x \simeq x - \frac{1}{6}x^3 + \frac{1}{120}x^5$$

we replace the powers of x by their equivalents in terms of Chebyshev polynomials, using Problem 21.45.

$$\sin x \simeq T_1 - \frac{1}{24}(3T_1 + T_3) + \frac{1}{1920}(10T_1 + 5T_3 + T_5) = \frac{169}{192}T_1 - \frac{5}{128}T_3 + \frac{1}{1920}T_5$$

The coefficients here are not exactly the a_k of Problem 21.46 since higher powers of x from the sine series would make further contributions to the T_1, T_3, and T_5 terms. But those contributions would be relatively small, particularly for the early T_k terms. For example, the x^5 term has altered the T_1 term by less than 1 percent, and the x^7 term would alter it by less than .01 percent. In contrast the x^5 term has altered the T_3 term by about 6 percent, though x^7 will contribute only about .02 percent more. This suggests that truncating our expansion will give us a close approximation to the least-squares cubic. Accordingly we take for our approximation

$$\sin x \simeq \frac{169}{192}T_1 - \frac{5}{128}T_3 \simeq .9974x - .1562x^3$$

The accuracy of this approximation may be estimated by noting that we have made two "truncation errors," first by using only three terms of the power series for $\sin x$ and second in dropping T_5. Both affect the fourth decimal place. Naturally, greater accuracy is available if we seek a least-squares polynomial of higher degree, but even the one we have has accuracy comparable to that of the fifth-degree Taylor polynomial with which we began. The errors of our present cubic and the Taylor cubic, obtained by dropping the x^5 term, are compared in Fig. 21-8. The Taylor cubic is superior near zero but the almost-equal-error property of the (almost) least-squares polynomial is evident and should be compared with $T_5(x)$.

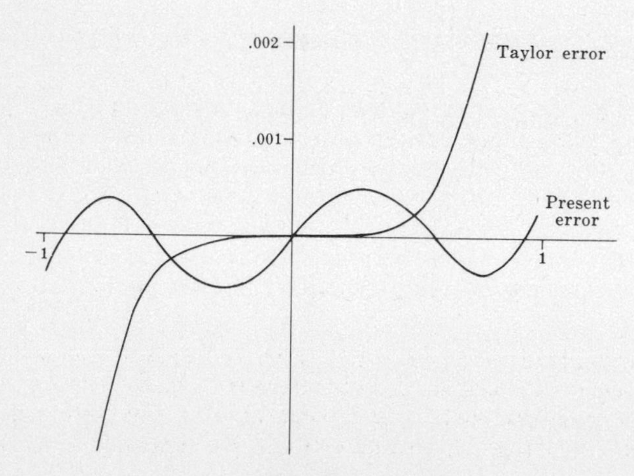

Fig. 21-8

21.51. Prove that for m and n less than N,

$$\sum_{i=0}^{N-1} T_m(x_i)T_n(x_i) = \begin{cases} 0 & m \neq n \\ N/2 & m = n \neq 0 \\ N & m = n = 0 \end{cases}$$

where $x_i = \cos A_i = \cos[(2i+1)\pi/2N]$, $i = 0, 1, \ldots, N-1$.

From the trigonometric definition of the Chebyshev polynomials, we find directly

$$\sum_{i=0}^{N-1} T_m(x_i)T_n(x_i) = \sum_{i=0}^{N-1} \cos mA_i \cos nA_i = \frac{1}{2}\sum_{i=0}^{N-1} [\cos(m+n)A_i + \cos(m-n)A_i]$$

Since $\cos ai = (\frac{1}{2}\sin\frac{1}{2}a)[\Delta \sin a(i-\frac{1}{2})]$ both cosine sums may be telescoped. It is simpler, however, to note that except when $m+n$ or $m-n$ is zero each sum vanishes by symmetry, the angles A_i being equally spaced between 0 and π. This already proves the result for $m \neq n$. If $m = n \neq 0$ the second sum contributes $N/2$, while if $m = n = 0$ both sums together total N. It should be noticed that the Chebyshev polynomials are orthogonal under summation as well as under integration. This is often a substantial advantage, since sums are far easier to compute than integrals of complicated functions, particularly when the factor $\sqrt{1-x^2}$ appears in the latter but not in the former.

21.52. What choice of coefficients a_k will minimize

$$\sum_{x_i} [y(x_i) - a_0T_0(x_i) - \cdots - a_mT_m(x_i)]^2$$

where the x_i are the arguments of the preceding problem?

With proper identification it follows directly from Problems 21.7 and 21.8 that the orthogonal projection $p = a_0T_0 + \cdots + a_mT_m$ determined by

$$a_k = \frac{\sum_i y(x_i)T_k(x_i)}{\sum_i [T_k(x_i)]^2}$$

provides the minimum. Using Problem 21.51 the coefficients are

$$a_0 = \frac{1}{N}\sum_i y(x_i) \qquad a_k = \frac{2}{N}\sum_i y(x_i)T_k(x_i) \qquad k = 1, \ldots, m$$

For $m = N-1$ we have the collocation polynomial for the N points $(x_i, y(x_i))$ and the minimum sum is zero.

21.53. Find the least-squares line for $y(t) = t^2$ over $(0, 1)$ by the method of Problem 21.52.

We have already found a line which minimizes the integral of Problem 21.46. To minimize the sum of Problem 21.52, choose $t = (x+1)/2$ as before. Suppose we use only two points, so that $N = 2$. These points will have to be $x_0 = \cos\pi/4 = 1/\sqrt{2}$ and $x_1 = \cos 3\pi/4 = -1/\sqrt{2}$. Then

$$a_0 = \frac{1}{2}\left[\frac{1}{8}(3+2\sqrt{2}) + \frac{1}{8}(3-2\sqrt{2})\right] = \frac{3}{8}$$

$$a_1 = \frac{1}{8}(3+2\sqrt{2})\left(\frac{1}{\sqrt{2}}\right) + \frac{1}{8}(3-2\sqrt{2})\left(-\frac{1}{\sqrt{2}}\right) = \frac{1}{2}$$

and the line is given by $p(x) = \frac{3}{8}T_0 + \frac{1}{2}T_1 = \frac{3}{8} + \frac{1}{2}x$. This is the same line as before and using a larger N would reproduce it again. The explanation of this is simply that y itself can be represented in the form $y = a_0T_0 + a_1T_1 + a_2T_2$ and, since the T_k are orthogonal relative to both integration and summation, the least-squares line in either sense is also available by *truncation*. (See the last paragraph of Problem 21.8.)

21.54. Find least-squares lines for $y(x) = x^3$ over $(-1, 1)$ by minimizing the sum of Problem 21.52.

In this problem the line we get will depend somewhat upon the number of points we use. First take $N = 2$, which means that we use $x_0 = -x_1 = 1/\sqrt{2}$ as before. Then

$$a_0 = \frac{1}{2}(x_0^3 + x_1^3) = 0 \qquad a_1 = x_0^4 + x_1^4 = \frac{1}{2}$$

Choosing $N=3$ we find $x_0=\sqrt{3}/2$, $x_1=0$, and $x_2=-\sqrt{3}/2$. This makes

$$a_0=\frac{1}{3}(x_0^3+x_1^3+x_2^3)=0 \qquad a_1=\frac{2}{3}(x_0^4+x_1^4+x_2^4)=\frac{3}{4}$$

Taking the general case of N points, we have $x_i=\cos A_i$ and

$$a_0=\frac{1}{N}\sum_{i=0}^{N-1}\cos^3 A_i=0$$

by the symmetry of the A_i in the first and second quadrants. Also,

$$a_1=\frac{2}{N}\sum_{i=0}^{N-1}\cos^4 A_i=\frac{2}{N}\sum_{i=0}^{N-1}\left(\frac{3}{8}+\frac{1}{2}\cos 2A_i+\frac{1}{8}\cos 4A_i\right)$$

Since the A_i are the angles $\pi/2N,\ 3\pi/2N,\ldots,(2N-1)\pi/2N$, the doubled angles are π/N, $3\pi/N,\ldots,(2N-1)\pi/N$ and these are symmetrically spaced around the entire circle. The sum of the $\cos 2A_i$ is therefore zero. Except when $N=2$, the sum of the $\cos 4A_i$ will also be zero so that $a_i=\frac{3}{4}$, for $N=2$. For N tending to infinity we thus have trivial convergence to the line $p(x)=3T_1/4=3x/4$.

If we adopt the minimum integral approach, then we find

$$a_0=\frac{1}{\pi}\int_{-1}^{1}\frac{x^3}{\sqrt{1-x^2}}\,dx=0 \qquad a_1=\frac{2}{\pi}\int_{-1}^{1}\frac{x^4}{\sqrt{1-x^2}}\,dx=\frac{3}{4}$$

which leads to the same line.

The present example may serve as further elementary illustration of the Problem 21.52 algorithm, but the result is more easily found and understood by noting that $y=x^3=\frac{3}{4}T_1+\frac{1}{4}T_3$ and once again appealing to the corollary in Problem 21.8 to obtain $3T_1/4$ or $3x/4$ by truncation. The truncation process fails for $N=2$ since then the polynomials T_0, T_1, T_2, T_3 are not orthogonal. (See Problem 21.51.)

21.55. Find least-squares lines for $y(x)=|x|$ over $(-1,1)$ by minimizing the sum of Problem 21.52.

With $N=2$ we quickly find $a_0=1/\sqrt{2}$, $a_1=0$. With $N=3$ the results $a_0=1/\sqrt{3}$, $a_1=0$ are just as easy. For arbitrary N,

$$a_0=\frac{1}{N}\sum_{i=0}^{N-1}|\cos A_i|=\frac{2}{N}\sum_{i=0}^{I}\cos A_i$$

where I is $(N-3)/2$ for odd N, and $(N-2)/2$ for even N. This trigonometric sum may be evaluated by telescoping or otherwise, with the result

$$a_0=\frac{\sin[\pi(I+1)/N]}{N\sin(\pi/2N)}$$

It is a further consequence of symmetry that $a_1=0$ for all N. For N tending to infinity it now follows that

$$\lim a_0=\lim\frac{1}{N\sin\pi/2N}=\frac{2}{\pi}$$

As more and more points are used, the limiting line is approached. Turning to the minimum integral approach, we of course anticipate this same line. The computation produces

$$a_0=\frac{1}{\pi}\int_{-1}^{1}\frac{|x|}{\sqrt{1-x^2}}\,dx=\frac{2}{\pi}$$

$$a_1=\frac{2}{\pi}\int_{-1}^{1}\frac{x\,|x|}{\sqrt{1-x^2}}\,dx=0$$

and so we are not disappointed. The limiting line is the solid line in Fig. 21-9.

21.56. Apply the method of the previous problems to the experimentally produced curve of Fig. 21-5.

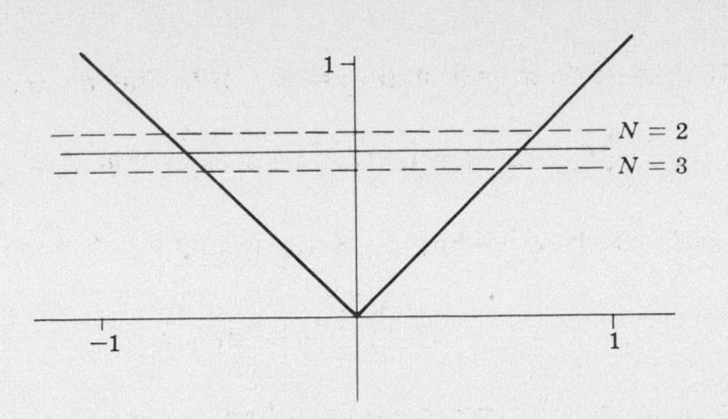

Fig. 21-9

For such a function, of unknown analytic character, any of our methods must involve discretization at some point. We have already chosen one discrete set of values of the function for use in Simpson's rule, thus maintaining at least in spirit the idea of minimizing an integral. We could have used the same equidistant set of arguments and minimized a sum. With the idea of obtaining a more nearly equal-ripple error, however, we now choose the arguments $x_i = \cos A_i = 2t_i - 1$ instead. With 11 points, the number used earlier, the arguments, $x_i = \cos A_i = \cos[(2i+1)\pi/22]$ and corresponding t_i as well as y_i values read from the curve are as follows:

x_i	.99	.91	.75	.54	.28	.00	−.28	−.54	−.75	−.91	−.99
t_i	1.00	.96	.88	.77	.64	.50	.36	.23	.12	.04	.00
y_i	.36	.33	.28	.24	.21	.25	.20	.12	.17	.13	.10

The coefficients become

$$a_0 = \frac{1}{11}\sum y_i \simeq .22 \qquad a_1 = \frac{2}{11}\sum x_i y_i \simeq .11$$

making the line $p(x) = .22 + .11x = .22t + .11$ which is almost indistinguishable from the earlier result. The data inaccuracies have not warranted the extra sophistication.

Supplementary Problems

21.57. The average scores reported by golfers of various handicaps on a par-four hole were as follows:

Handicap	6	8	10	12	14	16	18	20	22	24
Average	4.6	4.8	4.6	4.9	5.0	5.4	5.1	5.5	5.6	6.0

Find the least-squares line for this data.

21.58. Use the least-squares line of the preceding problem to smooth the reported data.

21.59. Estimate the rate at which the average score increases per unit handicap.

21.60. Find the least-squares parabola for the data of Problem 21.57. Does it differ noticeably from the line just found?

21.61. When the x_i and y_i are both subject to errors of about the same size, it has been argued that the sum of squares of perpendicular distances to a line should be minimized, rather than the sum of squares of vertical distances. Show that this requires minimizing

$$S = \frac{1}{1+M^2} \sum_{i=0}^{N} (y_i - Mx_i - B)^2$$

Then find the normal equations and show that M is determined by a quadratic equation.

21.62. Apply the method of the preceding problem to the data of Problem 21.57. Does the new line differ very much from the line found in that problem?

21.63. Find the least-squares line for the three points (x_0, y_0), (x_1, y_1), and (x_2, y_2) by the method of Problem 21.1. What is true of the signs of the three numbers $y(x_i) - y_i$?

21.64. Show that for the data

x_i	2.2	2.7	3.5	4.1
P_i	65	60	53	50

the introduction of $y = \log P$ and computation of the least-squares line for the (x_i, y_i) data pairs leads eventually to $P = 91.9x^{-.43}$.

21.65. Find a function of type $P = Ae^{Mx}$ for the data

x_i	1	2	3	4
P_i	60	30	20	15

21.66. Show that the least-squares parabola for seven points leads to the smoothing formula

$$y(x_k) \simeq y_k - \frac{1}{21}(9\delta^4 y_k + 2\delta^6 y_k)$$

by following the procedures of Problems 21.12 and 21.13.

21.67. Apply the preceding formula to smooth the center four y_i values of Table 21.1. Compare with the correct roots and note whether or not this formula yields better results than the five-point formula.

21.68. Use the seven-point parabola to derive the approximate differentiation formula

$$y'(x_k) \simeq \frac{1}{28h}(-3y_{k-3} - 2y_{k-2} - y_{k-1} + y_{k+1} + 2y_{k+2} + 3y_{k+3})$$

21.69. Apply the preceding formula to estimate $y'(x)$ for $x = 4$, 5, 6, and 7 from the y_i values of Table 21.1. How do the results compare with those obtained by the five-point parabola? (See Problem 21.19.)

21.70. The following are values of $y(x) = x^2$ with random errors of from $-.10$ to $.10$ added. (Errors were obtained by drawing cards from an ordinary pack with face cards removed, black meaning plus and red minus.) The correct values T_i are also included.

x_i	1.0	1.1	1.2	1.3	1.4	1.5	1.6	1.7	1.8	1.9	2.0
y_i	.98	1.23	1.40	1.72	1.86	2.17	2.55	2.82	3.28	3.54	3.92
T_i	1.00	1.21	1.44	1.69	1.96	2.25	2.56	2.89	3.24	3.61	4.00

Apply the smoothing formulas of Problem 21.13 and 21.15. Compare the RMS errors of the original and smoothed values.

21.71. Apply the differentiation formula of Problem 21.18, for the center seven arguments. Also apply the formula obtained from Stirling's polynomial (see Problem 21.19). Which produces better approximations to $y'(x) = 2x$? Note that in this example the "true" function is actually a parabola, so that except for the random errors which were introduced we would have exact results. Has the least-squares parabola penetrated through the errors to any extent and produced information about the true $y'(x)$?

21.72. What is the least-squares parabola for the data of Problem 21.70? Compare it with $y(x) = x^2$.

21.73. Use the formulas of Problem 21.20 to estimate $y'(x)$ near the ends of the data supply given in Problem 21.70.

21.74. Estimate $y''(x)$ from your computed $y'(x)$ values.

21.75. The following are values of $\sin x$ with random errors of $-.10$ to $.10$ added. Find the least-squares parabola and use it to compute smoothed values. Also apply the method of Problem 21.13 which uses a different least-squares parabola for each point to smooth the data. Which works best?

x	0	.2	.4	.6	.8	1.0	1.2	1.4	1.6
$\sin x$	−.09	.13	.44	.57	.64	.82	.97	.98	1.04

21.76. A simple and ancient smoothing procedure, which still finds use, is the *method of moving averages*. In this method each value y_i is replaced by the average of itself and nearby neighbors. For example, if two neighbors on each side are used, the formula is

$$p_i = \frac{1}{5}(y_{i-2} + y_{i-1} + y_i + y_{i+1} + y_{i+2})$$

where p_i is the smoothed substitute for y_i. Apply this to the data of the preceding problem. Devise a method for smoothing the end values for which two neighbors are not available on one side.

21.77. Apply the method of moving averages, using only one neighbor on each side, to the data of Problem 21.75. The formula for interior arguments will be

$$p_i = \frac{1}{3}(y_{i-1} + y_i + y_{i+1})$$

Devise a formula for smoothing the end values.

21.78. Apply the formula of the preceding problem to the values $y(x) = x^3$ below, obtaining the p_i values listed.

x_i	0	1	2	3	4	5	6	7
$y = x_i^3$	0	1	8	27	64	125	216	343
p_i		3	12	33	72	135	228	

Show that these p_i values belong to a different cubic function. Apply the moving average formula to the p_i values to obtain a second generation of smoothed values. Can you tell what happens as successive generations are computed, assuming that the supply of y_i values is augmented at both ends indefinitely?

21.79. Apply the method of moving averages to smooth the oscillating data below.

x_i	0	1	2	3	4	5	6	7	8
y_i	0	1	0	−1	0	1	0	−1	0

What happens if higher generations of smooth values are computed endlessly? It is easy to see that excessive smoothing can entirely alter the character of a data supply.

21.80. Use orthogonal polynomials to find the same least-squares line found in Problem 21.2.

21.81. Use orthogonal polynomials to find the same least-squares parabola found in Problem 21.10.

21.82. Use orthogonal polynomials to find the least-squares polynomial of degree four for the square root data of Problem 21.14. Use this single polynomial to smooth the data. Compute the RMS error of the smoothed values. Compare with those given in Problem 21.17.

21.83. The following are values of e^x with random errors of from −.10 to .10 added. Use orthogonal polynomials to find the least-squares cubic. How accurate is this cubic?

x	0	.1	.2	.3	.4	.5	.6	.7	.8	.9	1.0
y	.92	1.15	1.22	1.44	1.44	1.66	1.79	1.98	2.32	2.51	2.81

21.84. The following are values of the Bessel function $J_0(x)$ with random errors of from −.010 to .010 added. Use orthogonal polynomials to find a least-squares approximation. Choose the degree you feel appropriate. Then smooth the data and compare with the correct results which are also provided.

x	0	1	2	3	4	5	6	7	8	9	10
$y(x)$	.994	.761	.225	−.253	−.400	−.170	.161	.301	.177	−.094	−.240
Correct	1.00	.765	.224	−.260	−.397	−.178	.151	.300	.172	−.090	−.246

21.85. Find the least-squares line for $y(x) = x^2$ on the interval $(-1, 1)$.

21.86. Find the least-squares line for $y(x) = x^3$ on the interval $(-1, 1)$.

21.87. Find the least-squares parabola for $y(x) = x^3$ on the interval $(-1, 1)$.

21.88. Find approximately the least-squares parabola for the function in Fig. 21-10, evaluating the integrals by Simpson's rule. This curve should be imagined to be an experimental result which theory claims ought to have been a parabola.

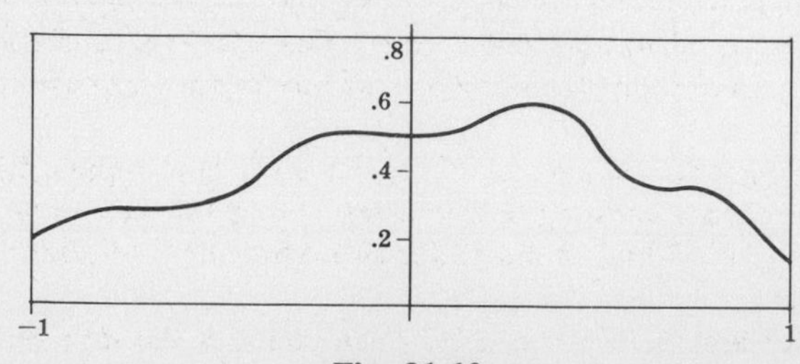

Fig. 21-10

21.89. Show that the Chebyshev series for $\arcsin x$ is

$$\arcsin x = \frac{4}{\pi}\left(T_1 + \frac{1}{9}T_3 + \frac{1}{25}T_5 + \frac{1}{49}T_7 + \cdots\right)$$

by evaluating the coefficient integrals directly. Truncate after T_3 to obtain the least-squares cubic for this function. Compute the actual error of this cubic and compare with the first omitted term (the T_5 term). Notice the (almost) equal-ripple behavior of the error.

21.90. Find the least-squares line for $y(x) = x^2$ on the interval $(-1, 1)$ with weight function $w(x) = 1/\sqrt{1-x^2}$. Compare this line with the one found in Problem 21.85. Which one has the equal-ripple property?

21.91. Find the least-squares parabola for $y(x) = x^3$ on the interval $(-1, 1)$ with weight function $w(x) = 1/\sqrt{1-x^2}$. Compare this with the parabola found in Problem 21.87.

21.92. Represent $y(x) = e^{-x}$ by terms of its power series through x^7. The error will be in the fifth decimal place for x near one. Rearrange the sum into Chebyshev polynomials. How many terms can then be dropped without seriously affecting the fourth decimal place? Rearrange the truncated polynomial into standard form. (This is another example of economization of a polynomial.)

21.93. Show that for $y(x) = T_n(x) = \cos(n \arccos x) = \cos nA$ it follows that $y'(x) = (n \sin nA)/(\sin A)$. Then show that $(1-x^2)y'' - xy' + n^2 y = 0$, which is the classical differential equation of the Chebyshev polynomials.

21.94. Show that $S_n(x) = \sin(n \arccos x)$ also satisfies the differential equation of Problem 21.93.

21.95. Let $U_n(x) = S_n(x)/\sqrt{1-x^2}$ and prove the recursion $U_{n+1}(x) = 2xU_n(x) - U_{n-1}(x)$.

21.96. Verify that $U_0(x) = 0$, $U_1(x) = 1$ and then apply the recursion to verify $U_2(x) = 2x$, $U_3(x) = 4x^2 - 1$, $U_4(x) = 8x^3 - 4x$, $U_5(x) = 16x^4 - 12x^2 + 1$, $U_6(x) = 32x^5 - 32x^3 + 6x$, $U_7(x) = 64x^6 - 80x^4 + 24x^2 - 1$.

21.97. Prove $T_{m+n}(x) + T_{m-n}(x) = 2T_m(x)T_n(x)$ and then put $m = n$ to obtain

$$T_{2n}(x) = 2T_n^2(x) - 1$$

21.98. Use the result of Problem 21.97 to find T_8, T_{16}, and T_{32}.

21.99. Prove $\frac{1}{n}T'_n = 2T_{n-1} + \frac{1}{n-2}T'_{n-2}$ and then deduce

$$T'_{2n+1} = 2(2n+1)(T_{2n} + T_{2n-2} + \cdots + T_2) + 1 \qquad T'_{2n} = 2(2n)(T_{2n-1} + T_{2n-3} + \cdots + T_1)$$

21.100. Prove $T_{2n+1} = x(2T_{2n} - 2T_{2n-2} + 2T_{2n-4} + \cdots \pm T_0)$.

21.101. Economize the result $\ln(1+x) \simeq x - \frac{1}{2}x^2 + \frac{1}{3}x^3 - \frac{1}{4}x^4 + \frac{1}{5}x^5$ by rearranging into Chebyshev polynomials and then retaining only the quadratic terms. Show that the final result $\ln(1+x) \simeq \frac{1}{32} + \frac{11}{8}x - \frac{3}{4}x^2$ has about the same accuracy as the fourth-degree part of the original approximation.

21.102. Economize the polynomial $y(x) = 1 + x + \frac{1}{2}x^2 + \frac{1}{6}x^3 + \frac{1}{24}x^4$, first representing it as a combination of Chebyshev polynomials, then truncating to two terms. Compare the result with $1 + x + \frac{1}{2}x^2$, considering both as approximations to e^x. Which is the better approximation? In what sense?

21.103. Show that the change of argument $x = 2t - 1$, which converts the interval to $(0, 1)$ in terms of t, also converts the Chebyshev polynomials into the following, which may be used instead of the classical polynomials if the intervals (0, 1) is felt to be more convenient:

$$T_0^*(x)=1 \quad T_1^*(x)=2t-1 \quad T_2^*(x)=8t^2-8t+1 \quad T_3^*(x)=32t^3-48t^2+18t-1 \quad \text{etc.}$$

Also prove the recursion $T_{n+1}^*(t)=(4t-2)T_n^*(t)-T_{n-1}^*(t)$.

21.104. Prove $\int T_0(x)\,dx = T_1(x)$, $\int T_1(x)\,dx = \frac{1}{4}T_2(x)$, and, for $n>1$,

$$\int T_n(x)\,dx = \frac{1}{2}\left[\frac{T_{n+1}(x)}{n+1} - \frac{T_{n-1}(x)}{n-1}\right]$$

21.105. Show that the same line found with $N=2$ in Problem 21.53 also appears for arbitrary N.

21.106. Use the method of Problem 21.52 to obtain a least-squares parabola for $y(x)=x^3$ over $(-1,1)$ choosing $N=3$. Show that the same result is obtained for arbitrary N and also by the method of minimizing the integral of Problem 21.91.

21.107. Find the least-squares parabolas for $y(x)=|x|$ over $(-1,1)$ and for arbitrary N. Also show that as N tends to infinity this parabola approaches the minimum integral parabola.

21.108. Apply the method of Problem 21.52 to the experimental data of Fig. 21-10. Use the result to compute smoothed values of $y(x)$ at $x=-1(.2)1$.

21.109. Smooth the following experimental data by fitting a least-squares polynomial of degree five:

t	0	5	10	15	20	25	30	35	40	45	50
y	0	.127	.216	.286	.344	.387	.415	.437	.451	.460	.466

21.110. The following table gives the number y of students who made a grade of x on an examination. To use these results as a standard norm, smooth the y numbers twice, using the smoothing formula

$$p=\frac{1}{35}(-3y_0+12y_1+17y_2+12y_3-3y_4)$$

It is assumed that $y=0$ for unlisted x values.

x	100	95	90	85	80	75	70	65	60	55	50	45
y	0	13	69	147	208	195	195	126	130	118	121	85

x	40	35	30	25	20	15	10	5	0
y	93	75	54	42	30	34	10	8	1

21.111. Find the least-squares polynomial of degree two for the following data. Then obtain smoothed values.

x	.78	1.56	2.34	3.12	3.81
y	2.50	1.20	1.12	2.25	4.28

Chapter 22

Min–Max Polynomial Approximation

DISCRETE DATA

The basic idea of min–max approximation by polynomials may be illustrated for the case of a *discrete data* supply x_i, y_i where $i = 1, \ldots, N$. Let $p(x)$ be a polynomial of degree n or less and let the amounts by which it misses our data points be $h_i = p(x_i) - y_i$. Let H be the largest of these "errors." The min–max polynomial is that particular $p(x)$ for which H is smallest. Min–max approximation is also called *Chebyshev approximation*. The principal results are as follows:

1. The **existence and uniqueness** of the min–max polynomial for any given value of n may be proved by the exchange method described below. The details will be provided for the case $n = 1$ only.
2. The **equal-error property** is the identifying feature of a min–max polynomial. Calling this polynomial $P(x)$, and the maximum error

$$E = \max |P(x_i) - y(x_i)|$$

we shall prove that $P(x)$ is the only polynomial for which $P(x_i) - y(x_i)$ takes the extreme values $\pm E$ at least $n + 2$ times, with alternating sign.

3. The **exchange method** is an algorithm for finding $P(x)$ through its equal-error property. Choosing some initial subset of $n + 2$ arguments x_i, an equal-error polynomial for these data points is found. If the maximum error of this polynomial over the subset chosen is also its overall maximum H, then it is $P(x)$. If not, some point of the subset is exchanged for an outside point and the process is repeated. Eventual convergence to $P(x)$ will be proved.

CONTINUOUS DATA

For continuous data $y(x)$ it is almost traditional to begin by recalling a classical theorem of analysis, known as the **Weierstrass theorem,** which states that for a continuous function $y(x)$ on an interval (a, b) there will be a polynomial $p(x)$ such that

$$|p(x) - y(x)| \leqq \epsilon$$

in (a, b) for arbitrary positive ϵ. In other words, there exists a polynomial which approximates $y(x)$ uniformly to any required accuracy. We prove this theorem using Bernstein polynomials, which have the form

$$B_n(x) = \sum_{k=0}^{n} p_{nk}\, y\left(\frac{k}{n}\right)$$

where $y(x)$ is a given function and

$$p_{nk} = \binom{n}{k} x^k (1 - x)^{n-k}$$

Our proof of the Weierstrass theorem involves showing that $\lim B_n(x) = y(x)$ uniformly for n tending to infinity. The rate of convergence of the Bernstein polynomials to $y(x)$ is often disappointing. Accurate uniform approximations are more often found in practice by min–max methods.

The essential facts of *min–max methods* somewhat parallel those for the discrete case.

1. The min–max approximation to $y(x)$, among all polynomials of degree n or less, minimizes the $\max |p(x) - y(x)|$ for the given interval (a, b).
2. It exists and is unique.

3. It has an equal-error property, being the only such polynomial for which $p(x) - y(x)$ takes extreme values of size E, with alternating sign, at $n+2$ or more arguments in (a, b). Thus the min–max polynomial can be identified by its equal-error property. In simple examples it may be displayed exactly. An example is the min–max line when $y''(x) > 0$. Here

$$P(x) = Mx + B$$

with
$$M = \frac{y(b) - y(a)}{b - a} \qquad B = \frac{y(a) + y(x_2)}{2} - \frac{(a + x_2)[y(b) - y(a)]}{2(b - a)}$$

and x_2 determined by
$$y'(x_2) = \frac{y(b) - y(a)}{b - a}$$

The three extreme points are a, x_2, and b. Ordinarily, however, the exact result is not within reach and an exchange method must be used to produce a polynomial which comes close to the equal-error behavior.

4. Series of Chebyshev polynomials, when truncated, often yield approximations having almost equal-error behavior. Such approximations are therefore almost min–max. If not entirely adequate by themselves, they may be used as inputs to the exchange method which then may be expected to converge more rapidly than it would from a more arbitrary start.

THE INFINITY NORM

The underlying theme of this chapter is to minimize the norm

$$\|y - p\|_\infty$$

where y represents the given data and p the approximating polynomial.

Solved Problems

DISCRETE DATA, THE MIN–MAX LINE

22.1. Show that for any three points (x_i, Y_i) with the arguments x_i distinct, there is exactly one straight line which misses all three points by equal amounts and with alternating signs. This is the *equal-error line* or Chebyshev line.

Let $y(x) = Mx + B$ represent an arbitrary line and let $h_i = y(x_i) - Y_i = y_i - Y_i$ be the "errors" at the three data points. An easy calculation shows that, since $y_i = Mx_i + B$, for any straight line at all

$$(x_3 - x_2)y_1 - (x_3 - x_1)y_2 + (x_2 - x_1)y_3 = 0$$

Defining $\beta_1 = x_3 - x_2$, $\beta_2 = x_3 - x_1$, $\beta_3 = x_2 - x_1$, the above equation becomes

$$\beta_1 y_1 - \beta_2 y_2 + \beta_3 y_3 = 0$$

We may take it that $x_1 < x_2 < x_3$ so that the three β's are positive numbers. We are to prove that there is one line for which

$$h_1 = h \qquad h_2 = -h \qquad h_3 = h$$

making the three errors of equal size and alternating sign. (This is what will be meant by an "equal-error" line.) Now, if a line having this property does exist, then

$$y_1 = Y_1 + h \qquad y_2 = Y_2 - h \qquad y_3 = Y_3 + h$$

and substituting above,
$$\beta_1(Y_1 + h) - \beta_2(Y_2 - h) + \beta_3(Y_3 + h) = 0$$

Solving for h,

$$h = -\frac{\beta_1 Y_1 - \beta_2 Y_2 + \beta_3 Y_3}{\beta_1 + \beta_2 + \beta_3}$$

This already proves that at most one equal-error line can exist and that it must pass through the three points $(x_1, Y_1 + h)$, $(x_2, Y_2 - h)$, $(x_3, Y_3 + h)$ for the value h just computed. Though normally one asks a line to pass through only two designated points, it is easy to see that in this special case the three points do fall on a line. The slopes of P_1P_2 and P_2P_3 (where P_1, P_2, P_3 are the three points taken from left to right) are

$$\frac{Y_2 - Y_1 - 2h}{x_2 - x_1} \quad \text{and} \quad \frac{Y_3 - Y_2 + 2h}{x_3 - x_2}$$

and using our earlier equations these are easily proved to be the same. So there is exactly one equal-error, or Chebyshev, line.

22.2. Find the equal-error line for the data points (0, 0), (1, 0), and (2. 1).

First we find $\beta_1 = 2 - 1 = 1$, $\beta_2 = 2 - 0 = 2$, $\beta_3 = 1 - 0 = 1$, and then compute

$$h = -\frac{(1)(0) - (2)(0) + (1)(1)}{1 + 2 + 1} = -\frac{1}{4}$$

The line passes through $(0, -\frac{1}{4})$, $(1, \frac{1}{4})$, and $(2, \frac{3}{4})$ and so has the equation $y(x) = \frac{1}{2}x - \frac{1}{4}$. The line and points appear in Fig. 22-1.

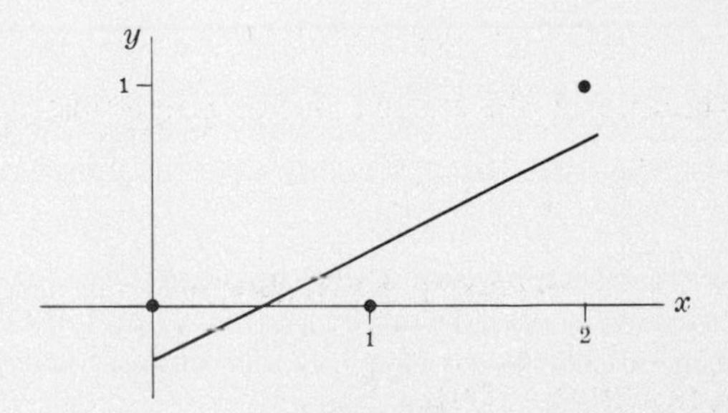

Fig. 22-1

22.3. Show that the equal-error line is also the min–max line for the three points (x_i, Y_i).

The errors of the equal-error line are $h, -h, h$. Let h_1, h_2, h_3 be the errors for any other line. Also let H be the largest of $|h_1|$, $|h_2|$, $|h_3|$. Then using our earlier formulas,

$$h = -\frac{\beta_1 Y_1 - \beta_2 Y_2 + \beta_3 Y_3}{\beta_1 + \beta_2 + \beta_3} = -\frac{\beta_1(y_1 - h_1) - \beta_2(y_2 - h_2) + \beta_3(y_3 - h_3)}{\beta_1 + \beta_2 + \beta_3}$$

where y_1, y_2, y_3 here refer to the "any other line." This rearranges to

$$h = -\frac{(\beta_1 y_1 - \beta_2 y_2 + \beta_3 y_3) - (\beta_1 h_1 - \beta_2 h_2 + \beta_3 h_3)}{\beta_1 + \beta_2 + \beta_3}$$

and the first term being zero we have a relationship between the h of the equal-error line and the h_1, h_2, h_3 of the other line,

$$h = \frac{\beta_1 h_1 - \beta_2 h_2 + \beta_3 h_3}{\beta_1 + \beta_2 + \beta_3}$$

Since the β's are positive, the right side of this equation will surely be increased if we replace h_1, h_2, h_3 by $H, -H, H$, respectively. Thus $|h| \leqq H$, and the maximum error size of the Chebyshev line, which is $|h|$, comes out no greater than that of any other line.

22.4. Show that no other line can have the same maximum error as the Chebyshev line, so that the min–max line is unique.

Suppose equality holds in our last result, $|h| = H$. This means that the substitution of $H, -H, H$ which produced this result has not actually increased the size of $\beta_1 h_1 - \beta_2 h_2 + \beta_3 h_3$. But this can be true only if h_1, h_2, h_3 themselves are all of equal size H and alternating sign, and these are the features which led us to the three points through which the Chebyshev line passes. Surely these are not two straight lines through these three points. This proves that the equality $|h| = H$ identifies the Chebyshev line. We have now proved that the equal-error line and the min–max line for three points are the same.

22.5. Illustrate the *exchange method* by applying it to the following data:

x_i	0	1	2	6	7
Y_i	0	0	1	2	3

We will prove shortly that there exists a unique min–max line for N points. The proof uses the exchange method, which is also an excellent algorithm for computing this line, and so this method will first be illustrated. It involves four steps.

Step 1. Choose any three of the data points. (A set of three data points will be called a triple. This step simply selects an initial triple. It will be changed in Step 4.)

Step 2. Find the Chebyshev line for this triple. The value h for this line will of course be computed in the process.

Step 3. Compute the errors at all data points for the Chebyshev line just found. Call the largest of these h_i values (in absolute value) H. If $|h| = H$ the search is over. The Chebyshev line for the triple in hand is the min–max line for the entire set of N points. (We shall prove this shortly.) If $|h| < H$ proceed to Step 4.

Step 4. This is the exchange step. Choose a new triple as follows. Add to the old triple a data point at which the greatest error size H occurs. Then discard one of the former points, in such a way that the remaining three have errors of alternating sign. (A moment's practice will show that this is always possible.) Return, with the new triple, to Steps 2 and 3.

To illustrate, suppose we choose for the initial triple

$$(0, 0) \qquad (1, 0) \qquad (2, 1)$$

consisting of the first three points. This is the triple of Problem 22.2, for which we have already found the Chebyshev line to be $y = \frac{1}{2}x - \frac{1}{4}$ with $h = -\frac{1}{4}$. This completes Steps 1 and 2. Proceeding to Step 3 we find the errors at all five data points to be $-\frac{1}{4}, \frac{1}{4}, -\frac{1}{4}, \frac{3}{4}, \frac{1}{4}$. This makes $H = h_4 = \frac{3}{4}$. This Chebyshev line is an equal-error line on its own triple but it misses the fourth data point by a larger amount. (See the dashed line in Fig. 22-2.)

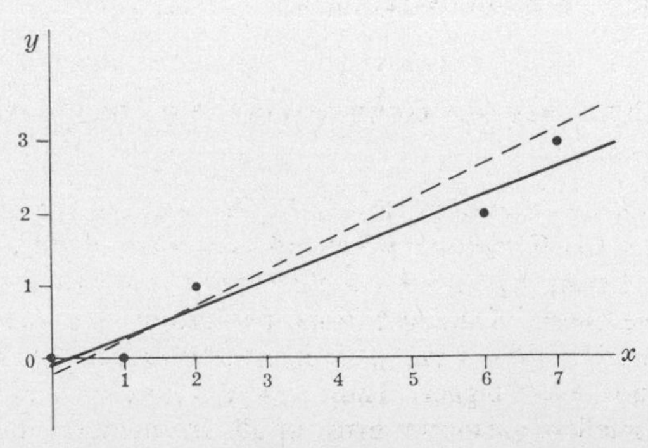

Fig. 22-2

Moving therefore to Step 4 we now include the fourth point and eliminate the first to obtain the new triple

$$(1, 0) \qquad (2, 1) \qquad (6, 2)$$

on which the errors of the old Chebyshev line do have the required alternation of sign $(\frac{1}{4}, -\frac{1}{4}, \frac{3}{4})$. With this triple we return to Step 2 and find a new Chebyshev line. The computation begins with

$$\beta_1 = 6 - 2 = 4 \qquad \beta_2 = 6 - 1 = 5 \qquad \beta_3 = 2 - 1 = 1$$

$$h = -\frac{(4)(0) - (5)(1) + (1)(2)}{4 + 5 + 1} = \frac{3}{10}$$

so that the line must pass through the three points $(1, \frac{3}{10})$, $(2, \frac{7}{10})$, and $(6, \frac{23}{10})$. This line is found to be $y = \frac{2}{5}x - \frac{1}{10}$. Repeating Step 3 we find the five errors $-\frac{1}{10}, \frac{3}{10}, -\frac{3}{10}, \frac{3}{10}, -\frac{3}{10}$; and since $H = \frac{3}{10} = |h|$, the job is done.

The Chebyshev line for the new triple is the min–max line for the entire point set. Its maximum error is $\frac{3}{10}$. The new line is shown solid in Fig. 22-2. Notice that the $|h|$ value of our new line $(\frac{3}{10})$ is larger than that of the first line $(\frac{1}{4})$. But over the entire point set the maximum error has been reduced from $\frac{3}{4}$ to $\frac{3}{10}$, and it is the min–max error. This will now be proved for the general case.

22.6. Prove that the condition $|h| = H$ in Step 3 of the exchange method will be satisfied eventually, so that the method will stop. (Conceivably we could be making exchanges forever.)

Recall that after any particular exchange the old Chebyshev line has errors of size $|h|, |h|, H$ on the new triple. Also recall that $|h| < H$ (or we would have stopped) and that the three errors alternate in sign. The Chebyshev line for this new triple is then found. Call its errors on this new triple $h^*, -h^*, h^*$. Returning to the formula for h in Problem 22.3, with the old Chebyshev line playing the role of "any other line," we have

$$h^* = \frac{\beta_1 h_1 - \beta_2 h_2 + \beta_3 h_3}{\beta_1 + \beta_2 + \beta_3}$$

where h_1, h_2, h_3 are the numbers h, h, H with alternating sign. Because of this alternation of sign all three terms in the numerator of this fraction have the same sign, so that

$$|h^*| = \frac{\beta_1 |h| + \beta_2 |h| + \beta_3 H}{\beta_1 + \beta_2 + \beta_3}$$

if we assume that the error H is at the third point, just to be specific. (It really makes no difference in which position it goes.) In any event, $|h^*| > |h|$ because $H > |h|$. The new Chebyshev line has a greater error size on its triple than the old one had on its triple. This result now gives excellent service. If it comes as a surprise, look at it this way. The old line gave excellent service ($h = \frac{1}{4}$ in our example) on its own triple, but poor service ($H = \frac{3}{4}$) elsewhere. The new line gave good service ($h = \frac{3}{10}$) on its own triple, and just as good service on the other points also.

We can now prove that the exchange method must come to a stop sometime. For there are only so many triples. And no triple is ever chosen twice, since as just proved the h values increase steadily. At some stage the condition $|h| = H$ will be satisfied.

22.7. Prove that the last Chebyshev line computed in the exchange method is the min–max line for the entire set of N points.

Let h be the equal-error value of the last Chebyshev line on its own triple. Then the maximum error size on the entire point set is $H = |h|$, or we would have proceeded by another exchange to still another triple and another line. Let $h_1, h_2, \ldots, h_N$ be the errors for any other line. Then $|h| < \max |h_i|$ where h_i is restricted to the three points of the last triple, because no line outperforms a Chebyshev line on its own triple. But then certainly $|h| < \max |h_i|$ for h_i unrestricted, for including the rest of the N points can only make the right side even bigger. Thus $H = |h| < \max |h_i|$ and the maximum error of the last Chebyshev line is the smallest maximum error of all. In summary, the min–max line for the set of N points is an equal-error line on a properly chosen triple.

22.8. Apply the exchange method to find the min–max line for the following data:

x_i	0	1	2	3	4	5	6	7	8	9	10	11	12	13	14	15
Y_i	0	1	1	2	1	3	2	2	3	5	3	4	5	4	5	6

x_i	16	17	18	19	20	21	22	23	24	25	26	27	28	29	30
Y_i	6	5	7	6	8	7	7	8	7	9	11	10	12	11	13

The number of available triples is $C(31, 3) = 4495$, so that finding the correct one might seem comparable to needle-hunting in haystacks. However, the exchange method wastes very little time on inconsequential triples. Beginning with the very poor triple at $x = (0, 1, 2)$ only three exchanges are necessary to produce the min–max line $y(x) = .38x - .29$ which has coefficients rounded off to two places. The successive triples with h and H values were as follows:

Triple at $x =$	(0, 1, 2)	(0, 1, 24)	(1, 24, 30)	(9, 24, 30)
h	.250	.354	−1.759	−1.857
H	5.250	3.896	2.448	1.857

Note that in this example no unwanted point is ever brought into the triple. Three points are needed, three exchanges suffice. Note also the steady increase of $|h|$, as forecast. The 31 points, the min–max line, and the final triple (dashed vertical lines show the equal errors) appear in Fig. 22-3.

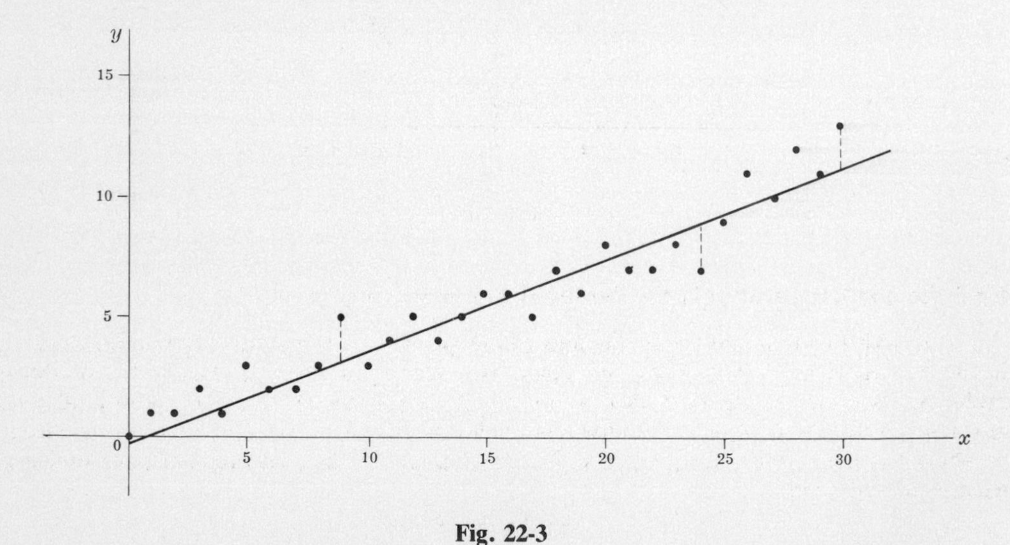

Fig. 22-3

DISCRETE DATA, THE MIN–MAX POLYNOMIAL

22.9. Extend the exchange method to find the min–max parabola for the data below.

x_i	−2	−1	0	1	2
y_i	2	1	0	1	2

The data are of course drawn from the function $y = |x|$ but this simple function will serve to illustrate how all the essential ideas of the exchange method carry over from the straight-line problems just treated to the discovery of a min–max polynomial. The proofs of the existence, uniqueness, and equal-error properties of such a polynomial are extensions of our proofs for the min–max line and will not be given. The algorithm now begins with the choice of an "initial quadruple" and we take the first four points, at $x = -2, -1, 0, 1$. For this quadruple we seek an equal-error parabola, say

$$p_1(x) = a + bx + cx^2$$

This means that we require $p(x_i) - y_i = \pm h$ alternately, or

$$\begin{aligned} a - 2b + 4c - 2 &= h \\ a - b + c - 1 &= -h \\ a \qquad\qquad - 0 &= h \\ a + b + c - 1 &= -h \end{aligned}$$

Solving these four equations, we find $a = \frac{1}{4}$, $b = 0$, $c = \frac{1}{2}$, $h = \frac{1}{4}$ so that $p_1(x) = \frac{1}{4} + \frac{1}{2}x^2$. This completes the equivalent of Steps 1 and 2, and we turn to Step 3 and compute the errors of our parabola at all five data points. They are $\frac{1}{4}, -\frac{1}{4}, \frac{1}{4}, -\frac{1}{4}, \frac{1}{4}$ so that the maximum error on the entire set ($H = \frac{1}{4}$) equals the maximum on our quadruple ($|h| = \frac{1}{4}$). The algorithm is ended and our first parabola is the min–max parabola. It is shown in Fig. 22-4.

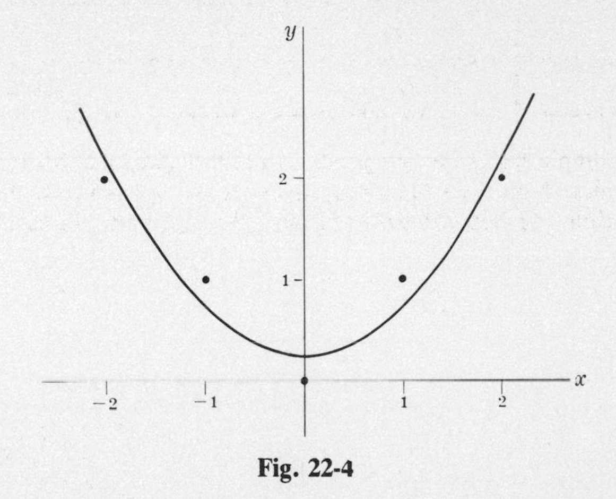

Fig. 22-4

22.10. Find the min–max parabola for the seven points $y = |x|$, $x = -3(1)3$.

This adds two more points at the ends of our previous data supply. Suppose we choose the same initial quadruple as before. Then we again have the equal-error parabola $p_1(x)$ of the preceding problem. Its errors at the new data points are $\frac{7}{4}$ so that now $H = \frac{7}{4}$ while $|h| = \frac{1}{4}$. Accordingly we introduce one of the new points into the quadruple and abandon $x = -2$. On the new quadruple the old parabola has the errors $-\frac{1}{4}, \frac{1}{4}, -\frac{1}{4}, \frac{7}{4}$ which do alternate in sign. Having made the exchange, a new equal-error parabola

$$p_2(x) = a_2 + b_2x + c_2x^2$$

must be found. Proceeding as in the previous problem we soon obtain the equal error $h_2 = -\frac{1}{3}$ and the parabola $p_2(x) = \frac{1}{3}(1 + x^2)$. Its errors at the seven data points are $\frac{1}{3}, -\frac{1}{3}, -\frac{1}{3}, \frac{1}{3}, -\frac{1}{3}, -\frac{1}{3}, \frac{1}{3}$ so that $H = |h| = \frac{1}{3}$ and the algorithm stops. The parabola $p_2(x)$ is the min–max parabola. The fact that all errors are of uniform size is a bonus, not characteristic of min–max polynomials generally, as the straight-line problems just solved show.

CONTINUOUS DATA, THE WEIERSTRASS THEOREM

22.11. Prove that $\sum_{k=0}^{n} p_{nk}^{(x)}(k - nx) = 0$ where $p_{nk}^{(x)} = \binom{n}{k} x^k (1 - x)^{n-k}$.

The binomial theorem for integers n and k,

$$(p+q)^n = \sum_{k=0}^{n} \binom{n}{k} p^k q^{n-k}$$

is an identity in p and q. Differentiating relative to p brings

$$n(p+q)^{n-1} = \sum_{k=0}^{n} \binom{n}{k} k p^{k-1} q^{n-k}$$

Multiplying by p and then setting $p = x$, $q = 1 - x$, this becomes $nx = \sum_{k=0}^{n} kp_{nk}^{(x)}$. Using the same p and q in the binomial theorem itself shows that $1 = \sum p_{nk}^{(x)}$ and so finally

$$\sum_{k=0}^{n} p_{nk}^{(x)}(k - nx) = nx - nx = 0$$

22.12. Prove also that $\sum_{k=0}^{n} p_{nk}^{(x)}(k-nx)^2 = nx(1-x)$.

A second differentiation relative to p brings

$$n(n-1)(p+q)^{n-2} = \sum_{k=0}^{n} \binom{n}{k} k(k-1) p^{k-2} q^{n-k}$$

Multiplying by p^2 and then setting $p = x$, $q = 1 - x$, this becomes

$$n(n-1)x^2 = \sum_{k=0}^{n} k(k-1) p_{nk}^{(x)}$$

from which we find $$\sum_{k=0}^{n} k^2 p_{nk}^{(x)} = n(n-1)x^2 + \sum_{k=0}^{n} kp_{nk}^{(x)} = n(n-1)x^2 + nx$$

Finally we compute

$$\begin{aligned}\sum_{k=0}^{n} p_{nk}^{(x)}(k-nx)^2 &= \sum k^2 p_{nk}^{(x)} - 2nx \sum kp_{nk}^{(x)} + n^2x^2 \sum p_{nk}^{(x)} \\ &= n(n-1)x^2 + nx - 2nx(nx) + n^2x^2 = nx(1-x)\end{aligned}$$

22.13. Prove that if $d > 0$ and $0 \leqq x \leqq 1$, then

$$\sum{}' p_{nk}^{(x)} \leqq \frac{x(1-x)}{nd^2}$$

where $\sum'$ is the sum over those integers k for which $|(k/n) - x| \geqq d$. (This is a special case of the famous Chebyshev inequality.)

Breaking the sum of the preceding problem into two parts

$$nx(1-x) = \sum{}' p_{nk}^{(x)}(k-nx)^2 + \sum{}'' p_{nk}^{(x)}(k-nx)^2$$

where $\sum''$ includes those integers k omitted in $\sum'$. But then

$$\begin{aligned}nx(1-x) &\geqq \sum{}' p_{nk}^{(x)}(k-nx)^2 \\ &\geqq \sum{}' p_{nk}^{(x)} n^2 d^2\end{aligned}$$

the first of these steps being possible since $\sum''$ is nonnegative and the second because in $\sum'$ we find $|k - nx| \geqq nd$. Dividing through by n^2d^2, we have the required result.

22.14. Derive these estimates for $\sum'$ and $\sum''$.

$$\sum{}' p_{nk}^{(x)} \leqq \frac{1}{4nd^2} \qquad \sum{}'' p_{nk}^{(x)} \geqq 1 - \frac{1}{4nd^2}$$

The function $x(1-x)$ takes its maximum at $x=\frac{1}{2}$ and so $0 \leqq x(1-x) \leqq \frac{1}{4}$ for $0 \leqq x \leqq 1$. The result for Σ' is thus an immediate consequence of the preceding problem. But then $\Sigma'' = 1 - \Sigma' \geqq 1 - (1/4nd^2)$.

22.15. Prove that if $f(x)$ is continuous for $0 \leqq x \leqq 1$, then $\lim \sum_{k=0}^{n} p_{nk}^{(x)} f(k/n) = f(x)$ uniformly as n tends to infinity.

This will prove the Weierstrass theorem, by exhibiting a sequence of polynomials

$$B_n(x) = \sum_{k=0}^{n} p_{nk}^{(x)} f\left(\frac{k}{n}\right)$$

which converges uniformly to $f(x)$. These polynomials are called the Bernstein polynomials for $f(x)$. The proof begins with the choice of an arbitrary positive number ϵ. Then for $|x'-x| < d$,

$$|f(x') - f(x)| < \frac{\epsilon}{2}$$

and d is independent of x by the uniform continuity of $f(x)$. Then with M denoting the maximum of $|f(x)|$, we have

$$\begin{aligned} |B_n(x) - f(x)| &= \left| \sum p_{nk}^{(x)} \left[f\left(\frac{k}{n}\right) - f(x) \right] \right| \\ &\leqq \sum{}' p_{nk}^{(x)} \left| f\left(\frac{k}{n}\right) - f(x) \right| + \sum{}'' p_{nk}^{(x)} \left| f\left(\frac{k}{n}\right) - f(x) \right| \\ &\leqq 2M \sum{}' p_{nk}^{(x)} + \frac{1}{2}\epsilon \sum{}'' p_{nk}^{(x)} \end{aligned}$$

with k/n in the Σ'' part playing the role of x'. The definition of Σ'' guarantees $|x'-x| < d$. Then

$$\begin{aligned} |B_n(x) - f(x)| &\leqq \left(\frac{2M}{4nd^2}\right) + \frac{1}{2}\epsilon \\ &\leqq \frac{1}{2}\epsilon + \frac{1}{2}\epsilon = \epsilon \end{aligned}$$

for n sufficiently large. This is the required result. Another interval than (0, 1) can be accommodated by a simple change of argument.

22.16. Show that in the case of $f(x) = x^2$, $B_n(x) = x^2 + x(1-x)/n$ so that Bernstein polynomials are not the best approximations of given degree to $f(x)$. [Surely the best quadratic approximation to $f(x) = x^2$ is x^2 itself.]

Since the sum $\sum k^2 p_{nk}^{(x)}$ was found in Problem 22.2,

$$B_n(x) = \sum_{k=0}^{n} p_{nk}^{(x)} f\left(\frac{k}{n}\right) = \sum_{k=0}^{n} \frac{p_{nk}^{(x)} k^2}{n^2} = \frac{1}{n^2}[n(n-1)x^2 + nx] = x^2 + \frac{x(1-x)}{n}$$

as required. The uniform convergence for n tending to infinity is apparent, but clearly $B_n(x)$ does not duplicate x^2. We now turn to a better class of uniform approximation polynomials.

CONTINUOUS DATA, THE CHEBYSHEV THEORY

22.17. Prove that if $y(x)$ is continuous for $a \leqq x \leqq b$, then there is a polynomial $P(x)$ of degree n or less such that $\max |P(x) - y(x)|$ on the interval (a, b) is a minimum. In other words, no other polynomial of this type produces a smaller maximum.

Let $p(x) = a_0 + a_1 x + \cdots + a_n x^n$ by any polynomial of degree n or less. Then

$$M(\bar{a}) = \max |p(x) - y(x)|$$

depends on the polynomial $p(x)$ chosen, that is, it depends upon the coefficient set $(a_0, a_1, \ldots, a_n)$ which we shall call $\bar{a}$ as indicated. Since $M(\bar{a})$ is a continuous function of $\bar{a}$ and nonnegative, it has a greatest lower bound. Call this bound L. What has to be proved is that for some particular coefficient set A, the coefficients of $P(x)$, the lower bound L is actually attained, that is, $M(A) = L$. By way of contrast, the function $f(t) = 1/t$ for positive t has greatest lower bound zero, but there is no argument t for which $f(t)$ actually attains this bound. The infinite range of t is of course the factor which allows this situation to occur. In our problem the coefficient set $\bar{a}$ also has unlimited range, but we now show that $M(A) = L$ nevertheless. To begin, let $a_i = Cb_i$ for $i = 0, 1, \ldots, n$ in such a way that $\sum b_i^2 = 1$. We may also write $\bar{a} = C\bar{b}$. Consider a second function

$$m(\bar{b}) = \max |b_0 + b_1 x + \cdots + b_n x^n|$$

where max refers as usual to the maximum of the polynomial on the interval (a, b). This is a continuous function on the unit sphere $\sum b_i^2 = 1$. On such a set (closed and bounded) a continuous function does assume its minimum value. Call this minimum μ. Plainly $\mu \geqq 0$. But the zero value is impossible since only $p(x) = 0$ can produce this minimum and the condition on the b_i temporarily excludes this polynomial. Thus $\mu > 0$. But then

$$m(\bar{a}) = \max |a_0 + a_1 x + \cdots + a_n x^n| = \max |p(x)| = Cm(\bar{b}) \geqq C\mu$$

Now returning to $M(\bar{a}) = \max |p(x) - y(x)|$, and using the fact that the absolute value of a difference exceeds the difference of absolute values, we find

$$\begin{aligned} M(\bar{a}) &\geqq m(\bar{a}) - \max |y(x)| \\ &\geqq C\mu - \max |y(x)| \end{aligned}$$

If we choose $C > (L + 1 + \max |y(x)|)/\mu = R$, then at once $M(\bar{a}) \geqq L + 1$. Recalling that L is the greatest lower bound of $M(\bar{a})$, we see that $M(\bar{a})$ is relatively large for $C > R$ and that its greatest lower bound under the constraint $C \leqq R$ will be this same number L. But this constraint is equivalent to $\sum a_i^2 \leqq R$, so that now it is again a matter of a continuous function $M(\bar{a})$ on a closed and bounded set (a solid sphere, or ball). On such a set the greatest lower bound is actually assumed, say at $\bar{a} = A$. Thus $M(A)$ is L, and $P(x)$ is a min–max polynomial.

22.18. Let $P(x)$ be a min–max polynomial approximation to $y(x)$ on the interval (a, b), among all polynomials of degree n or less. Let $E = \max |y(x) - P(x)|$, and assume $y(x)$ is not itself a polynomial of degree n or less, so that $E > 0$. Show that there must be at least one argument for which $y(x) - P(x) = E$, and similarly for $-E$. [We continue to assume $y(x)$ continuous.]

Since $y(x) - P(x)$ is continuous for $a \leqq x \leqq b$, it must attain either $\pm E$ somewhere. We are to prove

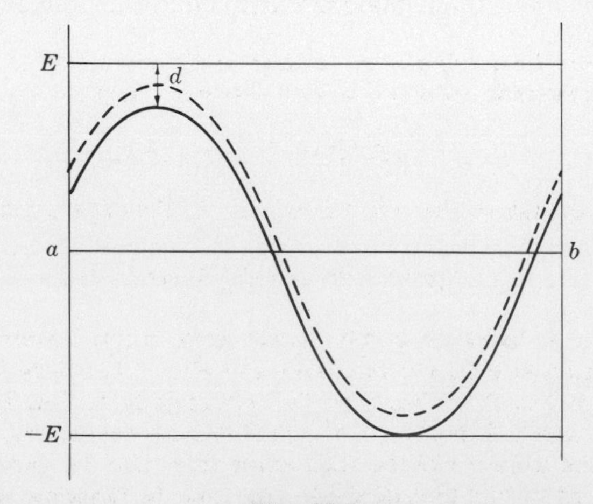

Fig. 22-5

that it must achieve both. Suppose that it did not equal E anywhere in (a, b). Then

$$\max\,[y(x) - P(x)] = E - d$$

where d is positive, and so

$$-E \leqq y(x) - P(x) \leqq E - d$$

But this can be written as

$$-E + \frac{1}{2}d \leqq y(x) - \left[P(x) - \frac{1}{2}d\right] \leqq E - \frac{1}{2}d$$

which flatly claims that $P(x) - \frac{1}{2}d$ approximates $y(x)$ with a maximum error of $E - \frac{1}{2}d$. This contradicts the original assumption that $P(x)$ itself is a min–max polynomial, with maximum error of E. Thus $y(x) - P(x)$ must equal E somewhere in (a, b). A very similar proof shows it must also equal $-E$. Figure 22.5 illustrates the simple idea of this proof. The error $y(x) - P(x)$ for the min–max polynomial cannot behave as shown solid, because raising the curve by $\frac{1}{2}d$ then brings a new error curve (shown dashed) with a smaller maximum absolute value of $E - \frac{1}{2}d$, and this is a contradiction.

22.19. Continuing the previous problem, show that, for $n = 1$, approximation by linear polynomials, there must be a third point at which the error $|y(x) - P(x)|$ of a min–max $P(x)$ assumes its maximum value E.

Let $y(x) - P(x) = E(x)$ and divide (a, b) into subintervals small enough so that for x_1, x_2 within any subinterval,

$$|E(x_1) - E(x_2)| \leqq \frac{1}{2}E$$

Since $E(x)$ is continuous for $a \leqq x \leqq b$, this can surely be done. In one subinterval, call it I_1, we know the error reaches E, say at $x = x_+$. It follows that throughout this subinterval,

$$|E(x) - E(x_+)| = |E(x) - E| \leqq \tfrac{1}{2}E$$

making $E(x) \geqq \frac{1}{2}E$. Similarly, in one subinterval, call it I_2, we find $E(x_-) = -E$, and therefore $|E(x)| \leqq -\frac{1}{2}E$. These two subintervals cannot therefore be adjacent and so we can choose a point u_1 between them. Suppose that I_1 is to the left of I_2. (The argument is almost identical for the reverse situation.) Then $u_1 - x$ has the same sign as $E(x)$ in each of the two subintervals discussed. Let $R = \max |u_1 - x|$ in (a, b).

Now suppose that there is no third point at which the error is $\pm E$. Then in all but the two subintervals just discussed we must have

$$\max |E(x)| < E$$

and since there are finitely many subintervals,

$$\max\,[\max |E(x)|] = E^* < E$$

Naturally $E^* \geqq \frac{1}{2}E$ since these subintervals extend to the endpoints of I_1 and I_2 where $|E(x)| \geqq \frac{1}{2}E$. Consider the following alteration of $P(x)$, still a linear polynomial:

$$P^*(x) = P(x) + \epsilon(u_1 - x)$$

If we choose ϵ small enough so that $\epsilon R < E - E^* \leqq \frac{1}{2}E$, then $P^*(x)$ becomes a better approximation than $P(x)$. For,

$$|y(x) - P^*(x) = |E(x) - \epsilon(u_1 - x)|$$

so that in I_1 the error is reduced but is still positive while in I_2 it is increased but remains negative; in both subintervals the error size has been reduced. Elsewhere, though the error size may grow, it cannot exceed $E^* + \epsilon R < E$, and so $P^*(x)$ has a smaller maximum error than $P(x)$. This contradiction shows that a third point with error $\pm E$ must exist. Figure 22-6 illustrates the simple idea behind this proof. The error curve $E(x)$ cannot behave like the solid curve (only two $\pm E$ points) because adding the linear correction term $\epsilon(u_1 - x)$ to $P(x)$ then diminishes the error by this same amount, leading to a new error curve (shown dashed) with smaller maximum absolute value.

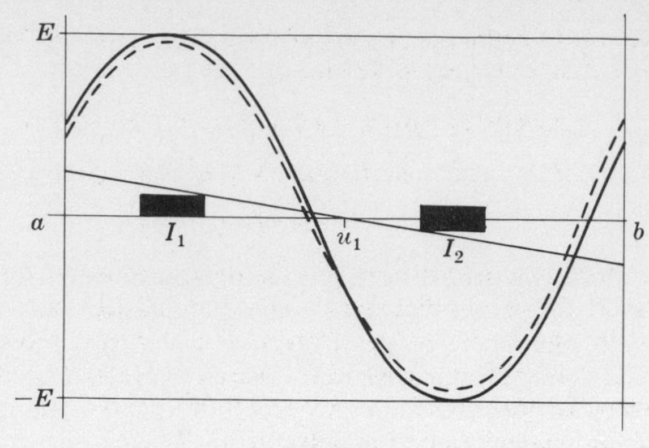

Fig. 22-6

22.20. Show that for the $P(x)$ of the previous problem there must be three points at which errors of size E and with alternating sign occur.

The proof of the previous problem is already sufficient. If, for example, the signs were $+, +, -$, then choosing u_1 between the adjacent $+$ and $-$ our $P^*(x)$ is again better than $P(x)$. The pattern $+, -, -$ is covered by exactly the same remark. Only the alternation of signs can avoid the contradiction.

22.21. Show that in the general case of the min–max polynomial of degree n or less, there must be $n+2$ points of maximum error size with alternating sign.

The proof is illustrated by treating the case $n = 2$. Let $P(x)$ be a min–max polynomial of degree two or less. By Problem 22.18 it must have at least two points of maximum error. The argument of Problems 22.19 and 22.20, with $P(x)$ now quadratic instead of linear but with no other changes, then shows that a third such point must exist and signs must alternate, say $+, -, +$ just to be definite. Now suppose that no fourth position of maximum error occurs. We repeat the argument of Problem 22.19, choosing two points u_1 and u_2 between the subintervals I_1, I_2, and I_3 in which the errors $\pm E$ occur, and using the correction term $\epsilon(u_1 - x)(u_2 - x)$, which agrees in sign with $E(x)$ in these subintervals. No other changes are necessary. The quadratic $P^*(x)$ will have a smaller maximum error than $P(x)$, and this contradiction proves that the fourth $\pm E$ point must exist. The alternation of sign is established by the same argument used in Problem 22.20, and the extension to higher values of n is entirely similar.

22.22. Prove that there is just one min–max polynomial for each n.

Suppose there were two, $P_1(x)$ and $P_2(x)$. Then

$$-E \leqq y(x) - P_1(x) \leqq E \qquad -E \leqq y(x) - P_2(x) \leqq E$$

Let $P_3(x) = \frac{1}{2}(P_1 + P_2)$. Then

$$-E \leqq y(x) - P_3(x) \leqq E$$

and P_3 is also a min–max polynomial. By Problem 22.21 there must be a sequence of $n+2$ points at which $y(x) - P_3(x)$ is alternately $\pm E$. Let $P_3(x_+) = E$. Then at x_+ we have $y - P_3 = E$, or

$$(y - P_1) + (y - P_2) = 2E$$

Since neither term on the left can exceed E, each must equal E. Thus $P_1(x_+) = P_2(x_-)$. Similarly $P_1(x_-) = P_2(x_-)$. The polynomials P_1 and P_2 therefore coincide at the $n+2$ points and so are identical. This proves the uniqueness of the min–max polynomial for each n.

22.23. Prove that a polynomial $p(x)$ of degree n or less, for which the error $y(x) - p(x)$ takes alternate extreme values of $\pm e$ on a set of $n+2$ points, must be the min–max polynomial.

This will show that only the min–max polynomial can have this *equal-error* feature, and it is useful in finding and identifying such polynomials. We have

$$\max |y(x) - p(x)| = e \geqq E = \max |y(x) - P(x)|$$

$P(x)$ being the unique min–max polynomial. Suppose $e > E$. Then since

$$P - p = (y - p) + (P - y)$$

we see that, at the $n + 2$ extreme points of $y - p$, the quantities $P - p$ and $y - p$ have the same sign. (The first term on the right equals e at these points and so dominates the second.) But the sign of $y - p$ alternates on this set, so the sign of $P - p$ does likewise. This is $n + 1$ alternations in all and means $n + 1$ zeros for $P - p$. Since $P - p$ is of degree n or less it must be identically zero, making $p = P$ and $E = e$. This contradicts our supposition of $e > E$ and leaves us with the only alternative, namely $e = E$. The polynomial $p(x)$ is thus the (unique) min–max polynomial $P(x)$.

CONTINUOUS DATA, EXAMPLES OF MIN–MAX POLYNOMIALS

22.24. Show that on the interval $(-1, 1)$ the min–max polynomial of degree n or less for $y(x) = x^{n+1}$ can be found by expressing x^{n+1} as a sum of Chebyshev polynomials and dropping the $T_{n+1}(x)$ term.

Let

$$x^{n+1} = a_0 T_0(x) + \cdots + a_n T_n(x) + a_{n+1} T_{n+1}(x) = p(x) + a_{n+1} T_{n+1}(x)$$

Then the error is

$$E(x) = x^{n+1} - p(x) = a_{n+1} T_{n+1}(x)$$

and we see that this error has alternate extremes of $\pm a_{n+1}$ at the $n + 2$ points where $T_{n+1} = \pm 1$. These points are $x_k = \cos [k\pi/(n + 1)]$, with $k = 0, 1, \ldots, n + 1$. Comparing coefficients of x^{n+1} on both sides above, we also find that $a_{n+1} = 2^{-n}$. [The leading coefficient of $T_{n+1}(x)$ is 2^n. See Problems 21.42 and 21.43.] The result of Problem 22.23 now applies and shows that $p(x)$ is the min–max polynomial, with $E = 2^{-n}$. As illustrations the sums in Problem 21.45 may be truncated to obtain

$$n = 1 \qquad x^2 \simeq \frac{1}{2} T_0 \qquad \text{Error} = \frac{T_2}{2}$$

$$n = 2 \qquad x^3 \simeq \frac{3}{4} T_1 \qquad \text{Error} = \frac{T_3}{4}$$

$$n = 3 \qquad x^4 \simeq \frac{1}{8}(3T_0 + 4T_2) \qquad \text{Error} = \frac{T_4}{8}$$

$$n = 4 \qquad x^5 \simeq \frac{1}{16}(10T_1 + 5T_3) \qquad \text{Error} = \frac{T_5}{16}$$

and so on. Note that in each case the min–max polynomial (of degree n or less) is actually of degree $n - 1$.

22.25. Show that in any series of Chebyshev polynomials $\sum_{i=0}^{\infty} a_i T_i(x)$ each partial sum S_n is the min–max polynomial of degree n or less for the next sum S_{n+1}. [The interval is again taken to be $(-1, 1)$.]

Just as in the previous problem, but with $y(x) = S_{n+1}(x)$ and $p(x) = S_n(x)$, we have

$$E(x) = S_{n+1}(x) - S_n(x) = a_{n+1} T_{n+1}(x)$$

The result of Problem 22.23 again applies. Note also, however, that $S_{n-1}(x)$ may not be the min–max polynomial of degree $n - 1$ or less, since $a_n T_n + a_{n+1} T_{n+1}$ is not necessarily an equal-ripple function. (It was in the previous problem, however, since a_n was zero.)

22.26. Use the result of Problem 22.24 to economize the polynomial $y(x) = x - \frac{1}{6}x^3 + \frac{1}{120}x^5$ to a cubic polynomial, for the interval $(-1, 1)$.

This was actually accomplished in Problem 21.50, but we may now view the result in a new light. Since

$$x - \frac{1}{6}x^3 + \frac{1}{120}x^5 = \frac{169}{192}T_1 - \frac{5}{128}T_3 + \frac{1}{1920}T_5$$

the truncation of the T_5 term leaves us with the min–max polynomial of degree four or less for $y(x)$, namely

$$P(x) = \frac{169}{192}x - \frac{5}{128}(4x^3 - 3x)$$

This is still only approximately the min–max polynomial of the same degree for $\sin x$. Further truncation, of the T_3 term, would not produce a min–max polynomial for $y(x)$, not exactly anyway.

22.27. Find the min–max polynomial of degree one or less, on the interval (a, b), for a function $y(x)$ with $y''(x) > 0$.

Let the polynomial be $P(x) = Mx + B$. We must find three points $x_1 < x_2 < x_3$ in (a, b) for which $E(x) = y(x) - P(x)$ attains its extreme values with alternate signs. This puts x_2 in the interior of (a, b) and requires $E'(x_2)$ to be zero, or $y'(x_2) = M$. Since $y'' > 0$, y' is strictly increasing and can equal M only once, which means that x_2 can be the only interior extreme point. Thus $x_1 = a$ and $x_3 = b$. Finally, by the equal-ripple property,

$$y(a) - P(a) = -[y(x_2) - P(x_2)] = y(b) - P(b)$$

Solving, we have

$$M = \frac{y(b) - y(a)}{b - a} \qquad B = \frac{y(a) + y(x_2)}{2} - \frac{(a + x_2)[y(b) - y(a)]}{2(b - a)}$$

with x_2 determined by $y'(x_2) = [y(b) - y(a)]/(b - a)$.

22.28. Apply the previous problem to $y(x) = -\sin x$ on the interval $(0, \pi/2)$.

We find $M = -2/\pi$ first; and then from $y'(x_2) = M$, $x_2 = \arccos(2/\pi)$. Finally,

$$B = -\frac{1}{2}\sqrt{1 - \frac{4}{\pi^2}} + \frac{1}{\pi}\arccos\frac{2}{\pi}$$

and from $P(x) = Mx + B$ we find

$$\sin x \simeq \frac{2x}{\pi} + \frac{1}{2}\sqrt{1 - \frac{4}{\pi^2}} + \frac{1}{\pi}\arccos\frac{2}{\pi}$$

the approximation being the min–max line.

22.29. Show that $P(x) = x^2 + \frac{1}{8}$ is the min–max *cubic* (or less) approximation to $y(x) = |x|$ over the interval $(-1, 1)$.

The error is $E(x) = |x| - x^2 - \frac{1}{8}$ and takes the extreme values $-\frac{1}{8}, \frac{1}{8}, -\frac{1}{8}, \frac{1}{8}, -\frac{1}{8}$ at $x = -1, -\frac{1}{2}, 0, \frac{1}{2}, 1$. These alternating errors of maximal size $E = \frac{1}{8}$ at $n + 2 = 5$ points guarantee (by Problem 22.23) that $P(x)$ is the min–max polynomial of degree $n = 3$ or less.

22.30. Use the function $y(x) = e^x$ on the interval $(-1, 1)$ to illustrate the *exchange method* for finding a min–max line.

The method of Problem 22.27 would produce the min–max line, but for a simple first illustration, we momentarily ignore that method and proceed by exchange, imitating the procedure of Problem 22.5.

Since we are after a line, we need $n+2=3$ points of maximum error $\pm E$. Try $x=-1, 0, 1$ for an initial triple. The corresponding values of $y(x)$ are about .368, 1, and 2.718. The equal-error line for this triple is easily found to be

$$p_1(x) \simeq 1.175x + 1.272$$

with errors $h = \pm .272$ on the triple. Off the triple, a computation of the error at intervals of .1 discovers a maximum error of size $H = .286$ (and negative) at $x = .2$. Accordingly we form a new triple, exchanging the old argument $x = 0$ for the new $x = .2$. This retains the alternation of error signs called for in Step 4 of the exchange method as presented earlier, and which we are now imitating. On the new triple $y(x)$ takes the values .368, 1.221, and 2.718 approximately. The equal-error line is found to be

$$p_2(x) = 1.175x + 1.264$$

with errors $h = \pm .278$ on the triple. Off the triple, anticipating maximum errors near $x = .2$, we check this neighborhood at intervals of .01 and find an error of .279 at $x = .16$. Since we are carrying only three places, this is the best we can expect. A shift to the triple $x = -1, .16, 1$ would actually reproduce $p_2(x)$.

Let us now see what the method of Problem 22.27 manages. With $a = -1$ and $b = 1$ it at once produces $M = (2.718 - .368)/2 = 1.175$. Then the equation $y'(x_2) = e^{x_2} = 1.175$ leads to $x_2 \simeq .16$, after which the result $B = 1.264$ is direct. The line is shown in Fig. 22-7 below, with the vertical scale compressed.

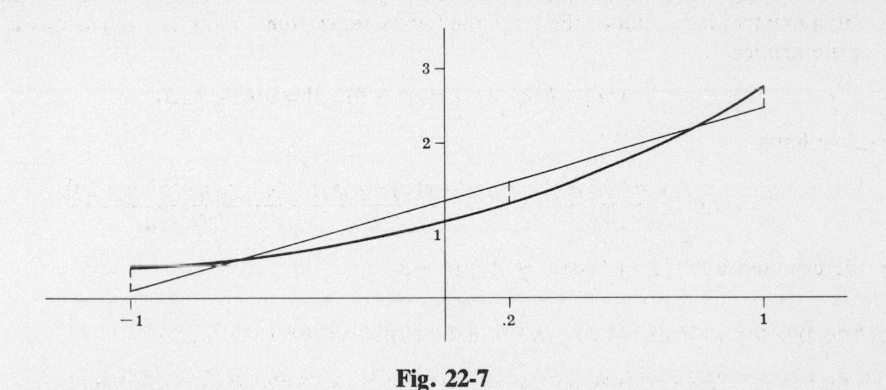

Fig. 22-7

22.31. Use the exchange method to find the min–max quadratic polynomial for $y(x) = e^x$ over $(-1, 1)$.

Recalling that truncation of a series of Chebyshev polynomials often leads to nearly equal ripple errors resembling the first omitted term, we take as our initial quadruple the four extreme points of $T_3(x)$, which are $x = \pm 1, \pm\frac{1}{2}$. The parabola which misses the four points

x	-1	$-\frac{1}{2}$	$\frac{1}{2}$	1
e^x	.3679	.6065	1.6487	2.7183

alternately by $\pm h$ proves to have its maximum error at $x = .56$. The new quadruple $(-1, -.5, .56, 1)$ then leads to a second parabola with maximum error at $x = -.44$. The next quadruple is $(-1, -.44, .56, 1)$ and proves to be our last. Its equal-ripple parabola is, to five decimal places,

$$p(x) = .55404x^2 + 1.13018x + .98904$$

and its maximum error both inside and outside the quadruple is $H = .04502$.

Supplementary Problems

DISCRETE DATA

22.32. Show that the least-squares line for the three data points of Problem 22.2 is $y(x) = \frac{1}{2}x - \frac{1}{6}$. Show that its errors at the data arguments are $\frac{1}{6}, -\frac{1}{3}, \frac{1}{6}$. The Chebyshev line was found to be $y(x) = \frac{1}{2}x - \frac{1}{4}$ with errors of $-\frac{1}{4}, \frac{1}{4}, -\frac{1}{4}$. Verify that the Chebyshev line does have the smaller maximum error and the least-squares line the smaller sum of errors squared.

22.33. Apply the exchange method to the average golf scores in Problem 21.2, producing the min–max line. Use this line to compute smoothed average scores. How do the results compare with those obtained by least squares?

22.34. Apply the exchange method to the data of Problem 21.5, obtaining the min–max line and then the corresponding exponential function $P(x) = Ae^{Mx}$.

22.35. Obtain a formula $y(x) = Mx + B$ for the Chebyshev line of an arbitrary triple (x_1, y_1), (x_2, y_2), (x_3, y_3). Such a formula could be useful in programming the exchange method for machine computation.

22.36. Show that if the arguments x_i are not distinct, then the min–max line may not be uniquely determined. For example, consider the three points $(0, 0)$, $(0, 1)$, and $(1, 0)$ and show that all lines between $y = \frac{1}{2}$ and $y = \frac{1}{2} - x$ have $H = \frac{1}{2}$. (See Fig. 22-8.)

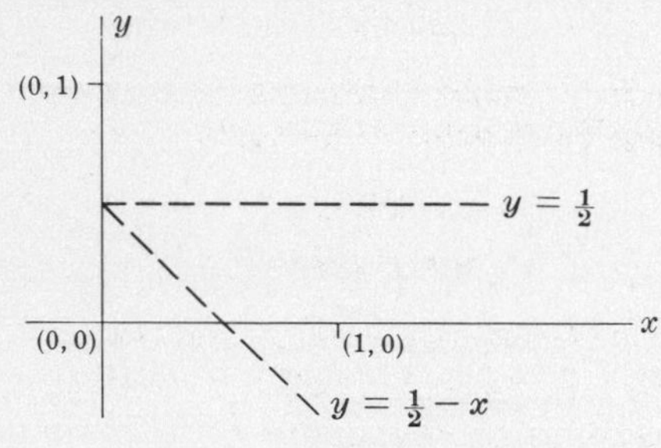

Fig. 22-8

22.37. Find the equal-error parabola for the four points $(0, 0)$, $(\pi/6, \frac{1}{2})$, $(\pi/3, \sqrt{3}/2)$, and $(\pi/2, 1)$ of the curve $y = \sin x$.

22.38. Find the min–max parabola for the five points $y = x^3$, $x = 0(\frac{1}{4})1$.

22.39. Use the exchange method to obtain the min–max parabola for the seven points $y = \cos x$, $x = 0(\pi/12)\pi/2$. What is the maximum error $|h|$ of this parabola? Compare its accuracy with that of the Taylor parabola $1 - \frac{1}{2}x^2$.

22.40. Extend the exchange method to obtain the min–max cubic polynomial for the seven points $y = \sin x$, $x = 0(\pi/12)\pi/2$. What is the maximum error $|h|$ of this cubic? Compare its accuracy with that of the Taylor cubic $x - \frac{1}{6}x^3$.

CONTINUOUS DATA

22.41. Find the min–max polynomial of degree five or less for $y(x) = x^6$ on the interval $(-1, 1)$. What is the error?

22.42. What is the min–max polynomial of degree two or less for $y(x) = T_0 + T_1 + T_2 + T_3$ and what is its error? Show that $T_0 + T_1$ is not, however, the min–max line for $y(x)$, by showing that the error of this approximation is not equal-ripple.

22.43. Find the min–max polynomial of degree five or less for $y(x) = 1 - \frac{1}{2}x^2 + \frac{1}{24}x^4 - \frac{1}{720}x^6$ and what is its error? [The interval is $(-1, 1)$.]

22.44. Apply Problem 22.27 to find the min–max line over $(0, \pi/2)$ for $y(x) = -\cos x$.

22.45. Does the method of Problem 22.27 work for $y(x) = |x|$ over $(-1, 1)$, or does the discontinuity in $y'(x)$ make the method inapplicable?

22.46. Use the exchange method to find the min–max line for $y(x) = \cos x$ over $(0, \pi/2)$. Work to three decimal places and compare with that found by another method in Problem 22.44.

22.47. Use the exchange method to find the min–max parabola for $y(x) = \cos x$ over $(0, \pi/2)$. [You may want to use the extreme points of $T_3(x)$, converted by a change of argument to the interval $(0, \pi/2)$, as an initial quadruple.]

22.48. Find a polynomial of minimum degree which approximates $y(x) = \cos x$ over $(0, \pi/2)$ with maximum error .005. Naturally, roundoff error will limit the precision to which the polynomial can be determined.

22.49. Prove that the min–max polynomial approximation to $f(x) = 0$, among all polynomials of degree n with leading coefficient 1, is $2^{1-n}T_n(x)$. The interval of approximation is taken to be $(-1, 1)$. This is covered by Problems 22.17 to 22.23, but carry out the details of the following historical argument. Let

$$p(x) = x^n + a_1x^{n-1} + \cdots + a_n$$

be any polynomial of the type described. Since $T_n(x) = \cos(n \arccos x)$, we have

$$\max |2^{1-n}T_n(x)| = 2^{1-n}$$

Notice that this polynomial takes its extreme values of $\pm 2^{1-n}$ alternately at the arguments $x_k = \cos k\pi/n$, where $k = 0, 1, \ldots, n$. Suppose that some polynomial $p(x)$ were such that

$$\max |p(x)| < 2^{1-n}$$

and let

$$P(x) = p(x) - 2^{1-n}T_n(x)$$

Then $P(x)$ is of degree $n-1$ or less and it does not vanish identically since this would require $\max |p(x)| = 2^{1-n}$. Consider the values $P(x_k)$. Since $p(x)$ is dominated by $2^{1-n}T_n(x)$ at these points, we see that the $P(x_k)$ have alternating signs. Being continuous, $P(x)$ must therefore have n zeros between the consecutive x_k. But this is impossible for a polynomial of degree $n-1$ or less which does not vanish identically. This proves that $\max |p(x)| \geqq 2^{1-n}$.

22.50. Values of $y(x) = e^{(t+2)/4}$ are given in the table below. Find the min–max parabola for this data. What is the min–max error?

x	-2	-1	0	1	2
$y(x)$	1.0000	1.2840	1.6487	2.1170	2.7183

22.51. What is the minimum degree of a polynomial approximation to e^x on the interval $(-1, 1)$ with maximum error .005 or less?

22.52. The Taylor series for $\ln(1 + x)$ converges so slowly that hundreds of terms would be needed for five-place accuracy over the interval $(0, 1)$. What is the maximum error of

$$p(x) = .999902x - .497875x^2 + .317650x^3 - .193761x^4 + .085569x^5 - .018339x^6$$

on this same interval?

22.53. Approximate $y(x) = 1 - x + x^2 - x^3 + x^4 - x^5 + x^6$ by a polynomial of minimum degree, with error not exceeding .005 in $(0, 1)$.

22.54. Continue the previous problem to produce a minimum degree approximation with error at most .1.

Chapter 23

Approximation by Rational Functions

COLLOCATION

Rational functions are quotients of polynomials and so constitute a much richer class of functions than polynomials. This greater supply increases the prospects for accurate approximation. Functions with poles, for instance, can hardly be expected to respond well to efforts at polynomial approximation, since polynomials do not have singularities. Such functions are a principal target of rational approximation. But even with nonsingular functions there are occasions when rational approximations may be preferred.

Two types of approximations will be discussed, the procedures resembling those used for polynomial approximation. Collocation at prescribed arguments is one basis for selecting a rational approximation, as it is for polynomials. *Continued fractions* and *reciprocal differences* are the main tools used. The continued fractions involved take the form

$$y(x) = y_1 + \cfrac{x - x_1}{\rho_1 + \cfrac{x - x_2}{\rho_2 - y_1 + \cfrac{x - x_3}{\rho_3 - \rho_1 + \cfrac{x - x_4}{\rho_4 - \rho_2}}}}$$

which may be continued further if required. It is not too hard to see that this particular fraction could be rearranged into the quotient of two quadratic polynomials, in other words, a rational function. The ρ coefficients are called **reciprocal differences**, and are to be chosen in such a way that collocation is achieved. For the present example we shall find that

$$\rho_1 = \frac{x_2 - x_1}{y_2 - y_1} \qquad \rho_2 - y_1 = \frac{x_3 - x_2}{\dfrac{x_3 - x_1}{y_3 - y_1} - \dfrac{x_2 - x_1}{y_2 - y_1}}$$

with similar expressions for ρ_3 and ρ_4. The term reciprocal difference is not unnatural.

MIN–MAX

Min–max rational approximations are also gaining an important place in applications. Their theory, including the equal-error property and an exchange algorithm, parallels that of the polynomial case. For example, a rational function

$$R(x) = \frac{1}{a + bx}$$

can be found which misses three specified data points (x_i, y_i) alternately by $\pm h$. This $R(x)$ will be the min–max rational function for the given points, in the sense that

$$\max |R(x_i) - y_i| = h$$

will be smaller than the corresponding maxima when $R(x)$ is replaced by other rational functions of the same form. If more than three points are specified, then an exchange algorithm identifies the min–max $R(x)$. The analogy with the problem of the min–max polynomial is apparent.

PADÉ APPROXIMATIONS

These take the form

$$R_{mn}(x) = \frac{P_m(x)}{Q_n(x)}$$

with P_m and Q_n polynomials of degree m and n, respectively. The normalization $Q_n(0)=1$ is customary. To approximate a given function $y(x)$, Padé suggested making y and R_{mn} agree in value at some specified point, together with their first N derivatives, where $N=m+n$. This provides $N+1$ conditions for determining the remaining $N+1$ coefficients of P_m and Q_n. The point in question is usually taken to be $x=0$, by a suitable translation of argument if necessary. The parallel with the Taylor polynomial of $y(x)$ at $x=0$ is evident and in fact the Taylor polynomial is R_{N0}. As it turns out, more accuracy is achieved for a given N by choosing $m=n+1$ or $m=n$, that is, by numerator and denominator polynomials of more or less equal degree.

Solved Problems

THE COLLOCATION RATIONAL FUNCTION

23.1. Find the rational function $y(x)=1/(a+bx)$ given that $y(1)=1$ and $y(3)=\frac{1}{2}$.

Substitution requires $a+b=1$ and $a+3b=2$, which force $a=b=\frac{1}{2}$. The required function is $y(x)=2/(1+x)$. This simple problem illustrates the fact that finding a rational function by collocation is equivalent to solving a set of *linear* equations for the unknown coefficients.

23.2. Also find rational functions $y_2(x)=Mx+B$ and $y_3(x)=c+d/x$ which have $y(1)=1$ and $y(3)=\frac{1}{2}$.

The linear function $y_2(x)=(5-x)/4$ may be found by inspection. For the other we need to satisfy the coefficient equations $c+d=1$, $3c+d=\frac{3}{2}$ and this means that $c=\frac{1}{4}$, $d=\frac{3}{4}$, making $y_3(x)=(x+3)/4x$. We now have three rational functions which pass through the three given points. Certainly there are others, but in a sense these are the simplest. At $x=2$ the three functions offer us the interpolated values $\frac{2}{3}$, $\frac{3}{4}$, and $\frac{5}{8}$. Inside the interval $(1,3)$ all three resemble each other to some extent. Outside they differ violently. (See Fig. 23-1.) The diversity of rational functions exceeds that of polynomials and it is very helpful to have knowledge of the type of rational function required.

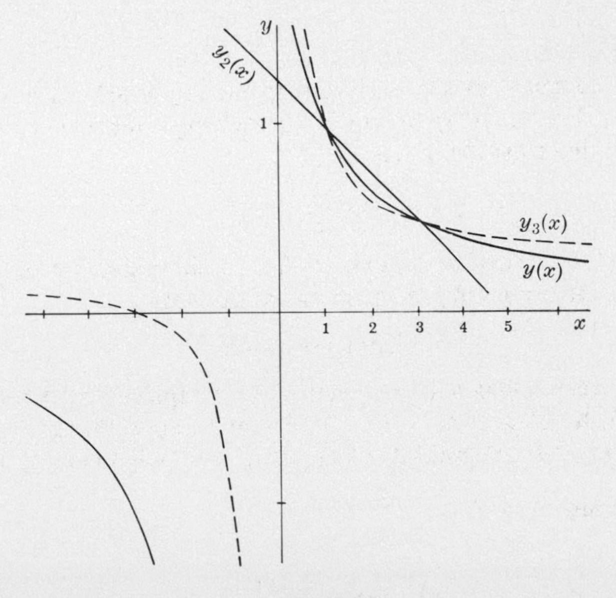

Fig. 23-1

23.3. Suppose it is known that $y(x)$ is of the form $y(x)=(a+bx^2)/(c+dx^2)$. Determine $y(x)$ by the requirements $y(0)=1$, $y(1)=\frac{2}{3}$, $y(2)=\frac{5}{9}$.

Substitution brings the linear system

$$a=c \qquad a+b=\frac{2}{3}(c+d) \qquad a+4b=\frac{5}{9}(c+4d)$$

Since only the ratio of the two polynomials is involved one coefficient may be taken to be 1, unless it later proves to be 0. Try $d=1$. Then one discovers that $a=b=c=\frac{1}{2}$, and $y(x)=(1+x^2)/(1+2x^2)$. Note that the rational function $y_2(x)=10/(10+6x-x^2)$ also includes these three points, and so does $y_3(x)=(x+3)/[3(x+1)]$.

CONTINUED FRACTIONS AND RECIPROCAL DIFFERENCES

23.4. Evaluate the continued fraction

$$y=1+\cfrac{x}{-3+\cfrac{x-1}{-\frac{2}{3}}} \qquad \text{at } x=0,\ 1, \text{ and } 2$$

Direct computation shows $y(0)=1$, $y(1)=\frac{2}{3}$, and $y(2)=\frac{5}{9}$. These are again the values of the previous problem. The point here is that the structure of a continued fraction of this sort makes these values equal to the successive "convergents" of the fraction, that is, the parts obtained by truncating the fraction before the x and $x-1$ terms and, of course, at the end. One finds easily that the fraction also rearranges into our $y_3(x)$.

23.5. Develop the connection between rational functions and continued fractions in the case

$$y(x)=\frac{a_0+a_1x+a_2x^2}{b_0+b_1x+b_2x^2}$$

We follow another historical path. Let the five data points (x_i, y_i) for $i=1,\dots,5$ be given. For collocation at these points,

$$a_0-b_0y+a_1x-b_1xy+a_2x^2-b_2x^2y=0$$

for each x_i, y_i pair. The determinant equation

$$\begin{vmatrix} 1 & y & x & xy & x^2 & x^2y \\ 1 & y_1 & x_1 & x_1y_1 & x_1^2 & x_1^2y_1 \\ 1 & y_2 & x_2 & x_2y_2 & x_2^2 & x_2^2y_2 \\ 1 & y_3 & x_3 & x_3y_3 & x_3^2 & x_3^2y_3 \\ 1 & y_4 & x_4 & x_4y_4 & x_4^2 & x_4^2y_4 \\ 1 & y_5 & x_5 & x_5y_5 & x_5^2 & x_5^2y_5 \end{vmatrix}=0$$

clearly has the required features. The second row is now reduced to 1, 0, 0, 0, 0, 0 by these operations:

Multiply column 1 by y_1 and subtract from column 2.

Multiply column 3 by y_1 and subtract from column 4.

Multiply column 5 by y_1 and subtract from column 6.

Multiply column 3 by x_1 and subtract from column 5.

Multiply column 1 by x_1 and subtract from column 3.

At this point the determinant has been replaced by the following substitute:

$$\begin{vmatrix} 1 & y-y_1 & x-x_1 & x(y-y_1) & x(x-x_1) & x^2(y-y_1) \\ 1 & 0 & 0 & 0 & 0 & 0 \\ 1 & y_2-y_1 & x_2-x_1 & x_2(y_2-y_1) & x_2(x_2-x_1) & x_2^2(y_2-y_1) \\ 1 & y_3-y_1 & x_3-x_1 & x_3(y_3-y_1) & x_3(x_3-x_1) & x_3^2(y_3-y_1) \\ 1 & y_4-y_1 & x_4-x_1 & x_4(y_4-y_1) & x_4(x_4-x_1) & x_4^2(y_4-y_1) \\ 1 & y_5-y_1 & x_5-x_1 & x_5(y_5-y_1) & x_5(x_5-x_1) & x_5^2(y_5-y_1) \end{vmatrix}$$

Expand this determinant by its second row and then

Divide row 1 by $y-y_1$.

Divide row i by y_i-y_1, for $i=2, 3, 4, 5$.

Introducing the symbol $\rho_1(xx_1)=\dfrac{x-x_1}{y-y_1}$, the equation may now be written as

$$\begin{vmatrix} 1 & \rho_1(xx_1) & x & x\rho_1(xx_1) & x^2 \\ 1 & \rho_1(x_2x_1) & x_2 & x_2\rho_1(x_2x_1) & x_2^2 \\ 1 & \rho_1(x_3x_1) & x_3 & x_3\rho_1(x_3x_1) & x_3^2 \\ 1 & \rho_1(x_4x_1) & x_4 & x_4\rho_1(x_4x_1) & x_4^2 \\ 1 & \rho_1(x_5x_1) & x_5 & x_5\rho_1(x_5x_1) & x_5^2 \end{vmatrix}=0$$

The operation is now repeated, to make the second row 1, 0, 0, 0, 0:

Multiply column 1 by $\rho_1(x_2x_1)$ and subtract from column 2.

Multiply column 3 by $\rho_1(x_2x_1)$ and subtract from column 4.

Multiply column 3 by x_2 and subtract from column 5.

Multiply column 1 by x_2 and subtract from column 3.

The determinant then has this form:

$$\begin{vmatrix} 1 & \rho_1(xx_1)-\rho_1(x_2x_1) & x-x_2 & x[\rho_1(xx_1)-\rho_1(x_2x_1)] & x(x-x_2) \\ 1 & 0 & 0 & 0 & 0 \\ 1 & \rho_1(x_3x_1)-\rho_1(x_2x_1) & x_3-x_2 & x[\rho_1(x_3x_1)-\rho_1(x_2x_1)] & x_3(x_3-x_2) \\ 1 & \rho_1(x_4x_1)-\rho_1(x_2x_1) & x_4-x_2 & x[\rho_1(x_4x_1)-\rho_1(x_2x_1)] & x_4(x_4-x_2) \\ 1 & \rho_1(x_5x_1)-\rho_1(x_2x_1) & x_5-x_2 & x[\rho_1(x_5x_1)-\rho_1(x_2x_1)] & x_5(x_5-x_2) \end{vmatrix}$$

Expand by the second row, and then

Divide row 1 by $\rho_1(xx_1)-\rho_1(x_2x_1)$.

Divide row i by $\rho_1(x_{i+1}x_1)-\rho_1(x_2x_1)$, for $i=2, 3, 4$.

An additional step is traditional at this point in order to assure a symmetry property of the ρ quantities to be defined. (See Problem 23.6.)

Multiply column 1 by y_1 and add to column 2.

Multiply column 3 by y_1 and add to column 4.

Introducing the symbol $\rho_2(xx_1x_2) = \dfrac{x - x_2}{\rho_1(xx_1) - \rho_1(x_2x_1)} + y_1$, the equation has now been reduced to

$$\begin{vmatrix} 1 & \rho_2(xx_1x_2) & x & x\rho_2(xx_1x_2) \\ 1 & \rho_2(x_3x_1x_2) & x_3 & x_3\rho_2(x_3x_1x_2) \\ 1 & \rho_2(x_4x_1x_2) & x_4 & x_4\rho_2(x_4x_1x_2) \\ 1 & \rho_2(x_5x_1x_2) & x_5 & x_5\rho_2(x_5x_1x_2) \end{vmatrix} = 0$$

Another similar reduction produces

$$\begin{vmatrix} 1 & \rho_3(xx_1x_2x_3) & x \\ 1 & \rho_3(x_4x_1x_2x_3) & x_4 \\ 1 & \rho_3(x_5x_1x_2x_3) & x_5 \end{vmatrix} = 0$$

where $$\rho_3(xx_1x_2x_3) = \frac{x - x_3}{\rho_2(xx_1x_2) - \rho_2(x_3x_1x_2)} + \rho_1(x_1x_2)$$

Finally, the last reduction manages

$$\begin{vmatrix} 1 & \rho_4(xx_1x_2x_3x_4) \\ 1 & \rho_4(x_5x_1x_2x_3x_4) \end{vmatrix} = 0$$

where $$\rho_4(xx_1x_2x_3x_4) = \frac{x - x_4}{\rho_3(xx_1x_2x_3) - \rho_3(x_4x_2x_3x_1)} + \rho_2(x_1x_2x_3)$$

We deduce that $\rho_4(xx_1x_2x_3x_4) = \rho_4(x_5x_1x_2x_3x_4)$. The various ρ_i's just introduced are called *reciprocal differences* of order i, and the equality of these fourth-order reciprocal differences is equivalent to the determinant equation with which we began and which identifies the rational function we are seeking.

The definitions of reciprocal differences now lead in a natural way to a continued fraction. We find successively

$$y = y_1 + \frac{x - x_1}{\rho_1(xx_1)} = y_1 + \cfrac{x - x_1}{\rho_1(x_2x_1) + \cfrac{x - x_2}{\rho_2(xx_1x_2) - y_1}}$$

$$= y_1 + \cfrac{x - x_1}{\rho_1(x_2x_1) + \cfrac{x - x_2}{\rho_2(x_3x_1x_2) - y_1 + \cfrac{x - x_3}{\rho_3(xx_1x_2x_3) - \rho_1(x_1x_2)}}}$$

$$= y_1 + \cfrac{x - x_1}{\rho_1(x_2x_1) + \cfrac{x - x_2}{\rho_2(x_3x_1x_2) - y_1 + \cfrac{x - x_3}{\rho_3(x_4x_1x_2x_3) - \rho_1(x_1x_2) + \cfrac{x - x_4}{\rho_4(x_5x_1x_2x_3x_4) - \rho_2(x_1x_2x_3)}}}}$$

where, in the last denominator, the equality of certain fourth differences, which was the culmination of our extensive determinant reduction, has finally been used. This is what makes the above continued fraction the required rational function. (Behind all these computations there has been the assumption that the data points do actually belong to such a rational function, and that the algebraic procedure will not break down at some point. See the problems for exceptional examples.)

23.6. Prove that reciprocal differences are symmetric.

For first-order differences it is at once clear that $\rho_1(x_1x_2) = \rho_1(x_2x_1)$. For second-order differences one verifies first that

$$\frac{x_3 - x_2}{\dfrac{x_3 - x_1}{y_3 - y_1} - \dfrac{x_2 - x_1}{y_2 - y_1}} + y_1 = \frac{x_3 - x_1}{\dfrac{x_3 - x_2}{y_3 - y_2} - \dfrac{x_1 - x_2}{y_1 - y_2}} + y_2 = \frac{x_2 - x_1}{\dfrac{x_2 - x_3}{y_2 - y_3} - \dfrac{x_1 - x_3}{y_1 - y_3}} + y_3$$

from which it follows that in $\rho_2(x_1x_2x_3)$ the x_i may be permuted in any way. For higher-order differences the proof is similar.

23.7. Apply reciprocal differences to recover the function $y(x) = 1/(1 + x^2)$ from the x, y data in the first two columns of Table 23.1.

Various reciprocal differences also appear in this table. For example, the entry 40 is obtained from the looped entries as follows:

$$\rho_3(x_2x_3x_4x_5) = \frac{4-1}{(-\frac{1}{25}) - (-\frac{1}{10})} + (-10) = 40$$

$$= \frac{x_5 - x_2}{\rho_2(x_3x_4x_5) - \rho_2(x_2x_3x_4)} + \rho_1(x_3x_4)$$

From the definition given in Problem 23.5 this third difference should be

$$\rho_3(x_2x_3x_4x_5) = \frac{x_2 - x_5}{\rho_2(x_2x_3x_4) - \rho_2(x_5x_3x_4)} + \rho_1(x_3x_4)$$

but by the symmetry property this is the same as what we have. The other differences are found in the same way.

Table 23.1

x	y				
0	1				
		-2			
①	$\frac{1}{2}$		-1		
		$-\frac{10}{3}$		0	
2	$\frac{1}{5}$		$-\frac{1}{10}$ (circled)		0
		-10 (circled)		40	
3	$\frac{1}{10}$		$-\frac{1}{25}$ (circled)		0
		$-\frac{170}{7}$		140	
④	$\frac{1}{17}$		$-\frac{1}{46}$		
		$-\frac{442}{9}$			
5	$\frac{1}{26}$				

The continued fraction is constructed from the top diagonal

$$y = 1 + \cfrac{x - 0}{-2 + \cfrac{x - 1}{-1 - 1 + \cfrac{x - 2}{0 - (-2) + \cfrac{x - 3}{0 - (-1)}}}}$$

and easily rearranges to the original $y(x) = 1/(1 + x^2)$. This test case merely illustrates the continued fractions algorithm.

By substituting successively the arguments $x = 0, 1, 2, 3, 4$ into this continued fraction it is easy to see that as the fraction becomes longer it absorbs the (x, y) data pairs one by one. This further implies that truncating the fraction will produce a rational collocation function for an initial segment of the data. The same remarks hold for the general case of Problem 23.5. It should also be pointed out that the zeros in the last column of the table cause the fraction to terminate without an $x - x_4$ term, but that the fraction in hand absorbs the (x_5, x_5) data pair anyway.

23.8. Use a rational approximation to interpolate for tan 1.565 from the data provided in Table 23.2.

The table also includes reciprocal differences through fourth order.

Table 23.2

x	$\tan x$				
1.53	24.498				
		.0012558			
1.54	32.461		−.033		
		.0006403		2.7279	
1.55	48.078		−.022		−.4167
		.0002245		1.7145	
1.56	92.631		−.0045		
		.0000086			
1.57	1255.8				

The interpolation then proceeds as follows:

$$\tan 1.565 \simeq 24.498 + \cfrac{1.565 - 1.53}{.0012558 + \cfrac{1.565 - 1.54}{-24.531 + \cfrac{1.565 - 1.55}{2.7266 + \cfrac{1.565 - 1.56}{-.3837}}}}$$

which works out to 172.552. This result is almost perfect, which is remarkable considering how terribly close we are to the pole of the tangent function at $x = \pi/2$. Newton's backward formula, using the same data, produces the value 433, so it is easy to see that our rational approximation is an easy winner. It is interesting to notice the results obtained by stopping at the earlier differences, truncating the fraction at its successive "convergents." Those results are

$$52.37 \qquad 172.36 \qquad 172.552$$

so that stopping at third and fourth differences we find identical values. This convergence is reassuring, suggesting implicitly that more data pairs and continuation of the fraction are unnecessary and that even the final data pair has served only as a check or safeguard.

23.9. It is possible that more than one rational function of the form in Problem 23.5 may include the given points. Which one will the continued fraction algorithm produce?

As the continued fraction grows it represents successively functions of the forms

$$a_0 + a_1 x \qquad \frac{a_0 + a_1 x}{b_0 + b_1 x} \qquad \frac{a_0 + a_1 x + a_2 x^2}{b_0 + b_1 x} \qquad \frac{a_0 + a_1 x + a_2 x^2}{b_0 + b_1 x + b_2 x^2} \qquad \cdots$$

Our algorithm chooses the simplest form (left to right) consistent with the data. See Problems 23.4, 23.18, and 23.19 for examples.

23.10. Given that $y(x)$ has a simple pole at $x = 0$ and is of the form used in Problem 23.5, determine it from these (x, y) points: $(1, 30)$, $(2, 10)$, $(3, 5)$, $(4, 3)$.

Such a function may be sought directly starting with

$$y(x) = \frac{1 + a_1 x + a_2 x^2}{b_1 x + b_2 x^2}$$

It may also be found by this slight variation of the continued fractions algorithm. The table of reciprocal differences

x	y				
1	30				
		$-\frac{1}{20}$			
2	10		$-\frac{10}{3}$		
		$-\frac{1}{5}$		$\frac{8}{5}$	
3	5		$-\frac{5}{3}$		0
		$-\frac{1}{2}$		1	
4	3		-3		
		0			
0	∞				

leads to the continued fraction

$$y = 30 + \cfrac{x-1}{-\cfrac{1}{20} + \cfrac{x-2}{-\cfrac{100}{3} + \cfrac{x-3}{\cfrac{33}{20} + \cfrac{x-4}{\frac{10}{3}}}}}$$

which collapses to $y(x) = 60/[x(x+1)]$.

MIN–MAX RATIONAL FUNCTIONS

23.11. How can a rational function $R(x) = 1/(a + bx)$ which misses the three points (x_1, y_1), (x_2, y_2), and (x_3, y_3) alternately by $\pm h$ be found?

The three conditions

$$y_i - \frac{1}{a + bx_i} = h, -h, h \qquad \text{for } i = 1, 2, 3$$

can be rewritten as

$$\begin{aligned} a(y_1 - h) + b(y_1 - h)x_1 - 1 &= 0 \\ a(y_2 + h) + b(y_2 + h)x_2 - 1 &= 0 \\ a(y_3 - h) + b(y_3 - h)x_3 - 1 &= 0 \end{aligned}$$

Eliminating a and b, we find that h is determined by the quadratic equation

$$\begin{vmatrix} y_1 - h & (y_1 - h)x_1 & -1 \\ y_2 + h & (y_2 + h)x_2 & -1 \\ y_3 - h & (y_3 - h)x_3 & -1 \end{vmatrix} = 0$$

Choosing the root with smaller absolute value, we substitute back and obtain a and b. (It is not hard to show that real roots will always exist.)

23.12. Apply the procedure of Problem 23.11 to these three points: $(0, .83)$, $(1, 1.06)$, $(2, 1.25)$.

The quadratic equation becomes $4h^2 - 4.12h - .130 = 0$ and the required root is $h = -.03$. The coefficients a and b then satisfy $.86a - 1 = 0$, $1.03a + 1.03b - 1 = 0$ and are $a \simeq 1.16$, $b \simeq -.19$.

23.13. Extending the previous problem, apply an exchange method to find a min–max rational function of the form $R = 1/(a + bx)$ for these points: (0, .83), (1, 1.06), (2, 1.25), (4, 4.15).

Our problem will be a close parallel to earlier exchange methods. Let the triple of the previous problem serve as initial triple. The equal-error rational function for this triple was found to be $R_1(x) = 1/(1.16 - .19x)$. At the four data points its errors may be computed to be $-.03$, .03, $-.03$, 1.65 and we see that $R_1(x)$ is very poor at $x = 4$. For a new triple we choose the last three points, to retain alternating error signs. The new quadratic equation is

$$6h^2 - 21.24h + 1.47 = 0$$

making $h = .07$. The new equations for a and b are

$$a + b = 1.010 \qquad a + 2b = .758 \qquad a + 4b = .245$$

making $a \simeq 1.265$ and $b \simeq -.255$. The errors at the four data points are now .04, .07, $-.07$, .07; and since no error exceeds the .07 of our present triple we stop, accepting

$$R_2(x) = \frac{1}{1.265 - .255x}$$

as the min–max approximation. This is the typical development of an exchange algorithm. Our result is of course accurate only to a point, but the data themselves are given to only two places so a greater struggle seems unwarranted. It is interesting to notice that the computation is quite sensitive. Rounding the third-digit 5s in our $R_2(x)$, for instance, can change $R_2(4)$ by almost half a unit. This sensitivity is due to the pole near $x = 5$. Both $R_1(x)$ and $R_2(x)$ are shown in Fig. 23-2.

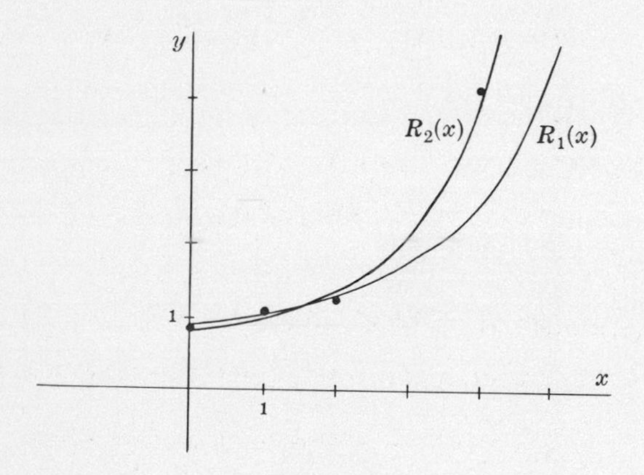

Fig. 23-2

23.14. The data points of the preceding problem were chosen by adding random "noise" of up to 5 percent to values of $y(x) = 4/(5 - x)$. Use $R_2(x)$ to compute smoothed values and compare with the correct values and the original data.

The required values are as follows, with entries at $x = 3$ added:

Original "noisy" data	.83	1.06	1.25	—	4.15
Values of $R_2(x)$	.79	.99	1.32	2.00	4.08
Correct values of $y(x)$	.80	1.00	1.33	2.00	4.00

Only the error at $x = 4$ is sizable and this has been reduced by almost half. The influence of the pole at $x = 5$ is evident. Approximation by means of polynomials would be far less successful.

23.15. Derive conditions on the coefficients such that the Padé rational function

$$R_{mn}(x) = \frac{P_m(x)}{Q_n(x)}$$

with

$$P_m(x) = a_0 + a_1x + a_2x^2 + \cdots + a_mx^m$$
$$Q_n(x) = 1 + b_1x + b_2x^2 + \cdots + b_nx^n$$

will satisfy

$$R_{mn}^{(k)}(0) = y^{(k)}(0) \qquad k = 0, 1, \ldots, N$$

for $N = m + n$, assuming that $y(x)$ has the series representation

$$y(x) = c_0 + c_1x + c_2x^2 + \cdots$$

We have

$$y(x) - R_{mn}(x) = \frac{\left(\sum_0^\infty c_ix^i\right)\left(\sum_0^n b_ix^i\right) - \sum_0^m a_ix^i}{\sum_0^n b_ix^i}$$

and will have achieved the required goal if the numerator on the right has no terms of lower degree than x^{N+1}. For this we need

$$a_0 = b_0c_0 \qquad a_1 = b_0c_1 + b_1c_0 \qquad a_2 = b_0c_2 + b_1c_1 + b_2c_0$$

and in general

$$a_j = \sum_{i=0}^{j} b_ic_{j-i} \qquad j = 0, 1, \ldots, N$$

subject to the constraints $b_0 = 1$ and

$$a_i = 0 \qquad \text{if } i > m$$
$$b_i = 0 \qquad \text{if } i > n$$

23.16. Apply the preceding problem to $y(x) = e^x$ with $m = n = 2$.

For this function we have $c_0 = 1$, $c_1 = 1$, $c_2 = \frac{1}{2}$, $c_3 = \frac{1}{6}$, $c_4 = \frac{1}{24}$, leading to these equations:

$$a_0 = 1 \qquad a_1 = 1 + b_1 \qquad a_2 = \frac{1}{2} + b_1 + b_2$$

$$0 = \frac{1}{6} + \frac{1}{2}b_1 + b_2 \qquad 0 = \frac{1}{24} + \frac{1}{6}b_1 + \frac{1}{2}b_2$$

Their solution is $a_0 = 1$, $a_1 = \frac{1}{2}$, $a_2 = \frac{1}{12}$, $b_1 = -\frac{1}{2}$, and $b_2 = \frac{1}{12}$. Substituting back we have finally

$$R_{22}(x) = \frac{12 + 6x + x^2}{12 - 6x + x^2}$$

for the Padé approximation. On the interval $(-1, 1)$ its absolute error ranges from zero at the center to .004 at $x = 1$. It is interesting to note that the approximation reflects a basic property of the exponential function, namely that replacing x by $-x$ produces the reciprocal.

23.17. For $y(x) = e^x$ it is clear that

$$R_{40} = 1 + x + \frac{1}{2}x^2 + \frac{1}{6}x^3 + \cdots$$

but use the method of Problem 23.15 to find $R_{04}(x)$.

The appropriate equations include $a_0 = 1$ and then the triangular system

$$0 = 1 + b_1$$

$$0 = \frac{1}{2} + b_1 + b_2$$

$$0 = \frac{1}{6} + \frac{1}{2}b_1 + b_2 + b_3$$

$$0 = \frac{1}{24} + \frac{1}{6}b_1 + \frac{1}{2}b_2 + b_3 + b_4$$

leading to the approximation

$$R_{04}(x) = \frac{1}{1 - x + \frac{1}{2}x^2 - \frac{1}{6}x^3 + \frac{1}{24}x^4}$$

of which the denominator is a five-term approximation to the reciprocal of $y(x)$. Presumably this could have been anticipated.

Over $(-1, 1)$ R_{04} is closer to e^x on the left half and farther from it on the right, relative to R_{40}. It is inferior all the way to R_{22} and this is generally true of Padé approximations. Those with m and n equal or nearly equal are the most accurate.

Supplementary Problems

23.18. Find directly, as in Problem 23.1, a function $y(x) = 1/(a + bx)$ such that $y(1) = 3$ and $y(3) = 1$. Will our method of continued fractions yield this function?

23.19. Find directly a function $y(x) = 1/(a + bx + cx^2)$ such that $y(0) = 1$, $y(1) = \frac{1}{2}$, and $y(10) = \frac{1}{4}$. Will our method of continued fractions yield this function?

23.20. Use the continued fractions method to find a rational function having the following values:

x	0	1	2	3	4
y	-1	0	$\frac{3}{5}$	$\frac{4}{5}$	$\frac{15}{17}$

23.21. Use the continued fractions method to find a rational function having the following values:

x	0	1	9	19
y	0	$\frac{1}{2}$	8.1	18.05

23.22. Find a rational function with these values:

x	0	1	$+\infty$
y	$\frac{1}{2}$	$\frac{2}{3}$	1

23.23. Find a rational function with these values:

x	0	1	2	4	∞
y	-2	$\pm\infty$	2	$\frac{6}{5}$	1

(The symbol $\pm\infty$ refers to a pole at which the function changes sign.)

23.24. Find a rational function with the values given below. Interpolate for $y(1.5)$. Where are the "poles" of this function?

x	0	± 1	± 2
y	$\frac{1}{2}$	1	$-\frac{1}{2}$

23.25. Find the min–max function

$$R(x)=\frac{1}{a+bx}$$

for $y(x)=x^2-1$ on the interval $(-1, 1)$.

23.26. Use an exchange method to find the min–max approximation $R(x)=1/(a+bx)$ to $y(x)=e^x$ on the interval $(0, 3)$.

23.27. Develop an exchange method for finding the min–max approximation $R(x)=(a+bx)/(1+dx)$ for a set of points (x_i, y_i) where $i=1, \ldots, N$. Apply it to the following data:

x	0	1	2	3	4	5
y	.38	.30	.16	.20	.12	.10

Use $R(x)$ to smooth the y values. How close do you come to $y(x)=1/(x+3)$ which was the parent function of this data, with random errors added?

23.28. Find a rational function which includes these points:

x	-1	0	1	2	3
y	∞	4	2	4	7

23.29. Find a rational function which includes these points:

x	-2	-1	0	1	2
y	$-\infty$	0	3	8	∞

23.30. Find a rational function which includes the following points. Does the function have any real poles?

x	-2	-1	0	1	2	3
y	$\frac{4}{3}$	2	2	$\frac{4}{3}$	$\frac{8}{7}$	$\frac{14}{13}$

23.31. Interpolate for $y(1.5)$ in the table below, using a rational approximation function.

x	1	2	3	4
y	57.298677	28.653706	19.107321	14.335588

23.32. Find a rational function, in the form of a cubic polynomial over a quadratic, including these points:

x	0	1	2	3	4	5
y	12	0	-4	-6	6	4

23.33. Work Problem 23.16 with $m = 3$, $n = 1$.

23.34. Work Problem 23.16 with $m = 1$, $n = 3$.

Chapter 24

Trigonometric Approximation

DISCRETE DATA

The sine and cosine functions share many of the desirable features of polynomials. They are easily computed, by rapidly convergent series. Their successive derivatives are again sines and cosines, the same then holding for integrals. They also have orthogonality properties, and of course periodicity, which polynomials do not have. The use of these familiar trigonometric functions in approximation theory is therefore understandable.

A trigonometric sum which collocates with a given data function at $2L+1$ prescribed arguments may be obtained in the form

$$y(x)=\frac{1}{2}a_0+\sum_{k=1}^{L}\left(a_k\cos\frac{2\pi}{2L+1}kx+b_k\sin\frac{2\pi}{2L+1}kx\right)$$

a slightly different form being used if the number of collocation arguments is even. An orthogonality property of these sines and cosines,

$$\sum_{x=0}^{N}\sin\frac{2\pi}{N+1}jx\sin\frac{2\pi}{N+1}kx=\begin{cases}0 & \text{if } j\neq k\\ (N+1)/2 & \text{if } j=k\neq 0\end{cases}$$

$$\sum_{x=0}^{N}\sin\frac{2\pi}{N+1}jx\cos\frac{2\pi}{N+1}kx=0$$

$$\sum_{x=0}^{N}\cos\frac{2\pi}{N+1}jx\cos\frac{2\pi}{N+1}kx=\begin{cases}0 & \text{if } j\neq k\\ (N+1)/2 & \text{if } j=k\neq 0,\ N+1\\ N+1 & \text{if } j=k=0,\ N+1\end{cases}$$

allows the coefficients to be easily determined as

$$a_k=\frac{2}{2L+1}\sum_{x=0}^{2L}y(x)\cos\frac{2\pi}{2L+1}kx \qquad k=0,1,\ldots,L$$

$$b_k=\frac{2}{2L+1}\sum_{x=0}^{2L}y(x)\sin\frac{2\pi}{2L+1}kx \qquad k=1,2,\ldots,L$$

These coefficients provide the unique collocation function of the form specified. For an even number of collocation arguments, say $2L$, the corresponding formula is

$$y(x)=\frac{1}{2}a_0+\sum_{k=1}^{L-1}\left(a_k\cos\frac{\pi}{L}kx+b_k\sin\frac{\pi}{L}kx\right)+\frac{1}{2}a_L\cos\pi x$$

with

$$a_k=\frac{1}{L}\sum_{x=0}^{2L-1}y(x)\cos\frac{\pi}{L}kx \qquad k=0,1,\ldots,L$$

$$b_k=\frac{1}{L}\sum_{x=0}^{2L-1}y(x)\sin\frac{\pi}{L}kx \qquad k=1,\ldots,L-1$$

Least-squares approximations for the same discrete data, using the same type of trigonometric sum, are obtained simply *by truncation of the collocation sum.* This is a famous and convenient result. As observed in Problem 21.8, it is true of other representations in terms of orthogonal functions. What is minimized here, in the case of $2L+1$ arguments, is

$$S=\sum_{x=0}^{2L}[y(x)-T_M(x)]^2$$

where $T_M(x)$ is the abbreviated sum (M being less than L)

$$T_M(x) = \frac{1}{2}A_0 + \sum_{k=1}^{M}\left(A_k \cos\frac{2\pi}{2L+1}kx + B_k \sin\frac{2\pi}{2L+1}kx\right)$$

The result just stated means that to minimize S we should choose $A_k = a_k$, $B_k = b_k$. The minimum value of S can be expressed as

$$S_{\min} = \frac{2L+1}{2}\sum_{k=M+1}^{L}(a_k^2 + b_k^2)$$

For $M = L$ this would be zero, which is hardly a surprise since then we have once again the collocation sum.

Periodicity is an obvious feature of trigonometric sums. If a data function $y(x)$ is not basically periodic, it may still be useful to construct a trigonometric approximation, provided we are concerned only with a finite interval. The given $y(x)$ may then be imagined extended outside this interval in a way which makes it periodic.

Odd and even functions are commonly used as extensions. An odd function has the property $y(-x) = -y(x)$. The classic example is $y(x) = \sin x$. For an odd function of period $P = 2L$, the coefficients of our trigonometric sum simplify to

$$a_k = 0 \qquad b_k = \frac{4}{P}\sum_{x=1}^{L-1} y(x)\sin\frac{2\pi}{P}kx$$

An even function has the property $y(-x) = y(x)$. The classic example is $y(x) = \cos x$. For an even function of period $P = 2L$, the coefficients become

$$a_k = \frac{2}{P}[y(0) + y(L)\cos k\pi] + \frac{4}{P}\sum_{x=1}^{L-1} y(x)\cos\frac{2\pi}{P}kx \qquad b_k = 0$$

These simplifications explain the popularity of odd and even functions.

CONTINUOUS DATA

Fourier series replace finite trigonometric sums when the data supply is continuous, much of the detail being analogous. For $y(x)$ defined over $(0, 2\pi)$, the series has the form

$$\frac{1}{2}\alpha_0 + \sum_{k=1}^{\infty}(\alpha_k \cos kt + \beta_k \sin kt)$$

A second orthogonality property of sines and cosines,

$$\int_0^{2\pi} \sin jt \sin kt\, dt = \begin{cases} 0 & \text{if } j \neq k \\ \pi & \text{if } j = k \neq 0 \end{cases}$$

$$\int_0^{2\pi} \sin jt \cos kt\, dt = 0$$

$$\int_0^{2\pi} \cos jt \cos kt\, dt = \begin{cases} 0 & \text{if } j \neq k \\ \pi & \text{if } j = k \neq 0 \\ 2\pi & \text{if } j = k = 0 \end{cases}$$

allows easy identification of the Fourier coefficients as

$$\alpha_k = \frac{1}{\pi}\int_0^{2\pi} y(t)\cos kt\, dt \qquad \beta_k = \frac{1}{\pi}\int_0^{2\pi} y(t)\sin kt\, dt$$

Since the series has period 2π, we must limit its use to the given interval $(0, 2\pi)$ unless $y(x)$ also happens to have this same period. Nonperiodic functions may be accommodated over a finite interval, if we imagine them extended as periodic. Again, odd and even extensions are the most common and in such cases the Fourier coefficients simplify much as above.

Fourier coefficients are related to collocation coefficients. Taking the example of an odd number of arguments we have, for example,

$$a_j = \frac{1}{L}\left[\frac{1}{2}y(0) + \frac{1}{2}y(2L) + \sum_{x=1}^{2L-1} y(x)\cos\frac{\pi}{L}jx\right]$$

which is the trapezoidal rule approximation to

$$\alpha_j = \frac{1}{L}\int_0^{2L} y(x)\cos\frac{\pi}{L}jx\,dx$$

in which a change of argument has been used to bring out the analogy.

Least-squares approximations for the case of continuous data are obtained by *truncation of the Fourier series.* This will minimize the integral

$$I = \int_0^{2\pi} [y(t) - T_M(t)]^2\,dt$$

where
$$T_M(t) = \frac{1}{2}A_0 + \sum_{k=1}^{M}(A_k\cos kt + B_k\sin kt)$$

In other words, to minimize I we should choose $A_k = \alpha_k$, $B_k = \beta_k$. The minimum value of I can be expressed as

$$I_{\min} = \pi\sum_{k=M+1}^{\infty}(\alpha_k^2 + \beta_k^2)$$

Convergence in the mean occurs under very mild assumptions on $y(t)$. This means that, for M tending to infinity, $I_{\min}$ has limit zero.

APPLICATIONS

The two major applications of trigonometric approximation in numerical analysis are

1. **Data smoothing.** Since least-squares approximations are so conveniently available by truncation, this application seems natural, the smoothing effect of the least-squares principle being similar to that observed for the case of polynomials.
2. **Approximate differentiation.** Here too the least-squares aspect of trigonometric approximation looms in the background. Sometimes the results of applying a formula such as

$$y(x) \simeq \frac{1}{10}[-2y(x-2) - y(x-1) + y(x+1) + 2y(x+2)]$$

derived earlier from a least-squares parabola, are further smoothed by the use of a trigonometric sum. The danger of oversmoothing, removing essential features of the target function, should be kept in mind.

COMPLEX FORMS

All the foregoing can also be represented in complex form. Trigonometric sums become

$$\sum_{j=-l}^{l} c_j e^{ijx}$$

where i is the imaginary unit. Because of the Euler formula

$$e^{ix} = \cos x + i\sin x$$

this is the equivalent of
$$\frac{a_0}{2}+\sum_{j=1}^{l}(a_j\cos jx+b_j\sin jx)$$
with
$$a_j=c_j+c_{-j}\qquad b_j=i(c_j-c_{-j})$$
The coefficients a_j, b_j, c_j may be real or complex. The Fourier series becomes
$$f(x)=\sum_{j=-\infty}^{\infty}f_je^{ijx}$$
with the Fourier coefficients
$$f_j=\frac{1}{2\pi}\int_0^{2\pi}f(x)e^{-ijx}\,dx$$
The finite sum
$$f_j^*=\frac{1}{N}\sum_{n=0}^{N-1}f(x_n)e^{-ijx_n}$$
where $x_n=2\pi n/N$ for $n=0$ to $N-1$, is an obvious approximation to f_j and is also the appropriate coefficient in the trigonometric sum which interpolates $f(x)$ at the data points x_n.
$$y(x)=\sum_{j=-l}^{l}f_j^*e^{ijx}$$
The f_j^* are essentially the elements of what is called a discrete Fourier transform. Given a vector V with components v_0 to v_{N-1}, the discrete Fourier transform of V may be defined as the vector V^T having components
$$v_j^T=\sum_{n=0}^{N-1}v_n\omega_N^{jn}$$
for $j=0$ to $j=N-1$ and ω_N an Nth root of 1.
$$\omega_N=e^{-2\pi i/N}$$
These various relationships will be explored in the problems.

What this means is that it is possible to compute approximations to the Fourier coefficients f_j by using discrete transforms. The use of Fast Fourier Transforms (FFT) has made such computations efficient even for rather large values of N. These coefficients are of interest in many applications, since they give the relative weights of the component terms in a complex periodic process.

Solved Problems

TRIGONOMETRIC SUMS BY COLLOCATION

24.1. Prove the orthogonality conditions
$$\sum_{x=0}^{N}\sin\frac{2\pi}{N+1}jx\sin\frac{2\pi}{N+1}kx=\begin{cases}0 & \text{if } j\neq k \text{ or } j=k=0\\ (N+1)/2 & \text{if } j=k\neq 0\end{cases}$$
$$\sum_{x=0}^{N}\sin\frac{2\pi}{N+1}jx\cos\frac{2\pi}{N+1}kx=0$$
$$\sum_{x=0}^{N}\cos\frac{2\pi}{N+1}jx\cos\frac{2\pi}{N+1}kx=\begin{cases}0 & \text{if } j\neq k\\ (N+1)/2 & \text{if } j=k\neq 0\\ N+1 & \text{if } j=k=0\end{cases}$$
for $j+k\leqq N$.

The proofs are by elementary trigonometry. As an example,

$$\sin\frac{2\pi}{N+1}jx\,\sin\frac{2\pi}{N+1}kx=\frac{1}{2}\left[\cos\frac{2\pi}{N+1}(j-k)x-\cos\frac{2\pi}{N+1}(j+k)x\right]$$

and each cosine sums to zero since the angles involved are symmetrically spaced between 0 and 2π, except when $j=k\neq 0$, in which case the first sum of cosines is $(N+1)/2$. The other two parts are proved in similar fashion.

24.2. For collocation at an odd number of arguments $x=0, 1, \ldots, N=2L$, the trigonometric sum may take the form

$$\frac{1}{2}a_0+\sum_{k=1}^{L}\left(a_k\cos\frac{2\pi}{2L+1}kx+b_k\sin\frac{2\pi}{2L+1}kx\right)$$

Use Problem 24.1 to determine the coefficients a_k and b_k.

To obtain a_j multiply by $\cos\frac{2\pi}{2L+1}jx$ and sum. We find

$$a_j=\frac{2}{2L+1}\sum_{x=0}^{2L}y(x)\cos\frac{2\pi}{2L+1}jx \qquad j=0, 1, \ldots, L$$

since all other terms on the right are zero. The factor $\frac{1}{2}$ in $y(x)$ makes this result true also for $j=0$. To obtain b_j we multiply $y(x)$ by $\sin\frac{2\pi}{2L+1}jx$ and sum, getting

$$b_j=\frac{2}{2L+1}\sum_{x=0}^{2L}y(x)\sin\frac{2\pi}{2L+1}jx \qquad j=1, 2, \ldots, L$$

Thus only one such expression can represent a given $y(x)$, the coefficients being uniquely determined by the values of $y(x)$ at $x=0, 1, \ldots, 2L$. Notice that this function will have period $N+1$.

24.3. Verify that, with the coefficients of Problem 24.2, the trigonometric sum does equal $y(x)$ for $x=0, 1, \ldots, 2L$. This will prove the existence of a unique sum of this type which collocates with $y(x)$ for these arguments.

Calling the sum $T(x)$ for the moment and letting x^* be any one of the $2L+1$ arguments, substitution of our formulas for the coefficients leads to

$$\begin{aligned}T(x^*)&=\frac{2}{2L+1}\sum_{x=0}^{2L}y(x)\left[\frac{1}{2}+\sum_{k=1}^{L}\left(\cos\frac{2\pi}{2L+1}kx\cos\frac{2\pi}{2L+1}kx^*+\sin\frac{2\pi}{2L+1}kx\sin\frac{2\pi}{2L+1}kx^*\right)\right]\\&=\frac{2}{2L+1}\sum_{x=0}^{2L}y(x)\left[\frac{1}{2}+\sum_{k=1}^{L}\cos\frac{2\pi}{2L+1}k(x-x^*)\right]\end{aligned}$$

in which the order of summation has been altered. The last sum is now written as

$$\sum_{k=1}^{L}\cos\frac{2\pi}{2L+1}k(x-x^*)=\frac{1}{2}\sum_{k=1}^{L}\cos\frac{2\pi}{2L+1}k(x-x^*)+\frac{1}{2}\sum_{k=L+1}^{2L}\cos\frac{2\pi}{2L+1}k(x-x^*)$$

which is possible because of the symmetry property

$$\cos\frac{2\pi}{2L+1}k(x-x^*)=\cos\frac{2\pi}{2L+1}(2L+1-k)(x-x^*)$$

of the cosine function. Filling in the $k=0$ term, we now find

$$T(x^*)=\frac{1}{2L+1}\sum_{x=0}^{2L}y(x)\left[\sum_{k=0}^{2L}\cos\frac{2\pi}{2L+1}k(x-x^*)\right]$$

But the term in brackets is zero by the orthogonality conditions unless $x = x^*$, when it becomes $2L+1$. Thus $T(x^*) = y(x^*)$, which was to be proved.

24.4. Suppose $y(x)$ is known to have the period 3. Find a trigonometric sum which includes the following data points and use it to interpolate for $y(\frac{1}{2})$ and $y(\frac{3}{2})$.

x	0	1	2
y	0	1	1

Using the formulas of Problem 24.2, we find

$$a_0 = \frac{2}{3}(0+1+1) = \frac{4}{3} \qquad a_1 = \frac{2}{3}\left(\cos\frac{2\pi}{3} + \cos\frac{4\pi}{3}\right) = -\frac{2}{3}$$

$$b_1 = \frac{2}{3}\left(\sin\frac{2\pi}{3} + \sin\frac{4\pi}{3}\right) = 0$$

24.5. For an even number of x arguments $(N+1 = 2L)$ the collocation sum is

$$y(x) = \frac{1}{2}a_0 + \sum_{k=1}^{L-1}\left(a_k \cos\frac{\pi}{L}kx + b_k \sin\frac{\pi}{L}kx\right) + \frac{1}{2}a_L \cos \pi x$$

with collocation at $x = 0, 1, \ldots, N$. The coefficients are found by an argument almost identical with that of Problems 24.1 and 24.2 to be

$$a_j = \frac{1}{L}\sum_{x=0}^{2L-1} y(x)\cos\frac{\pi}{L}jx \qquad j = 0, 1, \ldots, L$$

$$b_j = \frac{1}{L}\sum_{x=0}^{2L-1} y(x)\sin\frac{\pi}{L}jx \qquad j = 1, \ldots, L-1$$

Once again the function $y(x)$ is seen to have the period $N+1$. Apply these formulas to the data below, and then compute the maximum of $y(x)$.

x	0	1	2	3
y	0	1	1	0

We find $L = 2$ and then $a_0 = \frac{1}{2}(2) = 1$, $a_1 = \frac{1}{2}(-1) = -\frac{1}{2}$, $a_2 = \frac{1}{2}(-1+1) = 0$, $b_1 = \frac{1}{2}(1) = \frac{1}{2}$. The trigonometric sum is therefore

$$y(x) = \frac{1}{2} - \frac{1}{2}\cos\frac{1}{2}\pi x + \frac{1}{2}\sin\frac{1}{2}\pi x$$

The maximum of $y(x)$ is then found by standard procedures to be $y(\frac{3}{2}) = \frac{1}{2}(1+\sqrt{2})$.

TRIGONOMETRIC SUMS BY LEAST SQUARES, DISCRETE DATA

24.6. Determine the coefficients A_k and B_k so that the sum of squares

$$S = \sum_{x=0}^{2L}[y(x) - T_m(x)]^2 = \text{minimum}$$

where $T_m(x)$ is the trigonometric sum

$$T_m(x) = \frac{1}{2}A_0 + \sum_{k=1}^{M}\left(A_k \cos\frac{2\pi}{2L+1}kx + B_k \sin\frac{2\pi}{2L+1}kx\right)$$

and $M < L$.

Since by Problem 24.3 we have

$$y(x) = \frac{1}{2}a_0 + \sum_{k=1}^{L}\left(a_k \cos\frac{2\pi}{2L+1}kx + b_k \sin\frac{2\pi}{2L+1}kx\right)$$

the difference is

$$y(x) - T_m(x) = \frac{1}{2}(a_0 - A_0) + \sum_{k=1}^{M}\left[(a_k - A_k)\cos\frac{2\pi}{2L+1}kx + (b_k - B_k)\sin\frac{2\pi}{2L+1}kx\right]$$
$$+ \sum_{k=M+1}^{L}\left[a_k \cos\frac{2\pi}{2L+1}kx + b_k \sin\frac{2\pi}{2L+1}kx\right]$$

Squaring, summing over the arguments x, and using the orthogonality conditions,

$$S = \sum_{x=0}^{2L}[y(x) - T_m(x)]^2 = \frac{2L+1}{4}(a_0 - A_0)^2 + \frac{2L+1}{2}\sum_{k=1}^{M}[(a_k - A_k)^2 + (b_k - B_k)^2] + \frac{2L+1}{2}\sum_{k=M+1}^{L}(a_k^2 + b_k^2)$$

Only the first two terms depend upon the A_k and B_k, and since these terms are nonnegative the minimum sum can be achieved in only one way, by making these terms zero. Thus for a minimum,

$$A_k = a_k \qquad B_k = b_k$$

and we have the important result that truncation of the collocation sum $T(x)$ at $k = M$ produces the least-squares trigonometric sum $T_M(x)$. (This is actually another special case of the general result found in Problem 21.8.) We also find

$$S_{\min} = \frac{2L+1}{2}\sum_{k=M+1}^{L}(a_k^2 + b_k^2)$$

Since an almost identical computation shows that

$$\sum_{x=0}^{2L}[y(x)]^2 = \sum_{x=0}^{2L}[T(x)]^2 = \frac{2L+1}{4}a_0^2 + \frac{2L+1}{2}\sum_{k=1}^{L}(a_k^2 + b_k^2)$$

this may also be expressed in the form

$$S_{\min} = \sum_{x=0}^{2L}[y(x)^2] - \frac{2L+1}{4}a_0^2 - \frac{2L+1}{2}\sum_{k=1}^{M}(a_k^2 + b_k^2)$$

As M increases this sum steadily decreases, reaching zero for $M = L$, since then the least-squares and collocation sums are identical. A somewhat similar result holds for the case of an even number of x arguments.

24.7. Apply Problem 24.6 with $M = 0$ to the data of Problem 24.4.

Truncation leads to $T_0(x) = \frac{2}{3}$.

ODD OR EVEN PERIODIC FUNCTIONS

24.8. Suppose $y(x)$ has the period $P = 2L$, that is, $y(x + P) = y(x)$ for all x. Show that the formulas for a_j and b_j in Problem 24.5 may be written as

$$a_j = \frac{2}{P}\sum_{x=-L+1}^{L} y(x)\cos\frac{2\pi}{P}jx \qquad j = 0, 1, \ldots, L$$

$$b_j = \frac{2}{P}\sum_{x=-L+1}^{L} y(x)\sin\frac{2\pi}{P}jx \qquad j = 1, \ldots, L-1$$

Since the sine and cosine also have period P, it makes no difference whether the arguments $x = 0, \ldots, 2L-1$ or the arguments $-L+1, \ldots, L$ are used. Any such set of P consecutive arguments will lead to the same coefficients.

24.9. Suppose $y(x)$ has the period $P = 2L$ and is also an odd function, that is, $y(-x) = -y(x)$.

Prove that

$$a_j = 0 \qquad b_j = \frac{4}{P}\sum_{x=1}^{L-1} y(x)\sin\frac{2\pi}{P}jx$$

By periodicity, $y(0) = y(P) = y(-P)$. But since $y(x)$ is an odd function, $y(-P) = -y(P)$ also. This implies $y(0) = 0$. In the same way we find $y(L) = y(-L) = -y(L) = 0$. Then in the sum for a_j each remaining term at positive x cancels its mate at negative x, so that all a_j will be 0. In the sum for b_j the terms for x and $-x$ are identical and so we find b_j by doubling the sum over positive x.

24.10. Find a trigonometric sum $T(x)$ for the function of Problem 24.5, assuming it extended to an odd function of period $P = 6$.

By the previous problem all $a_j = 0$, and since $L = 3$,

$$b_1 = \frac{2}{3}\left(\sin\frac{\pi}{3} + \sin\frac{2\pi}{3}\right) = \frac{2}{\sqrt{3}} \qquad b_2 = \frac{2}{3}\left(\sin\frac{2\pi}{3} + \sin\frac{4\pi}{3}\right) = 0$$

making $T(x) = (2/\sqrt{3})\sin(\pi x/3)$.

24.11. If $y(x)$ has the period $P = 2L$ and is an even function, that is, $y(-x) = y(x)$, show that the formulas of Problem 24.8 become

$$a_j = \frac{2}{P}[y(0) + y(L)\cos j\pi] + \frac{4}{P}\sum_{x=1}^{L-1} y(x)\cos\frac{2\pi}{P}jx \qquad j = 0, 1, \ldots, L$$

$$b_j = 0$$

The terms for $\pm x$ in the formula for b_j cancel in pairs. In the a_j formula the terms for $x = 0$ and $x = L$ may be separated as above, after which the remaining terms come in matching pairs for $\pm x$.

24.12. Find a $T(x)$ for the function of Problem 24.5 assuming it extended to an even function of period 6. (This will make three representations of the data by trigonometric sums, but in different forms. See Problems 24.5 and 24.10.)

All b_j will be zero, and with $L = 3$ we find $a_0 = \frac{4}{3}$, $a_1 = 0$, $a_2 = -\frac{2}{3}$, $a_3 = 0$ making $T(x) = \frac{2}{3}(1 - \cos\frac{2}{3}\pi x)$.

CONTINUOUS DATA. THE FOURIER SERIES

24.13. Prove the orthogonality conditions

$$\int_0^{2\pi} \sin jt \sin kt\, dt = \begin{cases} 0 & \text{if } j \neq k \\ \pi & \text{if } j = k \neq 0 \end{cases}$$

$$\int_0^{2\pi} \sin jt \cos kt\, dt = 0$$

$$\int_0^{2\pi} \cos jt \cos kt\, dt = \begin{cases} 0 & \text{if } j \neq k \\ \pi & \text{if } j = k \neq 0 \\ 2\pi & \text{if } j = k = 0 \end{cases}$$

where $j, k = 0, 1, \ldots$ to infinity.

The proofs are elementary calculus. For example,

$$\sin jt \sin kt = \tfrac{1}{2}[\cos(j-k)t - \cos(j+k)t]$$

and each cosine integrates to zero since the interval of integration is a period of the cosine, except when $j = k \neq 0$, in which case the first integral becomes $\frac{1}{2}(2\pi)$. The other two parts are proved in similar fashion.

24.14. Derive the coefficient formulas

$$\alpha_j = \frac{1}{\pi}\int_0^{2\pi} y(t)\cos jt\,dt \qquad \beta_j = \frac{1}{\pi}\int_0^{2\pi} y(t)\sin jt\,dt$$

of the Fourier series
$$y(t) = \frac{1}{2}\alpha_0 + \sum_{k=1}^{\infty}(\alpha_k \cos kt + \beta_k \sin kt)$$

These are called the Fourier coefficients. As a matter of fact all such coefficients in sums or series of orthogonal functions are frequently called Fourier coefficients.

The proof follows a familiar path. Multiply $y(t)$ by $\cos jt$ and integrate over $(0, 2\pi)$. All terms but one on the right are zero and the formula for α_j emerges. The factor $\frac{1}{2}$ in the α_0 term makes the result true also for $j = 0$. To obtain β_j we multiply by $\sin jt$ and integrate. Here we are assuming that the series will converge to $y(t)$ and that term-by-term integration is valid. This is proved, under very mild assumptions about the smoothness of $y(t)$, in the theory of Fourier series. Clearly $y(t)$ must also have the period 2π.

24.15. Obtain the Fourier series for $y(t) = |t|$, $-\pi \leqq t \leqq \pi$.

Let $y(t)$ be extended to an even function of period 2π. (See solid curve in Fig. 24-1.) The limits of integration in our coefficient formulas may be shifted to $(-\pi, \pi)$ and we see that all $\beta_j = 0$. Also $\alpha_0 = \pi$; and for $j > 0$

$$\alpha_j = \frac{2}{\pi}\int_0^{\pi} t\cos jt\,dt = \frac{2(\cos j\pi - 1)}{\pi j^2}$$

Thus
$$y(t) = \frac{\pi}{2} - \frac{4}{\pi}\left(\cos t + \frac{\cos 3t}{3^2} + \frac{\cos 5t}{5^2} + \cdots\right)$$

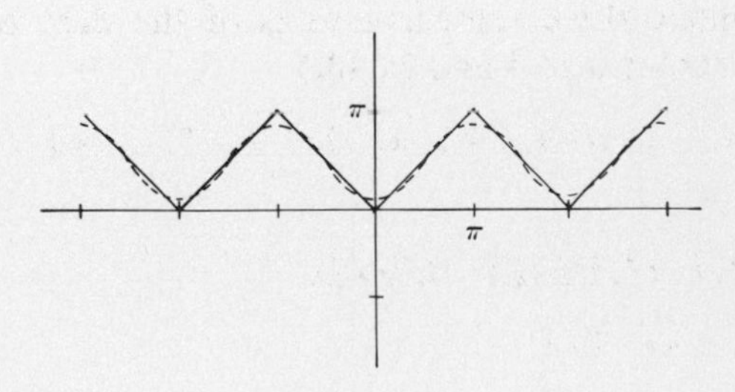

Fig. 24-1

24.16. Obtain the Fourier series for $y(t) = t$, $-\pi < t < \pi$.

We extend $y(t)$ to an odd function of period 2π. (See Fig. 24-2). Again shifting to limits $(-\pi, \pi)$ we find all $\alpha_j = 0$, and

$$\beta_j = \frac{2}{\pi}\int_0^{\pi} t\sin jt\,dt = \frac{2(-1)^{j-1}}{j}$$

Thus
$$y(t) = 2\left(\sin t - \frac{\sin 2t}{2} + \frac{\sin 3t}{3} - \frac{\sin 4t}{4} + \cdots\right)$$

Notice that the cosine series of Problem 24.15 converges more rapidly than the sine series. This is related to the fact that the $y(t)$ of that problem is continuous, while this one is not. The smoother $y(t)$ is, the more rapid the convergence. Notice also that at the points of discontinuity our sine series converges to zero, which is the average of the left and right extreme values (π and $-\pi$) of $y(t)$.

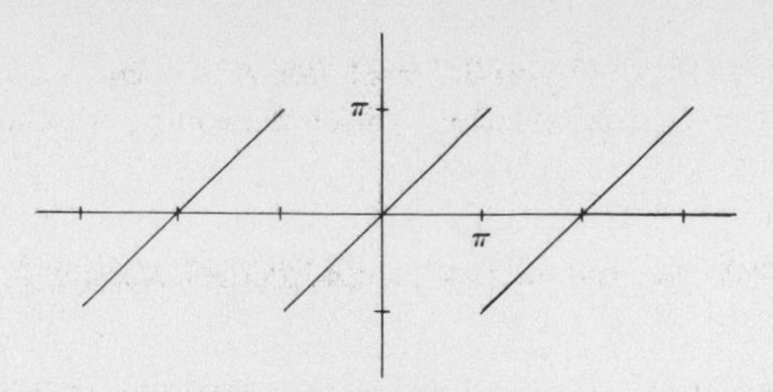

Fig. 24-2

24.17. Find the Fourier series for $y(t) = \begin{cases} t(\pi - t), & 0 \leqq t \leqq \pi \\ t(\pi + t), & -\pi \leqq t \leqq 0 \end{cases}$.

Extending the function to an odd function of period 2π, we have the result shown in Fig. 24-3. Notice that this function has no corners. At $t = 0$ its derivative is π from both sides, while both $y'(\pi)$ and $y'(-\pi)$ are $-\pi$ so that even the extended periodic function has no corners. This extra smoothness will affect the Fourier coefficients. Using limits $(-\pi, \pi)$ we again find all $\alpha_j = 0$, and

$$\beta_j = \frac{2}{\pi}\int_0^{\pi} t(\pi - t)\sin jt\,dt = \frac{2}{\pi}\int_0^{\pi} \frac{\pi - 2t}{j}\cos jt\,dt = \frac{4}{\pi j^2}\int_0^{\pi} \sin jt\,dt = \frac{4(1 - \cos j\pi)}{\pi j^3}$$

The series is therefore
$$y(t) = \frac{8}{\pi}\left(\sin t + \frac{\sin 3t}{3^3} + \frac{\sin 5t}{5^3} + \cdots\right)$$

The coefficients diminish as reciprocal cubes, which makes for very satisfactory convergence. The extra smoothness of the function has proved useful.

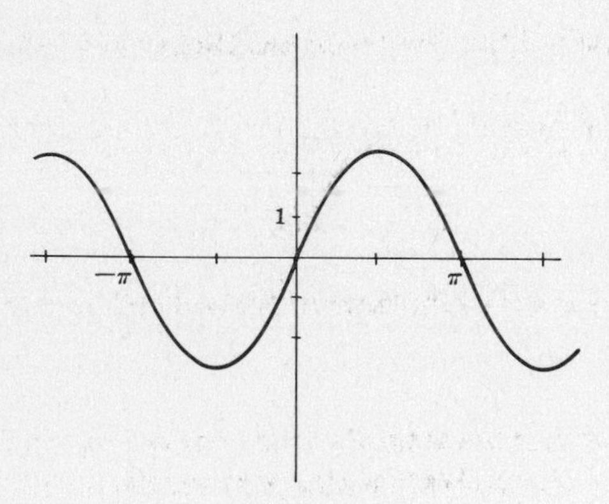

Fig. 24-3

24.18. Show that for the Bernoulli function

$$F_n(x) = B_n(x) \qquad 0 < x < 1 \qquad F_n(x \pm m) = F_n(x) \qquad m \text{ an integer}$$

$B_n(x)$ being a Bernoulli polynomial, the Fourier series is

$$F_n(x) = (-1)^{(n/2)+1}n!\left[\frac{2}{(2\pi)^n}\right]\sum_{k=1}^{\infty}\frac{\cos 2\pi kx}{k^n}$$

when n is even, and
$$F_n(x) = (-1)^{(n+1)/2}n!\left[\frac{2}{(2\pi)^n}\right]\sum_{k=1}^{\infty}\frac{\sin 2\pi kx}{k^n}$$

when n is odd. This result was used in Problem 17.28 of the chapter on sums and series.

Since $B_1(x) = x - \frac{1}{2}$, the series for $F_1(x)$ may be found directly from the coefficient formulas to be

$$F_1(x) = -\frac{1}{\pi}\left(\frac{\sin 2\pi x}{1} + \frac{\sin 4\pi x}{2} + \frac{\sin 6\pi x}{3} + \cdots\right)$$

Integrating, and recalling that

$$B_n'(x) = nB_{n-1}(x) \qquad \int_0^1 B_n(x)\,dx = 0 \qquad \text{for } n > 0$$

we soon find
$$F_2(x) = \frac{2\cdot 2!}{(2\pi)^2}\left(\frac{\cos 2\pi x}{1} + \frac{\cos 4\pi x}{2^2} + \frac{\cos 6\pi x}{3^2} + \cdots\right)$$

The next integration makes

$$F_3(x) = \frac{2\cdot 3!}{(2\pi)^3}\left(\frac{\sin 2\pi x}{1} + \frac{\sin 4\pi x}{2^3} + \frac{\sin 6\pi x}{3^3} + \cdots\right)$$

and an induction may be used to complete a formal proof. (Here it is useful to know that integration of a Fourier series term by term always produces the Fourier series of the integrated function. The analogous statement for differentiation is not generally true. For details see a theoretical treatment of Fourier series.

24.19. How are the collocation coefficients of Problem 24.5, or of Problem 24.2, related to the Fourier coefficients of Problem 24.14?

There are many ways of making the comparisons. One of the most interesting is to notice that in Problem 24.5, assuming $y(x)$ to have the period $P = 2L$, we may rewrite a_j as

$$a_j = \frac{1}{L}\left[\frac{1}{2}y(0) + \frac{1}{2}y(2L) + \sum_{x=1}^{2L-1} y(x)\cos\frac{\pi}{L}jx\right]$$

and this is the trapezoidal rule approximation to the Fourier coefficient

$$\alpha_j = \frac{1}{\pi}\int_0^{2\pi} y(t)\cos jt\,dt = \frac{1}{L}\int_0^{2L} y(x)\cos\frac{\pi}{L}jx\,dx$$

Similar results hold for b_j and β_j and for the coefficients in Problem 24.2. Since the trapezoidal rule converges to the integral for L becoming infinite, we see that the collocation coefficients converge upon the Fourier coefficients. (Here we may fix the period at 2π for convenience.) For an analogy with Chebyshev polynomials see Problems 21.53 to 21.55.

LEAST SQUARES, CONTINUOUS DATA

24.20. Determine the coefficients A_k and B_k so that the integral

$$I = \int_0^{2\pi} [y(t) - T_M(t)]^2\,dt$$

will be a minimum where $T_M(t) = \frac{1}{2}A_0 + \sum_{k=1}^{M} (A_k \cos kt + B_k \sin kt)$.

More or less as in Problem 24.6, we first find

$$y(t) - T_M(t) = \frac{1}{2}(\alpha_0 - A_0) + \sum_{k=1}^{M} [(\alpha_k - A_k)\cos kt + (\beta_k - B_k)\sin kt] + \sum_{k=M+1}^{\infty} (\alpha_k \cos kt + \beta_k \sin kt)$$

and then square, integrate, and use the orthogonality conditions to get

$$I = \frac{\pi}{2}(\alpha_0 - A)^2 + \pi\sum_{k=1}^{M} [(\alpha_k - A_k)^2 + (\beta_k - B_k)^2] + \pi\sum_{k=M+1}^{\infty} (\alpha_k^2 + \beta_k^2)$$

For a minimum we choose all $A_k = \alpha_k$, $B_k = \beta_k$ so that

$$I_{\min} = \pi \sum_{k=M+1}^{\infty} (\alpha_k^2 + \beta_k^2)$$

Again we have the important result that truncation of the Fourier series at $k = M$ produces the least-squares sum $T_M(t)$. (Once again this is a special case of Problem 21.8.) The minimum integral may be rewritten as

$$I_{\min} = \int_0^{2\pi} [y(t)]^2\, dt - \frac{1}{2}\pi\alpha_0^2 - \pi \sum_{k=1}^{M} (\alpha_k^2 + \beta_k^2)$$

As M increases, this diminishes; and it is proved in the theory of Fourier series that $I_{\min}$ tends to zero for M becoming infinite. This is called *convergence in the mean.*

24.21. Find the least-squares sum with $M = 1$ for the function $y(t)$ of Problem 24.15.

Truncation brings $T_1(t) = \pi/2 - (4/\pi)\cos t$. This function is shown dashed in Fig. 24-1. Notice that it smooths the corners of $y(t)$.

SMOOTHING BY FOURIER ANALYSIS

24.22. What is the basis of the Fourier analysis method for smoothing data?

If we think of given numerical data as consisting of the true values of a function with random errors superposed, the true functions being relatively smooth and the superposed errors quite unsmooth, then the examples in Problems 24.15 to 24.17 suggest a way of partially separating functions from error. Since the true function is smooth, its Fourier coefficients will decrease quickly. But the unsmoothness of the error suggests that its Fourier coefficients may decrease very slowly, if at all. The combined series will consist almost entirely of error, therefore, beyond a certain place. If we simply truncate the series at the right place, then we are discarding mostly error. There will still be error contributions in the terms retained. Since truncation produces a least-squares approximation, we may also view this method as least-squares smoothing.

24.23. Apply the method of the previous problem to the following data:

x	0	1	2	3	4	5	6	7	8	9	10
y	0	4.3	8.5	10.5	16.0	19.0	21.1	24.9	25.9	26.3	27.8

x	11	12	13	14	15	16	17	18	19	20
y	30.0	30.4	30.6	26.8	25.7	21.8	18.4	12.7	7.1	0

Assuming the function to be truly zero at both ends, we may suppose it extended to an odd function of period $P = 40$. Such a function will even have a continuous first derivative, which helps to speed convergence of Fourier series. Using the formulas of Problem 24.9, we now compute the b_j.

j	1	2	3	4	5	6	7	8	9	10
b_j	30.04	−3.58	1.35	−.13	−.14	−.43	.46	.24	−.19	.04

j	11	12	13	14	15	16	17	18	19	20
b_j	.34	.19	.20	−.12	−.36	−.18	−.05	−.37	.27	

The rapid decrease is apparent and we may take all b_j beyond the first three or four to be largely error effects. If four terms are used, we have the trigonometric sum

$$T(x) = 30.04 \sin \frac{\pi x}{20} - 3.58 \sin \frac{2\pi x}{20} + 1.35 \sin \frac{3\pi x}{20} - .13 \sin \frac{4\pi x}{20}$$

The values of this sum may be compared with the original data, which were actually values of $y(x) = x(400 - x^2)/100$ contaminated by artificially introduced random errors. (See Table 24.1). The RMS error of the given data was 1.06 and of the smoothed data .80.

Table 24.1

x	Given	Correct	Smoothed	x	Given	Correct	Smoothed
1	4.3	4.0	4.1	11	30.0	30.7	29.5
2	8.5	7.9	8.1	12	30.4	30.7	29.8
3	10.5	11.7	11.9	13	30.6	30.0	29.3
4	16.0	15.6	15.5	14	26.8	28.6	28.0
5	19.0	18.7	18.6	15	25.7	26.2	25.8
6	21.1	22.7	21.4	16	21.8	23.0	22.4
7	24.9	24.6	23.8	17	18.4	18.9	18.0
8	25.9	26.9	25.8	18	12.7	13.7	12.6
9	26.3	28.7	27.4	19	7.1	7.4	6.5
10	27.8	30.0	28.7	20			

24.24. Approximate the derivative $y'(x) = (400 - 3x^2)/100$ of the function in the preceding problem on the basis of the same given data.

First we shall apply the formula

$$y'(x) \simeq \frac{1}{10}[-2y(x-2) - y(x-1) + y(x-1) + 2y(x+2)]$$

derived earlier from the least-squares parabola for the five arguments $x - 2, \ldots, x + 2$. With similar formulas for the four end arguments, the results form the second column of Table 24.2. Using this local least-squares parabola already amounts to local smoothing of the original x, y data. We now attempt further overall smoothing by the Fourier method. Since the derivative of an odd function is even, the formula of Problem 24.11 is appropriate.

$$a_j = \frac{1}{20}[y'(0) + y'(20) \cos j\pi] + \frac{1}{10} \sum_{x=1}^{19} y'(x) \cos \frac{\pi}{20} jx$$

These coefficients may be computed to be

j	0	1	2	3	4	5	6	7	8	9	10
a_j	0	4.81	−1.05	.71	−.05	.05	−.20	.33	.15	.00	.06

j	11	12	13	14	15	16	17	18	19	20
a_j	.06	.06	−.03	.11	.06	.14	−.04	.16	−.09	.10

Again the sharp drop is noticeable. Neglecting all terms beyond $j = 4$, we have

$$y'(x) \simeq 4.81 \cos \frac{\pi x}{20} - 1.05 \cos \frac{2\pi x}{20} + .71 \cos \frac{3\pi x}{20} - .05 \cos \frac{4\pi x}{20}$$

Computing this for $x = 0, \ldots, 20$ produces the third column of Table 24.2. The last column gives the correct values. The RMS error in column 2, after local smoothing by a least-squares parabola, is .54, while the RMS error in column 3, after additional Fourier smoothing, is .39.

Table 24.2

x	Local	Fourier	Correct	x	Local	Fourier	Correct
0	5.3	4.4	4.0	11	1.1	.5	.4
1	4.1	4.4	4.0	12	−.1	−.1	−.3
2	3.8	4.1	3.9	13	−1.2	−.9	−1.1
3	3.7	3.8	3.7	14	−2.2	−1.8	−1.9
4	3.4	3.4	3.5	15	−2.9	−2.9	−2.8
5	3.4	3.0	3.2	16	−3.6	−4.0	−3.7
6	2.6	2.5	2.9	17	−4.6	−5.0	−4.7
7	1.9	2.1	2.5	18	−5.5	−5.8	−5.7
8	1.5	1.8	2.1	19	−7.1	−6.4	−6.8
9	1.2	1.4	1.6	20	−6.4	−6.6	−8.0
10	1.3	1.0	1.0				

COMPLEX FORMS

24.25. Prove the following orthogonality property of the functions e^{ijx}, e^{ikx} for j and k integers. The overbar denotes a complex conjugate.

$$\int_0^{2\pi} \overline{e^{ijx}} e^{ikx}\, dx = \begin{cases} 0 & \text{if } k \neq j \\ 2\pi & \text{if } k = j \end{cases}$$

The proof is elementary, the integral reducing at once to

$$\int_0^{2\pi} e^{i(k-j)}\, dx = \frac{1}{i(k-j)} e^{i(k-j)x} \Big|_0^{2\pi}$$

for $k \neq j$. But this is equal to one at both limits, hence zero. For $k = j$, the left side above is clearly 2π.

24.26. Derive the formula for Fourier coefficients in complex form.

The proof takes a familiar path. The Fourier series is

$$f(x) = \sum_{j=-\infty}^{\infty} f_j e^{ijx}$$

Multiplying by e^{ikx} and integrating brings

$$\int_0^{2\pi} f(x) e^{ikx}\, dx = \sum_{j=-\infty}^{\infty} f_j e^{ikx} e^{ijx}\, dx$$

and since all terms on the right vanish by orthogonality except the one for $j = k$, the required result is found.

$$f_k = \frac{1}{2\pi} \int_0^{2\pi} f(x) e^{-ikx}\, dx$$

24.27. Show that the functions e^{ijx_n}, e^{ikx_n} are orthogonal in the following sense.

$$\sum_{n=0}^{N-1} \overline{e^{ijx_n}} e^{ikx_n} = \begin{cases} N & \text{if } k = j \\ 0 & \text{if } k \neq j \end{cases}$$

Here as before $x_n = 2\pi n/N$.

We will find a geometric sum with ratio $r = e^{i(k-j)2\pi/N}$.

$$\sum_{n=0}^{N-1} \overline{e^{ijx_n}} e^{ikx_n} = \sum_{n=0}^{N-1} e^{i(k-j)x_n} = e^{i(k-j)x_0}(1 + r + r^2 + \cdots + r^{N-1})$$

For $j = k$ we have $r = 1$ and the sum is N. Otherwise the sum of the powers of r is $(1 - r^N)/(1 - r)$ by a familiar formula. But r^N is $e^{2\pi i(k-j)}$ which is 1, making the numerator zero and establishing the orthogonality.

24.28. Show that if $N = 2l + 1$ then the trigonometric sum

$$\sum_{j=-l}^{l} d_j e^{ijx}$$

must have coefficients $d_j = f_j^*$ if it is to collocate with the function $f(x)$ at $x_n = 2\pi n/N$.

Assume that collocation occurs, multiply by $\overline{e^{ikx_n}}$ and sum.

$$\sum_{n=0}^{N-1} f(x_n)\overline{e^{ikx_n}} = \sum_{n=0}^{N-1} \overline{e^{ikx_n}} \sum_{j=-l}^{l} d_j e^{ijx_n} = \sum_{j=-l}^{l} d_j \sum_{n=0}^{N-1} \overline{e^{ikx_n}} e^{ijx_n}$$

Again all terms on the right are zero except one, for $j = k$, and we have

$$\sum_{n=0}^{N-1} f(x_n) e^{ikx_n} = d_k(N) = f_k^* N$$

24.29. How are the coefficients f_j^* related to discrete Fourier transforms?

Let V be the vector with components $f(x_0), \ldots, f(x_{N-1})$. For $N = 2l + 1$ this makes V $(2l + 1)$-dimensional, as is the vector of coefficients f_j^* for the trigonometric sum

$$\sum_{j=-l}^{l} f_j^* e^{ijx}$$

in which

$$f_j^* = \frac{1}{N} \sum_{n=0}^{N-1} f(x_n) e^{-ijx_n}$$

for $j = -l$ to $j = l$. Comparing with

$$v_j^T = \sum_{n=0}^{N-1} v_n \omega_N^{jn} = \sum_{n=0}^{N-1} f(x_n) e^{-ijx_n}$$

where $x_n = 2\pi n/N$, and $j = 0$ to $j = N - 1$, the match is conspicuous. We do have one problem: the ranges of validity do not coincide. But we may deduce that where the ranges overlap, from $j = 0$ to $j = l$,

$$v_j^T = N f_j^* \qquad j = 0, \ldots, l$$

Now we observe that

$$v_{j+N}^T = \sum_{n=0}^{N-1} f(x_n) e^{-i(j+N)x_n} = \sum_{n=0}^{N-1} f(x_n) e^{-ijx_n}$$

for $j + N = 0, \ldots, N - 1$ or $j = -1, \ldots, -N$. Once again we have a match, this time for $j = -1$ to $j = -l$.

$$v_{j+N} = N f_j^* \qquad j = -l, \ldots, -1$$

Apart from the factor $1/N$ the components v_j^T do, therefore, match the coefficients f_j^*, though in a slightly scrambled order. Taking the v_j^T in their natural order v_0^T to v_{2l}^T it is easy to verify that the order of the coefficients will be this.

$$f_0^*, \ldots, f_l^* \qquad f_{-l}^*, \ldots, f_{-1}^*$$

24.30. Work through the details of the preceding problem for the simple example $V = (1, 0, -1)$.

Here $N = 3$ and $l = 1$.

$$3f_j^* = \sum_{n=0}^{2} f(x_n)e^{-ijx_n} = \sum_{n=0}^{2} f(x_n)\omega_3^{jn} = 1 - \omega_3^{2j}$$

This makes $\qquad 3f_{-1}^* = 1 - \omega_3 \qquad 3f_0^* = 0 \qquad 3f_1^* = 1 - \omega_3^2$

and we have the three coefficients directly. Turning to the transform,

$$v_j^T = \sum_{n=0}^{2} f(x_n)\omega_3^{jn} = 1 - \omega_3^{2j}$$

we find $\qquad v_0^T = 0 \qquad v_1^T = 1 - \omega_3^2 \qquad v_3^T = 1 - \omega_3$

and the correspondence discovered in Problem 24.29 is confirmed.

24.31. What is the central idea behind the Fast Fourier Transform?

When N is the product of integers, the numbers f_j^* prove to be closely interdependent. This interdependence can be exploited to substantially reduce the amount of computing required to generate these numbers.

24.32. Develop the FFT for the simplest case, when N is the product of two integers t_1 and t_2.

Let $j = j_1 + t_1 j_2$ and $n = n_2 + t_2 n_1$. Then for j_1, $n_1 = 0$ to $t_1 - 1$, and j_2, $n_2 = 0$ to $t_2 - 1$ both j and n run their required ranges 0 to $N - 1$. Now

$$\omega^{(j_1+t_1j_2)(n_2+t_2n_1)} = \omega_N^{j_1t_2n_1 + j_1n_2 + t_1j_2n_2}$$

since $t_1t_2 = N$ and $\omega_N^N = 1$. The transform can then be written as a double sum

$$v_j^T = \sum_{n_2=0}^{t_2-1} \sum_{n_1=0}^{t_1-1} v_n \omega_N^{j_1t_2n_1} \omega_N^{j_1n_2 + t_1j_2n_2}$$

This can also be arranged in a two-step algorithm.

$$F_1(j_1, n_2) = \sum_{n_1=0}^{t_1-1} v_n \omega_N^{j_1t_2n_1}$$

$$v_j^T = F_2(j_1, j_2) = \sum_{n_2=0}^{t_2-1} F_1(j_1, n_2)\omega_N^{j_1n_2 + t_1j_2n_2}$$

24.33. What is the gain in computing efficiency if the FFT of Problem 24.32 is used? In other words, just how fast is the Fast Fourier Transform?

To compute F_1 there are t_1 terms to be processed; to compute F_2 there are t_2. The total is $t_1 + t_2$. This must be done for each (j_1, n_2) and (j_1, j_2) pair, or N pairs. The final count is thus $N(t_1 + t_2)$ terms processed. The original form of the transform

$$v_j^T = \sum_{n=0}^{N-1} v_n \omega_N^{jn}$$

processed N terms for each j, a total of N^2 terms. The gain in efficiency, if measured by this standard, is thus

$$\frac{t_1 + t_2}{N}$$

and depends very much upon N. For a small data set, say $N = 12 = 3 \times 4$, the FFT will need about $\frac{7}{12}$ the computing time of a direct approach. This is hardly significant but points out the direction of things to come.

24.34. Run the FFT of Problem 24.32 for the following vector:

n	0	1	2	3	4	5		
v_n	0	1	1	0	−1	−1	0	...

The small scale of the problem, $N = 6$, makes it easy to see all the detail. Here $N = t_1 t_2 = 2 \times 3$ so we first find the F_1 values from

$$F_1(j_1, n_2) = \sum_{n_1=0}^{1} v_n \omega_6^{3j_1 n_1} \qquad n = n_2 + 3n_1$$

and they prove to be the following, with $\omega = \omega_6$.

$$F_1(0, 0) = v_0 + v_3 = 0 \qquad F_1(1, 0) = v_0 - v_3 = 0$$
$$F_1(0, 1) = v_1 + v_4 = 0 \qquad F_1(1, 1) = v_1 - v_4 = 2$$
$$F_1(0, 2) = v_2 + v_5 = 0 \qquad F_1(1, 2) = v_2 - v_5 = 2$$

Then

$$v_j^T = F_2(j_1, j_2) = \sum_{n_2=0}^{2} F_1(j_1, n_2)\omega^{j_1 n_2 + t_1 j_2 n_2}$$

leading to, since $j = j_1 + 2j_2$

$$v_0^T = F_2(0, 0) = v_0 + v_1 + v_2 + v_3 + v_4 + v_5 = 0$$
$$v_1^T = F_2(1, 0) = F_1(1, 0) + F_1(1, 1)\omega + F_1(1, 2)\omega^2 = 2\omega + 2\omega^2 = 2\sqrt{3}\, i$$
$$v_2^T = F_2(0, 1) = F_1(0, 0) + F_1(0, 1)\omega^2 + F_2(0, 2)\omega^4 = 0$$
$$v_3^T = F_2(1, 1) = F_1(1, 0) + F_1(1, 1)\omega^3 + F_1(1, 2)\omega^6 = 0$$

and similarly

$$v_4^T = F_2(0, 2) = 0$$
$$v_5^T = F_2(1, 2) = -2\sqrt{3}\, i$$

Note that Nt_1 terms were involved in computing the F_1 values and Nt_2 terms in getting F_2, a total of $12 + 18 = 30$ terms. The direct computation would have used 36 and would confirm the results just found. Also note the order of processing j_1, j_2 pairs. In programming language, the j_2 loop is external to the j_1 loop.

24.35. Extend the FFT of Problem 24.32 to the case $N = t_1 t_2 t_3$.

The details will suggest the way to generalization for still longer products. Let

$$j = j_1 + t_1 j_2 + t_1 t_2 j_3 \qquad n = n_3 + t_3 n_2 + t_3 t_2 n_1$$

and observe that of the nine possible power terms in

$$\omega_N^{(j_1 + t_1 j_2 + t_1 t_2 j_3)(n_3 + t_3 n_2 + t_3 t_2 n_1)}$$

three will contain the product $t_1 t_2 t_3$ and may be neglected since $\omega_N^N = 1$. The remaining six may be grouped as follows in the transform,

$$v_j^T = \sum_{n_3=0}^{t_3-1} \left[\sum_{n_2=0}^{t_2-1} \left(\sum_{n_1=0}^{t_1-1} v_n \omega_N^{j_1 t_3 t_2 n_1} \right) \omega_N^{(j_1+t_1 j_2) t_3 n_2} \right] \omega_N^{(j_1+t_1 j_2+t_1 t_2 j_3) n_3}$$

with n_1 appearing only in the inner sum and n_2 not appearing in the outer. As before, this triple sum can be expressed as an algorithm, this time having three steps.

$$F_1(j_1, n_2, n_3) = \sum_{n_1=0}^{t_1-1} v_n \omega_N^{j_1 t_3 t_2 n_1}$$

$$F_2(j_1, j_2, n_3) = \sum_{n_2=0}^{t_2-1} F_1(j_1, n_2, n_3) \omega_N^{(j_1+t_1 j_2) t_3 n_2}$$

$$v_j^T = F_3(j_1, j_2, j_3) = \sum_{n_3=0}^{t_3-1} F_2(j_1, j_2, n_3) \omega_N^{(j_1+t_1 j_2+t_1 t_2 j_3) n_3}$$

This is the required FFT.

24.36. Estimate the saving in computing time if this algorithm is used.

At each of the three steps the number of triples, such as (j_1, n_2, n_3), that must be processed is $t_1 t_2 t_3 = N$. In the sums we find the number of terms to be t_1, t_2, t_3 in turn. This makes a total of $N(t_1 + t_2 + t_3)$ terms altogether. The transform as defined still uses N^2 terms, so the efficiency of the FFT may be estimated as

$$\frac{t_1 + t_2 + t_3}{N}$$

If, for instance, $N = 1000 = 10 \times 10 \times 10$, then only 3 percent of the original 1,000,000 terms are needed.

24.37. Run the FFT algorithm of Problem 24.35 manually for this input vector.

n	0	1	2	3	4	5	6	7
v_n	1	$1+i$	i	$i-1$	-1	$-1-i$	$-i$	$1-i$

We have $N = 8 = 2 \times 2 \times 2$, making $j = j_1 + 2j_2 + 4j_3$ and $n = n_3 + 2n_2 + 4n_1$. The formula for F_1 is then

$$F_1(j_1, n_2, n_3) = \sum_{n_1=0}^{1} v_n \omega_8^{4 j_1 n_1}$$

and we have

$$\begin{aligned} F_1(0, 0, 0) &= v_0 + v_4 = 0 & F_1(1, 0, 0) &= v_0 + v_4\omega^4 = 2 \\ F_1(0, 0, 1) &= v_1 + v_5 = 0 & F_1(1, 0, 1) &= v_1 + v_5\omega^4 = 2 + 2i \\ F_1(0, 1, 0) &= v_2 + v_6 = 0 & F_1(1, 1, 0) &= v_2 + v_6\omega^4 = 2i \\ F_1(0, 1, 1) &= v_3 + v_7 = 0 & F_1(1, 1, 1) &= v_3 + v_7\omega^4 = 2i - 2 \end{aligned}$$

with ω_8 abbreviated to ω. Notice the $Nt_1 = 8 \times 2$ terms used. Next we use

$$F_2(j_1, j_2, n_3) = \sum_{n_2=0}^{1} F_1(j_1, n_2, n_3) \omega^{2(j_1+2j_2) n_2}$$

to compute

$$\begin{aligned} F_2(0, 0, 0) &= 0 & F_2(1, 0, 0) &= F_1(1, 0, 0) + F_1(1, 1, 0)\omega^2 = 4 \\ F_2(0, 0, 1) &= 0 & F_2(1, 0, 1) &= F_1(1, 0, 1) + F_1(1, 1, 1)\omega^2 = 4 + 4i \\ F_2(0, 1, 0) &= 0 & F_2(1, 1, 0) &= F_1(1, 0, 0) + F_1(1, 1, 0)\omega^6 = 0 \\ F_2(0, 1, 1) &= 0 & F_2(1, 1, 1) &= F_1(1, 0, 1) + F_1(1, 1, 1)\omega^6 = 0 \end{aligned}$$

and finally

$$v_j^T = F_3(j_1, j_2, j_3) = \sum_{n_3=0}^{1} F_2(j_1, j_2, n_3)\omega^{jn_3}$$

to get the transform

$$v_0^T = F_3(0,0,0) = F_2(0,0,0) + F_2(0,0,1) = 0$$
$$v_1^T = F_3(1,0,0) = F_2(1,0,0) + F_2(1,0,1)\omega = 4 + 4\sqrt{2}$$
$$v_2^T = F_3(0,1,0) = F_2(0,1,0) + F_2(0,1,1)\omega^2 = 0$$
$$v_3^T = F_3(1,1,0) = F_2(1,1,0) + F_2(1,1,1)\omega^3 = 0$$
$$v_4^T = F_3(0,0,1) = F_2(0,0,0) + F_2(0,0,1)\omega^4 = 0$$
$$v_5^T = F_3(1,0,1) = F_2(1,0,0) + F_2(1,0,1)\omega^5 = 4 - 4\sqrt{2}$$
$$v_6^T = F_3(0,1,1) = F_2(0,1,0) + F_2(0,1,1)\omega^6 = 0$$
$$v_7^T = F_3(1,1,1) = F_2(1,1,0) + F_2(1,1,1)\omega^7 = 0$$

A total of $N(t_1 + t_2 + t_3) = 48$ terms have been processed, only a slight saving from $N^2 = 64$ because of the problem's small scale.

24.38. The inverse discrete transform may be defined by

$$u_k^{-T} = \frac{1}{N}\sum_{j=0}^{N-1} u_j\omega^{-jk} = \frac{1}{N}\sum_{j=0}^{N-1} u_j e^{ikx_j}$$

Show that this definition does give an inverse relationship by inserting $u_j = v_j^T$ and discovering that $u_k^{-T} = v_k$. That is, the components of the original vector V are regained.

It may be useful to first rewrite the result of Problem 24.31 using

$$\omega^{jn} = e^{-ijx_n}$$

to obtain

$$\sum_{n=0}^{N-1} \omega^{jn}\omega^{-kn} = \begin{cases} N & \text{if } k = j \\ 0 & \text{if } k \neq j \end{cases}$$

for j, k in the interval $(0, N-1)$. Now

$$\frac{1}{N}\sum_{j=0}^{N-1} v_j^T\omega^{-jk} = \frac{1}{N}\sum_{j=0}^{N-1}\sum_{n=0}^{N-1} v_n\omega^{jn}\omega^{-jk} = \frac{1}{N}\sum_{n=0}^{N-1} v_n \sum_{j=0}^{N-1}\omega^{(n-k)j}$$

and the last sum being zero, unless n takes the value k, we soon have the anticipated v_k.

24.39. Invert the transform found in Problem 24.37.

The FFT could be used, but in view of the large number of zero components this is a good chance to proceed directly.

$$8u_0^{-T} = \sum_{j=0}^{7} v_j^T = 8 \qquad u_0^{-T} = 1 = v_0$$

$$8u_1^{-T} = \sum_{j=0}^{7} v_j^T\omega^{-j} = (4 + 4\sqrt{2})\omega^{-1} + (4 - 4\sqrt{2})\omega^{-5}$$
$$= 8(1+i) \qquad u_1^{-T} = 1 + i = v_2$$

$$8u_2^{-T} = \sum_{j=0}^{7} v_j^T\omega^{-2j} = (4 + 4\sqrt{2})\omega^{-2} + (4 - 4\sqrt{2})\omega^{-10}$$
$$= 8i \qquad u_2^{-T} = i = v_3$$

$$\vdots$$

The remaining components may be verified as Problem 24.63.

Supplementary Problems

24.40. Apply the method of Problem 24.2 to the data below.

x	0	1	2	3	4
y	0	1	2	1	0

24.41. Derive the coefficient formulas of Problem 24.5.

24.42. Apply the method of Problem 24.5 to the following data:

x	0	1	2	3	4	5
y	0	1	2	2	1	0

24.43. Use the result of Problem 24.6 to obtain least-squares sums $T_0(x)$ and $T_1(x)$ for the data of Problem 24.40.

24.44. Imitate the argument of Problem 24.6 to obtain a somewhat similar result for the case of an even number of x arguments.

24.45. Apply the preceding problem to the data of Problem 24.42.

24.46. Extend the data of Problem 24.40 to an odd function of period 8. Find a sum of sines to represent this function.

24.47. Extend the data of Problem 24.40 to an even function of period 8. Find a sum of cosines to represent this function.

24.48. Show that the Fourier series for $y(x) = |\sin x|$, the "fully rectified" sine wave, is

$$y(x) = \frac{4}{\pi}\left(\frac{1}{2} - \frac{\cos 2x}{1 \cdot 3} - \frac{\cos 4x}{3 \cdot 5} - \frac{\cos 6x}{5 \cdot 7} - \cdots\right)$$

24.49. Show that the Fourier series for $y(x) = x^2$ for x between $-\pi$ and π, and of period 2π, is

$$y(x) = \frac{\pi^2}{3} - 4\sum_{k=1}^{\infty} \frac{(-1)^{k-1}\cos kx}{k^2}$$

Use the result to evaluate the series $\sum_{k=1}^{\infty} (-1)^{k-1}/k^2$ and $\sum_{k=1}^{\infty} 1/k^2$.

24.50. Use the Fourier series of Problem 24.15 to evaluate $\sum_{k=1}^{\infty} 1/(2k-1)^2$.

24.51. Use the Fourier series of Problem 24.16 to show that $\pi/4 = 1 - \frac{1}{3} + \frac{1}{5} - \frac{1}{7} + \cdots$.

24.52. Use the series of Problem 24.17 to evaluate $1 - 1/3^3 + 1/5^3 - 1/7^3 + \cdots$.

24.53. What is the four-term least-squares trigonometric approximation to the function of Problem 24.48? What is the two-term least-squares approximation?

24.54. Apply Fourier smoothing to the following data, assuming that the end values are actually zero and extending the function as an odd function. Also try other methods of smoothing, or combinations of methods. Compare results with the correct values $y(x) = x(1-x)$ from which the given data were obtained by the addition of random errors of up to 20 percent. The arguments are $x = 0(.05)1$.

.00, .06, .10, .11, .14, .22, .22, .27, .28, .21, .22, .27, .21, .20, .19, .21, .19, .12, .08, .04, 00

24.55. Verify the coefficient relationships

$$a_j = c_j + c_{-j} \qquad b_j = i(c_j - c_{-j})$$

given in the introductory section, and the inverse relations

$$c_j = \frac{a_j - ib_j}{2} \qquad c_{-j} = \frac{a_j + ib_j}{2}$$

Deduce that if the a_j, b_j are real, then c_j and c_{-j} must be complex conjugates. Recalling that for the collocation trigonometric polynomial, we have $c_j = f_j^*$, and assuming a_j, b_j, and $f(x)$ all real, show that

$$a_j = 2\,\mathrm{Re}(f_j^*) = \frac{2}{N}\sum_{n=0}^{N-1} f(x_n)\cos jx_n$$

$$b_j = -2\,\mathrm{Im}(f_j^*) = \frac{2}{N}\sum_{n=0}^{N-1} f(x_n)\sin jx_n$$

24.56. Proceed as in Problem 24.30 using $V = (1, -1, 0)$.

24.57. Proceed as in Problem 24.34 using this vector V:

n	0	1	2	3	4	5
v_n	0	0	1	1	1	0

24.58. Proceed as in Problem 24.37 using this vector V:

n	0	1	2	3	4	5	6	7
v_n	1	$1+i$	0	$1-i$	0	$1+i$	0	$1-i$

24.59. Confirm the result of Problem 24.58 by applying the original transform

$$v_j^T = \sum_{n=0}^{N-1} v_n \omega_N^{jn}$$

24.60. Using elementary calculus show that if $t_1 t_2 = N$, then the minimum of $t_1 + t_2$ occurs for $t_1 = t_2$. Extend this result to the case $t_1 t_2 t_3 = N$. What is the implication for the FFT?

24.61. Invert the transform found in Problem 24.30.

24.62. Apply the FFT of Problem 24.32 to invert the output of Problem 24.34.

n	0	1	2	3	4	5
v_n^T	0	$2\sqrt{3}\,i$	0	0	0	$-2\sqrt{3}\,i$

24.63. Complete the inversion begun in Problem 24.39.

24.64. Make the same inversion using an FFT.

Chapter 25

Nonlinear Algebra

ROOTS OF EQUATIONS

The problem treated in this chapter is the ancient problem of finding roots of equations or of systems of equations. The long list of available methods shows the long history of this problem and its continuing importance. Which method to use depends upon whether one needs all the roots of a particular equation or only a few, whether the roots are real or complex, simple or multiple, whether one has a ready first approximation or not, and so on.

1. The **iterative method** solves $x = F(x)$ by the recursion

$$x_n = F(x_{n-1})$$

and converges to a root if $|F'(x)| \leqq L < 1$. The error $e_n = r - x_n$, where r is the exact root, has the property

$$e_n \simeq F'(r)e_{n-1}$$

so that each iteration reduces the error by a factor near $F'(r)$. If $F'(r)$ is near 1 this is slow convergence.

2. The Δ^2 **process can accelerate convergence** under some circumstances. It consists of the approximation

$$r \simeq x_{n+2} - \frac{(\Delta x_{n+1})^2}{\Delta^2 x_n}$$

which may be derived from the error property given above.

3. The **Newton method** obtains successive approximations

$$x_n = x_{n-1} - \frac{f(x_{n-1})}{f'(x_{n-1})}$$

to a root of $f(x) = 0$ and is unquestionably a very popular algorithm. If $f'(x)$ is complicated, the previous iterative method may be preferable, but Newton's method converges much more rapidly and usually gets the nod. The error e_n here satisfies

$$e_n \simeq -\frac{f''(r)}{2f'(r)} e_{n-1}^2$$

This is known as **quadratic convergence**, each error roughly proportional to the square of the previous error. The number of correct digits almost doubles with each iteration.

The square root iteration

$$x_n = \frac{1}{2}\left(x_{n-1} + \frac{Q}{x_{n-1}}\right)$$

is a special case of Newton's method, corresponding to $f(x) = x^2 - Q$. It converges quadratically to the positive square root of Q, for $Q > 0$.

The more general root-finding formula

$$x_n = x_{n-1} - \frac{x_{n-1}^p - Q}{px_{n-1}^{p-1}}$$

is also a special case of Newton's method. It produces a pth root of Q.

4. **Interpolation methods** use two or more approximations, usually some too small and some

too large, to obtain improved approximations to a root by use of collocation polynomials. The most ancient of these is based on linear interpolation between two previous approximations. It is called *regula falsi* and solves $f(x)=0$ by the iteration

$$x_n = x_{n-1} - \frac{(x_{n-1}-x_{n-2})f(x_{n-1})}{f(x_{n-1})-f(x_{n-2})}$$

The rate of convergence is between those of the previous two methods. A method based on quadratic interpolation between three previous approximations x_0, x_1, x_2 uses the formula

$$x_3 = x_2 - \frac{2C}{B \pm \sqrt{B^2-4AC}}$$

the expressions for A, B, C being given in Problem 25.18.

5. **Bernoulli's method** produces the *dominant root* of a real polynomial equation

$$a_0x^n + a_1x^{n-1} + \cdots + a_n = 0$$

provided a single dominant root exists, by computing a solution sequence of the difference equation

$$a_0x_k + a_1x_{k-1} + \cdots + a_nx_{k-n} = 0$$

and taking $\lim (x_{k+1}/x_k)$. The initial values $x_{-n+1} = \cdots = x_{-1} = 0$, $x_0 = 1$ are usually used. If a complex conjugate pair of roots is dominant, then the solution sequence is still computed, but the formulas

$$r^2 \simeq \frac{x_k^2 - x_{k+1}x_{k-1}}{x_{k-1}^2 - x_kx_{k-2}} \qquad -2r\cos\phi \simeq \frac{x_{k+1}x_{k-2} - x_{k-1}x_k}{x_{k-1}^2 - x_kx_{k-2}}$$

serve to determine the roots as $r_1, r_2 \simeq r(\cos\phi \pm i\sin\phi)$.

6. **Deflation** refers to the process of removing a known root from a polynomial equation, leading to a new equation of lower degree. Coupled with Bernoulli's method, this permits the discovery of next-dominant roots one after another. In practice it is found that continued deflation determines the smaller roots with diminishing accuracy. However, using the results obtained at each step as starting approximations for Newton's method often leads to accurate computation of all the roots.

7. The **quotient-difference algorithm** extends Bernoulli's method and may produce all roots of a polynomial equation, including complex conjugate pairs, simultaneously. It involves computing a table of quotients and differences (resembling a difference table) from which the roots are then deduced. The details are somewhat complicated and may be found in Problems 25.25 to 25.32.

8. **Sturm sequences** offer another historical approach to the real roots of an equation, again producing them more or less simultaneously. A Sturm sequence

$$f_0(x), f_1(x), \ldots, f_n(x)$$

meets five conditions as listed in Problem 25.33. These conditions assure that the number of real zeros of $f_0(x)$ in the interval (a, b) is precisely the difference between the number of sign changes in the sequence $f_0(a), f_1(a), \ldots, f_n(a)$ and the corresponding number in $f_0(b), f_1(b), \ldots, f_n(b)$. By choosing various intervals (a, b) the real zeros can therefore be located. When $f_0(x)$ is a polynomial, a suitable Sturm sequence may be found by using the Euclidean algorithm. Letting $f_1(x) = f_0'(x)$, the rest of the sequence is defined by

$$f_0(x) = f_1(x)L_1(x) - f_2(x)$$
$$f_1(x) = f_2(x)L_2(x) - f_2(x)$$
$$\cdots\cdots\cdots\cdots\cdots\cdots\cdots\cdots$$
$$f_{n-2}(x) = f_{n-1}(x)L_{n-1}(x) - f_n(x)$$

Like the deflation and quotient-difference methods, Sturm sequences can be used to obtain good starting approximations for Newton iterations, which then produce highly accurate roots at great speed.

SYSTEMS OF EQUATIONS AND OPTIMIZATION PROBLEMS

Systems of equations respond to generalizations of many of the previous methods and to other algorithms as well. We choose three.

1. The **iterative method**, for example, solves the pair of equations

$$x = F(x, y) \qquad y = G(x, y)$$

by the formulas $$x_n = F(x_{n-1}, y_{n-1}) \qquad y_n = G(x_{n-1}, y_{n-1})$$

assuming convergence of both the x_n and y_n sequences. Newton's method solves

$$f(x, y) = 0 \qquad g(x, y) = 0$$

through the sequences defined by

$$x_n = x_{n-1} + h_{n-1} \qquad y_n = y_{n-1} + k_{n-1}$$

with h_{n-1} and k_{n-1} determined by

$$f_x(x_{n-1}, y_{n-1})h_{n-1} + f_y(x_{n-1}, y_{n-1})k_{n-1} = -f(x_{n-1}, y_{n-1})$$
$$g_x(x_{n-1}, y_{n-1})h_{n-1} + g_y(x_{n-1}, y_{n-1})k_{n-1} = -g(x_{n-1}, y_{n-1})$$

More generally, the system

$$F(x) = 0$$

in which F, x, and 0 are vectors of n dimensions, may respond to the iteration

$$x^{(n)} = G(x^{(n-1)})$$

obtained by a rearrangement of the original system, with a suitable initial vector $x^{(0)}$. Or the Newton approach can be expressed in a compact vector-matrix form beginning with the Taylor series

$$F(x^{(n-1)} + h) = F(x^{(n-1)}) + J(x^{(n-1)})h + \cdots$$

ignoring the higher-order terms and setting the left side to the zero vector. The result is a linear system for h

$$J(x^{(n-1)})h = -F(x^{(n-1)})$$

which can even be written $$h = -J^{-1}(x^{(n-1)})F(x^{(n-1)})$$

The matrix J is called the **Jacobian** of F and has the elements

$$J_{ij} = \frac{\partial f_i}{\partial x_j}$$

where f_i and x_j are components of F and x. With an accurate initial approximation, and a cooperative F, the error decreases quadratically in the sense

$$\|x - x^{(n)}\| \leqq c\, \|x - x^{(n-1)}\|^2$$

but it must be pointed out that this quadratic convergence can be elusive. Finding sufficiently accurate first approximations is not always easy with systems of equations and Newton approximations sometimes wander about. In some cases it has been found that the shortened step

$$x^{(n)} = x^{(n-1)} + kn \qquad k < 1$$

does better, with k chosen to assure that the norm of F decreases.

$$\|F(x^{(n)})\| < \|F(x^{(n-1)})\|$$

In this way each step improves the situation. The device has been called the **damped Newton method**.

2. **Optimization methods** are based upon the idea that the system $F = 0$, or $f_i = 0$ for $i = 1, \ldots, n$, is solved whenever the function

$$S = f_1^2 + f_2^2 + \cdots + f_n^2$$

is minimized, since the minimum clearly occurs when all the f_i are zero. Direct methods for seeking this minimum, or descent methods, have been developed. For example, the two-dimensional problem (with a familiar change of notation)

$$f(x, y) = 0 \qquad g(x, y) = 0$$

is equivalent to minimizing this sum

$$S(x, y) = f^2 + g^2$$

Beginning at an initial approximation (x_0, y_0), we select the next approximation in the form

$$x_1 = x_0 - tS_{x0} \qquad y_1 = y_0 - tS_{y0}$$

where S_{x0} and S_{y0} are the components of the gradient vector of S at (x_0, y_0). Thus progress is in the direction of steepest descent and the algorithm is known as the **steepest descent algorithm**. The number t may be chosen to minimize S in this direction, though alternatives have been proposed. Similar steps then follow. The method is often used to provide initial approximations to the Newton method.

The above equivalence is, of course, often exploited in the opposite way. To optimize a function $f(x_1, \ldots, x_n)$, one looks for places where the gradient of f is zero

$$\text{grad}\,(f) = (f_1, f_2, \ldots, f_n) = (0, 0, \ldots, 0)$$

Here f_i denotes the partial derivative of f relative to x_i. The optimization is then attempted through the solution of the system of n nonlinear equations.

3. **Bairstow's method** produces complex roots of a real polynomial equation $p(x) = 0$ by applying the Newton method to a related system. More specifically, division of $p(x)$ by a quadratic polynomial suggests the identity

$$p(x) = (x^2 - ux - v)q(x) + r(x)$$

where $r(x)$ is a linear remainder

$$r(x) = b_{n-1}(u, v)(x - u) + b_n(u, v)$$

The quadratic divisor will be a factor of $p(x)$ if we can choose u and v so that

$$b_{n-1}(u, v) = 0 \qquad b_n(u, v) = 0$$

This is the system to which Newton's method is now applied. Once u and v are known, a complex pair of roots may be found by solving

$$x^2 - ux - v = 0$$

Solved Problems

THE ITERATIVE METHOD

25.1. Prove that if r is a root of $f(x)=0$ and if this equation is rewritten in the form $x=F(x)$ in such a way that $|F'(x)| \leqq L < 1$ in an interval I centered at $x=r$, then the sequence $x_n = F(x_{n-1})$ with x_0 arbitrary but in the interval I has $\lim x_n = r$.

First we find

$$|F(x)-F(y)| = |F'(\xi)(x-y)| \leqq L\,|x-y|$$

provided both x and y are close to r. Actually it is this Lipschitz condition rather than the more restrictive condition on $F'(x)$ which we need. Now

$$|x_n - r| = |F(x_{n-1}) - F(r)| \leqq L\,|x_{n-1} - r|$$

so that, since $L<1$, each approximation is at least as good as its predecessor. This guarantees that all our approximations are in the interval I, so that nothing interrupts the algorithm. Applying the last inequality n times, we have

$$|x_n - r| \leqq L^n\,|x_0 - r|$$

and since $L<1$, $\lim x_n = r$.

The convergence is illustrated in Fig. 25-1. Note that choosing $F(x_{n-1})$ as the next x_n amounts to following one of the horizontal line segments over to the line $y=x$. Notice also that in Fig. 25-2 the case $|F'(x)|>1$ leads to divergence.

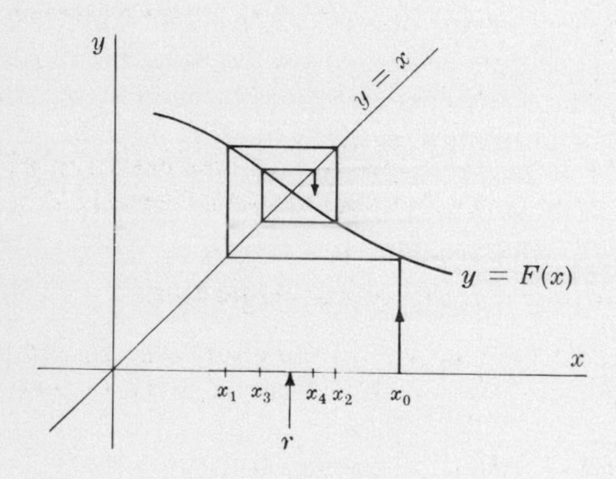

Fig. 25-1

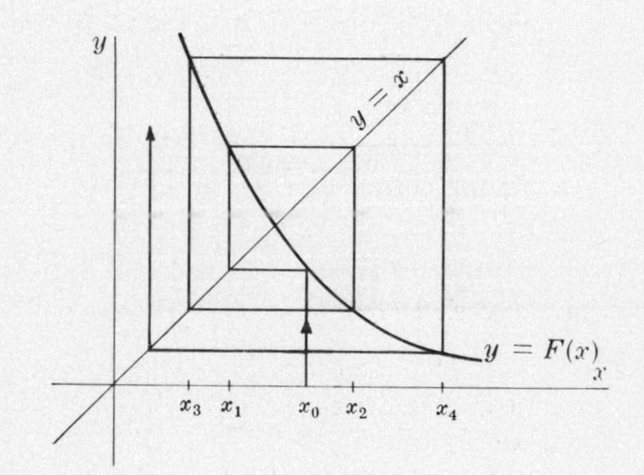

Fig. 25-2

25.2. In the year 1225 Leonardo of Pisa studied the equation

$$f(x) = x^3 + 2x^2 + 10x - 20 = 0$$

and produced $x = 1.368,808,107$. Nobody knows by what method Leonardo found this value but it is a remarkable result for his time. Apply the method of Problem 25.1 to obtain this result.

The equation can be put into the form $x=F(x)$ in many ways. We take $x = F(x) = 20/(x^2+2x+10)$ which suggests the iteration

$$x_n = \frac{20}{x_{n-1}^2 + 2x_{n-1} + 10}$$

With $x_0 = 1$ we find $x_1 = \frac{20}{13} \simeq 1.538461538$. Continuing the iteration produces the sequence of Table 25.1. Sure enough, on the twenty-fourth round Leonardo's value appears.

Table 25.1

n	x_n	n	x_n
1	1.538461538	13	1.368817874
2	1.295019157	14	1.368803773
3	1.401825309	15	1.368810031
4	1.354209390	16	1.368807254
5	1.375298092	17	1.368808486
6	1.365929788	18	1.368807940
7	1.370086003	19	1.368808181
8	1.368241023	20	1.368808075
9	1.369059812	21	1.368808122
10	1.368696397	22	1.368808101
11	1.368857688	23	1.368808110
12	1.368786102	24	1.368808107

25.3. Why is the convergence of the algorithm of the previous problem so slow?

The rate of convergence may be estimated from the relation

$$e_n = r - x_n = F(r) - F(x_{n-1}) = F'(\xi)(r - x_{n-1}) = F'(\xi)e_{n-1}$$

which compares the nth error e_n with the preceding error. As n increases we may take $F'(r)$ as an approximation to $F'(\xi)$, assuming the existence of this derivative. Then $e_n \simeq F'(r)e_{n-1}$. In our example,

$$F'(r) = -\frac{40(r+1)}{(r^2+2r+10)^2} \simeq -.44$$

making each error about $-.44$ times the one before it. This suggests that two or three iterations will be required for each new correct decimal place, and this is what the algorithm has actually achieved.

25.4. Apply the idea of extrapolation to the limit to accelerate the previous algorithm.

This idea may be used whenever information about the character of the error in an algorithm is available. Here we have the approximation $e_n \simeq F'(r)e_{n-1}$. Without knowledge of $F'(r)$ we may still write

$$r - x_{n+1} \simeq F'(r)(r - x_n)$$
$$r - x_{n+2} \simeq F'(r)(r - x_{n+1})$$

Dividing we find

$$\frac{r - x_{n+1}}{r - x_{n+2}} \simeq \frac{r - x_n}{r - x_{n+1}}$$

and solving for the root

$$r \simeq x_{n+2} - \frac{(x_{n+2} - x_{n+1})^2}{x_{n+2} - 2x_{n+1} + x_n} = x_{n+2} - \frac{(\Delta x_{n+1})^2}{\Delta^2 x_n}$$

This is often called the **Aitken Δ^2 process**.

25.5. Apply extrapolation to the limit to the computation of Problem 25.2.

Using x_{10}, x_{11}, and x_{12}, the formula produces

$$r \simeq 1.368786102 - \frac{(.000071586)^2}{-.000232877} \simeq 1.368808107$$

which is once again Leonardo's value. With this extrapolation, only half the iterations are needed. Using it earlier might have made still further economies by stimulating the convergence.

25.6. Using extrapolation to the limit systematically after each three iterations is what is known as **Steffensen's method**. Apply this to Leonardo's equation.

The first three approximations x_0, x_1, and x_2 may be borrowed from Problem 25.2. Aitken's formula is now used to produce x_3:

$$x_3 = x_2 - \frac{(x_2 - x_1)^2}{x_2 - 2x_1 + x_0} = 1.370813882$$

The original iteration is now resumed as in Problem 25.2 to produce x_4 and x_5:

$$x_4 = F(x_3) = 1.367918090 \qquad x_5 = F(x_4) = 1.369203162$$

Aitken's formula then yields x_6:

$$x_6 = x_5 - \frac{(x_5 - x_4)^2}{x_5 - 2x_4 + x_3} = 1.368808169$$

The next cycle brings the iterates

$$x_7 = 1.368808080 \qquad x_8 = 1.368808120$$

from which Aitken's formula manages $x_9 = 1.368808108$.

25.7. Show that other rearrangements of Leonardo's equation may not produce convergent sequences.

As an example we may take $x = (20 - 2x^2 - x^3)/10$ which suggests the iteration

$$x_n = \frac{20 - 2x_{n-1}^2 - x_{n-1}^3}{10}$$

Again starting with $x_0 = 1$, we are led to the sequence

$$\begin{array}{llll} x_1 \simeq 1.70 & x_3 \simeq 1.75 & x_5 \simeq 1.79 & x_7 \simeq 1.83 \\ x_2 \simeq .93 & x_4 \simeq .85 & x_6 \simeq .79 & x_8 \simeq .72 \end{array}$$

and so on. It seems clear that alternate approximations are headed in opposite directions. Comparing with Problem 25.1 we find that here $F'(r) = (-4r - 3r^2)/10 < -1$, confirming the computational evidence.

THE NEWTON METHOD

25.8. Derive the Newton iterative formula $x_n = x_{n-1} - \dfrac{f(x_{n-1})}{f'(x_{n-1})}$ for solving $f(r) = 0$.

Beginning with Taylor's formula

$$f(r) = f(x_{n-1}) + (r - x_{n-1})f'(x_{n-1}) + \frac{1}{2}(r - x_{n-1})^2 f''(\xi)$$

we retain the linear part, recall that $f(r) = 0$, and define x_n by putting it in place of the remaining r to obtain

$$0 = f(x_{n-1}) + (x_n - x_{n-1})f'(x_{n-1})$$

which rearranges at once into $r \simeq x_n = x_{n-1} - \dfrac{f(x_{n-1})}{f'(x_{n-1})}$.

25.9. What is the geometric interpretation of Newton's formula?

It amounts to using the tangent line to $y = f(x)$ at x_{n-1} in place of the curve. In Fig. 25-3 it can be

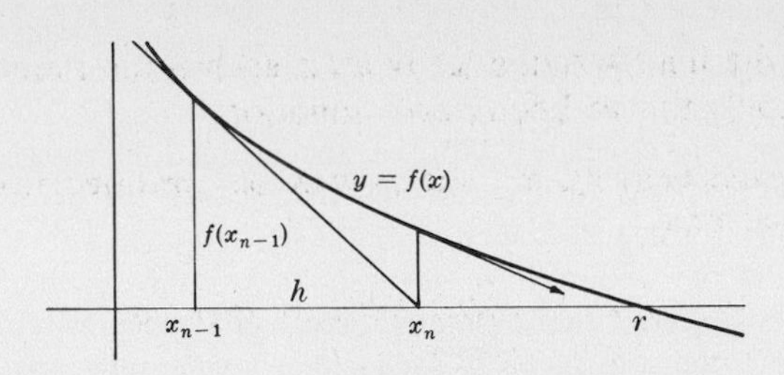

Fig. 25-3

seen that this leads to

$$\frac{f(x_{n-1})-0}{x_{n-1}-x_n}=f'(x_{n-1})$$

which is once again Newton's formula. Similar steps follow, as suggested by the arrow.

25.10. Apply Newton's formula to Leonardo's equation.

With $f(x)=x^3+2x^2+10x-20$ we find $f'(x)=3x^2+4x+10$, and the iterative formula becomes

$$x_n=x_{n-1}-\frac{x_{n-1}^3+2x_{n-1}^2+10x_{n-1}-20}{3x_{n-1}^2+4x_{n-1}+10}$$

Once more choosing $x_0=1$, we obtain the results in Table 25.2.

Table 25.2

n	1	2	3	4
x_n	1.411764706	1.369336471	1.368808189	1.368808108

The speed of convergence is remarkable. In four iterations we have essentially Leonardo's value. In fact, computation shows that

$$f(1.368808107)\simeq -.000000016$$
$$f(1.368808108)\simeq -.000000005$$

which suggests that the Newton result is the winner by a nose.

25.11. Explain the rapid convergence of Newton's iteration by showing that the convergence is "quadratic."

Recalling the equations of Problem 25.8 which led to the Newton formula,

$$f(r)=f(x_{n-1})+(r-x_{n-1})f'(x_{n-1})+\frac{1}{2}(r-x_{n-1})^2f''(\xi)$$

$$0=f(x_{n-1})+(x_n-x_{n-1})f'(x_{n-1})$$

we subtract to obtain

$$0=(r-x_n)f'(x_{n-1})+\frac{1}{2}(r-x_{n-1})^2f''(\xi)$$

or, letting $e_n=r-x_n$,

$$0=e_nf'(x_{n-1})+\frac{1}{2}e_{n-1}^2f''(\xi)$$

Assuming convergence, we replace both x_{n-1} and ξ by the root r and have

$$e_n\simeq -\frac{f''(r)}{2f'(r)}e_{n-1}^2$$

Each error is therefore roughly proportional to the square of the previous error. This means that the number of correct decimal places roughly doubles with each approximation and is what is called quadratic convergence. It may be compared with the slower, linear convergence in Problem 25.3, where each error was roughly proportional to the previous error. Since the error of our present x_3 is about .00000008, and $[f''(r)]/[2f'(r)]$ is about .3, we see that if we had been able to carry more decimal places in our computation the error of x_4 might have been about two units in the fifteenth place! This superb speed suggests that the Newton algorithm deserves a reasonably accurate first approximation to trigger it, and that its natural role is the conversion of such a reasonable approximation into an excellent one. In fact, other algorithms to be presented are better suited than Newton's for the "global" problem of obtaining first approximations to all the roots. Such methods usually converge very slowly, however, and it seems only natural to use them only as a source of reasonable first approximations, the Newton method then providing the polish. Such procedures are very popular and will be mentioned again as we proceed. It may also be noted that occasionally, given an inadequate first approximation, the Newton algorithm will converge at quadratic speed, but not to the root expected! Recalling the tangent line geometry behind the algorithm, it is easy to diagram a curve for which this happens, simply putting the first approximation near a maximum or minimum point.

25.12. Show that the formula for determining square roots,

$$x_n = \frac{1}{2}\left(x_{n-1} + \frac{Q}{x_{n-1}}\right)$$

is a special case of Newton's iteration.

With $f(x) = x^2 - Q$, it is clear that making $f(x) = 0$ amounts to finding a square root of Q. Since $f'(x) = 2x$, the Newton formula becomes

$$x_n = x_{n-1} - \frac{x_{n-1}^2 - Q}{2x_{n-1}} = \frac{1}{2}\left(x_{n-1} + \frac{Q}{x_{n-1}}\right)$$

25.13. Apply the square root iteration with $Q = 2$.

Choosing $x_0 = 1$, we find the results in Table 25.3. Notice once again the quadratic nature of the convergence. Each result has roughly twice as many correct digits as the one before it. Figure 25-4 illustrates the action. Since the first approximation was on the concave side of $y = x^2 - 2$, the next is on the other side of the root. After this the sequence is monotone, remaining on the convex side of the curve as tangent lines usually do.

Table 25.3

n	x_n
1	1.5
2	1.416 666 667
3	1.414 215 686
4	1.414 213 562
5	1.414 213 562

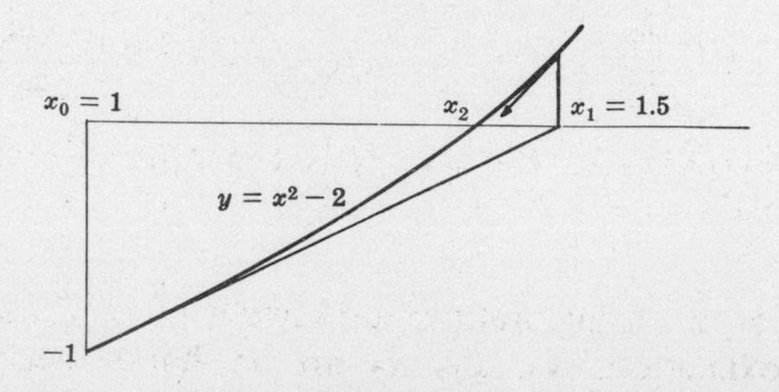

Fig. 25-4

25.14. Derive the iteration $x_n = x_{n-1} - \dfrac{x_{n-1}^p - Q}{px_{n-1}^{p-1}}$ for finding a pth root of Q.

With $f(x) = x^p - Q$ and $f'(x) = px^{p-1}$, the result is at once a special case of Newton's method.

25.15. Apply the preceding problem to find a cube root of 2.

With $Q = 2$ and $p = 3$, the iteration simplifies to $x_n = \dfrac{2}{3}\left(x_{n-1} + \dfrac{1}{x_{n-1}^2}\right)$.

Choosing $x_0 = 1$, we find $x_1 = \frac{4}{3}$ and then

$$x_2 = 1.263888889 \qquad x_3 = 1.259933493 \qquad x_4 = 1.259921049 \qquad x_5 = 1.259921049$$

The quadratic convergence is conspicuous.

INTERPOLATION METHODS

25.16. This ancient method uses two previous approximations and constructs the next approximation by making a linear interpolation between them. Derive the *regula falsi* (see Fig. 25-5),

$$c = a - \frac{(a-b)f(a)}{f(a) - f(b)}$$

The linear function

$$y = f(a) + \frac{f(a) - f(b)}{a - b}(x - a)$$

clearly has $y = f(x)$ at a and b. It vanishes at the argument c given in the *regula falsi*. This zero serves as our next approximation to the root of $f(x) = 0$, so effectively we have replaced the curve $y = f(x)$ by a linear collocation polynomial in the neighborhood of the root. It will also be noticed in Fig. 25-5 that the two given approximations a and b are on opposite sides of the exact root. Thus $f(a)$ and $f(b)$ have opposite signs. This opposition of signs is assumed when using *regula falsi*. Accordingly, having found c, to reapply *regula falsi* we use this c as either the new a or the new b, whichever choice preserves the opposition of signs. In Fig. 25-5, c would become the new a. In this way a sequence of approximations $x_0, x_1, x_2, \ldots$ may be generated, x_0 and x_1 being the original a and b.

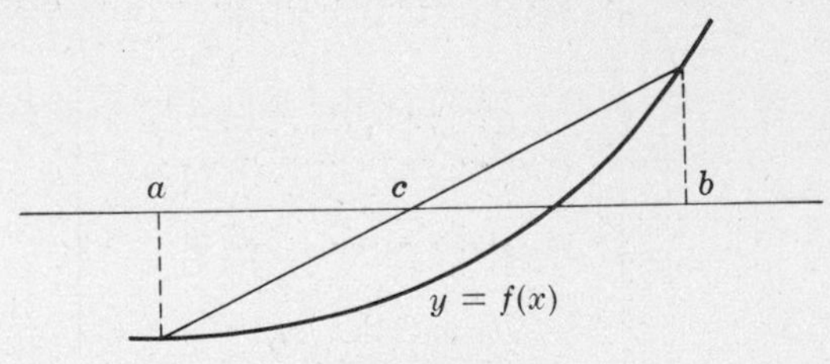

Fig. 25-5

25.17. Apply *regula falsi* to Leonardo's equation.

Choosing $x_0 = 1$ and $x_1 = 1.5$, the formula produces

$$x_2 = 1.5 - \frac{.5(2.875)}{9.875} \simeq 1.35 \qquad x_3 = 1.35 - \frac{(-.15)(-.3946)}{-3.2696} \simeq 1.368$$

and so on. The rate of convergence can be shown to be better than the rate in Problem 25.2 but not so good as that of Newton's method.

25.18. A natural next step is to use a quadratic interpolation polynomial rather than a linear one. Assuming three approximations x_0, x_1, x_2 are in hand, derive a formula for a new approximation x_3 which is a root of such a quadratic.

It is not hard to verify that the quadratic through the three points (x_0, y_0), (x_1, y_1), (x_2, y_2), where $y = f(x)$, can be written as

$$p(x) = \frac{x_1 - x_0}{x_2 - x_0}(Ah^2 + Bh + C)$$

where $h = x - x_2$ and A, B, C are

$$A = \frac{(x_1 - x_0)y_2 + (x_0 - x_2)y_1 + (x_2 - x_1)y_0}{(x_2 - x_1)(x_1 - x_0)^2}$$

$$B = \frac{(x_1 - x_0)(2x_2 - x_1 - x_0)y_2 - (x_2 - x_0)^2 y_1 + (x_2 - x_1)^2 y_0}{(x_2 - x_1)(x_1 - x_0)^2}$$

$$C = \frac{x_2 - x_0}{x_1 - x_0} y_2$$

Solving $p(x) = 0$ for h we find

$$h = -\frac{2C}{B \pm \sqrt{B^2 - 4AC}}$$

this form of the quadratic formula being chosen to avoid loss of significant digits during subtraction. Here the sign which makes the denominator larger in absolute value should be chosen. Then

$$x_3 = x_2 + h$$

becomes the next approximation and the process may be repeated with all subscripts advanced by one.

The method just described is what is known as **Muller's method** and has been found to converge to both real and complex roots. For the latter it is necessary, of course, to run the algorithm in complex arithmetic, but even with real roots, complex arithmetic is the wiser choice since traces of imaginary parts occasionally enter.

BERNOULLI'S METHOD

25.19. Prove that if the polynomial of degree n

$$p(x) = a_0 x^n + a_1 x^{n-1} + \cdots + a_n$$

has a single dominant zero, say r_1, then it may be found by computing a solution sequence for the difference equation of order n

$$a_0 x_k + a_1 x_{k-1} + \cdots + a_n x_{k-n} = 0$$

and taking $\lim (x_{k+1}/x_k)$.

This difference equation has $p(x) = 0$ for its characteristic equation and its solution can therefore be written as

$$x_k = c_1 r_1^k + c_2 r_2^k + \cdots + c_n r_n^k$$

If we choose initial values so that $c_1 \neq 0$, then

$$\frac{x_{k+1}}{x_k} = r_1 \frac{1 + (c_2/c_1)(r_2/r_1)^{k+1} + \cdots + (c_n/c_1)(r_n/r_1)^{k+1}}{1 + (c_2/c_1)(r_2/r_1)^k + \cdots + (c_n/c_1)(r_n/r_1)^k}$$

and since r_1 is the dominant root,

$$\lim \frac{r_i}{r_1} = 0 \qquad i = 2, 3, \ldots, n$$

making $\lim (x_{k+1}/x_k) = r_1$ as claimed. It can be shown using complex variable theory that the initial values $x_{-n+1} = \cdots = x_{-1} = 0$, $x_0 = 1$ will guarantee $c_1 \neq 0$.

25.20. Apply the Bernoulli method to the equation $x^4 - 5x^3 + 9x^2 - 7x + 2 = 0$.

The associated difference equation is

$$x_k - 5x_{k-1} + 9x_{k-2} - 7x_{k-3} + 2x_{k-4} = 0$$

and if we take the initial values $x_{-3} = x_{-2} = x_{-1} = 0$ and $x_0 = 1$, then the succeeding x_k are given in Table 25.4. The ratio x_{k+1}/x_k is also given. The convergence to $r = 2$ is slow, the rate of convergence of Bernoulli's method being linear. Frequently the method is used to generate a good starting approximation for Newton's or Steffensen's iteration, both of which are quadratic.

Table 25.4

k	x_k	x_{k+1}/x_k	k	x_k	x_{k+1}/x_k
1	5	3.2000	9	4,017	2.0164
2	16	2.6250	10	8,100	2.0096
3	42	2.3571	11	16,278	2.0056
4	99	2.2121	12	32,647	2.0032
5	219	2.1279	13	65,399	2.0018
6	466	2.0773	14	130,918	2.0010
7	968	2.0465	15	261,972	2.0006
8	1,981	2.0278	16	524,097	

25.21. Modify the Bernoulli method for the case in which a pair of complex conjugate roots are dominant.

Let r_1 and r_2 be complex conjugate roots. Then $|r_i| < |r_1|$ for $i = 3, \ldots, n$, since the r_1, r_2 pair is dominant. Using real starting values, the solution of the difference equation may be written as

$$x_k = c_1 r_1^k + c_2 r_2^k + \cdots + c_n r_n^k$$

where c_1 and c_2 are also complex conjugate. Let $r_1 = re^{i\phi} = \bar{r}_2$, $c_1 = ae^{i\theta} = \bar{c}_2$ with $r > 0$, $a > 0$, and $0 < \phi < \pi$ so that r_1 is the root in the upper half plane. Then

$$x_k = 2ar^k \cos(k\phi + \theta) + c_3 r_3^k + \cdots + c_n r_n^k$$

$$= 2ar^k\left[\cos(k\phi + \theta) + \frac{c_3}{2a}\left(\frac{r_3}{r}\right)^k + \cdots + \frac{c_n}{2a}\left(\frac{r_n}{r}\right)^k\right]$$

All terms except the first have limit zero; and so for large k, $x_k \simeq 2ar^k \cos(k\phi + \theta)$. We now use this result to determine r and ϕ. First we observe that

$$x_{k+1} - 2r\cos\phi\, x_k + r^2 x_{k-1} \simeq 0$$

as may be seen by substituting for x_k from the previous equation and using the identities for cosines of sums and differences. Reducing the subscripts, we also have

$$x_k - 2r\cos\phi\, x_{k-1} + r^2 x_{k-2} \simeq 0$$

Now solving these two simultaneously,

$$r^2 \simeq \frac{x_k^2 - x_{k+1}x_{k-1}}{x_{k-1}^2 - x_k x_{k-2}} \qquad -2r\cos\phi \simeq \frac{x_{k+1}x_{k-2} - x_{k-1}x_k}{x_{k-1}^2 - x_k x_{k-2}}$$

The necessary ingredients for determining r_1 and r_2 are now in hand.

25.22. Apply Bernoulli's method to Leonardo's equation.

The associated difference equation is $x_k = -2x_{k-1} - 10x_{k-2} + 20x_{k-3}$ and the solution sequence for initial values $x_{-2}=x_{-1}=0$, $x_0=1$ appears in Table 25.5. Some approximations to r^2 and $-2r\cos\phi$ also appear. The fluctuating $\pm$ signs are an indication that dominant complex roots are present. This may be seen by recalling the form of the x_k as given in Problem 25.21, namely $x_k \simeq 2ar^k \cos(k\phi+\theta)$. As k increases, the value of the cosine will vary between ± 1 in a somewhat irregular way which depends on the value of ϕ.

Table 25.5

k	x_k	k	x_k	r^2	$-2r\cos\phi$
1	−2	7	−2,608	14.6026	3.3642
2	−6	8	−32,464	14.6076	3.3696
3	52	9	147,488	14.6135	3.3692
4	−84	10	−22,496	14.6110	3.3686
5	−472	11	−2,079,168	14.6110	3.3688
6	2,824	12	7,333,056		

From the last approximations we find

$$r\cos\phi \simeq -1.6844 \qquad r\sin\phi = \pm\sqrt{r^2-(r\cos\phi)^2} \simeq \pm 3.4313$$

making the dominant pair of roots $r_1 r_2 \simeq -1.6844 \pm 3.4313i$. Since Leonardo's equation is cubic, these roots could also be found by using the real root found earlier to reduce to a quadratic equation. The Bernoulli method was not really needed in this case. The results found may be checked by computing the sum (-2) and product (20) of all the roots.

DEFLATION

25.23. Use the simple equation $x^4-10x^3+35x^2-50x+24=0$ to illustrate the idea of deflation.

The dominant root of this equation is exactly 4. Applying the factor theorem we remove the factor $x-4$ by division,

$$\begin{array}{rrrrr|l} 1 & -10 & 35 & -50 & 24 & \underline{4} \\ & 4 & -24 & 44 & -24 & \\ \hline 1 & -6 & 11 & -6 & 0 & \end{array}$$

The quotient is the cubic $x^3-6x^2+11x-6$ and we say that the original quartic polynomial has been deflated to this cubic. The dominant root of the cubic is exactly 3. Removing this factor,

$$\begin{array}{rrrr|l} 1 & -6 & 11 & -6 & \underline{3} \\ & 3 & -9 & 6 & \\ \hline 1 & -3 & 2 & 0 & \end{array}$$

we achieve a second deflation, to the quadratic x^2-3x+2 which may then be solved for the remaining roots 2 and 1. Or the quadratic may be deflated to the linear function $x-1$. The idea of deflation is that, one root having been found, the original equation may be exchanged for one of lower degree. Theoretically, a method for finding the dominant root of an equation, such as Bernoulli's method, could be used to find all the roots one after another, by successive deflations which remove each dominant root as it is found, and assuming no two roots are of equal size. Actually there are error problems which limit the use of this procedure, as the next problem suggests.

25.24. Show that if the dominant root is not known exactly, then the method of deflation may yield the next root with still less accuracy, and suggest a procedure for obtaining this second root to the same accuracy as the first.

Suppose, for simplicity, that the dominant root of the previous equation has been found correct to only two places to be 4.005. Deflation brings

1	−10	35	−50	24	$\underline{\vert 4.005}$
	4.005	−24.01	44.015	−23.97	
1	−5.995	10.99	−5.985	.03	

and the cubic $x^3 - 5.995x^2 + 10.99x - 5.985$. The dominant zero of this cubic (correct to two places) is 2.98. As far as the original quartic equation is concerned, this is incorrect in the last place. The natural procedure at this point is to use the 2.98 as the initial approximation for a Newton iteration, which would rapidly produce a root of the original equation correct to two places. A second deflation could then be made. In practice it is found that the smaller "roots" require substantial correction and that for polynomials of even moderate degree the result obtained by deflation may not be good enough to guarantee convergence of the Newton iteration to the desired root. Similar remarks hold when complex conjugate roots $a \pm bi$ are removed through division by the quadratic factor $x^2 - 2ax + a^2 + b^2$.

THE QUOTIENT-DIFFERENCE ALGORITHM

25.25. What is a quotient-difference scheme?

Given a polynomial $a_0x^n + a_1x^{n-1} + \cdots + a_n$ and the associated difference equation

$$a_0x_k + a_1x_{k-1} + \cdots + a_nx_{k-n} = 0$$

consider the solution sequence for which $x_{-n+1} = \cdots = x_{-1} = 0$ and $x_0 = 1$. Let $q_k^1 = x_{k+1}/x_k$ and $d_k^0 = 0$. Then define

$$q_k^{j+1} = \left(\frac{d_{k+1}^j}{d_k^j}\right) q_{k+1}^j \qquad d_k^j = q_{k+1}^j - q_k^j + d_{k+1}^{j-1}$$

where $j = 1, 2, \ldots, n-1$ and $k = 0, 1, 2, \ldots$. These various quotients (q) and differences (d) may be displayed as in Table 25.6. The definitions are easily remembered by observing the rhombus-shaped parts of the table. In a rhombus centered in a (q) column the sum of the SW pair equals the sum of the NE pair. In a rhombus centered in a (d) column the corresponding products are equal. These are the **rhombus rules**.

Table 25.6

	q_0^1						
0		d_0^1					
	q_1^1		q_0^2				
0		d_1^1		d_0^2			
	q_2^1		q_1^2		q_0^3		
0		d_2^1		d_1^2		d_0^3	
	q_3^1		q_2^2		q_1^3		q_0^4
0		d_3^1		d_2^2		d_1^3	
	q_4^1		q_3^2		q_2^3		q_1^4
0		d_4^1		d_3^2		d_2^3	
	q_5^1	.	q_4^2	.	q_3^3	.	q_2^4
	.	.	.	.	.	.	.
	.	.	.	.	.	.	.
	.		.		.		.

25.26. Compute the quotient-difference scheme for the polynomial $x^2 - x - 1$ associated with the Fibonacci sequence.

The results appear in Table 25.7.

Table 25.7

k	x_k	d_k^0	q_k^1	d_k^1	q_k^2	d_k^2
0	1	0				
			1.0000			
1	1	0		1.0000		
			2.0000		−1.0000	
2	2	0		−.5000		−.0001
			1.5000		−.5001	
3	3	0		.1667		−.0001
			1.6667		−.6669	
4	5	0		−.0667		.0005
			1.6000		−.5997	
5	8	0		.0250		.0007
			1.6250		−.6240	
6	13	0		−.0096		−.0082
			1.6154		−.6226	
7	21	0		.0037		
			1.6190			
8	34	0				

25.27. What is the first convergence theorem associated with the quotient-difference scheme?

Suppose no two zeros of the given polynomial have the same absolute value. Then

$$\lim q_k^j = r_j \qquad j = 1, 2, \ldots, n$$

for k tending to infinity, where $r_1, r_2, \ldots, r_n$ are in the order of diminishing absolute value. For $j = 1$ this is Bernoulli's result for the dominant root. For the other values of j the proof requires complex function theory and will be omitted. It has also been assumed here that none of the denominators involved in the scheme is zero. The convergence of the q's to the roots implies the convergence of the d's to zero. This may be seen as follows. By the first of the defining equations of Problem 25.25,

$$\frac{d_{k+1}^j}{d_k^j} = \frac{q_k^{j+1}}{q_{k+1}^j} \rightarrow \frac{r_{j+1}}{r_j} < 1$$

The d_k^j therefore converge geometrically to zero. The beginning of this convergence, in the present problem, is evident already in Table 25.7, except in the last column which will be discussed shortly. In this table the (q) columns should, by the convergence theorem, be approaching the roots $(1 \pm \sqrt{5})/2$ which are approximately 1.61803 and −.61803. Clearly we are closer to the first than to the second.

25.28. How can a quotient-difference scheme produce a pair of complex conjugate roots?

The presence of such roots may be indicated by (d) columns which do not converge to zero. Suppose the column of d_k^j entries does not. Then one forms the polynomial

$$p_j = x^2 - A_j x + B_j$$

where for k tending to infinity,

$$A_j = \lim (q_{k+1}^j + q_k^{j+1}) \qquad B_j = \lim q_k^j q_k^{j+1}$$

The polynomial will have the roots r_j and r_{j+1} which will be complex conjugates. Essentially, a quadratic factor of the original polynomial will have been found. Here we have assumed that the columns of d_k^{j-1} and d_k^{j+1} entries do converge to zero. If they do not, then more than two roots have equal absolute value and a more complicated procedure is needed. The details, and also the proofs of convergence claims just made, are given in *National Bureau of Standards Applied Mathematics Series,* vol. 49.

25.29. What is the row-by-row method of generating a quotient-difference scheme and what are its advantages?

The column-by-column method first introduced in Problem 25.25 is very sensitive to roundoff error. This is the explanation of the fact that the final column of Table 25.7 is not converging to zero as a (d) column should but instead shows the typical start of an error explosion. The following row-by-row method is less sensitive to error. Fictitious entries are supplied to fill out the top two rows of a quotient-difference scheme as follows, starting with the d_k^0 column and ending with d_k^n. Both of these boundary columns are to consist of zeros for all values of k. This amounts to forcing proper behavior of these boundary differences in an effort to control roundoff error effects.

	$-a_1/a_0$		0		0		0	
0		a_2/a_1		a_3/a_2		a_4/a_3		0

The rhombus rules are then applied, filling each new row in its turn. It can be shown that the same scheme found in Problem 25.25 will be developed by this method, assuming no errors in either procedure. In the presence of error the row-by-row method is more stable. Note that in this method it is not necessary to compute the x_k.

25.30. Apply the row-by-row method to the polynomial of the Fibonacci sequence, $x^2 - x - 1$.

The top rows are filled as suggested in the previous problem. The others are computed by the rhombus rules. Table 25.8 exhibits the results. The improved behavior in the last (q) column is apparent.

Table 25.8

k	d	q	d	q	d
		1		0	
1	0		1		0
		2		−1	
2	0		−.5000		0
		1.5000		−.5000	
3	0		.1667		0
		1.6667		−.6667	
4	0		−.0667		0
		1.6000		−.6000	
5	0		.0250		0
		1.6250		−.6250	
6	0		−.0096		0
		1.6154		−.6154	
7	0		.0037		0
		1.6191		−.6191	
8	0				0

25.31. Apply the quotient-difference algorithm to find all the roots of

$$x^4 - 10x^3 + 35x^2 - 50x + 24 = 0$$

The roots of this equation are exactly 1, 2, 3, and 4. No advance information about the roots is, however, required by this algorithm, so the equation serves as a simple test case. The quotient-difference scheme, generated by the method of Problem 25.29, appears as Table 25.9. Clearly the convergence is slow, but the expected pattern is emerging. The (d) columns seem headed for zero and

Table 25.9

k	d	q	d	q	d	q	d	q	d
		10		0		0		0	
1	0		−3.5000		−1.4286		−.4800		0
		6.5000		2.0714		.9486		.4800	
2	0		−1.1154		−.6542		−.2429		0
		5.3846		2.5326		1.3599		.7229	
3	0		−.5246		−.3513		−.1291		0
		4.8600		2.7059		1.5821		.8520	
4	0		−.2921		−.2054		−.0695		0
		4.5679		2.7926		1.7180		.9215	
5	0		−.1786		−.1264		−.0373		0
		4.3893		2.8448		1.8071		.9588	
6	0		−.1158		−.0803		−.0198		0
		4.2735		2.8803		1.8676		.9786	
7	0		−.0780		−.0521		−.0104		0
		4.1955		2.9062		1.9093		.9890	
8	0		−.0540		−.0342		−.0054		0
		4.1415		2.9260		1.9381		.9944	

the (q) columns for 4, 3, 2, 1 in that order. Probably it would be wise to switch at this point to Newton's method, which very quickly converts reasonable first approximations such as we now have into accurate results. The quotient-difference algorithm is often used for exactly this purpose, to prime the Newton iteration.

25.32. Apply the quotient-difference algorithm to Leonardo's equation.

Again using the row-by-row method, we generate the scheme displayed in Table 25.10.

Table 25.10

k	d	q	d	q	d	q	d
		−2		0		0	
1	0		5		-2		0
		3		−7		2	
2	0		−11.6667		.5714		0
		−8.6667		5.2381		1.4286	
3	0		7.0513		.1558		0
		−1.6154		−1.6574		1.2728	
4	0		7.2346		−.1196		0
		5.6192		−9.0116		1.3924	
5	0		−11.6022		.0185		0
		−5.9830		2.6091		1.3739	
6	0		5.0596		.0097		0
		−.9234		−2.4408		1.3642	

The convergence being slow, suppose we stop here. The second (d) column hardly seems headed for zero, suggesting that r_1 and r_2 are complex, as we already know anyway. The next (d) column does appear to be tending to zero, suggesting a real root which we know to be near 1.369. The Newton method would quickly produce an accurate root from the initial estimate of 1.3642 we now have here. Returning to the complex pair, we apply the procedure of Problem 25.28. From the first two (q) columns we compute

$$5.6192 - 9.0116 = -3.3924 \qquad (-1.6154)(-9.0116) \simeq 14.5573$$
$$-5.9830 + 2.6091 = -3.3739 \qquad (5.6192)(2.6091) \simeq 14.6611$$
$$-.9234 - 2.4408 = -3.3642 \qquad (-5.9830)(-2.4408) \simeq 14.6033$$

so that $A_1 \simeq -3.3642$ and $B_1 \simeq 14.6033$. The complex roots are therefore approximately given by $x^2 + 3.3642x + 14.6033 = 0$ which makes them $r_1, r_2 \simeq -1.682 \pm 3.431i$.

Newton's method using complex arithmetic could be used to improve these values, but an alternative procedure known as Bairstow's method will be presented shortly. Once again in this problem we have used the quotient-difference algorithm to provide respectable estimates of all the roots. A method which can do this should not be expected to converge rapidly, and the switch to a quadratically convergent algorithm at some appropriate point is a natural step.

STURM SEQUENCES

25.33. Define a Sturm sequence.

A sequence of functions $f_0(x), f_1(x), \ldots, f_n(x)$ which satisfy on an interval (a, b) of the real line the conditions:

1. Each $f_i(x)$ is continuous.
2. The sign of $f_n(x)$ is constant.
3. If $f_i(r) = 0$ then $f_{i-1}(r)$ and $f_{i+1}(r) \neq 0$.
4. If $f_i(r) = 0$ then $f_{i-1}(r)$ and $f_{i+1}(r)$ have opposite signs.
5. If $f_0(r) = 0$ then for h sufficiently small

$$\text{sign}\frac{f_0(r-h)}{f_1(r-h)} = -1 \qquad \text{sign}\frac{f_0(r+h)}{f_1(r+h)} = 1$$

is called a Sturm sequence.

25.34. Prove that the number of roots of the function $f_0(x)$ on the interval (a, b) is the difference between the number of changes of sign in the sequences $f_0(a), f_1(a), \ldots, f_n(a)$ and $f_0(b), f_1(b), \ldots, f_n(b)$.

As x increases from a to b the number of sign changes in the Sturm sequence can only be affected by one or more of the functions having a zero, since all are continuous. Actually only a zero of $f_0(x)$ can affect it. For, suppose $f_i(r) = 0$ with $i \neq 0, n$, then by properties 1, 3, and 4 the following sign patterns are possible for small h:

	f_{i-1}	f_i	f_{i+1}
$r-h$	+	±	−
r	+	0	−
$r+h$	+	±	−

or

	f_{i-1}	f_i	f_{i+1}
$r-h$	−	±	+
r	−	0	+
$r+h$	−	±	+

In all cases there is one sign change, so that moving across such a root does not affect the number of sign changes. By condition 2 the function $f_n(x)$ cannot have a zero, so we come finally to $f_0(x)$. By condition 5 we lose one sign change, between f_0 and f_1, as we move across the root r. This proves the theorem. One sees that the five conditions have been designed with this root-counting feature in mind.

25.35. If $f_0(x)$ is a polynomial of degree n with no multiple roots, how can a Sturm sequence for enumerating its roots be constructed?

Let $f_1(x)=f_0'(x)$ and then apply the Euclidean algorithm to construct the rest of the sequence as follows:

$$f_0(x)=f_1(x)L_1(x)-f_2(x)$$
$$f_1(x)=f_2(x)L_2(x)-f_3(x)$$
$$\cdots\cdots\cdots\cdots\cdots\cdots$$
$$f_{n-2}(x)=f_{n-1}(x)L_{n-1}(x)-f_n(x)$$

where $f_i(x)$ is of degree $n-i$ and the $L_i(x)$ are linear.

The sequence $f_0(x), f_1(x), \ldots, f_n(x)$ will be a Sturm sequence. To prove this we note first that all $f_i(x)$ are continuous, since f_0 and f_1 surely are. Condition 2 follows since f_n is a constant. Two consecutive $f_i(x)$ cannot vanish simultaneously since then all would vanish including f_0 and f_1 and this would imply a multiple root. This proves condition 3. Condition 4 is a direct consequence of our defining equations and 5 is satisfied since $f_1=f_0'$.

If the method were applied to a polynomial having multiple roots, then the simultaneous vanishing of all the $f_i(x)$ would give evidence of them. Deflation of the polynomial to remove multiplicities allows the method to be applied to find the simple roots.

25.36. Apply the method of Sturm sequences to locate all real roots of

$$x^4-2.4x^3+1.03x^2+.6x-.32=0$$

Denoting this polynomial $f_0(x)$, we first compute its derivative. Since we are concerned only with the signs of the various $f_i(x)$, it is often convenient to use a positive multiplier to normalize the leading coefficient. Accordingly we multiply $f_0'(x)$ by $\frac{1}{4}$ and take

$$f_1(x)=x^3-1.8x^2+.515x+.15$$

The next step is to divide f_0 by f_1. One finds the linear quotient $L_1(x)=x-.6$ which is of no immediate interest, and a remainder of $-.565x^2+.759x-.23$. A common error at this point is to forget that we want the *negative* of this remainder. Also normalizing, we have

$$f_2(x)=x^2-1.3434x+.4071$$

Dividing f_1 by f_2 brings a linear quotient $L_2(x)=x-.4566$ and a remainder whose negative, after normalizing, is

$$f_3(x)=x-.6645$$

Finally, dividing f_2 by f_3 we find the remainder to be $-.0440$. Taking the negative and normalizing, we may choose

$$f_4(x)=1$$

We now have our Sturm sequence and are ready to search out the roots. It is a simple matter to confirm the signs displayed in Table 25.11. They show that there is one root in the interval $(-1,0)$, one in $(1,2)$,

Table 25.11

	f_0	f_1	f_2	f_3	f_4	Changes
$-\infty$	+	−	+	−	+	4
-1	+	−	+	−	+	4
0	−	+	+	−	+	3
1	−	−	+	+	+	1
2	+	+	+	+	+	0
∞	+	+	+	+	+	0

and two roots in $(0, 1)$. Choosing more points within these intervals, all roots may be more precisely pinpointed. As with the quotient-difference algorithm, however, it is wise to shift at a certain point to a more rapidly convergent process such as Newton's. A method which provides first estimates of the locations of all real roots, as the Sturm method does, is uneconomical for the precise determination of any one root. In this example the roots prove to be $-.5$, $.5$, $.8$, and 1.6.

25.37. Show that Newton's method will produce all the roots of the equation in the previous problem provided sufficiently good initial approximations are obtained.

Figure 25-6 below exhibits the qualitative behavior of this polynomial. Clearly any first approximation $x_0 < -.5$ will lead to a sequence which converges upon this root, since such an x_0 is already on the convex side of the curve. Similarly any $x_0 > 1.6$ will bring convergence to the largest root. Roots that are close together ordinarily require accurate starting approximations. The simplicity of the roots in this example may be ignored in order to see how a more obscure pair might be separated. From the diagram it is apparent that an x_0 slightly below .5 will bring convergence to .5, while an x_0 slightly above .8 will bring convergence to .8, since in both cases we start on the convex side. Notice that starting with $x_0 = .65$, which is midway between two roots, means following an almost horizontal tangent line. Actually it leads to $x_1 \simeq 5$, after which convergence to the root at 1.6 would occur. This sort of thing can occur in a Newton iteration.

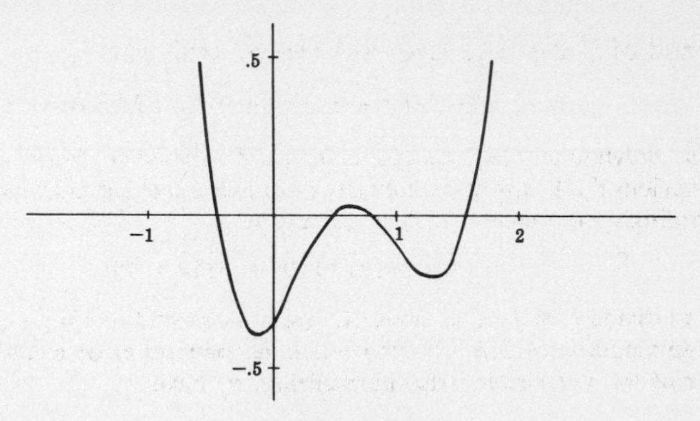

Fig. 25-6

SYSTEMS OF EQUATIONS, NEWTON'S METHOD

25.38. Derive the formulas for solving $f(x, y) = 0$, $g(x, y) = 0$,

$$x_n = x_{n-1} + h_{n-1}$$
$$y_n = y_{n-1} + k_{n-1}$$

where h and k satisfy

$$f_x(x_{n-1}, y_{n-1})h_{n-1} + f_y(x_{n-1}, y_{n-1})k_{n-1} = -f(x_{n-1}, y_{n-1})$$
$$g_x(x_{n-1}, y_{n-1})h_{n-1} + g_y(x_{n-1}, y_{n-1})k_{n-1} = -g(x_{n-1}, y_{n-1})$$

These formulas are known as the Newton method for solving two simultaneous equations.

Approximate f and g by the linear parts of their Taylor series for the neighborhood of (x_{n-1}, y_{n-1}):

$$f(x, y) \simeq f(x_{n-1}, y_{n-1}) + (x - x_{n-1})f_x(x_{n-1}, y_{n-1}) + (y - y_{n-1})f_y(x_{n-1}, y_{n-1})$$
$$g(x, y) \simeq g(x_{n-1}, y_{n-1}) + (x - x_{n-1})g_x(x_{n-1}, y_{n-1}) + (y - y_{n-1})g_y(x_{n-1}, y_{n-1})$$

This assumes that the derivatives involved exist. With (x, y) denoting an exact solution, both left sides vanish. Defining $x = x_n$ and $y = y_n$ as the numbers which make the right sides vanish, we have at once the equations required. This idea of replacing a Taylor series by its linear part is what led to the Newton method for solving a single equation in Problem 25.8.

25.39. Find the intersection points of the circle $x^2+y^2=2$ with the hyperbola $x^2-y^2=1$.

This particular problem can easily be solved by elimination. Addition brings $2x^2=3$ and $x \simeq \pm 1.2247$. Subtraction brings $2y^2=1$ and $y=\pm .7071$. Knowing the correct intersections makes the problem a simple test case for Newton's method. Take $x_0=1$, $y_0=1$. The formulas for determining h and k are

$$2x_{n-1}h_{n-1}+2y_{n-1}k_{n-1}=2-x_{n-1}^2-y_{n-1}^2$$
$$2x_{n-1}h_{n-1}-2y_{n-1}k_{n-1}=1-x_{n-1}^2+y_{n-1}^2$$

and with $n=1$ become $2h_0+2k_0=0$, $2h_0-2k_0=1$. Then $h_0=-k_0=\frac{1}{4}$, making

$$x_1=x_0+h_0=1.25 \qquad y_1=y_0+k_0=.75$$

The next iteration brings $2.5h_1+1.5k_1=-.125$, $2.5h_1-1.5k_1=0$ making $h_1=-.025$, $k_1=-.04167$ and

$$x_2=x_1+h_1=1.2250 \qquad y_2=y_1+k_1=.7083$$

A third iteration manages $2.45h_2+1.4167k_2=-.0024$, $2.45h_2-1.4167k_2=.0011$ making $h_2=-.0003$, $k_2=-.0012$ and

$$x_3=x_2+h_2=1.2247 \qquad y_3=y_2+k_2=.7071$$

The convergence to the correct results is evident. It can be proved that for sufficiently good initial approximations the convergence of Newton's method is quadratic. The idea of the method can easily be extended to any number of simultaneous equations.

25.40. Other iterative methods may also be generalized for simultaneous equations. For example, if our basic equations $f(x,y)=0$, $g(x,y)=0$ are rewritten as

$$x=F(x,y) \qquad y=G(x,y)$$

then under suitable assumptions on F and G, the iteration

$$x_n=F(x_{n-1},y_{n-1}) \qquad y_n=G(x_{n-1},y_{n-1})$$

will converge for sufficiently accurate initial approximations. Apply this method to the equations $x=\sin(x+y)$, $y=\cos(x-y)$.

These equations are already in the required form. Starting with the uninspired initial approximations $x_0=y_0=0$, we obtain the results given below. Convergence for such poor starting approximations is by no means the rule. Often one must labor long to find a convergent rearrangement of given equations and good first approximations.

n	0	1	2	3	4	5	6	7
x_n	0	0	.84	.984	.932	.936	.935	.935
y_n	0	1	.55	.958	1.000	.998	.998	.998

DESCENT METHODS AND OPTIMIZATION

25.41. What is the idea of a steepest descent algorithm?

A variety of *minimization methods* involves a function $S(x,y)$ defined in such a way that its minimum value occurs precisely where $f(x,y)=0$ and $g(x,y)=0$. The problem of solving these two equations simultaneously may then be replaced by the problem of minimizing $S(x,y)$. For example,

$$S(x,y)=[f(x,y)]^2+[g(x,y)]^2$$

surely achieves its minimum of zero wherever $f=g=0$. This is one popular choice of $S(x,y)$. The question of how to find such a minimum remains. The method of steepest descent begins with an initial approximation (x_0,y_0). At this point the function $S(x,y)$ decreases most rapidly in the direction of the vector

$$-\text{gradient } S(x, y)|_{x_0y_0} = [-S_x, -S_y]|_{x_0y_0}$$

Denoting this by $-\text{grad } S_0 = [-S_{x0}, -S_{y0}]$ for short, a new approximation (x_1, y_1) is now obtained in the form

$$x_1 = x_0 - tS_{x0} \qquad y_1 = y_0 - tS_{y0}$$

with t chosen so that $S(x_1, y_1)$ is a minimum. In other words, we proceed from (x_0, y_0) in the direction $-\text{grad } S_0$ until S starts to increase again. This completes one step and another is begun at (x_1, y_1) in the new direction $-\text{grad } S_1$. The process continues until, hopefully, the minimum point is found.

The process has been compared to a skier's return from a mountain to the bottom of the valley in a heavy fog. Unable to see his goal, he starts down in the direction of steepest descent and proceeds until his path begins to climb again. Then choosing a new direction of steepest descent, he makes a second run of the same sort. In a bowl-shaped valley ringed by mountains it is clear that this method will bring him gradually nearer and nearer to home. Figure 25-7 illustrates the action. The dashed lines are contour or level lines, on which $S(x, y)$ is constant. The gradient direction is orthogonal to the contour direction at each point, so we always leave a contour line at right angles. Proceeding to the minimum of $S(x, y)$ along this line means going to a point of tangency with a lower contour line. Actually it requires infinitely many steps of this sort to reach the minimum and a somewhat uneconomical zigzag path is followed.

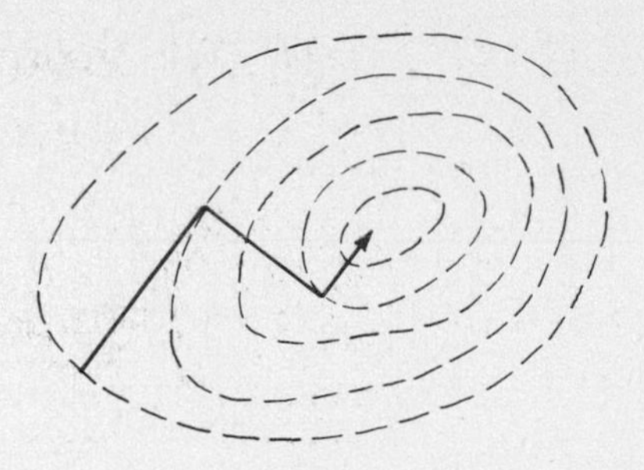

Fig. 25-7

25.42. Apply a method of steepest descent to solve the equation of Problem 25.40:

$$x = \sin(x + y) \qquad y = \cos(x - y)$$

Here we have

$$S = f^2 + g^2 = [x - \sin(x + y)]^2 + [y - \cos(x - y)]^2$$

making

$$\frac{1}{2}S_x = [x - \sin(x + y)][1 - \cos(x + y)] + [y - \cos(x - y)][\sin(x - y)]$$

$$\frac{1}{2}S_y = [x - \sin(x + y)][-\cos(x + y)] + [y - \cos(x - y)][1 - \sin(x - y)]$$

Suppose we choose $x_0 = y_0 = .5$. Then $-\text{grad } S_0 \simeq [.3, .6]$. Since a multiplicative constant can be absorbed in the parameter t, we may take

$$x_1 = .5 + t \qquad y_1 = .5 + 2t$$

The minimum of $S(.5 + t, .5 + 2t)$ is now to be found. Either by direct search or by setting $S'(t)$ to zero, we soon discover the minimum near $t = .3$, making $x_1 = .8$ and $y_1 = 1.1$. The value of $S(x_1, y_1)$ is about .04, so we proceed to a second step. Since $-\text{grad } S_1 \simeq [.5, -.25]$, we make our first right angle turn, choose

$$x_2 = .8 + 2t \qquad y_2 = 1.1 - t$$

and seek the minimum of $S(x_2, y_2)$. This proves to be near $t = .07$, making $x_2 = .94$ and $y_2 = 1.03$. Continuing in this way we obtain the successive approximations listed below. The slow convergence toward the result of Problem 25.40 may be noted. Slow convergence is typical of this method, which is often used to provide good starting approximations for the Newton algorithm.

x_n	.5	.8	.94	.928	.936	.934
y_n	.5	1.1	1.03	1.006	1.002	.998
S_n	.36	.04	.0017	.00013	.000025	.000002

The progress of the descent is suggested by path A in Fig. 25-8.

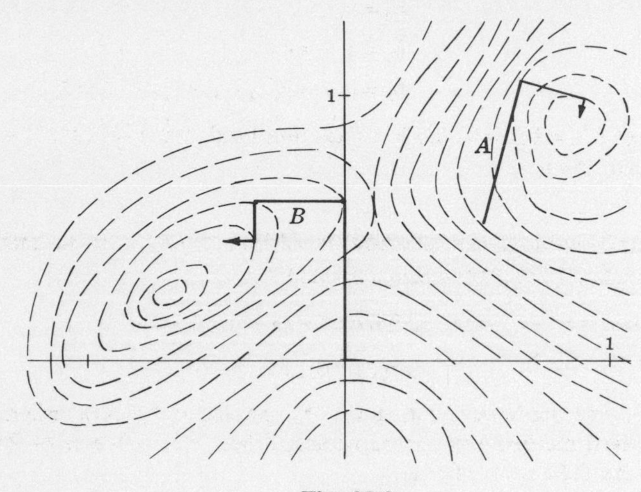

Fig. 25-8

25.43. Show that a steepest descent method may not converge to the required results.

Using the equations of the previous problem, suppose we choose the initial approximations $x_0 = y_0 = 0$. Then $-\text{grad } S_0 = [0, 2]$, so we take $x_1 = 0$ and $y_1 = t$. The minimum of $S(0, t)$ proves to be at $t = .55 = y_1$ with $S(x_1, y_1) = .73$. Computing the new gradient, we find $-\text{grad } S_1 \simeq [-.2, 0]$. This points us *westward*, away from the anticipated solution near $x = y = 1$. Succeeding steps find us traveling the path labeled B in Fig. 25-8. Our difficulty here is typical of minimization methods. There is a secondary valley near $x = -.75$, $y = .25$. Our first step has left us just to the west of the pass or saddle point between these two valleys. The direction of descent at (0, .55) is therefore westward and the descent into the secondary valley continues. Often a considerable amount of experimentation is necessary before a successful trail is found.

25.44. Generalize the idea of descent methods for the solution of optimization problems or of nonlinear systems.

The two principal questions are in what direction to proceed and how far to go. The formula

$$x^{(n)} = x^{(n-1)} + tu_{n-1}$$

keeps all options open, with $x^{(n-1)}$ the current approximation, u_{n-1} a unit vector in the next direction of search, and t the measure of how far to go. For steepest descent, u_{n-1} is the negative gradient vector. A wide variety of options have been proposed. Ideally perhaps one ought to follow a curve which is an orthogonal trajectory of the contour surfaces, on which f is constant, where f is the function being optimized. However, this leads to differential equations. Using steepest descent steps of equal length is equivalent to applying Euler's method for solving the differential equations. Even Newton's method might be viewed as a descent method, with tu_{n-1} equal to $-J^{-1}(x^{(n-1)})F(x^{(n-1)})$ in the notation used in the introduction.

QUADRATIC FACTORS. BAIRSTOW'S METHOD

25.45. Develop a recursion for the coefficients b_k in

$$q(x) = b_0x^{n-2} + \cdots + b_{n-2} \qquad r(x) = b_{n-1}(x-u) + b_n$$

when $q(x)$ and $r(x)$ are defined by

$$p(x) = a_0x^n + \cdots + a_n = (x^2 - ux - v)q(x) + r(x)$$

Multiplying out on the right and comparing the powers of x, we have

$$b_0 = a_0$$
$$b_1 = a_1 + ub_0$$
$$b_k = a_k + ub_{k-1} + vb_{k-2} \qquad k = 2, \ldots, n$$

If we artificially set $b_{-1} = b_{-2} = 0$, the last recursion holds for $k = 0, 1, \ldots n$. The b_k depend of course upon the numbers u and v.

25.46. How may the recursion of the previous problem be used to calculate $p(x)$ for a complex argument $x = a + bi$? (Assume the a_k are real.)

With $u = 2a$ and $v = -a^2 - b^2$, we have $x^2 - ux - v = 0$ so that

$$p(x) = b_{n-1}(x - 2a) + b_n$$

The advantage of this procedure is that the b_k are found by real arithmetic, so that no complex arithmetic occurs until the final step. In particular, if $b_{n-1} = b_n = 0$ then we have $p(x) = 0$. The complex conjugates $a \pm bi$ are then zeros of $p(x)$.

25.47. Develop Bairstow's method for using the Newton iteration to solve the simultaneous equations $b_{n-1}(u, v) = 0$, $b_n(u, v) = 0$.

To use Newton's iteration, as described in Problem 25.38, we need the partial derivatives of b_{n-1} and b_n relative to u and v. First taking derivatives relative to u, and letting $c_k = \partial b_{k+1}/\partial u$, we find $c_{-2} = c_{-1} = 0$, $c_0 = b_0$, $c_1 = b_1 + uc_0$, and then

$$c_k = b_k + uc_{k-1} + vc_{k-2}$$

The last result is actually valid for $k = 0, 1, \ldots, n-1$. Thus the c_k are computed from the b_k just as the b_k were obtained from the a_k. The two results we need are

$$\frac{\partial b_{n-1}}{\partial u} = c_{n-2} \qquad \frac{\partial b_n}{\partial u} = c_{n-1}$$

Similarly taking derivatives relative to v and letting $d_k = \partial b_{k+2}/\partial v$ we find $d_{-2} = d_{-1} = 0$, then $d_1 = b_1 + ud_0$, after which

$$d_k = b_k + ud_{k-1} + vd_{k-2}$$

The latter holds for $k = 0, 1, \ldots, n-2$. Since the c_k and d_k therefore satisfy the same recursion with the same initial conditions, we have proved $c_k = d_k$ for $k = 0, 1, \ldots, n-2$. In particular,

$$\frac{\partial b_{n-1}}{\partial v} = c_{n-3} \qquad \frac{\partial b_n}{\partial v} = c_{n-2}$$

and we are ready for Newton's iteration.

Suppose we have approximate roots $a \pm bi$ of $p(x) = 0$, and the associated quadratic factor $x^2 - ux - v$ of $p(x)$. This means we have approximate roots of $b_{n-1} = b_n = 0$ and are seeking improved approximations $u+h$, $v+k$. The corrections h and k are determined by

$$c_{n-2}h + c_{n-3}k = -b_{n-1}$$
$$c_{n-1}h + c_{n-2}k = -b_n$$

These are the central equations of Newton's iteration. Solving for h and k,

$$h = \frac{b_n c_{n-3} - b_{n-1}c_{n-2}}{c_{n-2}^2 - c_{n-1}c_{n-3}} \qquad k = \frac{b_{n-1}c_{n-1} - b_n c_{n-2}}{c_{n-2}^2 - c_{n-1}c_{n-3}}$$

25.48. Apply Bairstow's method to determine the complex roots of Leonardo's equation correct to nine places.

We have already found excellent initial approximations by the quotient-difference algorithm (see Problem 25.32): $u_0 \simeq -3.3642$, $v_0 \simeq -14.6033$. Our recursion now produces the following b_k and c_k:

k	0	1	2	3
a_k	1	2	10	−20
b_k	1	−1.3642	−.01386	−.03155
c_k	1	−4.7284	1.2901	

The formulas of Problem 25.47 then produce $h = -.004608$, $k = -.007930$ making

$$u_1 = u_0 + h = -3.368808 \qquad v_1 = v_0 + k = -14.611230$$

Repeating the process, we next find new b_k and c_k:

k	0	1	2	3
a_k	1	2	10	−20
b_k	1	−1.368808	.000021341	−.000103380
c_k	1	−4.737616	1.348910341	

These bring

$$h = -.000000108 \qquad k = -.000021852$$
$$u_2 = -3.368808108 \qquad v_2 = -14.611251852$$

Repeating the cycle once more finds $b_2 = b_3 = h = k = 0$ to nine places. The required roots are now

$$x_1, x_2 = \frac{1}{2}u \pm i\sqrt{-v - \frac{1}{4}u^2} = -1.684404054 \pm 3.431331350i$$

These may be further checked by computing the sum and product of all three roots and comparing with the coefficients of 2 and 20 in Leonardo's equation.

Supplementary Problems

25.49. Apply the method of Problem 25.1 to the equation $x = e^{-x}$ to find a root near $x = .5$. Show that starting with $x_0 = .5$, the approximations x_{10} and x_{11} agree to three places at .567.

25.50. Apply the Aitken acceleration to earlier approximations computed in the previous problem. When does it produce three-place accuracy?

25.51. Rewrite the equation $x^3 = x^2 + x + 1$ as $x = 1 + 1/x + 1/x^2$ and then use an iteration of the sort in Problem 25.1 to find a positive root.

25.52. Apply Newton's method to the equation of Problem 25.49. How many iterations are needed for three-place accuracy? For six-place accuracy?

25.53. Apply Newton's method to the equation of Problem 25.51.

25.54. Find the square root of 3 to six places.

25.55. Find the fifth root of 3 to six places.

25.56. Show that Newton's method applied to $f(x) = 1/x - Q = 0$ leads to the iteration $x_n = x_{n-1}(2 - Qx_{n-1})$ for producing reciprocals without division. Apply this iteration with $Q = e \simeq 2.7182818$, starting with $x_0 = .3$ and again starting with $x_0 = 1$. One of these initial approximations is not close enough to the correct result to produce a convergent sequence.

25.57. Apply *regula falsi* to the equation of Problem 25.49, starting with the approximations 0 and 1.

25.58. Apply the method of Problem 25.18 (quadratic interpolation) to the equation of Problem 25.49.

25.59. Apply the quadratic interpolation method to Leonardo's equation.

25.60. Use Bernoulli's method to find the dominant (real) root of the Fibonacci equation $x^2 - x - 1 = 0$.

25.61. Apply Bernoulli's method to the equation of Problem 25.31.

25.62. Apply Bernoulli's method to find a dominant pair of complex conjugate roots of

$$4x^4 + 4x^3 + 3x^2 - x - 1 = 0$$

25.63. Use the quotient-difference method to find all the roots of the equation of Problem 25.36.

25.64. Use the quotient-difference method to locate all the roots of the equation of Problem 25.62.

25.65. Use a Sturm sequence to show that $36x^6 + 36x^5 + 23x^4 - 13x^3 - 12x^2 + x + 1 = 0$ has only four real roots and to locate these four. Then apply Newton's method to pinpoint them.

25.66. Use a Sturm sequence to show that $288x^5 - 720x^4 + 694x^3 - 321x^2 + 71x - 6 = 0$ has five closely packed real roots. Apply Newton's method to determine these roots to six places.

25.67. Use the iterative method to find a solution of

$$x = .7 \sin x + .2 \cos y \qquad y = .7 \cos x - .2 \sin y$$

near (.5, .5).

25.68. Apply Newton's method to the system of the preceding problem.

25.69. Apply Newton's method to the system $x = x^2 + y^2$, $y = x^2 - y^2$ to find a solution near (.8, .4).

25.70. Apply the method of steepest descent to the system of the previous problem.

25.71. Apply the method of steepest descent to the system of Problem 25.67.

25.72. Given that 1 is an exact root of $x^3 - 2x^2 - 5x + 6 = 0$, find the other two roots by deflation to a quadratic equation.

25.73. Find all the roots of $x^4+2x^3+7x^2-11=0$ correct to six places using a deflation method supported by the Newton and Bairstow iterations.

25.74. Apply the Bairstow method to $x^4-3x^3+20x^2+44x+54=0$ to find a quadratic factor close to x^2+2x+2.

25.75. Find the largest root of $x^4-2.0379x^3-15.4245x^2+15.6696x+35.4936=0$.

25.76. Find two roots near $x=1$ of $2x^4+16x^3+x^2-74x+56=0$.

25.77. Find any real roots of $x^3=x+4$.

25.78. Find a small positive root of $x^{1.8632}=5.2171x-2.1167$.

25.79. Find a root near $x=2$ of $x=2\sin x$.

25.80. Find a complex pair of roots with negative real part for $x^4-3x^3+20x^2+44x+54=0$.

25.81. Find a solution of the system

$$x=\sin x\cosh y \qquad y=\cos x\sinh y$$

near $x=7$, $y=3$.

25.82. Solve the system $x^4+y^4-67=0$, $x^3-3xy^2+35=0$ near $x=2$, $y=3$.

25.83. Find the minimum for positive x of $y=(\tan x)/x^2$.

25.84. Where does the curve $y=e^{-x}\log x$ have an inflection point?

25.85. Find the smallest positive root of $1-x+\dfrac{x^2}{(2!)^2}-\dfrac{x^3}{(3!)^2}+\dfrac{x^4}{(4!)^2}-\cdots=0$.

25.86. Find the maximum value of $y(x)$ near $x=1$, given that $\sin(xy)=y-x$.

25.87. Find to twelve digits a root near 2 of $x^4-x=10$.

25.88. Find the smallest real root of $e^{-x}=\sin x$.

25.89. Split the fourth-degree polynomial $x^4+5x^3+3x^2-5x-9$ into quadratic factors.

25.90. Find a root near 1.5 of $x=\frac{1}{2}+\sin x$.

25.91. Find all the roots of $2x^3-13x^2-22x+3=0$.

25.92. Find a root near 1.5 of $x^6=x^4+x^3+1$.

25.93. Find two roots near $x=2$ of $x^4-5x^3-12x^2+76x-79=0$.

25.94. Show that the second-degree term is removed from the general cubic equation

$$x^3+ax^2+bx+c=0$$

by the translation $x=y-a/3$. See also the following problem.

25.95. In 1545 Cardano published this formula for solving the cubic equation $x^3 + bx + c = 0$. (Note the absence of a second-degree term.)

$$x = \left[-\frac{c}{2} + \sqrt{\left(\frac{c}{2}\right)^2 + \left(\frac{b}{3}\right)^3}\right]^{1/3} - \left[\frac{c}{2} + \sqrt{\left(\frac{c}{2}\right)^2 + \left(\frac{b}{3}\right)^3}\right]^{1/3}$$

Apply it to find at least the real root $x = 1$ of

$$x^3 + 3x - 4 = 0$$

Can it also manage the real root $x = 4$ of $x^3 - 15x - 4 = 0$?

Chapter 26

Linear Systems

SOLUTION OF LINEAR SYSTEMS

This may very well be the principal problem of numerical analysis. Much of applied mathematics reduces to a set of linear equations, or a linear system,

$$Ax = b$$

with the matrix A and vector b given and the vector x to be determined. An extensive set of algorithms have been developed for doing this, several of which will be presented. The variety of algorithms available indicates that the apparently elementary character of the problem is deceptive. There are numerous pitfalls.

Gaussian elimination is one of the oldest algorithms and still one of the most popular. It involves replacing equations by combinations of equations in such a way that a *triangular system* is obtained.

$$\begin{aligned} u_{11}x_1 + u_{12}x_2 + \cdots + u_{1n}x_n &= c_1 \\ u_{22}x_2 + \cdots + u_{2n}x_n &= c_2 \\ &\vdots \\ u_{nn}x_n &= c_n \end{aligned}$$

After this, the components of the vector x are easily found, one after the other, by a process called **back-substitution**. The last equation determines x_n, which is then substituted in the next-to-last equation to get x_{n-1}, and so on.

The Gauss algorithm also yields a factorization of the matrix A, in the form $A = LU$, where U is the upper triangular matrix shown above and L is a lower triangle with 1s on the diagonal. The algorithm can be used to prove the fundamental theorem of algebra, which deals with the question of whether or not a solution exists. The theorem guarantees a unique solution of $Ax = b$ precisely when the corresponding homogeneous system $Ax = 0$ has only the solution $x = 0$. Both systems, as well as the coefficient matrix A, are then called nonsingular. When $Ax = 0$ has solutions other than $x = 0$, both systems and the matrix A are singular. In this case $Ax = b$ will have either no solution at all or else an infinity of solutions. Singular systems occur in eigenvalue problems. If the methods of this chapter are applied inadvertently to a singular system, there is the curious possibility that unavoidable roundoff errors will alter it to an "almost identical" nonsingular system. A computed "solution" may then be produced where none actually exists.

Factorization methods convert A into products of the form LU or LDU, where L is zero above the main diagonal, U is zero below it, and D has only diagonal elements different from zero. The matrix L is called lower triangular and U is upper triangular. If L or U has all diagonal elements equal to 1, it is called unit triangular. The methods of Doolittle, Crout, Cholesky, and, as already mentioned, Gauss produce factorizations. When A has been factored in this way, the solution is easily accessible. Since

$$Ax = LUx = L(Ux) = Ly = b$$

we first solve $Ly = b$ for y and then $Ux = y$ for x. The first of these triangular systems responds to **forward-substitution,** and the second to back-substitution.

Iterative methods generate sequences of successive approximations to the solution vector x. The classic of this type is the Gauss–Seidel method, which reshapes the system $Ax = b$ in the form

$$\begin{aligned} x_1 &= \cdots \\ x_2 &= \cdots \\ &\vdots \\ x_n &= \cdots \end{aligned}$$

by solving the ith equation for x_i. An initial approximation to all the x_i allows each component to be corrected in its turn and when the cycle is complete to begin another cycle. A number of convergence theorems have been proved. The method is often used for sparse matrices A, in which many elements are zero.

Iterative refinement of an approximate solution $x^{(1)}$ using the *residual* vector r, defined by

$$r = b - Ax^{(1)}$$

is often a useful algorithm. Let e be the error

$$e = x - x^{(1)}$$

and observe that
$$Ae = Ax - Ax^{(1)} = b - (b - r) = r$$

Solving $Ae = r$ yields an approximation to e, say $e^{(1)}$, from which

$$x^{(2)} = x^{(1)} + e^{(1)}$$

manages a new approximation to the true solution x. The routine can be continued as long as it seems productive.

There are a wide variety of more elaborate iterative methods.

The error in a computed solution $x^{(c)}$ occurs for a combination of reasons. The input information may be imperfect, that is, the elements of A and b may contain error. There will almost surely be roundoff errors made during the course of the solution algorithm, probably millions of them in a large-scale problem. When a convergent iterative process is terminated, it is unlikely that the approximation in hand is the true solution. Estimates of the eventual error due to such sources can be made, and they are important, though often rather conservative. Backward error analysis is a useful tool in investigating the internal roundoff problem.

The character of the coefficient matrix A strongly influences error behavior. Nearly singular systems are extremely sensitive to even small errors in A and b and to internal roundoffs. The *condition* of A can be described numerically using the idea of a matrix norm, a high condition number meaning a nearly singular matrix and relatively poor error control. Such matrices are also called *ill-conditioned.* Sometimes poor condition will make itself known by erratic behavior of the algorithm. Unfortunately, this is not always true.

MATRIX INVERSION

Knowing the inverse of A would, of course, allow the system $Ax = b$ to be solved as a by-product, since

$$x = A^{-1}b$$

but this is usually an uneconomical route to the solution of a linear system. Complete knowledge of the elements of A^{-1} is required only in a few types of applications, notably statistical analysis. The methods just discussed for solving $Ax = b$ can be adapted to find inverses. Elimination, factorization, iteration, and an exchange method will be illustrated in the problems.

EIGENVALUE PROBLEMS

Eigenvalue problems require that we determine numbers λ such that the linear system $Ax = \lambda x$ will have solutions other than $x = 0$. These numbers are called **eigenvalues.** The corresponding solutions, or eigenvectors, are also of interest. Three general methods of approach will be presented.

1. The characteristic polynomial of a matrix A has as its zeros the eigenvalues of A. A direct procedure, resembling Gaussian elimination, for finding this polynomial will be included. To find its zeros, the methods of Chapter 25 may be used. With an eigenvalue in hand, substitution into $Ax = \lambda x$ yields a singular system. The value of some component of x may be specified and the reduced system solved by our methods for linear systems.

2. The power method generates the vectors

$$x^{(p)} = A^{(p)}V$$

with V a somewhat arbitrary initial vector, and produces the dominant eigenvalue with its eigenvector. For large values of p it proves that $x^{(p)}$ is close to an eigenvector corresponding to

$$\lambda = \frac{x^{(p)T}Ax^{(p)}}{x^{(p)T}x^{(p)}}$$

a formula known as the **Rayleigh quotient.** Modifications lead to the absolutely smallest and certain next-dominant eigenvectors.

An interesting variation uses the idea of shifting the eigenvalues to speed up the convergence of the power method. The inverse power method and inverse iteration are developments of this idea.

3. Reduction to canonical forms (simplified forms such as diagonal, triple diagonal, triangular, Hessenberg) is possible in many ways. When done by similarity transformations, the eigenvalues are not changed. The Jacobi method subjects a real, symmetric matrix to rotations based upon the submatrix

$$\begin{bmatrix} \cos\phi & -\sin\phi \\ \sin\phi & \cos\phi \end{bmatrix}$$

and leads to an almost diagonal form. The Givens method uses similar rotations and achieves a triple diagonal form in a finite number of steps. The QR method produces, under certain circumstances, a triangular matrix. The underlying idea of all these procedures is that eigenvalues of the canonical forms are found more easily.

COMPLEX SYSTEMS

Many of the methods used for real systems can be taken over for complex if a computer capable of complex arithmetic is available. If not, complex systems may be exchanged for equivalent, and larger, real systems. Thus, comparing real and imaginary parts of

$$(A + iB)(x + iy) = a + ib$$

leads to

$$\begin{bmatrix} A & -B \\ B & A \end{bmatrix}\begin{bmatrix} x \\ y \end{bmatrix} = \begin{bmatrix} a \\ b \end{bmatrix}$$

to which our real algorithms apply. The inversion problem

$$(A + iB)(C + iD) = I$$

responds to similar treatment. Eigenvalues can also be approached in this way.

Solved Problems

GAUSSIAN ELIMINATION

26.1. Solve by Gaussian elimination.

$$x_1 + \frac{1}{2}x_2 + \frac{1}{3}x_3 = 1$$

$$\frac{1}{2}x_1 + \frac{1}{3}x_2 + \frac{1}{4}x_3 = 0$$

$$\frac{1}{3}x_1 + \frac{1}{4}x_2 + \frac{1}{5}x_3 = 0$$

We begin by seeking the absolutely largest coefficient in column 1. Here it is in the top place. If this were not so, an interchange of rows would be made to arrange it. This largest element is called the *first pivot.* Now define

$$l_{21}=\frac{a_{21}}{a_{11}}=\frac{1}{2} \qquad l_{31}=\frac{a_{31}}{a_{11}}=\frac{1}{3}$$

and reduce the two lower coefficients in column 1 to zero in a familiar way, subtracting from the ith equation the product of l_{i1} by the first. Here is the result:

$$\begin{aligned} x_1+\frac{1}{2}x_2+\frac{1}{3}x_3&=1 \\ \frac{1}{12}x_2+\frac{1}{12}x_3&=-\frac{1}{2} \\ \frac{1}{12}x_2+\frac{4}{45}x_3&=-\frac{1}{3} \end{aligned}$$

This is the first modified system. The same procedure is now applied to the smaller system consisting of the two lower equations. Again the absolutely largest coefficient is already at the top of the leading column, so no interchange of rows is needed. We find

$$l_{32}=\frac{a_{32}^{(1)}}{a_{22}^{(1)}}=1$$

and so subtract from the third equation the product of l_{32} and the second equation. [The superscript (1) refers to the first modified system.] We then have

$$\begin{aligned} x_1+\frac{1}{2}x_2+\frac{1}{3}x_3&=1 \\ \frac{1}{12}x_2+\frac{1}{12}x_3&=-\frac{1}{2} \\ \frac{1}{180}x_3&=\frac{1}{6} \end{aligned}$$

and the triangular system is evident. The solution process is then completed by back-subsitution, which finds the components x_i from the bottom up, and in reverse order:

$$x_3=30 \qquad x_2=-36 \qquad x_1=9$$

26.2. Why is pivoting important?

Consider this extreme example:

$$\begin{aligned} 10^{-5}x_1+x_2&=1 \\ x_1+x_2&=2 \end{aligned}$$

The very small coefficient makes it clear that the solution ought to be quite close to $x_1=x_2=1$. Suppose we solve without pivoting and with the assumption that only four decimal places can be carried. Exact subtraction would yield the equation

$$(1-10^5)x_2=2-10^5$$

but with the restriction on decimal places we must settle for

$$10^5x_2=10^5$$

which still presents us with $x_2=1$. However, continuing the back-substitution we then face

$$10^{-5}x_1+1=1$$

making $x_1=0$ instead of the anticipated 1.

But now interchange the two equations, bringing the largest coefficient of column 1 into the pivot position:

$$x_1 + x_2 = 2$$
$$10^{-5}x_1 + x_2 = 1$$

Exact subtraction would now bring

$$(1 - 10^{-5})x_2 = 1 - 2(10^{-5})$$

which the same restrictions would round to $x_2 = 1$. This time the back-substitution manages

$$x_1 + 1 = 2$$

and $x_1 = 1$. Pivoting has made the difference between nonsense and a perfect result. Experience with many less dramatic systems has shown that pivoting is an important part of the elimination algorithm. The technique described is called *partial pivoting,* since the search for largest coefficient is limited to the immediate column. The value of a broader search, into other columns, and leading to column interchanges, is a matter of debate.

The example in hand may be used to illustrate a further point. Multiply the first equation by 10^5 to obtain

$$x_1 + 10^5 x_2 = 10^5$$
$$x_1 + \quad x_2 = 2$$

and make pivoting unnecessary. The usual subtraction manages

$$(1 - 10^5)x_2 = 2 - 10^5$$

when done exactly, but becomes $\quad -10^5 x_2 = -10^5$

after rounding. So $x_2 = 1$. But then $\quad x_1 = 10^5 - 10^5 = 0$

and we have the earlier "solution." The point is, even pivoting may not help when very large coefficients occur elsewhere. One way out of the difficulty might be to interchange columns, but an alternative is to *normalize* each equation, making the absolutely largest coefficient in each about the same. A popular way to do this is dividing each equation by its coefficient of greatest size. The "norm" of each equation will then be 1. In our example we would, of course, return to the original system. The lesson appears to be that the combination of normalization and partial pivoting has a good chance of yielding a good result.

26.3. Summarize the Gauss algorithm for the general n by n linear system.

Suppose that k steps of the type described in Problem 26.1 have been made, bringing the system to this form:

$$\begin{aligned}
u_{11}x_1 + u_{12}x_2 + \cdots + u_{1k}x_k + u_{1,k+1}x_{k+1} + \cdots + u_{1n}x_n &= b_1' \\
u_{22}x_2 + \cdots + u_{2k}x_k + u_{2,k+1}x_{k+1} + \cdots + u_{2n}x_n &= b_2' \\
&\vdots \\
u_{kk}x_k + u_{k,k+1}x_{k+1} + \cdots + u_{kn}x_n &= b_k' \\
a_{k+1,k+1}^{(k)}x_{k+1} + \cdots + a_{k+1,n}^{(k)}x_n &= b_{k+1}^{(k)} \\
&\vdots \\
a_{n,k+1}^{(k)}x_{k+1} + \cdots + a_{nn}^{(k)}x_n &= b_n^{(k)}
\end{aligned}$$

The top k equations are in their final form, with $u_{11}, \ldots, u_{kk}$ the first k pivots. In the remaining $n - k$ equations the coefficients bear the superscript (k) of this modified system. We next seek the $(k+1)$th pivot among the coefficients of x_{k+1} in the lower $n - k$ equations. It will be the absolutely largest and its equation will be interchanged with equation $k + 1$. With this new pivot in place, now called $u_{k+1,k+1}$, a new set of multipliers is found

$$l_{i,k+1} = \frac{a_{i,k+1}^{(k)}}{u_{k+1,k+1}} \qquad i = k+2, \ldots, n$$

and zeros are arranged under the new pivot by subtracting equations. Coefficient changes are governed by

$$a_{ij}^{(k+1)} = a_{ij}^{(k)} - l_{i,k+1}a_{k+1,j}^{(k)} \qquad k = 0, \ldots, n-2$$
$$b_i^{(k+1)} = b_i^{(k)} - l_{i,k+1}b_k^{(k)} \qquad j = k+2, \ldots, n$$
$$i = k+2, \ldots, n$$

with $k = 0$ referring to the original system. The back-substitution part of the algorithm is represented by

$$x_i = \frac{1}{u_{ii}}\left(b_i' - \sum_{j=i+1}^{n} u_{ij}x_j\right) \qquad i = n, \ldots, 1$$

26.4. What is the Gauss–Jordan variation?

Here zeros are generated both below and above each pivot, by further subtractions. The final matrix is thus diagonal rather than triangular and back-substitution is eliminated. The idea is attractive, but it involves more computing than the original algorithm and so is little used.

26.5. Estimate the amount of computing needed to carry out the Gauss algorithm for an n by n system.

Consider the reduction of the coefficient matrix A to triangular form. This is where the lion's share of the effort occurs. At the first step, $(n-1)^2$ modified coefficients are obtained. We further limit our attention to a count of such coefficients. In successive steps this number is reduced and the grand total will be

$$(n-1)^2 + (n-2)^2 + \cdots + 1$$

coefficients. By a well-known result of algebra this is equal to $(2n^3 - 3n^2 + n)/6$, from which the principal term $n^3/3$ is extracted as a simple measure of the computation's size. If $n = 100$, this number runs to six figures.

26.6. Apply Gaussian elimination to this system, assuming that a computer capable of carrying only two floating-point digits is to do the calculations.

$$x_1 + .67x_2 + .33x_3 = 2$$
$$.45x_1 + x_2 + .55x_3 = 2$$
$$.67x_1 + .33x_2 + x_3 = 2$$

With $l_{21} = .45$ and $l_{31} = .67$, the array below left summarizes the first stage of the process, and then with $l_{32} = -.17$ the array at the right shows the final triangularization.

1	.67	.33	2.0	1	.67	.33	2.0
0	.70	.40	1.1	0	.70	.40	1.1
0	−.12	.78	.7	0	0	.85	.89

Back-substitution now begins with

$$x_3 = \frac{.89}{.85} = 1.047$$

if we assume a double precision accumulator, but rounding to 1.0 in any event. Then

$$x_2 = \left(\frac{1}{.7}\right)(1.1 - .4) = 1.0$$
$$x_1 = 2 - .67 - .33 = 1.0$$

and the exact (1, 1, 1) solution has been found in spite of the severe limitations of the computer. This is because we have a very cooperative matrix. (See also Problem 26.20.)

26.7. **What is the connection between Gaussian elimination and factors of the coefficient matrix?**

Form matrices L and U as follows, using results of Problem 26.1:

$$L = \begin{bmatrix} 1 & 0 & 0 \\ l_{21} & 1 & 0 \\ l_{31} & l_{32} & 1 \end{bmatrix} = \begin{bmatrix} 1 & 0 & 0 \\ \frac{1}{2} & 1 & 0 \\ \frac{1}{3} & 1 & 1 \end{bmatrix}$$

$$U = \begin{bmatrix} u_{11} & u_{12} & u_{13} \\ 0 & u_{22} & u_{23} \\ 0 & 0 & u_{33} \end{bmatrix} = \begin{bmatrix} 1 & \frac{1}{2} & \frac{1}{3} \\ 0 & \frac{1}{12} & \frac{1}{12} \\ 0 & 0 & \frac{1}{180} \end{bmatrix}$$

Then

$$LU = \begin{bmatrix} 1 & \frac{1}{2} & \frac{1}{3} \\ \frac{1}{2} & \frac{1}{3} & \frac{1}{4} \\ \frac{1}{3} & \frac{1}{4} & \frac{1}{5} \end{bmatrix} = A$$

For a general proof of this factorization see the following problem.

26.8. **Show that if L is a lower triangular matrix with elements l_{ij} and $l_{ii} = 1$, and if U is an upper triangular matrix with elements u_{ij}, then $LU = A$.**

The proof involves some easy exercise with triangular matrices. Returning briefly to the opening example, define

$$S_1 = \begin{bmatrix} 1 & 0 & 0 \\ -\frac{1}{2} & 1 & 0 \\ -\frac{1}{3} & 0 & 1 \end{bmatrix} \qquad S = \begin{bmatrix} 1 & 0 & 0 \\ 0 & 1 & 0 \\ 0 & -1 & 1 \end{bmatrix}$$

and observe that the product S_1A effects Step 1 of the Gauss algorithm, as it applies to the left sides of the equations, while S_2S_1A then effects Step 2. This means that

$$S_2S_1A = U \qquad A = S_1^{-1}S_2^{-1}U = LU$$

with $L = S_1^{-1}S_2^{-1}$. Also note that

$$S_1^{-1} = \begin{bmatrix} 1 & 0 & 0 \\ \frac{1}{2} & 1 & 0 \\ \frac{1}{3} & 0 & 1 \end{bmatrix} \qquad S_2^{-1} = \begin{bmatrix} 1 & 0 & 0 \\ 0 & 1 & 0 \\ 0 & 1 & 0 \end{bmatrix}$$

so that inversions are achieved by changing the signs of the l_{ij} entries.

For the general problem assume at first that no interchanges will be needed. Define matrices

$$L_i = \begin{bmatrix} 1 & & & & & \\ & \ddots & & & & \\ & & 1 & & & \\ & & -l_{i+1,i} & \ddots & & \\ & & \vdots & & \ddots & \\ & & -l_{n,i} & & & 1 \end{bmatrix} \qquad i = 1, \ldots, n-1$$

with all other elements zero. As in the example, each of these effects one step of the elimination process, making

$$L_{n-1}L_{n-2}\cdots L_1A = U$$

This means that

$$A = L_1^{-1}\cdots L_{n-1}^{-1}U = LU$$

Since the product of lower triangles with diagonal 1s is itself of the same type we have our factorization. In addition, since each inversion is achieved by changing the signs of the l_{ij} entries, these are readily in hand and may be multiplied to rediscover

$$L = \begin{bmatrix} 1 & 0 & \cdots & & 0 \\ l_{21} & 1 & \cdots & & 0 \\ \cdots & \cdots & \cdots & \cdots & \cdots \\ l_{n1} & l_{n2} & \cdots & l_{n,n-1} & 1 \end{bmatrix}$$

Now suppose that some interchanges are to be made. Introduce the interchange matrices

$$I_{ij} = \begin{bmatrix} 1 & & & & & & \\ & \ddots & & & & & \\ & & 0 & 1 & & & \\ & & 1 & 0 & & & \\ & & & & \ddots & & \\ & & & & & & 1 \end{bmatrix} \begin{matrix} \\ \\ \text{row } i \\ \text{row } j \\ \\ \\ \end{matrix}$$

$$\begin{matrix} \text{col} & \text{col} \\ i & j \end{matrix}$$

The product $I_{ij}A$ will have rows i and j of A interchanged, while AI_{ij} has the corresponding columns interchanged. The elimination algorithm now uses a chain of I_{ij} interchanges and L_i operations, leading to this representation:

$$L_{n-1}I_{n-1,r_{n-1}}L_{n-2}I_{n-2,r_{n-2}}\cdots L_1L_{1,r_1}A = U$$

where the r_i are the rows containing the selected pivots. This can be rearranged as

$$(L_{n-1}L_{n-2}\cdots L_1)(I_{n-1,r_{n-1}}\cdots I_{1,r_1})A = U$$

or

$$L^{-1}PA = U \qquad PA = LU$$

with P the permutation matrix including the $n-1$ interchanges. Assuming A nonsingular, this means that there is a permutation of rows such that PA has an LU factorization. The uniqueness of this factorization will be evident from Problem 26.14.

26.9. Solve the system $Ax = b$ assuming an LU factorization has been done.

We have, since L, U, and P are in hand,

$$Ax = LUx = PAx = Pb$$

and letting $y = Ux$, first solve $Ly = Pb$ for y. This is easily done by forward-substitution. Then $Ux = y$ is solved by back-substitution. More specifically, and with p_i denoting an element of Pb, the system $Ly = Pb$ is

$$\begin{aligned} l_{11}y_1 \qquad\qquad\qquad\qquad &= p_1 \\ l_{21}y_1 + l_{22}y_2 \qquad\qquad\quad &= p_2 \\ &\vdots \\ l_{n1}y_1 + l_{n2}y_2 + \cdots + l_{nn}y_n &= p_n \end{aligned}$$

with all $l_{ii} = 1$. The solution by forward-substitution is clearly $y_1 = p_1$, $y_2 = p_2 - l_{21}y_1$, or more generally,

$$y_r = p_r - l_{r1}y_1 - \cdots - l_{r-1,r-1}y_{r-1}$$

for $r = 1, \ldots, n$. The backward-substitution is then achieved by the formula of Problem 26.3, modified only by the replacement of the vector b' by y:

$$x_i = \left(\frac{1}{u_{ii}}\right)(y_i - u_{i,i+1}x_{i+1} - \cdots - u_{in}x_n)$$

with $i = n, \ldots, 1$. The combination of factoring and forward-backward substitution is particularly useful if the system must be solved for more than one vector b.

26.10. What is a compact algorithm?

When Gaussian elimination was done by hand, many elements of A were copied many times. In a computer this would be equivalent to making liberal use of storage space. With large-scale systems it is advisable to be economical both of storage space and computer time. For this reason, compact algorithms have been devised. For example, as elimination proceeds, the lower triangle of matrix A is replaced by zeros. These storage locations may better be used to record successively the values l_{ij}, for $j < i$. At the end of the run the upper triangle of A will then have been replaced by U, and the lower triangle by L without its unit diagonal. And there is no need to store all the interchange matrices I_{ij}. It is enough to define initially a vector v with elements $(1, 2, 3, \ldots, n)$ and at each step to simply interchange the appropriate elements. If, for instance, the first pivot is in row 3, then $(3, 2, 1, 4, \ldots, n)$ records this. It is not necessary to physically interchange the rows, thus saving the time that would have been used for this maneuver. From the final v the permutation matrix P can be constructed if desired, or v itself used to permute the elements of vector b.

26.11. Apply the procedure of Problem 26.10 to this matrix

$$A = \begin{bmatrix} 0 & 1 & 2 & 3 \\ 3 & 0 & 1 & 2 \\ 2 & 3 & 0 & 1 \\ 1 & 2 & 3 & 0 \end{bmatrix}$$

The essential computations are displayed in Fig. 26-1. In three steps the original matrix is replaced by a four by four array containing all the information needed, except for the vector v which traces the interchanges.

At this point matrix A has been replaced by a triangular matrix in the LU factorization of PA. The vector v tells us that the triangle will be evident if we look at rows 2, 3, 4, 1 in that order. Indeed the unstarred elements are the factor U. The factor L can also be read by taking the starred elements in the same row order. As for the permutation matrix P, it is constructed by placing 1s in columns 2, 3, 4, 1 of

$$\begin{bmatrix} 0 & 1 & 2 & 3 \\ 3 & 0 & 1 & 2 \\ 2 & 3 & 0 & 1 \\ 1 & 2 & 3 & 0 \end{bmatrix}$$

The given matrix A $\qquad v = (1, 2, 3, 4)$

Identify the first pivot, 3.

$$\begin{bmatrix} 0^* & 1 & 2 & 3 \\ \textcircled{3} & 0 & 1 & 2 \\ \frac{2^*}{3} & 3 & -\frac{2}{3} & -\frac{1}{3} \\ \frac{1^*}{3} & 2 & \frac{8}{3} & -\frac{2}{3} \end{bmatrix}$$

Bring its row number to the first position in v. $v = (2, 1, 3, 4)$.
Compute and store the l_{i1} (starred).
Compute the nine new entries by subtractions (right of the solid line).

$$\begin{bmatrix} 0^* & \frac{1^*}{3} & \frac{20}{9} & \frac{28}{9} \\ 3 & 0 & 1 & 2 \\ \frac{2^*}{3} & \textcircled{3} & -\frac{2}{3} & -\frac{1}{3} \\ \frac{1^*}{3} & \frac{2^*}{3} & \frac{28}{9} & -\frac{4}{9} \end{bmatrix}$$

Identify the second pivot (column 2 and right of the solid line).
Bring its row number to second position in v (2, 3, 1, 4).
Compute the l_{i2} and store them (starred).
Compute the four new entries.

$$\begin{bmatrix} 0^* & \frac{1^*}{3} & \frac{5^*}{7} & \frac{24}{7} \\ 3 & 0 & 1 & 2 \\ \frac{2^*}{3} & 3 & -\frac{2}{3} & -\frac{1}{3} \\ \frac{1^*}{3} & \frac{2^*}{3} & \textcircled{\frac{28}{9}} & -\frac{4}{9} \end{bmatrix}$$

Identify the last pivot (column 3 and right of the solid line). Bring its row number to third position in v (2, 3, 4, 1).
Compute the l_{i3} and store them.
Compute the one new entry.

Fig. 26-1

an otherwise zero matrix, as follows:

$$P = \begin{bmatrix} 0 & 1 & 0 & 0 \\ 0 & 0 & 1 & 0 \\ 0 & 0 & 0 & 1 \\ 1 & 0 & 0 & 0 \end{bmatrix}$$

One may now calculate

$$PA = LU = \begin{bmatrix} 3 & 0 & 1 & 2 \\ 2 & 3 & 0 & 1 \\ 1 & 2 & 3 & 0 \\ 0 & 1 & 2 & 3 \end{bmatrix}$$

and so verify all steps taken.

26.12. Using the results of the preceding problem and given the vector b with components $(0, 1, 2, 3)$, solve $Ax = b$.

We use the suggestion in Problem 26.9. First either Pb or the vector v rearranges the components of b in the order $(1, 2, 3, 0)$. Although it is not necessary, suppose we display the system $Ly = Pb$ directly.

$$\begin{bmatrix} 1 & 0 & 0 & 0 \\ \frac{2}{3} & 1 & 0 & 0 \\ \frac{1}{3} & \frac{2}{3} & 1 & 0 \\ 0 & \frac{1}{3} & \frac{5}{7} & 1 \end{bmatrix} \begin{bmatrix} y_1 \\ y_2 \\ y_3 \\ y_4 \end{bmatrix} = \begin{bmatrix} 1 \\ 2 \\ 3 \\ 0 \end{bmatrix}$$

Forward-substitution then manages $y = (1, \frac{4}{3}, \frac{16}{9}, -\frac{12}{7})^T$. Turning to $Ux = y$ we face

$$\begin{bmatrix} 3 & 0 & 1 & 2 \\ 0 & 3 & -\frac{2}{3} & -\frac{1}{3} \\ 0 & 0 & \frac{28}{9} & -\frac{4}{9} \\ 0 & 0 & 0 & \frac{24}{7} \end{bmatrix} \begin{bmatrix} x_1 \\ x_2 \\ x_3 \\ x_4 \end{bmatrix} = \begin{bmatrix} 1 \\ \frac{4}{3} \\ \frac{16}{9} \\ -\frac{12}{7} \end{bmatrix}$$

from which comes $x = (\frac{1}{2}, \frac{1}{2}, \frac{1}{2}, -\frac{1}{2})^T$, which may be verified directly in $Ax = b$.

26.13. Prove the fundamental theorem of linear algebra.

We use the Gauss algorithm. If it can be continued to the end, producing a triangular system, then back-substitution will yield the unique solution. If all the b_i are zero, this solution has all zero components. This is already a principal part of the theorem. But suppose the algorithm cannot be continued to the anticipated triangular end. This happens only when at some point all coefficients below a certain level are zero. To be definite, say the algorithm has reached this point.

$$\begin{aligned} u_{11}x_1 + \cdots \qquad\qquad\qquad &= b_1' \\ u_{22}x_2 + \cdots \qquad\qquad &= b_2' \\ &\vdots \\ u_{kk}x_k + \cdots &= b_k' \\ 0 &= b_{k+1}^{(k)} \\ &\vdots \\ 0 &= b_n^{(k)} \end{aligned}$$

Then in the homogeneous case, where all the b's are zero, we may choose x_{k+1} to x_n as we please and then determine the other x_i. But in the general case, unless $b_{k+1}^{(k)}$ to $b_n^{(k)}$ are all zero, no solution is possible. If these b's do happen to be zero, then again we may choose x_{k+1} to x_n freely, after which the other x_i are determined. This is the content of the fundamental theorem.

FACTORIZATIONS

26.14. Determine the elements of matrices L and U such that $A = LU$ by a direct comparison of corresponding elements.

Assume that no interchanges will be necessary. Then we are to equate corresponding elements from the two sides of

$$\begin{bmatrix} l & 0 & 0 & \cdots & 0 \\ l_{21} & 1 & 0 & \cdots & 0 \\ l_{31} & l_{32} & 1 & \cdots & 0 \\ \cdots & \cdots & \cdots & \cdots & \cdots \\ l_{n1} & l_{n2} & l_{n3} & \cdots & 1 \end{bmatrix} \begin{bmatrix} u_{11} & u_{12} & u_{13} & \cdots & u_{1n} \\ 0 & u_{22} & u_{23} & \cdots & u_{2n} \\ 0 & 0 & u_{33} & \cdots & u_{3n} \\ \cdots & \cdots & \cdots & \cdots & \cdots \\ 0 & 0 & 0 & \cdots & u_{nn} \end{bmatrix} = \begin{bmatrix} a_{11} & \cdots & a_{1n} \\ a_{21} & \cdots & a_{2n} \\ a_{31} & \cdots & a_{3n} \\ \cdots & \cdots & \cdots \\ a_{n1} & \cdots & a_{nn} \end{bmatrix}$$

which amounts to n^2 equations in the n^2 unknowns l_{ij} and u_{ij}. The determination runs as follows. First multiply the top row of L by all columns of U to get

$$u_{1j} = a_{1j} \qquad j = 1, \ldots, n$$

Next multiply the rows of L (omit the first) by column 1 of U, finding $l_{i1}u_{11} = a_{i1}$, from which the l_{i1} follow.

$$l_{i1} = \frac{a_{i1}}{u_{11}} \qquad i = 2, \ldots, n$$

It is next the turn of the second row of L to multiply the columns of U (omit the first). The second row of U is then

$$u_{2j} = a_{2j} - l_{21}u_{1j} \qquad j = 2, \ldots, n$$

Now multiply the rows of L (omit the first two) by column 2 of U. All elements involved except the l_{i2} are in hand, so we solve for these.

$$l_{i2} = \frac{a_{i2} - l_{i1}u_{12}}{u_{22}} \qquad i = 3, \ldots, n$$

Continuing in this recursive way, we alternately find the rows of U to be

$$u_{rj} = a_{rj} - \sum_{k=1}^{r-1} l_{rk}u_{kj} \qquad j = r, \ldots, n$$

each row followed by the corresponding column of L.

$$l_{ir} = \frac{a_{ir} - \sum_{k=1}^{r-1} l_{ik}u_{kr}}{u_{rr}} \qquad i = r+1, \ldots, n$$

This procedure is called the **Doolittle algorithm.**

26.15. What is the Crout algorithm?

The Crout algorithm also produces a factorization of A, in the form $L'U'$, with U' having the diagonal of 1s and L' the general diagonal. Formulas for the elements of the factors may be found very much as in Problem 26.14, but it is of interest to note that, with D denoting the matrix of diagonal elements of our earlier U and zeros elsewhere,

$$A = LU = L(DD^{-1})U = (LD)(D^{-1}U) = L'U'$$

so the two factorizations are closely related.

26.16. Develop the Choleski method for factoring a real, symmetric, positive definite matrix.

Here we will find factors of the form

$$A = LL^T$$

the T denoting the transpose. The procedure is almost identical with that of Problem 26.14, with symmetry allowing us to consider only the lower triangle of A. The Hilbert matrix of order three can once again serve as a small-scale introduction.

$$\begin{bmatrix} l_{11} & 0 & 0 \\ l_{21} & l_{22} & 0 \\ l_{31} & l_{32} & l_{33} \end{bmatrix} \begin{bmatrix} l_{11} & l_{21} & l_{31} \\ 0 & l_{22} & l_{23} \\ 0 & 0 & l_{33} \end{bmatrix} = \begin{bmatrix} 1 & \frac{1}{2} & \frac{1}{3} \\ \frac{1}{2} & \frac{1}{3} & \frac{1}{4} \\ \frac{1}{3} & \frac{1}{4} & \frac{1}{5} \end{bmatrix}$$

The elements of L will be found from top to bottom and left to right.

$$\begin{aligned} l_{11}l_{11} &= 1 & l_{11} &= 1 \\ l_{21}l_{11} &= \frac{1}{2} & l_{21} &= \frac{1}{2} \\ l_{21}^2 + l_{22}^2 &= \frac{1}{3} & l_{22} &= \frac{1}{\sqrt{12}} \\ l_{31}l_{11} &= \frac{1}{3} & l_{31} &= \frac{1}{3} \\ l_{31}l_{21} + l_{32}l_{22} &= \frac{1}{4} & l_{32} &= \frac{1}{\sqrt{12}} \\ l_{31}^2 + l_{32}^2 + l_{33}^2 &= \frac{1}{5} & l_{33} &= \frac{1}{\sqrt{180}} \end{aligned}$$

The computation is again recursive, each line having only one unknown.

Because of the way the algorithm develops, should we now wish to extend our effort to the Hilbert matrix of order four, it is only necessary to border L with a new bottom row and fourth column.

$$LL^T = \begin{bmatrix} 1 & 0 & 0 & 0 \\ \frac{1}{2} & \frac{1}{\sqrt{12}} & 0 & 0 \\ \frac{1}{3} & \frac{1}{\sqrt{12}} & \frac{1}{\sqrt{180}} & 0 \\ l_{41} & l_{42} & l_{43} & l_{44} \end{bmatrix} L^T = \begin{bmatrix} 1 & \frac{1}{2} & \frac{1}{3} & \frac{1}{4} \\ \frac{1}{2} & \frac{1}{3} & \frac{1}{4} & \frac{1}{5} \\ \frac{1}{3} & \frac{1}{4} & \frac{1}{5} & \frac{1}{6} \\ \frac{1}{4} & \frac{1}{5} & \frac{1}{6} & \frac{1}{7} \end{bmatrix}$$

We then find

$$\begin{aligned} l_{41}l_{11} &= \frac{1}{4} & l_{41} &= \frac{1}{4} \\ l_{41}l_{21} + l_{42}l_{22} &= \frac{1}{5} & l_{42} &= \frac{3\sqrt{3}}{20} \end{aligned}$$

and so on to $l_{43} = \sqrt{5}/20$ and $l_{44} = \sqrt{7}/140$.

The algorithm can be summarized in the equations

$$\sum_{j=1}^{i-1} l_{rj}l_{ij} + l_{ri}l_{ii} = a_{ri} \qquad i = 1, \ldots, r-1$$

$$\sum_{j=1}^{r-1} l_{rj}^2 + l_{rr}^2 = a_{rr}$$

to be used for $r = 1, \ldots, n$ in turn.

ERRORS AND NORMS

26.17. What is a condition number of a matrix A?

It is a measure of how trustworthy the matrix is in computations. For a given norm, we define the condition number as

$$C(A) = \|A\| \cdot \|A^{-1}\|$$

and observe, using Problem 1.34, that $C(I) = 1$, where I is the identity matrix. Moreover, using Problem 1.38,

$$C(A) = \|A\| \cdot \|A^{-1}\| \geqq \|I\| = 1$$

so the identity matrix has the lowest condition number.

26.18. Suppose the vector b of the system $Ax = b$ contains input errors. Estimate the influence of such errors on the solution vector x.

Rewrite the system as

$$Ax_e = b + e$$

and combine with $Ax = b$ to obtain

$$A(x_e - x) = e \qquad x_e - x = A^{-1}e$$

from which it follows that, using Problem 1.60,

$$\|x - x_e\| \leqq \|A^{-1}\| \cdot \|e\|$$

To convert this to a relative error estimate, we have, from $Ax = b$,

$$\|A\| \cdot \|x\| \geqq \|b\| \qquad \|x\| \geqq \frac{\|b\|}{\|A\|}$$

and finally

$$\frac{\|x_e - x\|}{\|x\|} \leqq \|A\| \cdot \|A^{-1}\| \frac{\|e\|}{\|b\|} = C(A) \frac{\|e\|}{\|b\|}$$

in which the condition number of A appears.

Similarly from

$$\|e\| \leqq \|A\| \cdot \|x_e - x\| \qquad \text{and} \qquad \|A^{-1}\| \cdot \|b\| \geqq \|x\|$$

we find

$$\frac{\|e\|}{C(A)\,\|b\|} \leqq \frac{\|x_e - x\|}{\|x\|}$$

giving us both a lower and an upper bound for the relative error.

26.19. Suppose the matrix A of the system $Ax = b$ contains input errors. Estimate the influence of such errors on the solution vector x.

Write the system as

$$(A + E)x_e = b$$

and combine with $Ax = b$ to obtain

$$A(x_e - x) = -Ex_e$$

leading to

$$\|x_e - x\| \leqq \|A^{-1}\| \cdot \|E\| \cdot \|x_e\|$$

$$\frac{\|x_e - x\|}{\|x_e\|} \leqq \|A^{-1}\| \cdot \|A\| \cdot \frac{\|E\|}{\|A\|} = C(A) \frac{\|E\|}{\|A\|}$$

which estimates the error relative to the solution x_e. Here again the condition number of A appears.

Here and in the preceding problem it measures how much the input errors are inflated.

An estimate relative to the solution x can also be found. One such estimate is this:

$$\frac{\|x_e - x\|}{\|x\|} \leqq \frac{C(A)(\|E\|/\|A\|)}{1 - C(A)(\|E\|/\|A\|)}$$

26.20. Rework the opening example (Problem 26.1) under the assumption that a computer carrying only two floating-point digits is to do the computations.

The system now takes the form

$$1.0x_1 + .50x_2 + .33x_3 = 1.0$$
$$.50x_1 + .33x_2 + .25x_3 = 0$$
$$.33x_1 + .25x_2 + .20x_3 = 0$$

and with $l_{21} = .5$ and $l_{31} = .33$ is soon converted to

$$.08x_2 + .09x_3 = -.50$$
$$.09x_2 + .09x_3 = -.33$$

with the first equation retained as is. Here we may as well complete the triangularization by simply subtracting what we have.

$$.01x_2 = .17$$

Now back-substitution manages $x_2 = 17$, $x_3 = -21$, $x_1 = -.6$, and a "solution" vector $(-.6, 17, -21)$. Comparing with the correct $(9, -36, 30)$ we see no resemblance whatsoever. The point is, the matrix of this system is a junior member of a notorious family, the Hilbert matrices. Coupling this with the severe limitations of our computer has led to a grotesque result.

In Problem 26.42 the inverse matrix will be found to be

$$\begin{bmatrix} 9 & -36 & 30 \\ -36 & 192 & -180 \\ 30 & -180 & 180 \end{bmatrix}$$

in which the large elements should be noted. The maximum norm is $36 + 192 + 180 = 408$, making a condition number of

$$C(A) = \|A\| \cdot \|A^{-1}\| = \frac{11}{6}(408) = 748$$

By Problem 26.19 we now have the estimate

$$\frac{\|x_e - x\|}{\|x_e\|} \leqq 748\left(\frac{.005}{\frac{11}{6}}\right) = 2.04$$

suggesting a relative error of 200 percent. Clearly the computation was naive. At least four digits are needed.

By way of contrast, recall the cooperative matrix of Problem 26.6 which permitted an exact solution to be found even by a two-digit computer. For that matrix the maximum norm is 2 and the inverse also has norm near 2. The condition number is then near 4 and we estimate

$$\frac{\|x_e - x\|}{\|x_e\|} \leqq 4\left(\frac{.005}{1}\right) = .02$$

or a maximum error of 2 percent.

26.21. What is the "nearest singular matrix" theorem?

Suppose A is nonsingular and B singular. Then, by the fundamental theorem of linear algebra, there exists a vector $x \neq 0$ satisfying $Bx = 0$. For this x

$$\|Ax\| = \|Ax - Bx\| = \|(A - B)x\| \leqq \|A - B\| \cdot \|x\|$$

and since $x = A^{-1}Ax$, we also have $\|x\| \leqq \|A^{-1}\| \cdot \|Ax\|$

Since A is nonsingular, we cancel the factor $\|Ax\|$ and have

$$\frac{1}{\|A - B\|} \leqq \|A^{-1}\|$$

which is the required theorem.

Its message is that the size of the inverse matrix of A is at least the reciprocal of the "distance" of A from the nearest singular matrix B. If A is nearly singular, then A^{-1} will have a large norm. If A is normalized, in the sense $\|A\| = 1$, the condition number will also be large.

As a corollary we have the following intuitive result. If B is "close enough" to the nonsingular A, in the sense that $1/\|A - B\|$ is larger than $\|A^{-1}\|$, then B is nonsingular also.

26.22. Use the theorem of Problem 26.21 to estimate the condition of the matrix of this system, presented earlier in Problem 1.13.

$$x_1 + x_2 = 1$$
$$1.1x_1 + x_2 = 2$$

The point is that A^{-1}, required for the condition number, is not always easy to find with accuracy. Though this is not true here, we observe that the matrix of coefficients is close to the singular matrix

$$B = \begin{bmatrix} 1 & 1 \\ 1 & 1 \end{bmatrix}$$

and find, using maximum norms, $\|A\| = 2.1$, $\|A - B\| = .1$, so that

$$\|A^{-1}\| \geqq \frac{1}{.1} = 10 \qquad C(A) \geqq (2.1)(10) - 21$$

26.23. Estimate the error caused by using $1.01x_2$ in place of x_2 in the second equation in Problem 26.22.

The error matrix is $E = \begin{bmatrix} 0 & 0 \\ 0 & .01 \end{bmatrix}$ with maximum norm .01. Thus

$$\frac{\|x_e - x\|}{\|x_e\|} \leqq C(A)\frac{\|E\|}{\|A\|} \leqq 21\left(\frac{.01}{2.1}\right) = .1$$

which is our estimate. For an input error of 1 percent we anticipate an output error of 10 percent. This inflation is due to the ill-condition of A, as measured by $C(A)$.

Solving the system directly, we find $x = (10, -9)$ and $x_e = (11, -10)$. This makes $\|x_e - x\| = 1$ and $\|x_e\| = 11$, for a relative error of .09. So the 10 percent inflation is just about realized.

26.24. The many intermediate computations that are made in solving a linear system make roundoff error an important factor. How can this error be estimated?

Backward error analysis has produced the only real successes in this difficult area. It shows that the cumulative effect of roundoffs can be estimated by considering the substitute system $(A + E)x = b$, where E is a perturbation of A. It then finds bounds for the elements of E. The error in x can then be estimated by the formula of Problem 26.19. The details are far from trivial but have been carried through for most of the solution algorithms. The full story must be sought in the literature, but a simplified approach leading to the partially satisfactory bound

$$\max |e_{ij}| \leqq n\Delta[\max |a_{ij}| + (3 + n\Delta) \max |b_{ij}|]$$

is offered in Problems 26.113 to 26.117. Here Δ depends upon the unit roundoff and the b_{ij} upon the computed factors L and U of the given matrix A.

The somewhat deeper estimate

$$\|E\| \leqq (1.06 \max |u_{ij}|)(3n^2 + n^3)2^{-p}$$

may be easier to apply. For example, if A has order ten ($n = 10$), and the equivalent of eight decimal places are carried ($2^{-p} = 10^{-8}$), and we make the crude guess of ten for the first factor, then we find

$$\|E\| \leqq (1.3)10^{-4}$$

suggesting that perhaps half the digits being carried may no longer be significant. The estimate is, of course, conservative, since it ignores the fact that errors often cancel one another to some extent.

26.25. How does the condition of the coefficient matrix A enter into the roundoff error estimation process?

Recalling Problem 26.19, the relative error of the solution is bounded by

$$\frac{\|x_e - x\|}{\|x_e\|} \leqq C(A)\frac{\|E\|}{\|A\|}$$

where E is now the perturbation of A due to internal roundoffs. For a normalized A, the relative error in x_e is thus the product of two factors, the condition of A and the norm of E.

26.26. If double precision arithmetic is available, how much does it improve the roundoff situation?

By the formula in Problem 26.24, if the factor 2^{-p} can be reduced from 10^{-8} to 10^{-16}, eight additional decimal figures will be gained, surely a significant improvement. But there is a side effect. A large-scale system uses a lot of computer storage space, even at single precision. Doubling the precision may just burst the seams. There is a compromise, similar to the one described in Problem 19.48, where the motivation was computing time rather than storage space. Instead of doing and storing everything in double precision, limit this higher level of activity to the numerous inner product evaluations which permeate these algorithms. Once calculated, their values can be stored in single precision, making just one roundoff where there might have been n. Only a modest programming effort is needed to incorporate this feature, and the reward can be dramatic.

26.27. The **residual** of an approximate solution x_e is defined as the vector

$$r = b - Ax_e$$

and gives the amount by which each equation of the linear system fails to be satisfied. How is the residual related to the error of x_e?

Since $Ax = b$ for the exact solution x, we have

$$r = A(x - x_e) \qquad x - x_e = A^{-1}r$$

and, using Problem 1.37,

$$\frac{\|r\|}{\|A\|} \leqq \|x - x_e\| \leqq \|A^{-1}\| \cdot \|r\|$$

From $Ax = b$ we have similarly

$$\|A^{-1}\| \cdot \|b\| \geqq \|x\| \geqq \frac{\|b\|}{\|A\|}$$

so dividing corresponding elements leads to the required result.

$$\frac{1}{C(A)}\frac{\|r\|}{\|b\|} \leqq \frac{\|x - x_e\|}{\|x\|} \leqq C(A)\frac{\|r\|}{\|b\|}$$

The relative error of x_e is bounded above and below by multiples of the relative residual, the multipliers involving the condition number of A. If $C(A)$ is near 1, then the relative error is close to the relative residual, which is, of course, readily available. If, however, $C(A)$ is large, there may be good reason to suspect inaccuracy in x_e even though r may be small. In other words, if A is ill-conditioned, the system may be nearly satisfied by an x_e containing large error. On the optimistic side, and looking primarily at the left half of the above inequality, when $C(A)$ is large, even a large residual still allows the error $x - x_e$ to be small, though the probability of this happening is likely to be rather small too.

26.28. What is the method of iterative refinement?

Let $h = x - x_e$ and rewrite the equation $A(x - x_e) = r$ of the preceding problem as

$$Ah = r$$

This system has the same coefficient matrix as the original. If A has been factored, or the steps of Gaussian elimination retained in some way, it is solved with relatively little cost. With h in hand, one computes

$$x = x_e + h$$

and has a new, and presumably better, approximation to the true solution. New residuals may now be calculated and the process repeated as long as seems fruitful. This is the idea of iterative refinement. If double precision arithmetic is available, this is an excellent opportunity to use it.

ITERATIVE METHODS

26.29. Illustrate the Gauss–Seidel iteration for solving linear systems using the following well-known example. A dog is lost in a square maze of corridors (Fig. 26-2). At each intersection it chooses a direction at random and proceeds to the next intersection, where it again chooses at random and so on. What is the probability that a dog starting at intersection i will eventually emerge on the south side?

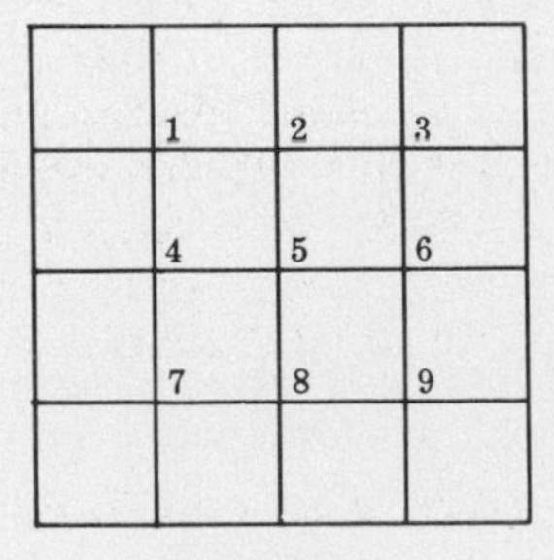

Fig. 26-2

Suppose there are just nine interior intersections, as shown. Let P_1 stand for the probability that a dog starting at intersection 1 will eventually emerge on the south side. Let $P_2, \ldots, P_9$ be similarly defined. Assuming that at each intersection he reaches, a dog is as likely to choose one direction as another, and that having reached any exit his walk is over, probability theory then offers the following nine equations for the P_k:

$$P_1 = \frac{1}{4}(0 + 0 + P_2 + P_4) \qquad P_4 = \frac{1}{4}(P_1 + 0 + P_5 + P_7) \qquad P_7 = \frac{1}{4}(P_4 + 0 + P_8 + 1)$$

$$P_2 = \frac{1}{4}(0 + P_1 + P_3 + P_5) \qquad P_5 = \frac{1}{4}(P_2 + P_4 + P_6 + P_8) \qquad P_8 = \frac{1}{4}(P_5 + P_7 + P_9 + 1)$$

$$P_3 = \frac{1}{4}(0 + P_2 + 0 + P_6) \qquad P_6 = \frac{1}{4}(P_3 + P_5 + 0 + P_9) \qquad P_9 = \frac{1}{4}(P_6 + P_8 + 0 + 1)$$

Leaving the equations in this form, we choose initial approximations to the P_k. It would be possible to make intelligent guesses here, but suppose we choose the uninspired initial values $P_k = 0$ for all k.

Taking the equations in the order listed we compute second approximations, one by one. First P_1 comes out zero. And so do $P_2, P_3, \ldots, P_6$. But then we find

$$P_7 = \frac{1}{4}(0+0+0+1) = \frac{1}{4} \qquad P_8 = \frac{1}{4}\left(0+\frac{1}{4}+0+1\right) = \frac{5}{16} \qquad P_9 = \frac{1}{4}\left(0+\frac{5}{16}+0+1\right) = \frac{21}{64}$$

and the second approximation to each P_k is in hand. Notice that in computing P_8 and P_9, the newest approximations to P_7 and P_8 respectively have been used. There seems little point in using more antique approximations. This procedure leads to the correct results more rapidly. Succeeding approximations are now found in the same way, and the iteration continues until no further changes occur in the required decimal places. Working to three places, the results of Table 26.1 are obtained. Note that P_5 comes out .250, which means that one-fourth of the dogs starting at the center should emerge on the south side. From the symmetry this makes sense. All nine values may be substituted back into the original equations as a further check, to see if the residuals are small.

Table 26.1

Iteration	P_1	P_2	P_3	P_4	P_5	P_6	P_7	P_8	P_9
0	0	0	0	0	0	0	0	0	0
1	0	0	0	0	0	0	.250	.312	.328
2	0	0	0	.062	.078	.082	.328	.394	.328
3	.016	.024	.027	.106	.152	.127	.375	.464	.398
4	.032	.053	.045	.140	.196	.160	.401	.499	.415
5	.048	.072	.058	.161	.223	.174	.415	.513	.422
6	.058	.085	.065	.174	.236	.181	.422	.520	.425
7	.065	.092	.068	.181	.244	.184	.425	.524	.427
8	.068	.095	.070	.184	.247	.186	.427	.525	.428
9	.070	.097	.071	.186	.249	.187	.428	.526	.428
10	.071	.098	.071	.187	.250	.187	.428	.526	.428

In this example of the Gauss–Seidel method each of the nine equations comes to us in the form

$$P_i = \cdots$$

and is used to update the approximation to P_i using the most recent values of the other components. It is worth noting that in each equation the unknown on the left side has the dominant coefficient.

26.30. Develop the Gauss–Seidel method for a general linear system.

The algorithm is applied most often to systems $Ax = b$ for which the diagonal elements of A are dominant. In any event, one should arrange by row and column interchanges that larger elements fall along the diagonal, to the extent that this is possible. The ith equation of the system is then solved for x_i in terms of the other unknowns. If we use the symbol $x_i^{(k)}$ to represent the kth approximation to x_i, then the algorithm proceeds as in the example.

$$x_1^{(1)} = \frac{b_1 - a_{12}x_2^{(0)} - \cdots - a_{1n}x_n^{(0)}}{a_{11}}$$

$$x_2^{(1)} = \frac{b_2 - a_{21}x_1^{(1)} - a_{23}x_3^{(0)} - \cdots - a_{2n}x_n^{(0)}}{a_{22}}$$

$$x_3^{(1)} = \frac{b_3 - a_{31}x_1^{(1)} - a_{32}x_2^{(1)} - a_{34}x_4^{(0)} - \cdots - a_{3n}x_n^{(0)}}{a_{33}}$$

$$\vdots$$

the superscript (0) denoting an initial approximation. More generally we have for the kth approximation to x_i

$$x_i^{(k)} = \frac{b_i - \sum_{j=1}^{i-1} a_{ij}x_j^{(k)} - \sum_{j=i+1}^{n} a_{ij}x_j^{(k-1)}}{a_{ii}}$$

in which the first sum uses kth approximations to x_j having $j<i$, while the second uses $(k-1)$th approximations to x_j with $j>i$. Here $i=1,\ldots,n$ and $k=1,\ldots$.

26.31. Express the Gauss–Seidel algorithm in matrix form.

First the matrix A is split into

$$A = L + D + U$$

where L and U are lower and upper triangles with zero elements on the diagonal. The general formula for Problem 26.30 can then be written as

$$x^{(k)} = D^{-1}(b - Lx^{(k)} - Ux^{(k-1)})$$

which can be solved for $x^{(k)}$. First

$$(I + D^{-1}L)x^{(k)} = D^{-1}b - D^{-1}Ux^{(k-1)}$$

which leads to
$$x^{(k)} = (I + D^{-1}L)^{-1}(D^{-1}b - D^{-1}Ux^{(k-1)})$$

or
$$x^{(k)} = -(D+L)^{-1}Ux^{(k-1)} + (D+L)^{-1}b$$

26.32. What is a stationary matrix iteration?

A matrix iteration of the form

$$x^{(k)} = M_k x^{(k-1)} + C_k b$$

is called stationary if M_k and C_k are independent of k. The iteration then becomes

$$x^{(k)} = Mx^{(k-1)} + Cb$$

The Gauss–Seidel method is stationary, with this M and C.

$$M = -(D+L)^{-1}U \qquad C = (D+L)^{-1}$$

26.33. Discuss the convergence of matrix iterations.

First we ask that the exact solution of $Ax = b$ be a fixed point of the iteration. That is, we substitute $x = A^{-1}b$ for both the input and output approximations in

$$x^{(k)} = M_k x^{(k-1)} + C_k b$$

and have
$$x = A^{-1}b = M_k A^{-1}b + C_k b = M_k x + C_k b$$

This is to hold for all vectors b, so we equate coefficients.

$$A^{-1} = M_k A^{-1} + C_k$$
$$I = M_k + C_k A$$

Now define $e^{(k)}$ as the error of the kth approximation.

$$e^{(k)} = x - x^{(k)}$$

Then
$$e^{(k)} = x - M_k x^{(k-1)} - C_k b$$
$$= M_k(x - x^{(k-1)}) = M_k e^{(k-1)}$$

which shows that it is the matrices M_k that control error behavior. Using this result repeatedly,

$$e^{(k)} = M_k M_{k-1} \cdots M_1 e^{(0)}$$

where $e^{(0)}$ is the initial error. For a stationary iteration this becomes

$$e^{(k)} = M^k e^{(0)}$$

26.34. Prove that the Gauss–Seidel iteration converges for an arbitrary initial vector $x^{(0)}$, if the matrix A is positive definite, symmetric.

Because of the symmetry, $A = L + D + L^T$, which makes

$$M = -(D+L)^{-1}L^T$$

If λ and v are an eigenvalue and eigenvector of M, then

$$(D+L)^{-1}L^T v = -\lambda v$$
$$L^T v = -\lambda(D+L)v$$

Premultiplying by the conjugate transpose of v (denoted v^*)

$$v^* L^T v = -v^* \lambda (D+L)v$$

and then adding $v^*(D+L)v$ to both sides

$$v^* A v = (1-\lambda)v^*(D+L)v$$

since $A = L + D + L^T$. But the conjugate transpose of v^*Av is v^*Av, so the same must be true for the right side of this last equation. Thus, with $\bar{\lambda}$ denoting the conjugate of λ,

$$\begin{aligned}(1-\bar{\lambda})v^*(D+L)^T v &= (1-\lambda)v^*(D+L)v \\ &= (1-\lambda)(v^*Dv + v^*Lv) \\ &= (1-\lambda)(v^*Dv - \bar{\lambda}v^*(D+L)^T v)\end{aligned}$$

Combining terms

$$(1-|\lambda|^2)v^*(D+L)^T v = (1-\lambda)v^*Dv$$

multiplying both sides by $(1-\bar{\lambda})$, and doing a little algebra we have finally

$$(1-|\lambda|^2)v^*Av = |1-\lambda|^2 v^*Dv$$

But both v^*Av and v^*Dv are nonnegative and λ cannot equal 1 (since this would lead back to $Av = 0$), so

$$|\lambda|^2 < 1$$

placing all eigenvalues within the unit circle and guaranteeing that $\lim M^k = 0$. Thus $e^{(k)}$ has limit zero for any $e^{(0)}$.

26.35. How can an acceleration method be applied to the Gauss–Seidel iteration?

Since $e^{(k)} = Me^{(k-1)}$, we anticipate that errors may diminish in a constant ratio, much as in Problem 25.4. The extrapolation to the limit idea then suggests itself. Here it would take the form

$$x_i = x_i^{(k+2)} - \frac{\Delta x_i^{(k+1)}}{\Delta^2 x_i^{(k)}}$$

for $i = 1, \ldots, n$. The superscripts denote three successive approximations.

For example, using the center column of Table 26-1, in which we know the correct value to be .250, the errors in rows 4 to 8 are 54, 27, 14, 6, and 3 in the third decimal place. This is very close to a steady reduction by one-half. Suppose we try extrapolation to the limit using the three entries below, together with the corresponding differences as given.

$$\begin{array}{ccc} .196 & & \\ & .027 & \\ .223 & & -.014 \\ & .013 & \\ .236 & & \end{array}$$

We find

$$P_5 = .236 - \frac{(.013)^2}{-.014} = .248$$

which is in the right direction if not especially dramatic.

26.36. What are relaxation and overrelaxation methods?

The central idea is to use residuals as indicators of how to correct approximations already in hand. For example, the iteration

$$x^{(k)} = x^{(k-1)} + (b - Ax^{(k-1)})$$

has the character of a relaxation method. It has been found that giving extra weight to the residual can speed convergence, leading to overrelaxation formulas such as

$$x^{(k)} = x^{(k-1)} + w(b - Ax^{(k-1)})$$

with $w > 1$. Other variations of the idea have also been used.

26.37. Adapt the overrelaxation method to accelerate the convergence of Gauss–Seidel.

The natural adaptation is

$$x^{(k)} = x^{(k-1)} + w[b - Lx^{(k)} - (D + U)x^{(k-1)}]$$

with $A = L + D + U$ as before. We take $w = 1.2$, $x^{(0)} = 0$, and try once more the problem of the dog in the maze. We find zeros generated as earlier until

$$P_7^{(1)} = P_7^{(0)} + 1.2\left(.250 + \frac{1}{4}P_4^{(1)} - P_7^{(0)} + \frac{1}{4}P_8^{(0)}\right) = .300$$

$$P_8^{(1)} = P_8^{(0)} + 1.2\left(.250 + \frac{1}{4}P_5^{(1)} + \frac{1}{4}P_7^{(1)} - P_8^{(0)} + \frac{1}{4}P_9^{(0)}\right) = .390$$

$$P_9^{(1)} = P_9^{(0)} + 1.2\left(.250 + \frac{1}{4}P_6^{(1)} + \frac{1}{4}P_8^{(1)} - P_9^{(0)}\right) = .418$$

Succeeding approximations are found in the same way and are listed in Table 26.2. Notice that about half as many iterations are now needed.

Table 26.2

Iteration	P_1	P_2	P_3	P_4	P_5	P_6	P_7	P_8	P_9
0	0	0	0	0	0	0	0	0	0
1	0	0	0	0	0	0	.300	.390	.418
2	0	0	0	.090	.144	.169	.384	.506	.419
3	.028	.052	.066	.149	.234	.182	.420	.520	.427
4	.054	.096	.071	.183	.247	.187	.427	.526	.428
5	.073	.098	.071	.188	.251	.187	.428	.527	.428
6	.071	.098	.071	.187	.250	.187	.428	.526	.428

MATRIX INVERSION

26.38. Extend the Gauss elimination algorithm to produce the inverse of the coefficient matrix A, that is, the A^{-1} such that $AA^{-1}=I$.

Taking once again the system of Problem 26.1, we simply treat three b vectors simultaneously. The starting point is the array

$$\begin{array}{cccccc} 1 & \frac{1}{2} & \frac{1}{3} & 1 & 0 & 0 \\ \frac{1}{2} & \frac{1}{3} & \frac{1}{4} & 0 & 1 & 0 \\ \frac{1}{3} & \frac{1}{4} & \frac{1}{5} & 0 & 0 & 1 \end{array}$$

the left half of which is A and the right half I. The first Gaussian step now leads to this new array.

$$\begin{array}{cccccc} 1 & \frac{1}{2} & \frac{1}{3} & 1 & 0 & 0 \\ 0 & \frac{1}{12} & \frac{1}{12} & -\frac{1}{2} & 1 & 0 \\ 0 & \frac{1}{12} & \frac{4}{45} & -\frac{1}{3} & 0 & 1 \end{array}$$

Here the method is modified slightly by reducing the next pivot to 1, a multiplication by 12 performing this service.

$$\begin{array}{cccccc} 1 & \frac{1}{2} & \frac{1}{3} & 1 & 0 & 0 \\ 0 & 1 & 1 & -6 & 12 & 0 \\ 0 & 0 & \frac{1}{180} & \frac{1}{6} & -1 & 1 \end{array}$$

The second step has also been performed to triangularize the system. At this point back-substitution could be used to solve three separate systems, each involving one of the last three column vectors. Instead, however, we extend the second Gaussian step. Continuing with the second row as pivotal row, we subtract half of it from row 1 to create one more zero:

$$\begin{array}{cccccc} 1 & 0 & -\frac{1}{6} & 4 & -6 & 0 \\ 0 & 1 & 1 & -6 & 12 & 0 \\ 0 & 0 & \frac{1}{180} & \frac{1}{6} & -1 & 1 \end{array}$$

The third Gaussian step then follows, after reducing the last pivot to 1. The purpose of this step is to create zeros above the new pivot. The final array then appears.

$$\begin{array}{cccccc} 1 & 0 & 0 & 9 & -36 & 30 \\ 0 & 1 & 0 & -36 & 192 & -180 \\ 0 & 0 & 1 & 30 & -180 & 180 \end{array}$$

Since we have actually solved three linear systems of the form $Ax = b$, with vectors $b^T = (1, 0, 0)$, $(0, 1, 0)$, and $(0, 0, 1)$ in turn, it is clear that the last three columns now contain A^{-1}. The original array was (A, I). The final array is (I, A^{-1}). The same process can be applied to other matrices A, row or column interchanges being made if required. If such interchanges are made, they must be restored at the completion of the algorithm.

26.39. Assuming that the matrix A has been factored as $A = LU$, how can A^{-1} be found from the factors?

Since $A^{-1} = U^{-1}L^{-1}$, the question is one of inverting triangular matrices. Consider L and seek an inverse in the same form.

$$\begin{bmatrix} 1 & 0 & 0 & \cdots & 0 \\ l_{21} & 1 & 0 & \cdots & 0 \\ l_{31} & l_{32} & 1 & \cdots & 0 \\ \cdots & \cdots & \cdots & \cdots & \cdots \\ l_{n1} & l_{n2} & l_{n3} & \cdots & 1 \end{bmatrix} \begin{bmatrix} 1 & 0 & 0 & \cdots & 0 \\ c_{21} & 1 & 0 & \cdots & 0 \\ c_{31} & c_{32} & 1 & \cdots & 0 \\ \cdots & \cdots & \cdots & \cdots & \cdots \\ c_{n1} & c_{n2} & c_{n3} & \cdots & 1 \end{bmatrix} = LL^{-1} = I$$

The validity of the assumption will be clear as we proceed. Now match the elements of the two sides, much as in the Choleski factorization algorithm, top to bottom and left to right. We find

$$\begin{aligned} l_{21} + c_{21} &= 0 & c_{21} &= -l_{21} \\ l_{31} + l_{32}c_{21} + c_{31} &= 0 & c_{31} &= -(l_{31} + l_{32}c_{21}) \\ l_{32} + c_{32} &= 0 & c_{32} &= -l_{32} \\ l_{41} + l_{42}c_{21} + l_{43}c_{31} + c_{41} &= 0 & c_{41} &= -(l_{41} + l_{42}c_{21} + l_{43}c_{31}) \\ l_{42} + l_{43}c_{32} + c_{42} &= 0 & c_{42} &= -(l_{42} + l_{43}c_{32}) \\ l_{43} + c_{43} &= 0 & c_{43} &= -l_{43} \\ &\vdots \end{aligned}$$

The elements are determined recursively, the general formula being

$$c_{ij} = -\sum_{k=j}^{i-1} l_{ik}c_{kj} \qquad \begin{matrix} i = 2, \ldots, n \\ j = 1, \ldots, i-1 \end{matrix}$$

All diagonal elements are 1.

The inversion of U is similar. Assuming the inverse to be an upper triangle, with elements d_{ij}, we proceed from bottom to top and right to left, finding

$$d_{ii} = \frac{1}{u_{ii}} \qquad i = n, \ldots, 1$$

and

$$d_{ij} = \frac{-1}{u_{ii}} \sum_{k=i+1}^{j} u_{ik}d_{kj} \qquad \begin{matrix} i = n, \ldots, 1 \\ j = n, \ldots, i+1 \end{matrix}$$

26.40. Apply the method of the preceding problem to the matrix of Problem 26.11.

In that problem the factorization

$$PA = LU = \begin{bmatrix} 1 & 0 & 0 & 0 \\ \frac{2}{3} & 1 & 0 & 0 \\ \frac{1}{3} & \frac{2}{3} & 1 & 0 \\ 0 & \frac{1}{3} & \frac{5}{7} & 1 \end{bmatrix} \begin{bmatrix} 3 & 0 & 1 & 2 \\ 0 & 3 & -\frac{2}{3} & -\frac{1}{3} \\ 0 & 0 & \frac{28}{9} & -\frac{4}{9} \\ 0 & 0 & 0 & \frac{24}{7} \end{bmatrix}$$

was made. Applying the above recursions, we now have

$$L^{-1} = \frac{1}{63}\begin{bmatrix} 63 & 0 & 0 & 0 \\ -42 & 63 & 0 & 0 \\ 7 & -42 & 63 & 0 \\ 9 & 9 & -45 & 63 \end{bmatrix} \qquad U^{-1} = \frac{1}{168}\begin{bmatrix} 56 & 0 & -18 & -35 \\ 0 & 56 & 12 & 7 \\ 0 & 0 & 54 & 7 \\ 0 & 0 & 0 & 49 \end{bmatrix}$$

from which there comes eventually

$$(PA)^{-1} = U^{-1}L^{-1} = \frac{1}{24}\begin{bmatrix} 7 & 1 & 1 & -5 \\ -5 & 7 & 1 & 1 \\ 1 & -5 & 7 & 1 \\ 1 & 1 & -5 & 7 \end{bmatrix}$$

To produce the ultimate A^{-1}, we use $A^{-1} = (PA)^{-1}P$ and recall that postmultiplication by a permutation matrix P rearranges the columns. Referring back to the earlier problem, it is found that the above columns should be taken in the order 4, 1, 2, 3.

26.41 Derive the formula for making an *exchange step* in a linear system.

Let the linear system be $Ax = b$, or

$$\sum_{k=1}^{n} a_{ik}x_k = b_i \qquad i = 1, \ldots, n$$

The essential ingredients may be displayed as in this array for $n = 3$.

	x_1	x_2	x_3
b_1	a_{11}	a_{12}	a_{13}
b_2	a_{21}	a_{22}	a_{23}
b_3	a_{31}	a_{32}	a_{33}

We proceed to exchange one of the "dependent" variables (say b_2) with one of the independent variables (say x_3). Solving the second equation for x_3, $x_3 = (b_2 - a_{21}x_1 - a_{22}x_2)/a_{23}$. This requires that the *pivot* coefficient a_{23} not be zero. Substituting the expression for x_3 in the remaining two equations brings

$$b_1 = a_{11}x_1 + a_{12}x_2 + \frac{a_{13}(b_2 - a_{21}x_1 - a_{22}x_2)}{a_{23}}$$

$$b_3 = a_{31}x_1 + a_{32}x_2 + \frac{a_{33}(b_2 - a_{21}x_1 - a_{22}x_2)}{a_{23}}$$

The array for the new system, after the exchange, is as follows.

	x_1	x_2	b_2
b_1	$a_{11} - \frac{a_{13}a_{21}}{a_{23}}$	$a_{12} - \frac{a_{13}a_{22}}{a_{23}}$	$\frac{a_{13}}{a_{23}}$
x_3	$-\frac{a_{21}}{a_{23}}$	$-\frac{a_{22}}{a_{23}}$	$\frac{1}{a_{23}}$
b_3	$a_{31} - \frac{a_{33}a_{21}}{a_{23}}$	$a_{32} - \frac{a_{33}a_{22}}{a_{23}}$	$\frac{a_{33}}{a_{23}}$

This may be summarized in four rules:

1. The pivot coefficient is replaced by its reciprocal.
2. The rest of the pivot column is divided by the pivot coefficient.

3. The rest of the pivot row is divided by the pivot coefficient with a change of sign.
4. Any other coefficient (say a_{lm}) is replaced by $a_{lm} - \dfrac{a_{lk}a_{im}}{a_{ik}}$ where a_{ik} is the pivot coefficient.

26.42. Illustrate the *exchange method* for finding the inverse matrix.

Once again we take the matrix of Problem 26.1.

	x_1	x_2	x_3
b_1	1	$\frac{1}{2}$	$\frac{1}{3}$
b_2	$\frac{1}{2}$	$\frac{1}{3}$	$\frac{1}{4}$
b_3	$\frac{1}{3}$	$\frac{1}{4}$	$\frac{1}{5}$

For error control it is the practice to choose the largest coefficient for the pivot, in this case 1. Exchanging b_1 and x_1, we have this new array:

	b_1	x_2	x_3
x_1	1	$-\frac{1}{2}$	$-\frac{1}{3}$
b_2	$\frac{1}{2}$	$\frac{1}{12}$	$\frac{1}{12}$
b_3	$\frac{1}{3}$	$\frac{1}{12}$	$\frac{4}{45}$

Two similar exchanges of b_3 and x_3, then of b_2 and x_2, lead to the two arrays shown below. In each case the largest coefficient in a b row and an x column is used as pivot.

	b_1	x_2	b_3
x_1	$\frac{9}{4}$	$-\frac{3}{16}$	$-\frac{15}{4}$
b_2	$\frac{3}{16}$	$\frac{1}{192}$	$\frac{15}{16}$
x_3	$-\frac{15}{4}$	$-\frac{15}{16}$	$\frac{45}{4}$

	b_1	b_2	b_3
x_1	9	-36	30
x_2	-36	192	-180
x_2	30	-180	180

Since what we have done is to exchange the system $b = Ax$ for the system $x = A^{-1}b$, the last matrix is A^{-1}.

26.43. Derive the formula $A^{-1} = (I + R + R^2 + \cdots)B$ where $R = I - BA$.

The idea here is that B is an approximate inverse of A, so that the residual R has small elements. A few terms of the series involved may therefore be enough to produce a much better approximation to A^{-1}. To derive the formula note first that $(I - R)(I + R + R^2 + \cdots) = I$ provided the matrix series is convergent. Then $I + R + R^2 + \cdots = (I - R)^{-1}$ and so

$$(I + R + R^2 + \cdots)B = (I - R)^{-1}B = (BA)^{-1}B = A^{-1}B^{-1}B$$

which reduces to A^{-1}.

26.44. Apply the formula of the preceding problem to the matrix

$$A = \begin{bmatrix} 1 & 10 & 1 \\ 2 & 0 & 1 \\ 3 & 3 & 2 \end{bmatrix}$$

assuming only a three-digit computer is available. Since any computer carries only a limited number of digits, this will again illustrate the power of a method of successive corrections.

First we apply Gaussian elimination to obtain a first approximation to the inverse. The three steps, using the largest pivot available in each case, appear below along with the approximate inverse B which results from two interchanges of rows, bringing the bottom row to the top.

$$\begin{matrix} .1 & 1 & .1 & .1 & 0 & 0 \\ 2.0 & 0 & 1.0 & 0 & 1 & 0 \\ 2.7 & 0 & 1.7 & -.3 & 0 & 1 \end{matrix} \qquad \begin{matrix} 0 & 1 & .037 & .111 & 0 & -.0371 \\ 0 & 0 & -.260 & .222 & 1 & -.742 \\ 1 & 0 & .630 & -.111 & 0 & .371 \end{matrix}$$

Step 1 Step 2

$$\begin{matrix} 0 & 1 & 0 & .143 & .143 & -.143 \\ 0 & 0 & 1 & -.854 & -3.85 & 2.85 \\ 1 & 0 & 0 & .427 & 2.43 & -1.43 \end{matrix} \qquad \begin{bmatrix} .427 & 2.43 & -1.43 \\ .143 & .143 & -.143 \\ -.854 & -3.85 & 2.85 \end{bmatrix}$$

Step 3 The Matrix B

Next we easily compute

$$R = I - BA = \begin{bmatrix} .003 & .020 & .003 \\ 0 & -.001 & 0 \\ .004 & -.010 & .004 \end{bmatrix}$$

after which RB, $B + RB$, $R^2B = R(RB)$, and $B + RB + R^2B$ are found in that order. (Notice that because the elements in R^2B are so small, a factor of 10,000 has been introduced for simplicity in presentation.)

$$\begin{bmatrix} .001580 & -.001400 & .001400 \\ -.000143 & -.000143 & .000143 \\ -.003140 & -.007110 & .007110 \end{bmatrix} \qquad \begin{bmatrix} .428579 & 2.428600 & -1.428600 \\ .142857 & .142857 & -.142857 \\ -.857138 & -3.857110 & 2.857110 \end{bmatrix}$$

RB $B + RB$

$$\begin{bmatrix} -.07540 & -.28400 & .28400 \\ .00143 & .00143 & -.00143 \\ -.04810 & -.32600 & .32600 \end{bmatrix} \qquad \begin{bmatrix} .4285715 & 2.4285716 & -1.4285716 \\ .1428571 & .1428571 & -.1428571 \\ -.8571428 & -3.8571426 & 2.8571426 \end{bmatrix}$$

$10^4 \cdot R(RB)$ $B + RB + R^2B$

Notice that except in the additive processes, only three significant digits have been carried. Since the exact inverse is

$$A^{-1} = \frac{1}{7}\begin{bmatrix} 3 & 17 & -10 \\ 1 & 1 & -1 \\ -6 & -27 & 20 \end{bmatrix}$$

it can be verified that $B + RB + R^2B$ is at fault only in the seventh decimal place. More terms of the series formula would bring still further accuracy. This method can often be used to improve the result of inversion by Gaussian elimination, since that algorithm is far more sensitive to roundoff error accumulation.

DETERMINANTS

26.45. Determinants are no longer used extensively in the solution of linear systems, but continue to have application in other ways. Direct evaluation of a determinant of order n would require the computation of $n!$ terms, which is prohibitive except for small n. What is the alternative?

From the properties of determinants, no step in a Gaussian elimination alters the determinant of the coefficient matrix except normalization and interchanges. If these were not performed, the determinant is available by multiplication of the diagonal elements after triangularization. For the matrix of Problem 26.1 the determinant is, therefore, a quick $(\frac{1}{12})(\frac{1}{180}) = \frac{1}{2160}$. This small value is another indication of the troublesome character of the matrix.

Determinants can also be found from the factorization $PA = LU$. Since $A = P^{-1}LU$ we have

$$\det(A) = \det(P^{-1})\det(L)\det(U) = (-1)^p \det(U)$$

where p is the number of interchanges represented by the permutation matrix P, or P^{-1}. For the matrix of Problem 26.11

$$\det(U) = 3(3)\left(\frac{28}{9}\right)\left(\frac{24}{7}\right) = 96$$

while $\det(P)$ is easily found to be -1. (Or recall that three interchanges were made during factorization, making $p = 3$.) Thus

$$\det(A) = -96$$

EIGENVALUE PROBLEMS, THE CHARACTERISTIC POLYNOMIAL

26.46. What are eigenvalues and eigenvectors of a matrix A?

A number λ for which the system $Ax = \lambda x$ or $(A - \lambda I)x = 0$ has a nonzero solution vector x is called an eigenvalue of the system. Any corresponding nonzero solution vector x is called an eigenvector. Clearly, if x is an eigenvector then so is Cx for any number C.

26.47. Find the eigenvalues and eigenvectors of the system

$$\begin{aligned} (2-\lambda)x_1 - x_2 \qquad &= 0 \\ -x_1 + (2-\lambda)x_2 - x_3 &= 0 \\ -x_2 + (2-\lambda)x_3 &= 0 \end{aligned}$$

which arises in various physical settings, including the vibration of a system of three masses connected by springs.

We illustrate the method of finding the *characteristic polynomial* directly and then obtaining the eigenvalues as roots of this polynomial. The eigenvectors are then found last. The first step is to take linear combinations of equations much as in Gaussian elimination, until only the x_3 column of coefficients involves λ. For example, if E_1, E_2, and E_3 denote the three equations, then $-E_2 + \lambda E_3$ is the equation

$$x_1 - 2x_2 + (1 + 2\lambda - \lambda^2)x_3 = 0$$

Calling this E_4, the combination $E_1 - 2E_2 + \lambda E_4$ becomes

$$4x_1 - 5x_2 + (2 + \lambda + 2\lambda^2 - \lambda^3)x_3 = 0$$

These last two equations together with E_3 now involve λ in only the x_3 coefficients.

The second step of the process is to triangularize this system by the Gauss elimination algorithm or its equivalent. With this small system we may take a few liberties as to pivots, retain

$$\begin{aligned} x_1 - 2x_2 + (1 + 2\lambda - \lambda^2)x_3 &= 0 \\ -x_2 + (2-\lambda)x_3 &= 0 \end{aligned}$$

as our first two equations and soon achieve

$$(4 - 10\lambda + 6\lambda^2 - \lambda^3)x_3 = 0$$

to complete the triangularization. To satisfy the last equation we must avoid making $x_3 = 0$, because this at once forces $x_2 = x_1 = 0$ and we do not have a nonzero solution vector. Accordingly we must require

$$4 - 10\lambda + 6\lambda^2 - \lambda^3 = 0$$

This cubic is the *characteristic polynomial,* and the eigenvalues must be its zeros since in no other way can we obtain a nonzero solution vector. By methods of an earlier chapter we find those eigenvalues to be $\lambda_1 = 2 - \sqrt{2}$, $\lambda_2 = 2$, $\lambda_3 = 2 + \sqrt{2}$ in increasing order.

The last step is to find the eigenvectors, but with the system already triangularized this involves no more than back-substitution. Taking λ_1 first, and recalling that eigenvectors are determined only to an arbitrary multiplier so that we may choose $x_3 = 1$, we find $x_2 = \sqrt{2}$ and then $x_1 = 1$. The other eigenvectors are found in the same way, using λ_2 and λ_3. The final results are

λ	x_1	x_2	x_3
$2 - \sqrt{2}$	1	$\sqrt{2}$	1
2	-1	0	1
$2 + \sqrt{2}$	1	$-\sqrt{2}$	1

In this case the original system of three equations has three distinct eigenvalues, to each of which there corresponds one independent eigenvector. This is the simplest, but not the only, possible outcome of an eigenvalue problem. It should be noted that the present matrix is both real and symmetric. For a real, symmetric $n \times n$ matrix an important theorem of algebra states that

(*a*) All eigenvalues are real, though perhaps not distinct.

(*b*) n independent eigenvectors always exist.

This is not true of all matrices. It is fortunate that many of the matrix problems which computers currently face are real and symmetric.

26.48. To make the algorithm for direct computation of the characteristic polynomial more clear, apply it to this larger system:

$$\begin{aligned}
E_1&: \quad (1-\lambda)x_1 + x_2 + x_3 + x_4 = 0 \\
E_2&: \quad x_1 + (2-\lambda)x_2 + 3x_3 + 4x_4 = 0 \\
E_3&: \quad x_1 + 3x_2 + (6-\lambda)x_3 + 10x_4 = 0 \\
E_4&: \quad x_1 + 4x_2 + 10x_3 + (20-\lambda)x_4 = 0
\end{aligned}$$

Calling these equations E_1, E_2, E_3, E_4, the combination $E_1 + 4E_2 + 10E_3 + \lambda E_4$ is

$$15x_1 + 39x_2 + 73x_3 + (117 + 20\lambda - \lambda^2)x_4 = 0$$

and is our second equation in which all but the x_4 term are free of λ. We at once begin triangularization by subtracting $15E_4$ to obtain

$$E_5: \quad -21x_2 - 77x_3 + (-183 + 35\lambda - \lambda^2)x_4 = 0$$

The combination $-21E_2 - 77E_3 + \lambda E_5$ becomes

$$-98x_1 - 273x_2 - 525x_3 + (-854 - 183\lambda + 35\lambda^2 - \lambda^3)x_4 = 0$$

and is our third equation in which all but the x_4 term are free of λ. The triangularization continues by blending this last equation with E_4 and E_5 to obtain

$$E_6: \quad 392x_3 + (1449 - 1736\lambda + 616\lambda^2 - 21\lambda^3)x_4 = 0$$

Now the combination $392E_3 + \lambda E_6$ is formed,

$$392x_1 + 1176x_2 + 2352x_3 + (3920 + 1449\lambda - 1736\lambda^2 + 616\lambda^3 - 21\lambda^4)x_4 = 0$$

and the triangularization is completed by blending this equation with E_4, E_5, and E_6 to obtain

E_7: $$(1 - 29\lambda + 72\lambda^2 - 29\lambda^3 + \lambda^4)x_4 = 0$$

The system E_4, E_5, E_6, E_7 is now the triangular system we have been aiming for. To avoid the zero solution vector, λ must be a zero of $1 - 29\lambda + 72\lambda^2 - 29\lambda^3 + \lambda^4$ which is the characteristic polynomial. Finding these zeros and the corresponding eigenvectors will be left as a problem. The routine just used can be generalized for larger systems.

26.49. Illustrate the use of the Cayley–Hamilton theorem for finding the characteristic equation of a matrix.

Writing the equation as

$$f(\lambda) = \lambda^n + c_1\lambda^{n-1} + \cdots + c_{n-1}\lambda + c_n = 0$$

the Cayley–Hamilton theorem states that the matrix A itself satisfies this equation. That is,

$$f(A) = A^n + c_1A^{n-1} + \cdots + c_{n-1}A + c_nI = 0$$

where the right side is now the zero matrix. This comes to n^2 equations for the n coefficients c_i so there is substantial redundance.

Take, for example, the Fibonacci matrix $F = \begin{bmatrix} 1 & 1 \\ 1 & 0 \end{bmatrix}$. Since $F^2 = \begin{bmatrix} 2 & 1 \\ 1 & 1 \end{bmatrix}$, we have

$$\begin{bmatrix} 2 & 1 \\ 1 & 1 \end{bmatrix} + c_1\begin{bmatrix} 1 & 1 \\ 1 & 0 \end{bmatrix} + c_2\begin{bmatrix} 1 & 0 \\ 0 & 1 \end{bmatrix} = \begin{bmatrix} 0 & 0 \\ 0 & 0 \end{bmatrix}$$

or

$$2 + c_1 + c_2 = 0$$

$$1 + c_1 = 0 \qquad 1 + c_2 = 0$$

with the second of these repeated. The familiar equation $\lambda^2 = \lambda + 1$ is again in hand. (See Problems 18.24 and 26.128.)

Or consider the permutation matrix P with

$$P = \begin{bmatrix} 0 & 0 & 1 \\ 1 & 0 & 0 \\ 0 & 1 & 0 \end{bmatrix} \qquad P^2 = \begin{bmatrix} 0 & 1 & 0 \\ 0 & 0 & 1 \\ 1 & 0 & 0 \end{bmatrix} \qquad P^3 = \begin{bmatrix} 1 & 0 & 0 \\ 0 & 1 & 0 \\ 0 & 0 & 1 \end{bmatrix}$$

which leads quickly to the set

$$1 + c_3 = 0 \qquad c_1 = 0 \qquad c_2 = 0$$

repeated twice. The characteristic equation is $\lambda^3 - 1 = 0$.

Several devices have been suggested for selecting a suitable subset of the available n^2 equations. One such device calls for computing

$$f(A)v = 0$$

for an appropriate vector v, and solving this system.

26.50. Prove Gerschgorin's theorem, which states that every eigenvalue of a matrix A falls within one of the complex circles having centers at a_{ii} and radii

$$R_i = \sum_{j \neq i} a_{ij}$$

with $i = 1, \ldots, n$.

Let x_i be the component of largest magnitude of one of the eigenvectors of A. From the ith equation of the system $(A - \lambda I)x = 0$, we have

$$(a_{ii} - \lambda)x_i = -\sum_{j \neq i} a_{ij}x_j$$

$$|a_{ii} - \lambda| \leqq \sum_{j \neq i} |a_{ij}| \left|\frac{x_j}{x_i}\right| \leqq \sum_{j \neq i} |a_{ij}|$$

which is the theorem.

26.51. What does the Gerschgorin theorem tell us about the eigenvalues of a permutation matrix which has a single 1 in each row and column, with zeros elsewhere?

The circles either have center at 0 with radius 1, or center at 1 with radius 0. All eigenvalues lie within a unit of the origin. For example, the eigenvalues of

$$\begin{bmatrix} 0 & 0 & 1 \\ 1 & 0 & 0 \\ 0 & 1 & 0 \end{bmatrix}$$

are the cube roots of 1. In particular, the eigenvalues of the identity matrix must be within the circle having center at 1 and radius 0.

26.52. The Gerschgorin theorem is especially useful for matrices having a dominant diagonal. Apply it to this matrix.

$$\begin{bmatrix} 4 & -1 & -1 & 0 \\ -1 & 4 & -1 & -1 \\ -1 & -1 & 4 & -1 \\ 0 & -1 & -1 & 4 \end{bmatrix}$$

All the eigenvalues must fall inside the circle with center at 4 and radius 3. By the symmetry, they must also be real.

THE POWER METHOD

26.53. What is the power method for producing the dominant eigenvalue and eigenvector of a matrix?

Assume that the matrix A is of size $n \times n$ with n independent eigenvectors $V_1, V_2, \ldots, V_n$ and a truly dominant eigenvalue λ_1: $|\lambda_1| > |\lambda_2| \geqq \cdots \geqq |\lambda_n|$. Then an arbitrary vector V can be expressed as a combination of eigenvectors,

$$V = a_1V_1 + a_2V_2 + \cdots + a_nV_n$$

It follows that

$$AV = a_1AV_1 + a_2AV_2 + \cdots + a_nAV_n = a_1\lambda_1V_1 + a_2\lambda_2V_2 + \cdots + a_n\lambda_nV_n$$

Continuing to multiply by A we arrive at

$$A^pV = a_1\lambda_1^pV_1 + a_2\lambda_2^pV_2 + \cdots + a_n\lambda_n^pV_n = \lambda_1^p\left[a_1V_1 + a_2\left(\frac{\lambda_2}{\lambda_1}\right)^pV_2 + \cdots + a_n\left(\frac{\lambda_n}{\lambda_1}\right)^pV_n\right]$$

provided $a_1 \neq 0$. Since λ_1 is dominant, all terms inside the brackets have limit zero except the first term. If we take the ratio of any corresponding components of $A^{p+1}V$ and A^pV, this ratio should therefore have limit λ_1. Moreover, $\lambda_1^{-p}A^pV$ will converge to the eigenvector a_1V_1.

26.54. Apply the power method to find the dominant eigenvalue and eigenvector of the matrix used in Problem 26.47.

$$A = \begin{bmatrix} 2 & -1 & 0 \\ -1 & 2 & -1 \\ 0 & -1 & 2 \end{bmatrix}$$

Choose the initial vector $V = (1, 1, 1)$. Then $AV = (1, 0, 1)$ and $A^2V = (2, -2, 2)$. It is convenient here to divide by 2, and in the future we continue to divide by some suitable factor to keep the numbers reasonable. In this way we find

$$A^7V = c(99, -140, 99) \qquad A^8V = c(338, -478, 338)$$

where c is some factor. The ratios of components are

$$\frac{338}{99} \simeq 3.41414 \qquad \frac{478}{140} \simeq 3.41429$$

and we are already close to the correct $\lambda_1 = 2 + \sqrt{2} \simeq 3.414214$. Dividing our last output vector by 338, it becomes $(1, -1.41420, 1)$ approximately and this is close to the correct $(1, -\sqrt{2}, 1)$ found in Problem 26.47.

26.55. What is the Rayleigh quotient and how may it be used to find the dominant eigenvalue?

The Rayleigh quotient is x^TAx/x^Tx, where T denotes the transpose. If $Ax = \lambda x$ this collapses to λ. If $Ax \simeq \lambda x$ then it is conceivable that the Rayleigh quotient is approximately λ. Under certain circumstances the Rayleigh quotients for the successive vectors generated by the power method converge to λ_1. For example, let x be the last output vector of the preceding problem, $(1, -1.41420, 1)$. Then

$$Ax = (3.41420, -4.82840, 3.41420) \qquad x^TAx = 13.65672 \qquad x^Tx = 3.99996$$

and the Rayleigh quotient is 3.414214 approximately. This is correct to six decimal places, suggesting that the convergence to λ_1 here is more rapid than for ratios of components.

26.56. Assuming all eigenvalues are real, how may the other extreme eigenvalue be found?

If $Ax = \lambda x$, then $(A - qI)x = (\lambda - q)x$. This means that $\lambda - q$ is an eigenvalue of $A - qI$. By choosing q properly, perhaps $q = \lambda_1$, we make the other extreme eigenvalue dominant and the power method can be applied. For the matrix of Problem 26.55 we may choose $q = 4$ and consider

$$A - 4I = \begin{bmatrix} -2 & -1 & 0 \\ -1 & -2 & -1 \\ 0 & -1 & -2 \end{bmatrix}$$

Again taking $V = (1, 1, 1)$ we soon find the Rayleigh quotient -3.414214 for the vector $(1, 1.41421, 1)$ which is essentially $(A - 4I)^8V$. Adding 4 we have .585786 which is the other extreme eigenvalue $2 - \sqrt{2}$ correct to six places. The vector is also close to $(1, \sqrt{2}, 1)$, the correct eigenvector.

26.57. How may the absolutely smallest eigenvalue be found by the power method?

If $Ax = \lambda x$, then $A^{-1}x = \lambda^{-1}x$. This means that the absolutely smallest eigenvalue of A can be found as the reciprocal of the dominant λ of A^{-1}. For the matrix of Problem 26.55 we first find

$$A^{-1} = \frac{1}{4}\begin{bmatrix} 3 & 2 & 1 \\ 2 & 4 & 2 \\ 1 & 2 & 3 \end{bmatrix}$$

Again choosing $V = (1, 1, 1)$ but now using A^{-1} instead of A, we soon find the Rayleigh quotient 1.707107 for the vector $(1, 1.41418, 1)$. The reciprocal quotient is .585786 so that we again have this eigenvalue and vector already found in Problems 26.47 and 26.56. Finding A^{-1} is ordinarily no simple task, but this method is sometimes the best approach to the absolutely smallest eigenvalue.

26.58. How may a next dominant eigenvalue be found by a suitable choice of starting vector V?

Various algorithms have been proposed, with varying degrees of success. The difficulty is to sidetrack the dominant eigenvalue itself and to keep it sidetracked. Roundoff errors have spoiled several theoretically sound methods by returning the dominant eigenvalue to the main line of the computation and obscuring the next dominant, or limiting the accuracy to which this runnerup can be determined. For example, suppose that in the argument of Problem 26.53 it could be arranged that the starting vector V is such that a_1 is zero. Then λ_1 and V_1 never actually appear, and if λ_2 dominates the remaining eigenvalues it assumes the role formerly played by λ_1 and the same reasoning proves convergence to λ_2 and V_2. With our matrix of Problem 26.54 this can be nicely illustrated. Being real and symmetric, this matrix has the property that its eigenvectors are orthogonal. (Problem 26.47 allows a quick verification of this.) This means that $V_1^T V = a_1 V_1^T V_1$ so that a_1 will be zero if V is orthogonal to V_1. Suppose we take $V = (-1, 0, 1)$. This is orthogonal to V_1. At once we find $AV = (-2, 0, 2) = 2V$, so that we have the exact $\lambda_2 = 2$ and $V_2 = (-1, 0, 1)$. However, our choice of starting vector here was fortunate.

It is almost entertaining to watch what happens with a reasonable but not so fortunate V, say $V = (0, 1, 1.4142)$ which is also orthogonal to V_1 as required. Then we soon find $A^3V \simeq 4.8(-1, .04, 1.20)$ which is something like V_2 and from which the Rayleigh quotient yields the satisfactory $\lambda_2 \simeq 1.996$. After this, however, the computation deteriorates and eventually we come to $A^{20}V \simeq c(1, -1.419, 1.007)$ which offers us good approximations once again to λ_1 and V_1. Roundoff errors have brought the dominant eigenvalue back into action. By taking the trouble to alter each vector A^pV slightly, to make it orthogonal to V_1, a better result can be achieved. Other devices also have been attempted using several starting vectors.

26.59. Develop the inverse power method.

This is an extension of the eigenvalue shift used in Problem 26.56. If A has eigenvalues λ_i, then $A - tI$ and $(A - tI)^{-1}$ have eigenvalues $\lambda_i - t$ and $(\lambda_i - t)^{-1}$, respectively. Applying the power method as in Problem 26.53, but using $(A - tI)^{-1}$ in place of A, we have

$$(A - tI)^{-p}V = a_1(\lambda_1 - t)^{-p}V_1 + \cdots + a_n(\lambda_n - t)^{-p}V_n$$

If t is near an eigenvalue λ_k, then the term $a_k(\lambda_k - t)^{-p}V_k$ will dominate the sum, assuming that $a_k \neq 0$ and λ_k is an isolated eigenvalue. The powers being computed will then lead to an eigenvalue of A, because all these matrices have the same eigenvectors. This is the basis of the inverse power method.

An interesting variation of this idea uses a sequence of values t_i. Given an initial approximation to an eigenvector, say $x^{(0)}$, compute successively

$$t_{i+1} = \frac{x^{(i)T}Ax^{(i)}}{x^{(i)T}x^{(i)}} \qquad x^{(i+1)} = c_{i+1}(A - t_{i+1}I)^{-1}x^{(i)}$$

the t_{i+1} being Rayleigh quotient estimates to λ_k and the $x^{(i+1)}$ approximations to V_k. Convergence has been proved under various hypotheses. The factor c_{i+1} is chosen to make $\|x^{(i+1)}\| = 1$ for some norm.

It is not actually necessary to compute the inverse matrix. What is needed is the vector $w^{(i+1)}$ defined by

$$w^{(i+1)} = (A - t_{i+1}I)^{-1}x^{(i)}$$

so it is more economical to get it by solving the system

$$(A - t_{i+1}I)w^{(i+1)} = x^{(i)}$$

for this vector. Then $x^{(i+1)} = c_{i+1}w^{(i+1)}$. As the sequence develops, the matrices $A - t_{i+1}I$ will approach singularity, suggesting that the method may have a perilous character, but with attention to normalization and pivoting, accurate results can be obtained.

26.60. What is inverse iteration?

Given an accurate approximation to an eigenvalue of A, inverse iteration is a fast way to obtain the corresponding eigenvector. Let t be an approximation to λ, obtained from the characteristic polynomial or other method which produces eigenvalues only. Then $A - tI$ is near singular, but still has a factorization

$$P(A - tI) = LU \qquad A - tI = P^{-1}LU$$

as in Problem 26.8. Just as in the preceding problem, we begin an iteration with

$$(A - tI)x^{(1)} = P^{-1}LUx^{(1)} = x^{(0)}$$

using an $x^{(0)}$ with a nonzero component in the direction of x, the eigenvector corresponding to λ. The choice $x^{(0)} = P^{-1}L(1, 1, \ldots, 1)^T$ has sometimes been suitable, or what is the same thing,

$$Ux^{(1)} = (1, 1, \ldots, 1)^T$$

26.61. Apply inverse iteration to the matrix of Problem 26.47, using .586 as an approximation to the eigenvalue $2 - \sqrt{2}$. Since the eigenvector $x = (1, \sqrt{2}, 1)$ has already been found, this will serve as a small-scale illustration of the method's potential.

To start, we need the factors L and U, which prove to be the following:

$$L = \begin{bmatrix} 1 & 0 & 0 \\ -.70721 & 1 & 0 \\ 0 & -1.4148 & 1 \end{bmatrix} \qquad U = \begin{bmatrix} 1.414 & -1 & 0 \\ 0 & .7068 & -1 \\ 0 & 0 & -.0008 \end{bmatrix}$$

In this example $P = I$. The solution of $Ux^{(1)} = (1, 1, \ldots, 1)^T$, found by back-substitution, is $x^{(1)} = (1250, 1767, -1250)^T$, after which

$$LUx^{(2)} = x^{(1)}$$

yields $x^{(2)} = (31{,}319, 44{,}273, 31{,}265)^T$ to five figures. Normalizing then brings the approximate eigenvector $(1, 1.414, .998)^T$.

REDUCTION TO CANONICAL FORMS

26.62. A basic theorem of linear algebra states that a real symmetric matrix A has only real eigenvalues and that there exists a real orthogonal matrix Q such that $Q^{-1}AQ$ is diagonal. The diagonal elements are then the eigenvalues and the columns of Q are the eigenvectors. Derive the Jacobi formulas for producing this orthogonal matrix Q.

In the Jacobi method Q is obtained as an infinite product of "rotation" matrices of the form

$$Q_1 = \begin{bmatrix} \cos\phi & -\sin\phi \\ \sin\phi & \cos\phi \end{bmatrix}$$

all other elements being identical with those of the unit matrix I. If the four entries shown are in positions (i, i), (i, k), (k, i), and (k, k), then the corresponding elements of $Q_1^{-1}AQ_1$ may easily be computed to be

$$b_{ii} = a_{ii}\cos^2\phi + 2a_{ik}\sin\phi\cos\phi + a_{kk}\sin^2\phi$$

$$b_{ki} = b_{ik} = (a_{kk} - a_{ii})\sin\phi\cos\phi + a_{ik}(\cos^2\phi - \sin^2\phi)$$

$$b_{kk} = a_{ii}\sin^2\phi - 2a_{ik}\sin\phi\cos\phi + a_{kk}\cos^2\phi$$

Choosing ϕ such that $\tan 2\phi = 2a_{ik}/(a_{ii} - a_{kk})$ then makes $b_{ik} = b_{ki} = 0$. Each step of the Jacobi algorithm therefore makes a pair of off-diagonal elements zero. Unfortunately the next step, while it creates a new pair of zeros, introduces nonzero contributions to formerly zero positions. Nevertheless, successive matrices of the form $Q_2^{-1}Q_1^{-1}AQ_1Q_2$, and so on, approach the required diagonal form and $Q = Q_1Q_2\cdots$.

26.63. Apply Jacobi's method to $A=\begin{bmatrix} 2 & -1 & 0 \\ -1 & 2 & -1 \\ 0 & -1 & 2 \end{bmatrix}$.

With $i=1$, $k=2$ we have $\tan 2\phi = -2/0$ which we interpret to mean $2\phi = \pi/2$. Then $\cos\phi = \sin\phi = 1/\sqrt{2}$ and

$$A_1 = Q_1^{-1}AQ_1 = \begin{bmatrix} \frac{1}{\sqrt{2}} & \frac{1}{\sqrt{2}} & 0 \\ -\frac{1}{\sqrt{2}} & \frac{1}{\sqrt{2}} & 0 \\ 0 & 0 & 1 \end{bmatrix} \begin{bmatrix} 2 & -1 & 0 \\ -1 & 2 & -1 \\ 0 & -1 & 2 \end{bmatrix} \begin{bmatrix} \frac{1}{\sqrt{2}} & -\frac{1}{\sqrt{2}} & 0 \\ \frac{1}{\sqrt{2}} & \frac{1}{\sqrt{2}} & 0 \\ 0 & 0 & 1 \end{bmatrix} = \begin{bmatrix} 1 & 0 & -\frac{1}{\sqrt{2}} \\ 0 & 3 & -\frac{1}{\sqrt{2}} \\ -\frac{1}{\sqrt{2}} & -\frac{1}{\sqrt{2}} & 2 \end{bmatrix}$$

Next we take $i=1$, $k=3$ making $\tan 2\phi = -\sqrt{2}/(-1) = \sqrt{2}$. Then $\sin\phi \simeq .45969$, $\cos\phi \simeq .88808$ and we compute

$$A_2 = Q_2^{-1}A_1Q_2 = \begin{bmatrix} .88808 & 0 & .45969 \\ 0 & 1 & 0 \\ -.45969 & 0 & .88808 \end{bmatrix} A_1 \begin{bmatrix} .88808 & 0 & -.45969 \\ 0 & 1 & 0 \\ .45969 & 0 & .88808 \end{bmatrix}$$

$$= \begin{bmatrix} .63398 & -.32505 & 0 \\ -.32505 & 3 & -.62797 \\ 0 & -.62797 & 2.36603 \end{bmatrix}$$

The convergence of the off-diagonal elements toward zero is not startling, but at least the decrease has begun. After nine rotations of this sort we achieve

$$A_9 = \begin{bmatrix} .58578 & .000000 & .000000 \\ .00000 & 2.00000 & .00000 \\ .00000 & .00000 & 3.41421 \end{bmatrix}$$

in which the eigenvalues found earlier have reappeared. We also have

$$Q \simeq Q_1Q_2\cdots Q_9 = \begin{bmatrix} .50000 & .70710 & .50000 \\ .70710 & .00000 & -.70710 \\ .50000 & -.70710 & .50000 \end{bmatrix}$$

in which the eigenvectors are also conspicuous.

26.64. What are the three main parts of Givens' variation of the Jacobi rotation algorithm for a real symmetric matrix?

In the first part of the algorithm rotations are used to reduce the matrix to triple-diagonal form, only the main diagonal and its two neighbors being different from zero. The first rotation is in the $(2,3)$ plane, involving the elements a_{22}, a_{23}, a_{32}, and a_{33}. It is easy to verify that such a rotation, with ϕ determined by $\tan\phi = a_{13}/a_{12}$, will replace the a_{13} (and a_{31}) elements by 0. Succeeding rotations in the $(2,i)$ planes then replace the elements a_{1i} and a_{i1} by zero, for $i=4,\ldots,n$. The ϕ values are determined by $\tan\phi = a_{1i}/a'_{12}$, where a'_{12} denotes the current occupant of row 1, column 2. Next it is the turn of the elements $a_{24},\ldots,a_{2n}$ which are replaced by zeros by rotations in the $(3,4),\ldots,(3,n)$ planes. Continuing in this way a matrix of triple-diagonal form will be achieved, since no zero that we have worked to create will be lost in a later rotation. This may be proved by a direct computation and makes the Givens' reduction finite whereas the Jacobi diagonalization is an infinite process.

The second step involves forming the sequence

$$f_0(\lambda)=1 \qquad f_i(\lambda)=(\lambda-\alpha_i)f_{i-1}(\lambda)-\beta_{i-1}^2 f_{i-2}(\lambda)$$

where the α's and β's are the elements of our new matrix

$$B=\begin{bmatrix} \alpha_1 & \beta_1 & 0 & \cdots & 0 \\ \beta_1 & \alpha_2 & \beta_2 & \cdots & 0 \\ 0 & \beta_2 & \alpha_3 & \cdots & 0 \\ \cdots & \cdots & \cdots & \cdots & \beta_{n-1} \\ 0 & 0 & 0 & \beta_{n-1} & \alpha_n \end{bmatrix}$$

and $\beta_0=0$. These $f_i(\lambda)$ prove to be the determinants of the principal minors of the matrix $\lambda I-B$, as may be seen from

$$f_i(\lambda)=\begin{vmatrix} \lambda-\alpha_1 & -\beta_1 & 0 & \cdots & 0 \\ -\beta_1 & \lambda-\alpha_2 & -\beta_2 & \cdots & 0 \\ 0 & -\beta_2 & \lambda-\alpha_3 & \cdots & 0 \\ \cdots & \cdots & \cdots & \cdots & -\beta_{i-1} \\ \cdots & \cdots & \cdots & -\beta_{i-1} & \lambda-\alpha_i \end{vmatrix}$$

by expanding along the last column,

$$f_i(\lambda)=(\lambda-\alpha_i)f_{i-1}(\lambda)+\beta_{i-1}D$$

where D has only the element $-\beta_{i-1}$ in its bottom row and so equals $D=-\beta_{i-1}f_{i-2}(\lambda)$. For $i=n$ we therefore have in $f_n(\lambda)$ the characteristic polynomial of B. Since our rotations do not alter the polynomial, it is also the characteristic polynomial of A.

Now, if some β_i are zero, the determinant splits into two smaller determinants which may be treated separately. If no β_i is zero, the sequence of functions $f_i(\lambda)$ proves to be a Sturm sequence (with the numbering reversed from the order given in Problem 25.33). Consequently the number of eigenvalues in a given interval may be determined by counting variations of sign.

Finally, the third step involves finding the eigenvectors. Here the diagonal nature of B makes Gaussian elimination a reasonable process for obtaining its eigenvectors U_j directly (deleting one equation and assigning some component the arbitrary value of 1). The corresponding eigenvectors of A are then $V_j=QU_j$ where Q is once again the product of our rotation matrices.

26.65. Apply the Givens method to the Hilbert matrix of order three.

$$H=\begin{bmatrix} 1 & \frac{1}{2} & \frac{1}{3} \\ \frac{1}{2} & \frac{1}{3} & \frac{1}{4} \\ \frac{1}{3} & \frac{1}{4} & \frac{1}{5} \end{bmatrix}$$

For this small matrix only one rotation is needed. With $\tan\phi=\frac{2}{3}$, we have $\cos\phi=3/\sqrt{13}$ and $\sin\phi=2/\sqrt{13}$. Then

$$Q=\frac{1}{\sqrt{13}}\begin{bmatrix} \sqrt{13} & 0 & 0 \\ 0 & 3 & -2 \\ 0 & 2 & 3 \end{bmatrix} \qquad B=Q^{-1}HQ=\begin{bmatrix} 1 & \frac{\sqrt{13}}{6} & 0 \\ \frac{\sqrt{13}}{6} & \frac{34}{65} & \frac{9}{260} \\ 0 & \frac{9}{260} & \frac{2}{195} \end{bmatrix}$$

and we have our triple diagonal matrix. The Sturm sequence consists of

$$f_0(\lambda)=1 \qquad f_1(\lambda)=\lambda-1 \qquad f_2(\lambda)=\left(\lambda-\frac{34}{65}\right)(\lambda-1)-\frac{13}{16}$$

$$f_3(\lambda)=\left(\lambda-\frac{2}{195}\right)f_2(\lambda)-\frac{81}{67{,}600}(\lambda-1)$$

which lead to the ± signs shown in Table 26.3. There are two roots between 0 and 1 and a third between 1 and 1.5. Iterations then locate these more precisely at .002688, .122327, and 1.408319. The eigenvalue so close to zero is another indication of the near singularity of this matrix.

Table 26.3

	f_0	f_1	f_2	f_3	**Changes**
0	+	−	+	−	3
1	+	0	−	−	1
1.5	+	+	+	+	0

To find the eigenvector for λ_1, we solve $BU_1=\lambda_1 U_1$ and soon discover $u_1=1$, $u_2=-1.6596$, $u_3=7.5906$ to be one possibility. Finally

$$V_1=QU_1=(1,\ -5.591,\ 5.395)^T$$

which can be normalized as desired. Eigenvectors for the other two eigenvalues respond to the same process.

26.66. A similarity transformation of A is defined by $M^{-1}AM$, for any nonsingular matrix M. Show that such a transformation leaves the eigenvalues unchanged.

Since $Ax=\lambda x$ implies

$$MAM^{-1}(Mx)=\lambda(Mx)$$

we have at once that λ is an eigenvalue of MAM^{-1} with corresponding eigenvector Mx. The orthogonal transformations used in the Jacobi and Givens methods are special cases of similarity transformations.

26.67. Show that a matrix having all distinct eigenvalues, and corresponding independent eigenvectors, can be reduced to diagonal form by a similarity transformation.

Form the matrix M by using the eigenvectors of A as columns. It follows that

$$AM=MD$$

where D is diagonal and has the eigenvalues along its diagonal. Because the eigenvectors are linearly independent, M^{-1} exists and

$$M^{-1}AM=D$$

as required. This classic theorem on the reduction of matrices to special, or canonical, form has questionable computational value, since to find M appears to presuppose the solution of the entire problem.

26.68. What is a Hessenberg matrix?

It is a matrix in which either the upper or the lower triangle is zero except for the elements adjacent to the main diagonal. If the upper triangle has the zeros, the matrix is a lower Hessenberg, and vice versa. Here are two small Hessenbergs, the second being also triple diagonal since it is symmetric:

$$\begin{bmatrix} 1 & 1 & 1 & 1 \\ 1 & 1 & 1 & 1 \\ 0 & 1 & 1 & 1 \\ 0 & 0 & 1 & 1 \end{bmatrix} \qquad \begin{bmatrix} 1 & 1 & 0 \\ 1 & 1 & 1 \\ 0 & 1 & 1 \end{bmatrix}$$

26.69. Show that a matrix A can be reduced to Hessenberg form by Gaussian elimination and a similarity transformation.

Suppose we take an upper Hessenberg as our goal. The required zeros in the lower triangle can be generated column by column in $n-2$ stages. Assume $k-1$ such stages finished, and denote the new elements by a'_{ij}. The zeros for column k are then arranged as follows:

(*a*) From the elements $a'_{k+1,k}, \ldots, a'_{nk}$ find the absolutely largest and interchange its row with row $k+1$. This is the partial pivoting step and can be achieved by premultiplying the current matrix A' by an interchange matrix $I_{r,k+1}$ as introduced in Problem 26.8.

(*b*) Calculate the multipliers

$$c_{jk} = -\frac{a''_{jk}}{a_{k+1,k}} \qquad j = k+2, \ldots, n$$

(the double prime referring to elements after the interchange). Add c_{jk} times row $k+1$ to row j. This can be done for all the j simultaneously by premultiplying the current matrix A'' by a matrix G_k similar to the L_i of Problem 26.8.

$$G_k = \begin{bmatrix} 1 & & & & & \\ & \ddots & & & & \\ & & 1 & & & \\ & & -c_{k+2,k} & 1 & & \\ & & \vdots & & \ddots & \\ & & -c_{nk} & & & 1 \end{bmatrix} \quad \text{row } k+2$$

$$\text{col. } k+1$$

This is the Gaussian step.

(*c*) Postmultiply the current matrix by the inverses of $I_{r,k+1}$ and G_k. This is the similarity step. Of course, $I_{r,k+1}$ is its own inverse, while that of G_k is found by changing the signs of the c elements. This completes the kth stage of the reduction, which can be summarized by

$$G_k I_{r,k+1} A' I_{r,k+1} G_k^{-1}$$

with A' the input from the preceding stage, or A itself if $k=1$.

The steps a, b, and c are carried out for $k = 1, \ldots, n-2$ and it is easy to discover that the target zeros of any stage are retained.

26.70. Apply the algorithm of the preceding problem to this matrix:

$$\begin{bmatrix} 0 & 1 & 2 & 3 \\ 2 & 3 & 0 & 1 \\ 3 & 0 & 1 & 2 \\ 1 & 2 & 3 & 0 \end{bmatrix}$$

All the essentials appear in Fig. 26-3, the two stages side by side. Remember that as a premultiplier,

$$I_{23}A \quad \begin{matrix} 0 & 1 & 2 & 3 \\ 3 & 0 & 1 & 2 \\ 2 & 3 & 0 & 1 \\ 1 & 2 & 3 & 0 \end{matrix} \qquad I_{34}A' \quad \begin{matrix} 0 & \frac{11}{3} & 1 & 3 \\ 3 & \frac{5}{3} & 0 & 2 \\ 0 & \frac{34}{9} & 2 & \frac{-2}{3} \\ 0 & \frac{11}{9} & 3 & \frac{-1}{3} \end{matrix}$$

$$G_1I_{23}A \quad \begin{matrix} 0 & 1 & 2 & 3 \\ 3 & 0 & 1 & 2 \\ 0 & 3 & \frac{-2}{3} & \frac{-1}{3} \\ 0 & 2 & \frac{8}{3} & \frac{-2}{3} \end{matrix} \qquad G_2I_{34}A' \quad \begin{matrix} 0 & \frac{11}{3} & 1 & 3 \\ 3 & \frac{5}{3} & 0 & 2 \\ 0 & \frac{34}{9} & 2 & \frac{-2}{3} \\ 0 & 0 & \frac{40}{17} & \frac{-2}{17} \end{matrix}$$

$$G_1I_{23}AI_{23} \quad \begin{matrix} 0 & 2 & 1 & 3 \\ 3 & 1 & 0 & 2 \\ 0 & \frac{-2}{3} & 3 & \frac{-1}{3} \\ 0 & \frac{8}{3} & 2 & \frac{-2}{3} \end{matrix} \qquad G_2I_{34}A'I_{34} \quad \begin{matrix} 0 & \frac{11}{3} & 3 & 1 \\ 3 & \frac{5}{3} & 2 & 0 \\ 0 & \frac{34}{9} & \frac{-2}{3} & 2 \\ 0 & 0 & \frac{-2}{17} & \frac{40}{17} \end{matrix}$$

$$\begin{matrix} G_1I_{23}AI_{23}G_1^{-1} \\ (=A') \end{matrix} \quad \begin{matrix} 0 & \frac{11}{3} & 1 & 3 \\ 3 & \frac{5}{3} & 0 & 2 \\ 0 & \frac{11}{9} & 3 & \frac{-1}{3} \\ 0 & \frac{34}{9} & 2 & \frac{-2}{3} \end{matrix} \qquad G_2I_{34}A'I_{34}G_2^{-1} \quad \begin{matrix} 0 & \frac{11}{3} & \frac{113}{34} & 1 \\ 3 & \frac{5}{3} & 2 & 0 \\ 0 & \frac{34}{9} & \frac{-1}{51} & 2 \\ 0 & 0 & \frac{186}{289} & \frac{40}{17} \end{matrix}$$

$$I_{23} \quad \begin{matrix} 1 & 0 & 0 & 0 \\ 0 & 0 & 1 & 0 \\ 0 & 1 & 0 & 0 \\ 0 & 0 & 0 & 1 \end{matrix} \qquad I_{34} \quad \begin{matrix} 1 & 0 & 0 & 0 \\ 0 & 1 & 0 & 0 \\ 0 & 0 & 0 & 1 \\ 0 & 0 & 1 & 0 \end{matrix}$$

$$G_1 \quad \begin{matrix} 1 & 0 & 0 & 0 \\ 0 & 1 & 0 & 0 \\ 0 & \frac{-2}{3} & 1 & 0 \\ 0 & \frac{-1}{3} & 0 & 1 \end{matrix} \qquad G_2 \quad \begin{matrix} 1 & 0 & 0 & 0 \\ 0 & 1 & 0 & 0 \\ 0 & 0 & 1 & 0 \\ 0 & 0 & \frac{-11}{34} & 1 \end{matrix}$$

Fig. 26-3

$I_{r,k+1}$ swaps rows but as its own inverse and postmultiplier it swaps columns. The given matrix A is not symmetric so the result is Hessenberg but not triple diagonal. The matrix M of the similarity transformation MAM^{-1} is $G_2 I_{34} G_1 I_{23}$.

26.71. What is the QR method of finding eigenvalues?

Suppose we have an upper Hessenberg matrix H and can factor it as

$$H = QR$$

with Q orthogonal and R an upper (or right?) triangle. In the algorithm to come what we actually find first is

$$Q^T H = R$$

by reducing H to triangular form through successive rotations. Define

$$H^{(2)} = RQ = Q^T HQ$$

and note that $H^{(2)}$ will have the same eigenvalues as H, because of the theorem in Problem 26.66. (Since Q is orthogonal, $Q^T = Q^{-1}$.) It turns out that $H^{(2)}$ is also Hessenberg, so the process can be repeated to generate $H^{(k+1)}$ from $H^{(k)}$, with H serving as $H^{(1)}$ and $k = 1, \ldots$. The convergence picture is fairly complicated, but under various hypotheses the diagonal elements approach the eigenvalues while the lower triangle approaches zero. (Of course, the R factor at each stage is upper triangular, but in forming the product RQ, to recover the original eigenvalues, subdiagonal elements become nonzero again.) This is the essential idea of the QR method, the eventual annihilation of the lower triangle.

26.72. How can the matrix $Q^{(k)}$, required for the kth stage of the QR method, be found? That is, find $Q^{(k)}$ such that

$$H^{(k+1)} = Q^{(k)T} H^{(k)} Q^{(k)}$$

for $k = 1, \ldots$.

One way of doing this uses rotations, very much as in the Givens method presented in Problem 26.64. Since we are assuming that H is upper Hessenberg, it is only the elements $h_{i+1,i}$ that need our attention, for $i = 1, \ldots, n-1$. But $h_{i+1,i}$ can be replaced by zero using the rotation

$$S_i^T = \begin{bmatrix} 1 & & & & & \\ & \ddots & & & & \\ & & \cos\phi & \sin\phi & & \\ & & -\sin\phi & \cos\phi & & \\ & & & & \ddots & \\ & & & & & 1 \end{bmatrix} \begin{matrix} \\ \\ \text{row } i \\ \text{row } i+1 \\ \\ \\ \end{matrix}$$

$$\qquad\qquad \text{col. } i \qquad \text{col. } i+1$$

and calculating $S_i^T H$, provided $\tan\phi = h_{i+1,i}/h_{i,i}$. (It is easier to let $\sin\phi = ch_{i+1,i}$, $\cos\phi = ch_{i,i}$ and choose c to make the sum of squares 1.) Then the product of these rotations

$$Q^T = S_{n-1}^T \cdots S_1^T$$

is what we need. The same argument applies for any stage, so the superscript (k) has been suppressed here.

26.73. How has the idea of eigenvalue shifting, presented in Problem 26.56, been applied to accelerate the convergence of the QR algorithm?

Instead of factoring the matrix H, we try instead the reduction

$$Q^T(H - pI) = R$$

for some suitable value of p. The factorization $H - pI = QR$ is thus implied. Then

$$Q^T(H - pI)Q = RQ = H^{(2)} - pI$$

exhibits the reversed product which is central to the method and also defines $H^{(2)}$. But then

$$H^{(2)} = Q^T(H - pI)Q + pI = Q^T HQ$$

so $H^{(2)}$ again has the same eigenvalues as H. With $H^{(2)}$ in hand, we are ready to begin the next iteration. It would be nice to choose p near an eigenvalue, but in the absence of such inside information, the following alternative is recommended. Find the eigenvalues of the 2 by 2 submatrix in the lower right corner of the current H and set p equal to the one closest to h_{nn}, assuming these eigenvalues real. If they are complex, set p to their common real part.

26.74. Given the midget Hessenberg matrix

$$H = \begin{bmatrix} 4 & 2 & 1 \\ 0 & 1 & 0 \\ 0 & 2 & 3 \end{bmatrix}$$

find the eigenvalues by the QR method.

It is easy to discover that the eigenvalues are the diagonal elements 4, 1, 3. But it is also interesting to watch the QR method perform the triangularization. Choosing a shift of 3, we compute

$$H - 3I = \begin{bmatrix} 1 & 2 & 1 \\ 0 & -2 & 0 \\ 0 & 2 & 0 \end{bmatrix}$$

which will need just one rotation to reach triangular form.

$$S^T = \frac{1}{\sqrt{2}} \begin{bmatrix} \sqrt{2} & 0 & 0 \\ 0 & -1 & 1 \\ 0 & -1 & -1 \end{bmatrix} \qquad S^T(H - 3I) = \frac{1}{\sqrt{2}} \begin{bmatrix} \sqrt{2} & 2\sqrt{2} & \sqrt{2} \\ 0 & 4 & 0 \\ 0 & 0 & 0 \end{bmatrix}$$

Postmultiplication by S then completes the similarity transformation.

$$S^T(H - 3I)S = \frac{1}{2} \begin{bmatrix} 2 & -\sqrt{2} & -3\sqrt{2} \\ 0 & -4 & -4 \\ 0 & 0 & 0 \end{bmatrix}$$

Finally we add $3I$ and have

$$H^{(2)} = \begin{bmatrix} 4 & -\frac{\sqrt{2}}{2} & -\frac{3\sqrt{2}}{2} \\ 0 & 1 & -2 \\ 0 & 0 & 3 \end{bmatrix}$$

the triangular form having been preserved. Ordinarily this will not happen, and several stages such as the above will be needed.

26.75. Apply the QR method to the Hessenberg matrix

$$H = \begin{bmatrix} 4 & 1 & 1 & 1 \\ 1 & 4 & 1 & 1 \\ 0 & 1 & 4 & 1 \\ 0 & 0 & 1 & 4 \end{bmatrix}$$

for which the exact eigenvalues are 6, 4, 3, and 3.

A substantial number of rotation cycles eventually reduced this matrix to the following triangle:

$$\begin{bmatrix} 5.99997 & 1.50750 & -.17830 & .29457 \\ & 3.99997 & -.44270 & .22152 \\ & & 3.00098 & -.60302 \\ & & & 2.99895 \end{bmatrix}$$

in which the eigenvalues are evident along the diagonal. For larger jobs a saving in computing time would be realized by a reduction of the order when one of the subdiagonal elements becomes zero. Here it was entertaining simply to watch the lower triangle slowly vanish. Using the above approximate eigenvalues, the corresponding vectors were found directly and matched the correct $(3, 3, 2, 1)$, $(-1, -1, 0, 1)$, and $(0, 0, -1, 1)$ to three decimal places more or less. There is no fourth eigenvector.

26.76. Apply the QR method to the triple diagonal matrix

$$\begin{bmatrix} 4 & 1 & 0 & 0 \\ 1 & 4 & 1 & 0 \\ 0 & 1 & 4 & 1 \\ 0 & 0 & 1 & 4 \end{bmatrix}$$

and then use the results obtained to "guess" the correct eigenvalues.

Once again the rotation cycles were allowed to run their course, with this result. Off-diagonal elements were essentially zero.

$$\begin{bmatrix} 5.618031 & & & \\ & 4.618065 & & \\ & & 3.381945 & \\ & & & 2.381942 \end{bmatrix}$$

Since the given matrix was symmetric, both the lower and upper triangles have become zero, leaving the eigenvalues quite conspicuous. Taking the largest, a direct calculation of the eigenvector managed

$$(1.00002,\ 1.61806,\ 1.61806,\ 1)$$

the fourth component having been fixed in advance. Guessing that this ought to have been $(1, x, x, 1)$ leads quickly to the equations

$$\lambda = x + 4 \qquad x^2 - x - 1 = 0$$

the second of which is familiar by its connection with Fibonacci numbers. The root $x = (1 + \sqrt{5})/2$ is now paired with $\lambda = (9 + \sqrt{5})/2$, while $x = (1 - \sqrt{5})/2$ is paired with $\lambda = (9 - \sqrt{5})/2$ giving us two of the exact solutions. The other two are found similarly.

COMPLEX SYSTEMS

26.77. How can the problem of solving a system of complex equations be replaced by that of solving a real system?

This is almost automatic, since complex numbers are equal precisely when their real and imaginary parts are equal. The equation

$$(A+iB)(x+iy)=a+ib$$

is at once equivalent to $\quad Ax-By=a \qquad Ay+Bx=b$

and this may be written in matrix form as

$$\begin{bmatrix} A & -B \\ B & A \end{bmatrix}\begin{pmatrix} x \\ y \end{pmatrix}=\begin{pmatrix} a \\ b \end{pmatrix}$$

A complex $n \times n$ system has been replaced by a real $2n \times 2n$ system, and any of our methods for real systems may now be used. It is also possible to replace this real system by two systems

$$(B^{-1}A+A^{-1}B)x=B^{-1}a+A^{-1}b$$
$$(B^{-1}A+A^{-1}B)y=B^{-1}b-A^{-1}a$$

of size $n \times n$ with identical coefficient matrices. This follows from

$$(B^{-1}A+A^{-1}B)x=B^{-1}(Ax-By)+A^{-1}(Bx+Ay)=B^{-1}a+A^{-1}b$$
$$(B^{-1}A+A^{-1}B)y=B^{-1}(Ay+Bx)+A^{-1}(By-Ax)=B^{-1}b-A^{-1}a$$

Using these smaller systems slightly shortens the overall computation.

26.78. Reduce the problem of inverting a complex matrix to that of inverting real matrices.

Let the given matrix be $A+iB$ and its inverse $C+iD$. We are to find C and D such that $(A+iB)(C+iD)=I$. Suppose A is nonsingular so that A^{-1} exists. Then

$$C=(A+BA^{-1}B)^{-1} \qquad D=-A^{-1}B(A+BA^{-1}B)^{-1}$$

as may be verified by direct substitution. If B is nonsingular, then

$$C=B^{-1}A(AB^{-1}A+B)^{-1} \qquad D=-(AB^{-1}A+B)^{-1}$$

as may be verified by substitution. If both A and B are nonsingular, the two results are of course identical. In case both A and B are singular, but $(A+iB)$ is not, then a more complicated procedure seems necessary. First a real number t is determined such that the matrix $E=A+tB$ is nonsingular. Then, with $F=B-tA$, we find $E+iF=(1-it)(A+iB)$ and so

$$(A+iB)^{-1}=(1-it)(E+iF)^{-1}$$

This can be computed by the first method since E is nonsingular.

26.79. Extend Jacobi's method for finding eigenvalues and vectors to the case of a Hermitian matrix.

We use the fact that a Hermitian matrix H becomes diagonalized under a unitary transformation, that is, $U^{-1}HU$ is a diagonal matrix. The matrices H and U have the properties $\bar{H}^T=H$ and $\bar{U}^T=U^{-1}$. The matrix U is to be obtained as an infinite product of matrices of the form

$$U_1=\begin{bmatrix} \cos\phi & -\sin\phi\, e^{-i\theta} \\ \sin\phi\, e^{i\theta} & \cos\phi \end{bmatrix}$$

all other elements agreeing with those of I. The four elements shown are in positions (i, i), (i, k), (k, i), and (k, k). If the corresponding elements of H are

$$H=\begin{bmatrix} a & b-ic \\ b+ic & d \end{bmatrix}$$

then the (i, k) and (k, i) elements of $U^{-1}HU$ will have real and imaginary parts equal to zero,

$$(d-a)\cos\phi\sin\phi\cos\theta + b\cos^2\phi - b\sin^2\phi\cos 2\theta - c\sin^2\phi\sin 2\theta = 0$$
$$(a-d)\cos\phi\sin\phi\sin\theta - c\cos^2\phi + b\sin^2\phi\sin 2\theta - c\sin^2\phi\cos 2\theta = 0$$

if ϕ and θ are chosen so that

$$\tan\theta = \frac{c}{b} \qquad \tan 2\phi = \frac{2(b\cos\theta + c\sin\theta)}{a-d}$$

This type of rotation is applied iteratively as in Problem 26.62 until all off-diagonal elements have been made satisfactorily small. The (real) eigenvalues are then approximated by the resulting diagonal elements, and the eigenvectors by the columns of $U = U_1U_2U_3\cdots$.

26.80. How may the eigenvalues and vectors of a general complex matrix be found? Assume all eigenvalues are distinct.

As a first step we obtain a unitary matrix U such that $U^{-1}AU = T$ where T is an upper triangular matrix, all elements below the main diagonal being zero. Once again U is to be obtained as an infinite product of rotation matrices of the form U_1 shown in the preceding problem, which we now write as

$$U_1 = \begin{bmatrix} x & -\bar{y} \\ y & x \end{bmatrix}$$

The element in position (k, i) of $U_1^{-1}AU_1$ is then

$$a_{ki}x^2 + (a_{kk} - a_{ii})xy - a_{ik}y^2$$

To make this zero we let $y = Cx$, $x = 1/\sqrt{1+|C|^2}$ which automatically assures us that U_1 will be unitary, and then determine C by the condition $a_{ik}C^2 + (a_{ii} - a_{kk})C - a_{ki} = 0$ which makes

$$C = \frac{1}{2a_{ik}}\left[(a_{kk} - a_{ii}) \pm \sqrt{(a_{kk} - a_{ii})^2 + 4a_{ik}a_{ki}}\right]$$

Either sign may be used, preferably the one that makes $|C|$ smaller. Rotations of this sort are made in succession until all elements below the main diagonal are essentially zero. The resulting matrix is

$$T = U^{-1}AU = \begin{bmatrix} t_{11} & t_{12} & \cdots & t_{1n} \\ 0 & t_{22} & \cdots & t_{2n} \\ \cdots & \cdots & \cdots & \cdots \\ 0 & 0 & \cdots & t_{nn} \end{bmatrix}$$

where $U = U_1U_2\cdots U_N$. The eigenvalues of both T and A are the diagonal elements t_{ii}.

We next obtain the eigenvectors of T, as the columns of

$$W = \begin{bmatrix} 1 & w_{12} & w_{13} & \cdots & w_{1n} \\ 0 & 1 & w_{23} & \cdots & w_{2n} \\ 0 & 0 & 1 & \cdots & \cdots \\ \cdots & \cdots & \cdots & \cdots & \cdots \\ 0 & 0 & 0 & \cdots & w_{nn} \end{bmatrix}$$

The first column is already an eigenvector belonging to t_{11}. To make the second column an eigenvector belonging to t_{22} we require $t_{11}w_{12} + t_{12} = t_{22}w_{12}$ or $w_{12} = t_{12}/(t_{22} - t_{11})$ assuming $t_{11} \neq t_{22}$. Similarly, to make the third column an eigenvector we need

$$w_{23} = \frac{t_{23}}{t_{33} - t_{22}} \qquad w_{13} = \frac{t_{12}w_{23} + t_{13}}{t_{33} - t_{11}}$$

In general the w_{ik} are found from the recursion

$$w_{ik} = \sum_{j=i+1}^{k} \frac{t_{ij} w_{jk}}{t_{kk} - t_{ii}}$$

with $i = k-1, k-2, \ldots, 1$ successively. Finally the eigenvectors of A itself are available as the columns of UW.

Supplementary Problems

26.81. Apply the Gauss elimination algorithm to find the solution vector of this system:

$$\begin{aligned} w + 2x - 12y + 8z &= 27 \\ 5w + 4x + 7y - 2z &= 4 \\ -3w + 7x + 9y + 5z &= 11 \\ 6w - 12x - 8y + 3z &= 49 \end{aligned}$$

26.82. Apply the method of Problem 26.10 to find the solution vector of this system:

$$\begin{aligned} 33x_1 + 16x_2 + 72x_3 &= 359 \\ -24x_1 - 10x_2 - 57x_3 &= 281 \\ -8x_1 - 4x_2 - 17x_3 &= 85 \end{aligned}$$

26.83. Suppose it has been found that the system

$$\begin{aligned} 1.7x_1 + 2.3x_2 - 1.5x_3 &= 2.35 \\ 1.1x_1 + 1.6x_2 - 1.9x_3 &= -.94 \\ 2.7x_1 - 2.2x_2 + 1.5x_3 &= 2.70 \end{aligned}$$

has a solution near $(1, 2, 3)$. Apply the method of Problem 26.28 to obtain an improved approximation.

26.84. Apply Gaussian elimination to the system which follows, computing in rational form so that no roundoff errors are introduced, and so getting an exact solution. The coefficient matrix is the Hilbert matrix of order four.

$$\begin{aligned} x_1 + \frac{1}{2}x_2 + \frac{1}{3}x_3 + \frac{1}{4}x_4 &= 1 \\ \frac{1}{2}x_1 + \frac{1}{3}x_2 + \frac{1}{4}x_3 + \frac{1}{5}x_4 &= 0 \\ \frac{1}{3}x_1 + \frac{1}{4}x_2 + \frac{1}{5}x_3 + \frac{1}{6}x_4 &= 0 \\ \frac{1}{4}x_1 + \frac{1}{5}x_2 + \frac{1}{6}x_3 + \frac{1}{7}x_4 &= 0 \end{aligned}$$

26.85. Repeat the preceding problem with all coefficients replaced by decimals having three significant digits. Retain only three significant digits throughout the computation. How close do your results come to the exact solution of the preceding problem? (The Hilbert matrices of higher order are extremely troublesome even when many decimal digits can be carried.)

26.86. Apply the Gauss–Seidel iteration to the following system:

$$\begin{aligned} -2x_1 + x_2 \qquad\qquad &= -1 \\ x_1 - 2x_2 + x_3 \qquad &= \ \ 0 \\ x_2 - 2x_3 + x_4 &= \ \ 0 \\ x_3 - 2x_4 &= \ \ 0 \end{aligned}$$

Start with the approximation $x_k = 0$ for all k, rewriting the system with each equation solved for its diagonal unknown. After making several iterations can you guess the correct solution vector? This problem may be interpreted in terms of a random walker, who takes each step to left or right at random along the line of Fig. 26-4. When he reaches an end he stops. Each x_k value represents his probability of reaching the left end from position k. We may define $x_0 = 1$ and $x_5 = 0$, in which case each equation has the form $x_{k-1} - 2x_k + x_{k+1} = 0$, $k = 1, \ldots, 4$.

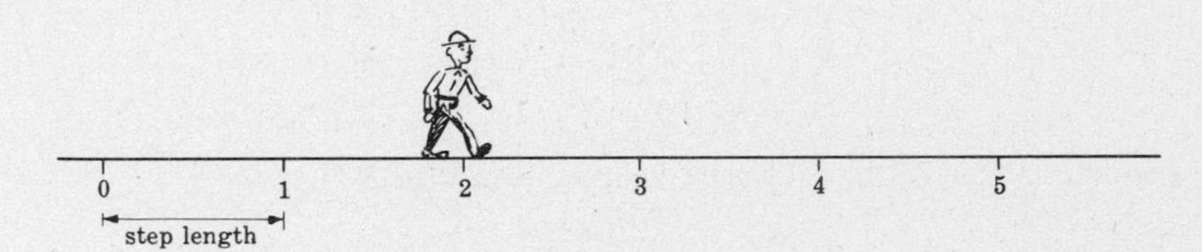

Fig. 26-4

26.87. Does overrelaxation speed convergence toward the exact solution of Problem 26.86?

26.88. Apply the Gauss–Seidel method to the system

$$x_k = \frac{3}{4}x_{k-1} + \frac{1}{4}x_{k+1} \qquad k = 1, \ldots, 19$$

$$x_0 = 1 \qquad x_{20} = 0$$

which may be interpreted as representing a random walker who moves to the left three times as often as to the right, on a line with positions numbered 0 to 20.

26.89. The previous problem is a boundary value problem for a difference equation. Show that its exact solution is $x_k = 1 - (3^k - 1)/(3^{20} - 1)$. Compute these values for $k = 0(1)20$ and compare with the results found by the iterative algorithm.

26.90. Apply overrelaxation to the same system. Experiment with values of w. Does underrelaxation ($w < 1$) look promising for this system?

26.91. Apply any of our methods to the following system:

$$\begin{aligned} x_1 + x_2 + x_3 + x_4 + x_5 &= 1 \\ x_1 + 2x_2 + 3x_3 + 4x_4 + 5x_5 &= 0 \\ x_1 + 3x_2 + 6x_3 + 10x_4 + 15x_5 &= 0 \\ x_1 + 4x_2 + 10x_3 + 20x_4 + 35x_5 &= 0 \\ x_1 + 5x_2 + 15x_3 + 35x_4 + 70x_5 &= 0 \end{aligned}$$

26.92. Invert the coefficient matrix of Problem 26.81 by the elimination algorithm of Problem 26.38.

26.93. Invert the same matrix by the exchange method.

26.94. Invert the coefficient matrix of Problem 26.86 by any of our methods.

26.95. Try to invert the Hilbert matrix of order four using three-digit arithmetic.

26.96. Try to invert Wilson's matrix. Invert the inverse. How close do you come to the original?

$$\begin{bmatrix} 10 & 7 & 8 & 7 \\ 7 & 5 & 6 & 5 \\ 8 & 6 & 10 & 9 \\ 7 & 5 & 9 & 10 \end{bmatrix}$$

26.97. Apply the method of Problem 26.43 to the matrix of Problem 26.82. Does it appear to converge toward the exact inverse?

$$A^{-1} = \frac{1}{6}\begin{bmatrix} -58 & -16 & -192 \\ 48 & 15 & 153 \\ 16 & 4 & 54 \end{bmatrix}$$

26.98. Evaluate the determinant of the coefficient matrix of Problem 26.81.

26.99. Evaluate the determinant of the coefficient matrix of Problem 26.82.

26.100. What is the determinant of the Hilbert matrix of order four?

26.101. Apply the method of Problem 26.48 to find the eigenvalues and eigenvectors of $Ax = \lambda x$ where A is the Hilbert matrix of order three. Use rational arithmetic and obtain the exact characteristic polynomial.

26.102. Referring to Problem 26.101, apply the same method to

$$\begin{aligned}
(2-\lambda)x_1 - x_2 \qquad\qquad\qquad\qquad\qquad\qquad &= 0 \\
-x_1 + (2-\lambda)x_2 - x_3 \qquad\qquad\qquad\qquad &= 0 \\
-x_2 + (2-\lambda)x_3 - x_4 \qquad\qquad &= 0 \\
-x_3 + (2-\lambda)x_4 - x_5 &= 0 \\
-x_4 + (2-\lambda)x_5 &= 0
\end{aligned}$$

26.103. Use the power method to find the dominant eigenvalue and eigenvector of the matrix

$$A = \begin{bmatrix} 2 & -1 & 0 & 0 \\ -1 & 2 & -1 & 0 \\ 0 & -1 & 2 & -1 \\ 0 & 0 & -1 & 2 \end{bmatrix}$$

26.104. Use the power method to find the dominant eigenvalue and eigenvector of the Hilbert matrix of order three.

26.105. Apply Jacobi's method to the Hilbert matrix of order three.

26.106. Apply Jacobi's method to the matrix of Problem 26.103.

26.107. Apply Givens' method to the matrix of Problem 26.103.

26.108. Apply Givens' method to the Hilbert matrix of order four.

26.109. Solve the system

$$\begin{aligned} x_1 + ix_2 \qquad &= 1 \\ -ix_1 + x_2 + ix_3 &= 0 \\ -ix_2 + x_3 &= 0 \end{aligned}$$

by the method of Problem 26.77.

26.110. Apply the method of Problem 26.78 to invert the coefficient matrix in Problem 26.109.

26.111. Apply Jacobi's method, as outlined in Problem 26.79, to find the eigenvalues and vectors for the coefficient matrix of Problem 26.109.

26.112. Apply the algorithm of Problem 26.80 to the matrix $A = \begin{bmatrix} 1 & i & -1 \\ i & 1 & i \\ -1 & i & 1 \end{bmatrix}$.

26.113. Assuming that a matrix A has an LU factorization, we have the formulas of Problem 26.14 for determining the factor elements.

$$u_{rj} = a_{rj} - l_{r1}u_{1j} - l_{r2}u_{2j} - \cdots - l_{r,r-1}u_{r-1,j} \qquad j \geqq r$$
$$u_{rr}l_{ir} = a_{ir} - l_{i1}u_{1r} - l_{i2}u_{2r} - \cdots - l_{i,r-1}u_{r-1,r} \qquad i > r$$

Suppose these are computed from left to right. With primes denoting computed values, subject to roundoff error, the calculation of u'_{rj} then begins like this. (See Problems 1.22 and 1.23.)

$$a_{rj}(1+E) - l'_{r1}u'_{1j}(1+E)^{(2)}$$

Each E represents a roundoff error, probably a different error at each appearance, and the superscript is not a power but only a count of the number of different $(1+E)$ factors. This device will shorten some otherwise lengthy expressions. Continuing,

$$a_{rj}(1+E)^{(2)} - l'_{r1}u'_{1j}(1+E)^{(3)} - l'_{r2}u'_{2j}(1+E)^{(2)}$$

until ultimately we obtain the computed u'_{rj}:

$$u'_{rj} = a_{rj}(1+E)^{(r-1)} - l'_{r1}u'_{1j}(1+E)^{(r)} - \cdots - l'_{r,r-1}u'_{r-1,j}(1+E)^{(2)}$$

Show that the corresponding expression for the computed l'_{ir} is as follows:

$$u'_{rr}l'_{ir}(1+E) = a_{ir}(1+E)^{(r-1)} - l'_{i1}u'_{1r}(1+E)^{(r)} - \cdots - l'_{i,r-1}u'_{r-1,r}(1+E)^{(2)}$$

26.114. Define Δ_2 by

$$(1+E_1)(1+E_2) = 1 + 2\Delta_2$$

and note that

$$|\Delta_2| = \left|\frac{1}{2}(E_1 + E_2 + E_1E_2)\right| \leqq u + \frac{1}{2}u^2$$

with u the maximum roundoff error. Show similarly that with Δ_3 defined by $(1+E_1)(1+E_2)(1+E_3) = 1 + 3\Delta_3$ the bound $u + u^2 + \frac{1}{3}u^3$ exists, and that more generally we may write

$$(1+E)^{(n)} = 1 + n\Delta_n$$

with Δ_n bounded by $[(1+u)^n - 1]/n$.

26.115. Combine the results of the preceding two problems to obtain (with Δ an appropriate Δ_k)

$$\begin{aligned} l'_{rk}u'_{kj} - a_{rj} &= a_{rj}(j-1)\Delta - l_{i1}u_{1j}j\Delta - \cdots - l_{ij}u_{jj}\Delta \qquad r > j \\ &= a_{rj}(r-1)\Delta - l_{i1}u_{1j}r\Delta - \cdots - l_{rr}u_{rj}\Delta \qquad r \leqq j \end{aligned}$$

and note that this is equivalent in matrix form to

$$L'U' = A + F$$

with the elements of F as shown on the right-hand side above. This shows that the factorization $L'U'$ is exact for the perturbed matrix $A + F$.

26.116. Show that the elements of the matrix F of the preceding problem do not exceed in absolute value $n\Delta$ times the combined terms of A and $L'U'$. That is,

$$|f_{ij}| \leqq n\Delta(|a_{ij}| + b_{ij})$$

where Δ bounds all the Δ_k involved and b_{ij} is computed from the absolute elements of the ith row of L' and the jth column of U'. This estimate of the effect of internal roundoffs depends strongly upon the b_{ij}. These may be computed after the factorization has been made. Here n is the order of the original matrix A. In a way we may deduce that the overall error is a modest multiple of the maximum roundoff, provided that n is not too large and the b_{ij} cooperative.

26.117. The formulas for forward- and back-substitution, derived in Problem 26.9

$$y_r = b_r - l_{r1}y_1 - \cdots - l_{r-1,r-1}y_{r-1}$$
$$u_{ii}x_i = y_i - u_{i,i+1}x_{i+1} - \cdots - u_{in}x_n$$

have the same form as those just analyzed for roundoff error propagation. Reasoning much as in the preceding problem, one may obtain this equation for the computed y'

$$(L' + G)y' = b$$

where $|g_{ij}| \leqq n\Delta\,|l'_{ij}|$, and then to $\qquad (U' + H)x' = y'$

for the computed solution itself. Here $|h_{ij}| \leqq n\Delta\,|u_{ij}|$.

By combining these results with that of the preceding problem, show that

$$(A + E)x' = b$$

with E a blend of F, G, H, L, and U. Further deduce the estimate

$$|e_{ij}| \leqq n\Delta[|a_{ij}| + (3 + n\Delta)\,|b_{ij}|]$$

with b_{ij} as defined earlier.

26.118. Apply the algorithm of Problem 26.80 to the real but nonsymmetric matrix

$$A = \begin{bmatrix} 1 & 2 & 3 \\ 1 & 3 & 5 \\ 1 & 4 & 7 \end{bmatrix}$$

26.119. Solve the system

$$\begin{aligned} 6.4375x_1 + 2.1849x_2 - 3.7474x_3 + 1.8822x_4 &= 4.6351 \\ 2.1356x_1 + 5.2101x_2 + 1.5220x_3 - 1.1234x_4 &= 5.2131 \\ -3.7362x_1 + 1.4998x_2 + 7.6421x_3 + 1.2324x_4 &= 5.8665 \\ 1.8666x_1 - 1.1104x_2 + 1.2460x_3 + 8.3312x_4 &= 4.1322 \end{aligned}$$

26.120. Find all the eigenvalues of this system:

$$\begin{aligned} 4x + 2y + z &= \lambda x \\ 2x + 4y + 2z &= \lambda y \\ x + 2y + 4z &= \lambda z \end{aligned}$$

26.121. Find all the eigenvalues and eigenvectors of this system:

$$\begin{bmatrix} 4 & 2 & 2 \\ 2 & 5 & 1 \\ 2 & 1 & 6 \end{bmatrix} \begin{pmatrix} x_1 \\ x_2 \\ x_3 \end{pmatrix} = \lambda \begin{pmatrix} x_1 \\ x_2 \\ x_3 \end{pmatrix}$$

26.122. Invert Pascal's matrix.

$$\begin{bmatrix} 1 & 1 & 1 & 1 & 1 \\ 1 & 2 & 3 & 4 & 5 \\ 1 & 3 & 6 & 10 & 15 \\ 1 & 4 & 10 & 20 & 35 \\ 1 & 5 & 15 & 35 & 70 \end{bmatrix}$$

26.123. Invert the following matrix:

$$\begin{bmatrix} 1 & \frac{1}{3} & \frac{1}{5} \\ \frac{1}{3} & \frac{1}{5} & \frac{1}{7} \\ \frac{1}{5} & \frac{1}{7} & \frac{1}{9} \end{bmatrix}$$

26.124. Invert the following matrix:

$$\begin{bmatrix} 5+i & 4+2i \\ 10+3i & 8+6i \end{bmatrix}$$

26.125. Find the largest eigenvalue of

$$\begin{bmatrix} 25 & -41 & 10 & -6 \\ -41 & 68 & -17 & 10 \\ 10 & -17 & 5 & -3 \\ -6 & 10 & -3 & 2 \end{bmatrix}$$

to three places.

26.126. Find the largest eigenvalue of

$$\begin{bmatrix} 8 & -5i & 3-2i \\ 5i & 3 & 0 \\ 3+2i & 0 & 2 \end{bmatrix}$$

and the corresponding eigenvector.

26.127. Find the two extreme eigenvalues of $\begin{bmatrix} 9 & 10 & 8 \\ 10 & 5 & -1 \\ 8 & -1 & 3 \end{bmatrix}$.

26.128. Show that the characteristic polynomial for the matrix

$$\begin{bmatrix} 1 & 1 \\ 1 & 0 \end{bmatrix}$$

is $\lambda^2 - \lambda - 1$ and note the relationship with Fibonacci numbers as encountered in Problem 18.23 and elsewhere. What is the characteristic polynomial for the more general "Fibonacci" matrix of order n?

Find its eigenvalues by any of our methods.

$$F_n = \begin{bmatrix} 1 & 1 & 1 & \cdots & 1 & 1 \\ 1 & 0 & 0 & \cdots & 0 & 0 \\ 0 & 1 & 0 & \cdots & 0 & 0 \\ 0 & 0 & 1 & \cdots & 0 & 0 \\ \cdots & \cdots & \cdots & \cdots & \cdots & \cdots \\ 0 & 0 & 0 & \cdots & 1 & 0 \end{bmatrix}$$

Given some initial vector x, what are the vectors $F_n^p x$ for $p = 2, \ldots$?

26.129. Apply the QR method to this Hessenberg matrix:

$$\begin{bmatrix} 2 & 1 & .5 & .1 \\ 1 & 3 & 1 & .5 \\ 0 & 1 & 2 & 1 \\ 0 & 0 & .5 & 1 \end{bmatrix}$$

26.130. Apply the QR method to this triple diagonal matrix:

$$\begin{bmatrix} 2.5 & -2.0 & 0 & 0 \\ -2.0 & 3.5 & 1.5 & 0 \\ 0 & 1.5 & 2.5 & -1.0 \\ 0 & 0 & -1.0 & 1.5 \end{bmatrix}$$

26.131. Rotating a square a quarter turn clockwise can be simulated by applying the permutation matrix R to the vector $(1, 2, 3, 4)^T$. (See Fig. 26-5.) Reflection in the vertical (dashed) line can be simulated using the matrix V. The eigenvalues of R are easily found to be $1, i, -1, -i$, while those of V are $1, 1, -1, -1$. Both matrices are of Hessenberg type. Will the QR algorithm of Problem 26.73 be convergent in either case?

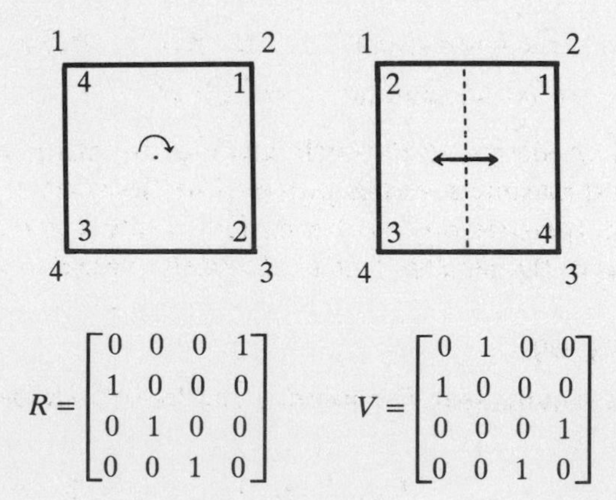

Fig. 26-5

Chapter 27

Linear Programming

THE BASIC PROBLEM

A linear programming problem requires that a linear function

$$H = c_1x_1 + \cdots + c_nx_n$$

be minimized (or maximized) subject to constraints of the form

$$a_{i1}x_1 + \cdots + a_{in}x_n \leqq b_i \qquad 0 \leqq x_j$$

where $i = 1, \ldots, m$ and $j = 1, \ldots, n$. In vector form the problem may be written as

$$H(x) = c^Tx = \text{minimum} \qquad Ax \leqq b, \qquad 0 \leqq x$$

An important theorem of linear programming states that the required minimum (or maximum) occurs at an *extreme feasible point*. A point $(x_1, \ldots, x_n)$ is called feasible if its coordinates satisfy all $n + m$ constraints, and an extreme feasible point is one where at least n of the constraints actually become equalities. The introduction of slack variables $x_{n+1}, \ldots, x_{n+m}$ converts the constraints to the form

$$a_{i1}x_1 + a_{i2}x_2 + \cdots + a_{in}x_n + x_{n+i} = b_i$$

for $i = 1, \ldots, m$. It allows an extreme feasible point to be identified as one at which n or more variables (including slack variables) are zero. This is a great convenience. In special cases more than one extreme feasible point may yield the required minimum, in which case other feasible points also serve the purpose. A minimum point of H is called a *solution point*.

The **simplex method** is an algorithm for starting at some extreme feasible point and, by a sequence of exchanges, proceeding systematically to other such points until a solution point is found. This is done in a way which steadily reduces the value of H. The exchange process involved is essentially the same as that presented in the previous chapter for matrix inversion.

The **duality theorem** is a relationship between the solutions of the two problems

$$c^Tx = \text{minimum} \qquad Ax \geqq b, \qquad 0 \leqq x$$
$$y^Tb = \text{maximum} \qquad y^TA \leqq c^T \qquad 0 \leqq y$$

which are known as dual problems, and which involve the same a_{ij}, b_i, and c_j numbers. The corresponding minimum and maximum values prove to be the same, and application of the simplex method to either problem (presumably to the easier of the two) allows the solutions of both problems to be extracted from the results. This is obviously a great convenience.

TWO RELATED PROBLEMS

1. **Two-person games** require that R choose a row and C choose a column of the following "payoff" matrix:

$$\begin{bmatrix} a_{11} & a_{12} & \cdots & a_{1n} \\ a_{21} & a_{22} & \cdots & a_{2n} \\ \cdots & \cdots & \cdots & \cdots \\ a_{m1} & a_{m2} & \cdots & a_{mn} \end{bmatrix}$$

The element a_{ij} where the selected row and column cross, determines the amount which R must then pay to C. Naturally C wishes to maximize his expected winnings while R wishes to minimize his expected losses. These conflicting viewpoints lead to dual linear programs which may be solved by the simplex method. The solutions are called *optimal strategies* for the two players.

2. **Overdetermined systems** of linear equations, in which there are more equations than unknowns and no vector x can satisfy the entire system, may be treated as linear programming problems in which we seek the vector x which in some sense has minimum error. The details appear in Chapter 28.

Solved Problems

THE SIMPLEX METHOD

27.1. Find x_1 and x_2 satisfying the inequalities

$$0 \leqq x_1 \qquad 0 \leqq x_2 \qquad -x_1 + 2x_2 \leqq 2 \qquad x_1 + x_2 \leqq 4 \qquad x_1 \leqq 3$$

and such that the function $F = x_2 - x_1$ is maximized.

Since only two variables are involved it is convenient to interpret the entire problem geometrically. In an x_1, x_2 plane the five inequalities constrain the point (x_1, x_2) to fall within the shaded region of Fig. 27-1. In each case the equality sign corresponds to (x_1, x_2) being on one of the five linear boundary segments. Maximizing F subject to these constraints is equivalent to finding that line of slope 1 having the largest y intercept and still intersecting the shaded region. It seems clear that the required line L_1 is $1 = x_2 - x_1$ and the intersection point $(0, 1)$. Thus, for a maximum, $x_1 = 0$, $x_2 = 1$, $F = 1$.

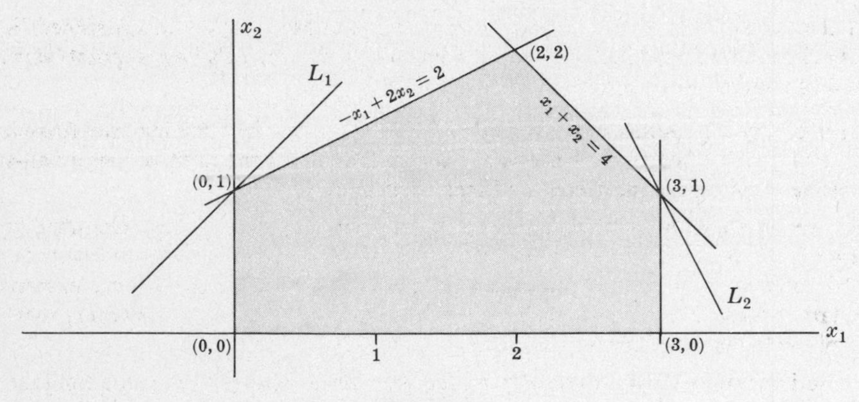

Fig. 27-1

27.2. With the same inequality constraints as in Problem 27.1, find (x_1, x_2) such that $G = 2x_1 + x_2$ is a maximum.

We now seek the line of slope -2 and having the largest y intercept while still intersecting the shaded region. This line L_2 is $7 = 2x_1 + x_2$ and the required point has $x_1 = 3$, $x_2 = 1$. (See Fig. 27-1.)

27.3. Find y_1, y_2, y_3 satisfying the constraints

$$0 \leqq y_1 \qquad 0 \leqq y_2 \qquad 0 \leqq y_3 \qquad y_1 - y_2 - y_3 \leqq 1 \qquad -2y_1 - y_2 \leqq -1$$

and minimizing $H = 2y_1 + 4y_2 + 3y_3$.

Interpreting the entire problem geometrically, we find that the five inequalities constrain the point (y_1, y_2, y_3) to fall within the region pictured in Fig. 27-2. This region is unbounded in the positive y_1, y_2, y_3 directions, but is otherwise bounded by portions of five planes, shown shaded. These planes

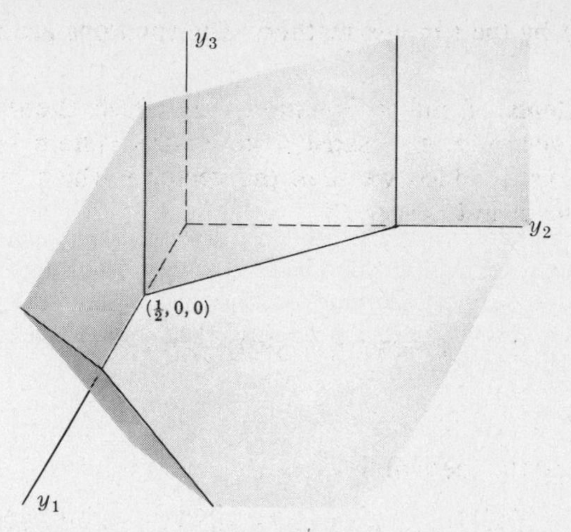

Fig. 27-2

correspond to equality holding in our five constraints. Minimizing H subject to these constraints is equivalent to finding a plane with normal vector $(2, 4, 3)$ having smallest intercepts and still intersecting the given region. It is easy to discover that this plane is $1 = 2y_1 + 4y_2 + 3y_3$ and the intersection point is $(\frac{1}{2}, 0, 0)$.

27.4. List three principal features of linear programming problems and their solutions which are illustrated by the previous problems.

Let the problem be to find a point x with coordinates $(x_1, x_2, \ldots, x_n)$ subject to the constraints $0 \leqq x$, $Ax \leqq b$ and minimizing a function $H(x) = c^T x = \sum c_i x_i$. Calling a point which meets all the constraints a *feasible point* (if any such exists), then:

1. The set of feasible points is convex, that is, the line segment joining two feasible points consists entirely of feasible points. This is due to the fact that each constraint defines a half-space and the set of feasible points is the intersection of these half-spaces.
2. There are certain *extreme feasible points,* the vertices of the convex set, identified by the fact that at least n of the constraints become equalities at these points. In the two-dimensional examples, exactly $n = 2$ boundary segments meet at such vertices. In the three-dimensional example, exactly three boundary planes meet at each such vertex. For $n \geqq 3$ it is possible, however, that more planes (or hyperplanes) come together at a vertex.
3. The solution point is always an extreme feasible point. This is due to the linearity of the function $H(x)$ being minimized. (It is possible that two extreme points are solutions, in which case the entire edge joining them consists of solutions, etc.)

These three features of linear programming problems will not be proved here. They are also true if $H(x)$ is to be maximized, or if the constraints read $Ax \geqq b$.

27.5. What is the general idea behind the *simplex method* for solving linear programs?

Since the solution occurs at an extreme feasible point, we may begin at some such point and compute the value of H. We then exchange this extreme point for its mate at the other end of an edge, in such a way that a smaller (in the case of a minimum problem) H value is obtained. The process of exchange and edge-following continues until H can no longer be decreased. This exchange algorithm is known as the simplex method. The details are provided in the following problem.

27.6. Develop the simplex method.

Let the problem be

$$0 \leqq x \qquad Ax \leqq b \qquad H(x) = c^T x = \text{minimum}$$

We first introduce *slack variables* $x_{n+1}, \ldots, x_{n+m}$ to make

$$\begin{gathered} a_{11}x_1 + a_{12}x_2 + \cdots + a_{1n}x_n + x_{n+1} = b_1 \\ a_{21}x_1 + a_{22}x_2 + \cdots + a_{2n}x_n + x_{n+2} = b_2 \\ \cdots\cdots\cdots\cdots\cdots\cdots\cdots\cdots\cdots \\ a_{m1}x_1 + a_{m2}x_2 + \cdots + a_{mn}x_n + x_{n+m} = b_m \end{gathered}$$

Notice that these slack variables, like the other x_i, must be nonnegative. The use of slack variables allows us to identify an extreme feasible point in another way. Since equality in $Ax \leqq b$ now corresponds to a slack variable being zero, an extreme point becomes one where at least n of the variables $x_1, \ldots, x_{n+m}$ are zero. Or said differently, at an extreme feasible point at most m of these variables are nonzero. The matrix of coefficients has become

$$\begin{bmatrix} a_{11} & a_{12} & \cdots & a_{1n} & 1 & 0 & \cdots & 0 \\ a_{21} & a_{22} & \cdots & a_{2n} & 0 & 1 & \cdots & 0 \\ \cdots & \cdots & \cdots & \cdots & \cdots & \cdots & \cdots & \cdots \\ a_{m1} & a_{m2} & \cdots & a_{mn} & 0 & 0 & \cdots & 1 \end{bmatrix}$$

the last m columns corresponding to the slack variables. Let the columns of this matrix be called $v_1, v_2, \ldots, v_{n+m}$. The linear system can then be written as

$$x_1v_1 + x_2v_2 + \cdots + x_{n+m}v_{n+m} = b$$

Now suppose that we know an extreme feasible point. For simplicity we will take it that $x_{m+1}, \ldots, x_{m+n}$ are all zero at this point so that $x_1, \ldots, x_m$ are the (at most m) nonzero variables. Then

$$x_1v_1 + x_2v_2 + \cdots + x_mv_m = b \tag{1}$$

and the corresponding H value is

$$H_1 = x_1c_1 + x_2c_2 + \cdots + x_mc_m \tag{2}$$

Assuming the vectors $v_1, \ldots, v_m$ linearly independent, all $n + m$ vectors may be expressed in terms of this basis:

$$v_j = v_{1j}v_1 + \cdots + v_{mj}v_m \qquad j = 1, \ldots, n+m \tag{3}$$

Also define
$$h_j = v_{1j}c_1 + \cdots + v_{mj}c_m - c_j \qquad j = 1, \ldots, n+m \tag{4}$$

Now, suppose we try to reduce H_1 by including a piece px_k, for $k > m$ and p positive. To preserve the constraint we multiply (3) for $j = k$ by p, which is still to be determined, and subtract from (1) to find

$$(x_1 - pv_{1k})v_1 + (x_2 - pv_{2k})v_2 + \cdots + (x_m - pv_{mk})v_m + pv_k = b$$

Similarly from (2) and (4) the new value of H will be

$$(x_1 - pv_{1k})c_1 + (x_2 - pv_{2k})c_2 + \cdots + (x_m - pv_{mk})c_m + pc_k = H_1 - ph_k$$

The change will be profitable only if $h_k > 0$. In this case it is optimal to make p as large as possible without a coefficient $x_i - pv_{ik}$ becoming negative. This suggests the choice

$$p = \min_i \frac{x_i}{v_{ik}} = \frac{x_l}{v_{lk}}$$

the minimum being taken over terms with positive v_{ik} only. With this choice of p the coefficient of c_l becomes zero, the others are nonnegative, and we have a new extreme feasible point with H value

$$H_1' = H_1 - ph_k$$

which is definitely smaller than H_1. We also have a new basis, having exchanged the basis vector v_l for the new v_k. The process is now repeated until all h_j are negative, or until for some positive h_k no v_{ik} is positive. In the former case the present extreme point is as good as any adjacent extreme point, and it can further be shown that it is as good as any other adjacent or not. In the latter case p may be arbitrarily large and there is no minimum for H.

Before another exchange can be made all vectors must be represented in terms of the new basis. Such exchanges have already been made in our section on matrix inversion but the details will be repeated. The vector v_l is to be replaced by the vector v_k. From

$$v_k = v_{1k}v_1 + \cdots + v_{mk}v_m$$

we solve for v_l and substitute into (3) to obtain the new representation

$$v_j = v'_{1j}v_1 + \cdots + v'_{l-1,j}v_{l-1} + v'_{kj}v_k + v'_{l+1,j}v_{l+1} + \cdots + v'_{mj}v_m$$

where

$$v'_{ij} = \begin{cases} v_{ij} - \dfrac{v_{lj}}{v_{lk}} v_{ik} & \text{for } i \neq l \\ \dfrac{v_{ij}}{v_{ik}} & \text{for } i = l \end{cases}$$

Also, substituting for v_l in (1) brings

$$x'_1v_1 + \cdots + x'_{l-1}v_{l-1} + x'_kv_k + x'_{l+1}v_{l+1} + \cdots + x'_mv_m = b$$

where

$$x'_i = \begin{cases} x_i - \dfrac{x_l}{v_{lk}} v_{ik} & \text{for } i \neq l \\ \dfrac{x_i}{v_{lk}} & \text{for } i = l \end{cases}$$

Furthermore, a short calculation proves

$$h'_j = v'_{1j}c_1 + \cdots + v'_{mj}c_m - c_j = h_j - \frac{v_{lj}}{v_{lk}} h_k$$

and we already have

$$H'_1 = H_1 - \frac{x_l}{v_{lk}} h_k$$

This entire set of equations may be summarized compactly by displaying the various ingredients as follows:

$$\begin{bmatrix} x_1 & v_{11} & v_{12} & \cdots & v_{1,n+m} \\ x_2 & v_{21} & v_{22} & \cdots & v_{2,n+m} \\ \cdots & \cdots & \cdots & \cdots & \cdots \\ x_m & v_{m1} & v_{m2} & \cdots & v_{m,n+m} \\ H_1 & h_1 & h_2 & \cdots & h_{n+m} \end{bmatrix}$$

Calling v_{lk} the *pivot,* all entries in the pivot row are divided by the pivot, the pivot column becomes zero except for a 1 in the pivot position, and all other entries are subjected to what was formerly called the *rectangle rule*. This will now be illustrated in a variety of examples.

27.7. Solve Problem 27.1 by the simplex method.

After introducing slack variables, the constraints are

$$\begin{aligned} -x_1 + 2x_2 + x_3 \qquad\qquad &= 2 \\ x_1 + x_2 \qquad + x_4 \qquad &= 4 \\ x_1 \qquad\qquad\qquad + x_5 &= 3 \end{aligned}$$

with all five variables required to be nonnegative. Instead of maximizing $x_2 - x_1$ we will minimize $x_1 - x_2$. Such a switch between minimum and maximum problems is always available to us. Since the origin is an extreme feasible point, we may choose $x_1 = x_2 = 0$, $x_3 = 2$, $x_4 = 4$, $x_5 = 3$ to start. This is very convenient since it amounts to choosing v_3, v_4, and v_5 as our first basis which makes all $v_{ij} = a_{ij}$. The starting display is therefore the following:

Basis	b	v_1	v_2	v_3	v_4	v_5
v_3	2	−1	②	1	0	0
v_4	4	1	1	0	1	0
v_5	3	1	0	0	0	1
	0	−1	1	0	0	0

Comparing with the format in Problem 27.6, one finds the six vectors b and $v_1, \ldots, v_5$ forming the top three rows, and the numbers $H, h_1, \ldots, h_5$ in the bottom row. Only h_2 is positive. This determines the pivot column. In this column there are two positive v_{i2} numbers, but 2/2 is less than 4/1 and so the pivot is $v_{12} = 2$. This number has been circled. The formulas of the previous problem now apply to produce a new display. The top row is simply divided by 2, and all other entries are subjected to the rectangle rule:

Basis	b	v_1	v_2	v_3	v_4	v_5
v_2	1	$-\frac{1}{2}$	1	$\frac{1}{2}$	0	0
v_4	3	$\frac{3}{2}$	0	$-\frac{1}{2}$	1	0
v_5	3	1	0	0	0	1
	−1	$-\frac{1}{2}$	0	$-\frac{1}{2}$	0	0

The basis vector v_3 has been exchanged for v_2 and all vectors are now represented in terms of this new basis. But more important for this example, no h_j is now positive so the algorithm stops. The minimum of $x_1 - x_2$ is -1 (making the maximum of $x_2 - x_1$ equal to 1 as before). This minimum is achieved for $x_2 = 1$, $x_4 = 3$, $x_5 = 3$ as the first column shows. The constraints then make $x_1 = 0$, $x_3 = 0$ which we anticipate since the x_j not corresponding to basis vectors should always be zero. The results $x_1 = 0$, $x_2 = 1$ correspond to our earlier geometrical conclusions. Notice that the simplex algorithm has taken us from the extreme point $(0, 0)$ of the set of feasible points to the extreme point $(0, 1)$ which proves to be the solution point. (See Fig. 27-1.)

27.8. Solve Problem 27.2 by the simplex method.

Slack variables and constraints are the same as in the previous problem. We shall minimize $H = -2x_1 - x_2$. The origin being an extreme point, we may start with this display:

Basis	b	v_1	v_2	v_3	v_4	v_5
v_3	2	−1	2	1	0	0
v_4	4	1	1	0	1	0
v_5	3	①	0	0	0	1
	0	2	1	0	0	0

Both h_1 and h_2 are positive, so we have a choice. Selecting $h_1 = 2$ makes v_{13} the pivot, since 3/1 is less than 4/1. This pivot has been circled. Exchanging v_5 for v_1 we have a new basis, a new extreme point, and a new display.

Basis	b	v_1	v_2	v_3	v_4	v_5
v_3	5	0	2	1	0	1
v_4	1	0	①	0	1	−1
v_1	3	1	0	0	0	1
	−6	0	1	0	0	−2

Now we have no choices. The new pivot has been circled and means that we exchange v_4 for v_2 with the following result:

v_3	3	0	0	1	−2	3
v_2	1	0	1	0	1	−1
v_1	3	1	0	0	0	1
	−7	0	0	0	−1	−1

Now no h_j is positive, so we stop. The minimum is −7, which agrees with the maximum of 7 for $2x_1 + x_2$ found in Problem 27.2. The solution point is at $x_1 = 3$, $x_2 = 1$ which also agrees with the result found in Problem 27.2. The simplex method has led us from (0, 0) to (3, 0) to (3, 1). The other choice available to us at the first exchange would have led us around the feasible set in the other direction.

27.9. Solve Problem 27.3 by the simplex method.

With slack variables the constraints become

$$y_1 - y_2 - y_3 + y_4 \quad = 1$$
$$-2y_1 - y_2 \quad + y_5 = -1$$

all five variables being required to be positive or zero. This time, however, the origin ($y_1 = y_2 = y_3 = 0$) is not a feasible point, as Fig. 27-2 shows and as the enforced negative value $y_5 = -1$ corroborates. We cannot therefore follow the starting procedure of the previous two examples based on a display such as this:

Basis	b	v_1	v_2	v_3	v_4	v_5
v_4	1	1	−1	−1	1	0
v_5	−1	−2	−1	0	0	1

The negative value $y_5 = -1$ in the b column cannot be allowed. Essentially our problem is that we do not have an extreme feasible point to start from. A standard procedure for finding such a point, even for a much larger problem than this, is to introduce an *artificial basis*. Here it will be enough to alter the second constraint, which contains the negative b component, to

$$-2y_1 - y_2 + y_5 - y_6 = -1$$

One new column may now be attached to our earlier display.

Basis	b	v_1	v_2	v_3	v_4	v_5	v_6
v_4	1	1	−1	−1	1	0	0
v_5	−1	−2	−1	0	0	1	−1

But an extreme feasible point now corresponds to $y_4 = y_6 = 1$, all other y_j being zero. This makes it natural to exchange v_5 for v_6 in the basis. Only a few sign changes across the v_6 row are required.

Basis	b	v_1	v_2	v_3	v_4	v_5	v_6
v_4	1	1	−1	−1	1	0	0
v_6	1	②	1	0	0	−1	1
	W	$2W-2$	$W-4$	−3	0	$-W$	0

The last row of this starting display will now be explained.

Introducing the artificial basis has altered our original problem, unless we can be sure that y_6 will eventually turn out to be zero. Fortunately this can be arranged, by changing the function to be minimized from $H = 2y_1 + 4y_2 + 3y_3$ as it was in Problem 27.2 to

$$H^* = 2y_1 + 4y_2 + 3y_3 + Wy_6$$

where W is such a large positive number that for a minimum we will surely have to make y_6 equal to zero. With these alterations we have a starting H value of W. The numbers h_j may also be computed and the last row of the starting display is as shown.

We now proceed in normal simplex style. Since W is large and positive we have a choice of two positive h_j values. Choosing h_1 leads to the circled pivot. Exchanging v_6 for v_1 brings a new display from which the last column has been dropped since v_6 is of no further interest:

v_4	$\frac{1}{2}$	0	$-\frac{3}{2}$	-1	1	$\frac{1}{2}$
v_1	$\frac{1}{2}$	1	$\frac{1}{2}$	0	0	$-\frac{1}{2}$
	1	0	-3	-3	0	-1

Since no h_j is positive we are already at the end. The minimum is 1, which agrees with our geometrical conclusion of Problem 27.3. Moreover, from the first column we find $y_1 = \frac{1}{2}$, $y_4 = \frac{1}{2}$ with all other y_j equal to zero. This yields the minimum point $(\frac{1}{2}, 0, 0)$ also found in Problem 27.3.

27.10. Minimize the function $H = 2y_1 + 4y_2 + 3y_3$ subject to the constraints $y_1 - y_2 - y_3 \leqq -2$, $-2y_1 - y_2 \leqq -1$, all y_j being positive or zero.

Slack variables and an artificial basis convert the constraints to

$$\begin{aligned} y_1 - y_2 - y_3 + y_4 \quad - y_6 \quad &= -2 \\ -2y_1 - y_2 \quad + y_5 \quad - y_7 &= -1 \end{aligned}$$

and much as in the preceding problem we soon have this starting display:

Basis	b	v_1	v_2	v_3	v_4	v_5	v_6	v_7
v_6	2	-1	1	1	-1	0	1	0
v_7	1	2	(1)	0	0	-1	0	1
	$3W$	$W-2$	$2W-4$	$W-3$	$-W$	$-W$	0	0

The function to be minimized is

$$H^* = 2y_1 + 4y_2 + 3y_3 + Wy_6 + Wy_7$$

and this determines the last row. There are various choices for pivot and we choose the one circled. This leads to a new display by exchanging v_7 for v_2 and dropping the v_7 column.

v_6	1	-3	0	(1)	-1	1	1
v_2	1	2	1	0	0	-1	0
	$W+4$	$-3W+6$	0	$W-3$	$-W$	$W-4$	0

A new pivot has been circled and the final display follows:

v_3	1	-3	0	1	-1	1
v_2	1	2	1	0	0	-1
	7	-3	0	0	-3	-1

The minimum of H^* and H is 7, and it occurs at $(0, 1, 1)$.

THE DUALITY THEOREM

27.11. What is the *duality theorem* of linear programming?

Consider these two linear programming problems:

Problem A	**Problem B**
$c^Tx = \text{minimum}$	$y^Tb = \text{maximum}$
$x \geqq 0$	$y \geqq 0$
$Ax \geqq b$	$y^TA \leqq c^T$

They are called dual problems because of the many relationships between them, such as the following:

1. If either problem has a solution then the other does also and the minimum of c^Tx equals the maximum of y^Tb.
2. For either problem the solution vector is found in the usual way. The solution vector of the dual problem may then be obtained by taking the slack variables in order, assigning those in the final basis the value zero, and giving each of the others the corresponding value of $-h_j$.

These results will not be proved here but will be illustrated using our earlier examples. The duality makes it possible to obtain the solution of both Problems A and B by solving either one.

27.12. Show that Problems 27.1 and 27.3 are dual problems and verify the two relationships claimed in Problem 27.11.

A few minor alterations are involved. To match Problems 27.1 and A we minimize $x_1 - x_2$ instead of maximizing $x_2 - x_1$. The vector c^T is then $(1, -1)$. The constraints are rewritten as

$$x_1 - 2x_2 \geqq -2 \qquad -x_1 - x_2 \geqq -4 \qquad -x_1 \geqq -3$$

which makes

$$A = \begin{bmatrix} 1 & -2 \\ -1 & -1 \\ -1 & 0 \end{bmatrix} \qquad b = \begin{bmatrix} -2 \\ -4 \\ -3 \end{bmatrix}$$

For Problem B we then have

$$y^TA = \begin{bmatrix} y_1 - y_2 - y_3 \\ -2y_1 - y_2 \end{bmatrix} \leqq \begin{bmatrix} 1 \\ -1 \end{bmatrix}$$

which are the constraints of Problem 27.3. The condition $y^Tb = \text{maximum}$ is also equivalent to

$$y^T(-b) = 2y_1 + 4y_2 + 3y_3 = \text{minimum}$$

so that Problems 27.3 and B have also been matched. The extreme values for both problems proved to be 1, which verifies relationship 1 of Problem 27.11. From the final simplex display in Problem 27.7 we obtain $x^T = (0, 1)$ and $y^T = (\frac{1}{2}, 0, 0)$ while from the computations of Problem 27.9 we find $y^T = (\frac{1}{2}, 0, 0)$ and $x^T = (0, 1)$, verifying relationship 2.

27.13. Verify that Problems 27.2 and 27.10 are duals.

The matrix A and vector b are the same as in Problem 27.12. However, we now have $c^T = (-2, -1)$. This matches Problem 27.2 with Problem A and Problem 27.10 with Problem B. The final display of Problem 27.8 yields $x^T = (3, 1)$ and $y^T = (0, 1, 1)$ and the same results come from Problem 27.10. The common minimum of c^Tx and maximum of y^Tb is -7.

SOLUTION OF TWO-PERSON GAMES

27.14. Show how a two-person game may be made equivalent to a linear program.

Let the payoff matrix, consisting of positive numbers a_{ij}, be

$$A = \begin{bmatrix} a_{11} & a_{12} & a_{13} \\ a_{21} & a_{22} & a_{23} \\ a_{31} & a_{32} & a_{33} \end{bmatrix}$$

by which we mean that when player R has chosen row i of this matrix and player C has (independently) chosen column j, a payoff of amount a_{ij} is then made from R to C. This constitutes one play of the game. The problem is to determine the best strategy for each player in the selection of rows or columns. To be more specific, let C choose the three columns with probabilities p_1, p_2, p_3, respectively. Then

$$p_1, p_2, p_3 \geqq 0 \qquad \text{and} \qquad p_1 + p_2 + p_3 = 1$$

Depending on R's choice of row, C now has one of the following three quantities for his expected winnings:

$$P_1 = a_{11}p_1 + a_{12}p_2 + a_{13}p_3$$
$$P_2 = a_{21}p_1 + a_{22}p_2 + a_{23}p_3$$
$$P_3 = a_{31}p_1 + a_{32}p_2 + a_{33}p_3$$

Let P be the least of these three numbers. Then, no matter how R plays, C will have expected winnings of at least P on each play and therefore asks himself how this amount P can be maximized. Since all the numbers involved are positive, so is P; and we obtain an equivalent problem by letting

$$x_1 = \frac{p_1}{P} \qquad x_2 = \frac{p_2}{P} \qquad x_3 = \frac{p_3}{P}$$

and minimizing

$$F = x_1 + x_2 + x_3 = \frac{1}{P}$$

The various constraints may be expressed as $x_1, x_2, x_3 \geqq 0$ and

$$a_{11}x_1 + a_{12}x_2 + a_{13}x_3 \geqq 1$$
$$a_{21}x_1 + a_{22}x_2 + a_{23}x_3 \geqq 1$$
$$a_{31}x_1 + a_{32}x_2 + a_{33}x_3 \geqq 1$$

This is the type A problem of our duality theorem with $c^T = b^T = (1, 1, 1)$.

Now look at things from R's point of view. Suppose he chooses the three rows with probabilities q_1, q_2, q_3, respectively. Depending on C's choice of column he has one of the following quantities as his expected loss:

$$q_1a_{11} + q_2a_{21} + q_3a_{31} \leqq Q$$
$$q_1a_{12} + q_2a_{22} + q_3a_{32} \leqq Q$$
$$q_1a_{13} + q_2a_{23} + q_3a_{33} \leqq Q$$

where Q is the largest of the three. Then, no matter how C plays, R will have expected loss of no more than Q on each play. Accordingly he asks how this amount Q can be minimized. Since $Q > 0$, we let

$$y_1 = \frac{q_1}{Q} \qquad y_2 = \frac{q_2}{Q} \qquad y_3 = \frac{q_3}{Q}$$

and consider the equivalent problem of maximizing

$$G = y_1 + y_2 + y_3 = \frac{1}{Q}$$

The constraints are $y_1, y_2, y_3 \geqq 0$ and

$$y_1a_{11} + y_2a_{21} + y_3a_{31} \leqq 1$$
$$y_1a_{12} + y_2a_{22} + y_3a_{32} \leqq 1$$
$$y_1a_{13} + y_2a_{23} + y_3a_{33} \leqq 1$$

This is the type B problem of our duality theorem with $c^T = b^T = (1, 1, 1)$. We have discovered that R's problem and C's problem are duals. This means that the maximum P and minimum Q values will be the same, so that both players will agree on the average payment which is optimal. It also means that the optimal strategies for both players may be found by solving just one of the dual programs. We choose R's problem since it avoids the introduction of an artificial basis.

The same arguments apply for payoff matrices of other sizes. Moreover, the requirement that all a_{ij} be positive can easily be removed since, if all a_{ij} are replaced by $a_{ij} + a$, then P and Q are replaced by $P + a$ and $Q + a$. Thus only the value of the game is changed, not the optimal strategies. Examples will now be offered.

27.15. Find optimal strategies for both players and the optimal payoff for the game with matrix

$$A = \begin{bmatrix} 0 & 1 & 2 \\ 1 & 0 & 1 \\ 1 & 2 & 0 \end{bmatrix}$$

Instead we minimize the function $-G = -y_1 - y_2 - y_3$ subject to the constraints

$$\begin{aligned} y_2 + y_3 + y_4 \qquad\qquad &= 1 \\ y_1 \quad + 2y_3 \qquad + y_5 \quad &= 1 \\ 2y_1 + y_2 \qquad\qquad + y_6 &= 1 \end{aligned}$$

all y_j including the slack variables y_4, y_5, y_6 being nonnegative. Since the origin is an extreme feasible point, we have this starting display:

Basis	b	v_1	v_2	v_3	v_4	v_5	v_6
v_4	1	0	1	1	1	0	0
v_5	1	1	0	2	0	1	0
v_6	1	②	1	0	0	0	1
	0	1	1	1	0	0	0

Using the indicated pivots we make three exchanges as follows:

v_4	1	0	1	1	1	0	0
v_5	$\frac{1}{2}$	0	$-\frac{1}{2}$	②	0	1	$-\frac{1}{2}$
v_1	$\frac{1}{2}$	1	$\frac{1}{2}$	0	0	0	$\frac{1}{2}$
	$-\frac{1}{2}$	0	$\frac{1}{2}$	1	0	0	$-\frac{1}{2}$

v_4	$\frac{3}{4}$	0	$\frac{5}{4}$	0	1	$-\frac{1}{2}$	$\frac{1}{4}$
v_3	$\frac{1}{4}$	0	$-\frac{1}{4}$	1	0	$\frac{1}{2}$	$-\frac{1}{4}$
v_1	$\frac{1}{2}$	1	$\frac{1}{2}$	0	0	0	$\frac{1}{2}$
	$-\frac{3}{4}$	0	$\frac{3}{4}$	0	0	$-\frac{1}{2}$	$-\frac{1}{4}$

v_2	$\frac{3}{5}$	—	—	—	—	—	—
v_3	$\frac{2}{5}$	—	—	—	—	—	—
v_1	$\frac{1}{5}$	—	—	—	—	—	—
	$-\frac{6}{5}$	0	0	0	$-\frac{3}{5}$	$-\frac{1}{5}$	$-\frac{2}{5}$

From the final display we deduce that the optimal payoff, or value of the game, is $\frac{5}{6}$. the optimal strategy for R can be found directly by normalizing the solution $y_1 = \frac{1}{5}$, $y_2 = \frac{3}{5}$, $y_3 = \frac{2}{5}$. The probabilities q_1, q_2, q_3 must be proportional to these y_j but must sum to 1. Accordingly,

$$q_1 = \frac{1}{6} \qquad q_2 = \frac{3}{6} \qquad q_3 = \frac{2}{6}$$

To obtain the optimal strategy for C we note that there are no slack variables in the final basis so that putting the $-h_j$ in place of the (nonbasis) slack variables,

$$x_1 = \frac{3}{5} \qquad x_2 = \frac{1}{5} \qquad x_3 = \frac{2}{5}$$

Normalizing brings

$$p_1 = \frac{3}{6} \qquad p_2 = \frac{1}{6} \qquad p_3 = \frac{2}{6}$$

If either player uses the optimal strategy for mixing his choices the average payoff will be $\frac{5}{6}$. To make the game fair, all payoffs could be reduced by this amount, or C could be asked to pay this amount before each play is made.

27.16. Find the optimal strategy for each player and the optimal payoff for the game with matrix

$$A = \begin{bmatrix} 0 & 3 & 4 \\ 1 & 2 & 1 \\ 4 & 3 & 0 \end{bmatrix}$$

Notice that the center element is both the maximum in its row and the minimum in its column. It is also the smallest row maximum and the largest column minimum. Such a *saddle point* identifies a game with *pure strategies*. The simplex method leads directly to this result using the saddle point as pivot. The starting display is as follows:

Basis	b	v_1	v_2	v_3	v_4	v_5	v_6
v_4	1	0	1	4	1	0	0
v_5	1	3	②	3	0	1	0
v_6	1	4	1	0	0	0	1
	0	1	1	1	0	0	0

One exchange is sufficient:

v_4	$\frac{1}{2}$	—	—	—	—	—	—
v_2	$\frac{1}{2}$	—	—	—	—	—	—
v_6	$\frac{1}{2}$	—	—	—	—	—	—
	$-\frac{1}{2}$	$-\frac{1}{2}$	0	$-\frac{1}{2}$	0	$-\frac{1}{2}$	0

The optimal payoff is the negative reciprocal of $-\frac{1}{2}$, that is, the pivot element 2. The optimal strategy for R is found directly. Since $y_1 = 0$, $y_2 = \frac{1}{2}$, $y_3 = 0$, we normalize to obtain the pure strategy

$$q_1 = 0 \qquad q_2 = 1 \qquad q_3 = 0$$

Only the second row should ever be used. The strategy for C is found through the slack variables. Since v_4 and v_6 are in the final basis we have $x_1 = x_3 = 0$, and finally $x_2 = -h_5 = \frac{1}{2}$. Normalizing, we have another pure strategy

$$p_1 = 0 \qquad p_2 = 1 \qquad p_3 = 0$$

Supplementary Problems

27.17. Make a diagram showing all points which satisfy the following constraints simultaneously:

$$0 \leqq x_1 \qquad 0 \leqq x_2 \qquad x_1 + 2x_2 \leqq 4 \qquad -x_1 + x_2 \leqq 1 \qquad x_1 + x_2 \leqq 3$$

27.18. What are the five extreme feasible points for the previous problem? At which extreme point does $F = x_1 - 2x_2$ take its minimum value and what is that minimum? At which extreme point does this function take its maximum value?

27.19. Find the minimum of $F = x_1 - 2x_2$ subject to the constraints of Problem 27.17 by applying the simplex method. Do you obtain the same value and the same extreme feasible point as by the geometrical method?

27.20. What is the dual of Problem 27.19? Show by using the final simplex display obtained in that problem that the solution of the dual is the vector $y_1 = \frac{1}{3}$, $y_2 = \frac{4}{3}$, $y_3 = 0$.

27.21. Find the maximum of $F = x_1 - 2x_2$ subject to the constraints of Problem 27.17 by applying the simplex method. (Minimize $-F$.) Do you obtain the same results as by the geometrical method?

27.22. What is the dual of Problem 27.21? Find its solution from the final simplex display of that problem.

27.23. Solve the dual of Problem 27.19 directly by the simplex method, using one extra variable for an artificial basis. The constraints should then read

$$\begin{aligned} -y_1 + y_2 - y_3 + y_4 \qquad\qquad &= 1 \\ -2y_1 - y_2 - y_3 \qquad + y_5 - y_6 &= -2 \end{aligned}$$

with y_4 and y_5 the slack variables. The function $H = 4y_1 + y_2 + 3y_3$ is to be minimized. From the final display recover both the solution of the dual and of Problem 27.19 itself.

27.24. Minimize $F = 2x_1 + x_2$ subject to the constraints

$$3x_1 + x_2 \geqq 3 \qquad 4x_1 + 3x_2 \geqq 6 \qquad x_1 + 2x_2 \geqq 2$$

all x_j being nonnegative. (The solution finds $x_1 = \frac{3}{5}$, $x_2 = \frac{6}{5}$.)

27.25. Show geometrically that for a minimum of $F = x_1 - x_2$ subject to the constraints of Problem 27.17 there will be infinitely many solution points. Where are they? Show that the simplex method produces one extreme solution point directly and that it also produces another if a final exchange of v_3 and v_1 is made even though the corresponding h_j value is zero. The set of solution points is the segment joining these extreme points.

27.26. Minimize $F = x_1 + x_4$ subject to the constraints

$$\begin{aligned} 2x_1 + 2x_2 + x_3 &\leqq 7 \qquad & x_2 + x_4 &\geqq 1 \\ 2x_1 + x_2 + 2x_3 &\leqq 4 \qquad & x_2 + x_3 + x_4 &= 3 \end{aligned}$$

all x_j being nonnegative. (The minimum is zero and it occurs for more than one feasible point.)

27.27. Find optimal strategies and payoff for the game

$$A = \begin{bmatrix} 1 & 3 \\ 4 & 2 \end{bmatrix}$$

using the simplex method. [The payoff is 2.5, the strategy for R being $(\frac{1}{2}, \frac{1}{2})$ and that for C being $(\frac{1}{4}, \frac{3}{4})$.]

27.28. Solve the game with matrix

$$A = \begin{bmatrix} 0 & 3 & -4 \\ 3 & 0 & 5 \\ -4 & 5 & 0 \end{bmatrix}$$

showing the optimal payoff to be $\frac{10}{7}$, the optimal strategy for R to be $(\frac{5}{14}, \frac{4}{7}, \frac{1}{14})$, and that for C to be the same.

27.29. Solve the following game by the simplex method:

$$A = \begin{bmatrix} 0 & 0 & 1 & 1 \\ 1 & 1 & -2 & -2 \\ 1 & -2 & 1 & -2 \\ -2 & 3 & -2 & 3 \end{bmatrix}$$

27.30. Find the min–max cubic polynomial for the following function. What is the min–max error and where is it attained?

x	-2	-1.5	-1	$-.5$	0	$.5$	1	1.5	2
$y(x)$	5	5	4	2	1	3	7	10	12

27.31. Find the min–max quadratic polynomial for

$$y(x) = \frac{1}{1 + (4.1163x)^2} \qquad x = 0(.01)1$$

as well as the min–max error and the arguments at which it is attained.

27.32. What is the result of seeking a cubic approximation to the function of the preceding problem? How can this be forecast from the results of that problem?

27.33. Maximize $x_1 - x_2 + 2x_3$ subject to

$$\begin{aligned} x_1 + x_2 + 3x_3 + x_4 &\leqq 5 \\ x_1 \qquad + x_3 - 4x_4 &\leqq 2 \end{aligned}$$

and all $x_k \geqq 0$.

27.34. Solve the dual of the preceding problem.

27.35. Maximize $2x_1 + x_2$ subject to

$$x_1 - x_2 \leqq 2 \qquad x_1 + x_2 \leqq 6 \qquad x_1 + 2x_2 \leqq A$$

and all $x_k \geqq 0$. Treat the cases $A = 0, 3, 6, 9, 12$.

27.36. Use linear programming to find optimum strategies for both players in the following game:

$$\begin{bmatrix} -6 & 4 \\ 4 & -2 \end{bmatrix}$$

27.37. Solve as a linear program the game with payoff matrix

$$\begin{bmatrix} 3 & 1 \\ 2 & 3 \end{bmatrix}.$$

Chapter 28

Overdetermined Systems

NATURE OF THE PROBLEM

An overdetermined system of linear equations takes the form

$$Ax = b$$

the matrix A having more rows than columns. Ordinarily no solution vector x will exist, so that the equation as written is meaningless. The system is also called inconsistent. Overdetermined systems arise in experimental or computational work whenever more results are generated than would be required if precision were attainable. In a sense, a mass of inexact, conflicting information becomes a substitute for a few perfect results and one hopes that good approximations to the exact results can somehow be squeezed from the conflict.

TWO METHODS OF APPROACH

The two principal methods involve the *residual vector*

$$R = Ax - b$$

Since R cannot ordinarily be reduced to the zero vector, an effort is made to choose x in such a way that r is minimized in some sense.

1. The **least-squares solution** of an overdetermined system is the vector x which makes the sum of the squares of the components of the residual vector a minimum. In vector language we want

$$R^T R = \text{minimum}$$

For m equations and n unknowns, with $m > n$, the type of argument used in Chapter 21 leads to the *normal equations*

$$\begin{aligned} (a_1, a_1)x_1 + \cdots + (a_1, a_n)x_n &= (a_1, b) \\ \cdots\cdots\cdots\cdots\cdots\cdots\cdots \\ (a_n, a_1)x_1 + \cdots + (a_n, a_n)x_n &= (a_n, b) \end{aligned}$$

which determine the components of x. Here

$$(a_i, a_j) = a_{1i}a_{1j} + \cdots + a_{mi}a_{mj}$$

is the scalar product of two column vectors of A.

2. The **Chebyshev or min–max solution** is the vector x for which the absolutely largest component of the residual vector is a minimum. That is, we try to minimize

$$r = \max(|r_1|, \ldots, |r_m|)$$

where the r_i are the components of R. For $m = 3$, $n = 2$ this translates into the set of constraints

$$\begin{aligned} a_{11}x_1 + a_{12}x_2 - b_1 &\leqq r & \qquad -a_{11}x_1 - a_{12}x_1 + b_1 &\leqq r \\ a_{21}x_1 + a_{22}x_2 - b_2 &\leqq r & \qquad -a_{21}x_1 - a_{22}x_2 + b_2 &\leqq r \\ a_{31}x_1 + a_{32}x_2 - b_3 &\leqq r & \qquad -a_{31}x_1 - a_{32}x_2 + b_3 &\leqq r \end{aligned}$$

with r to be minimized. This now transforms easily into a linear programming problem. Similar linear programs solve the case of arbitrary m and n.

Solved Problems

LEAST-SQUARES SOLUTION

28.1. Derive the *normal equations* for finding the least-squares solution of an overdetermined system of linear equations.

Let the given system be

$$a_{11}x_1 + a_{12}x_2 = b_1$$
$$a_{21}x_1 + a_{22}x_2 = b_2$$
$$a_{31}x_1 + a_{32}x_2 = b_3$$

This involves only the two unknowns x_1 and x_2 and is only slightly overdetermined, but the details for larger systems are almost identical. Ordinarily we will not be able to satisfy all three of our equations. The problem as it stands probably has no solution. Accordingly we rewrite it as

$$a_{11}x_1 + a_{12}x_2 - b_1 = r_1$$
$$a_{21}x_1 + a_{22}x_2 - b_2 = r_2$$
$$a_{31}x_1 + a_{32}x_2 - b_3 = r_3$$

the numbers r_1, r_2, r_3 being called residuals, and look for the numbers x_1, x_2 which make $r_1^2 + r_2^2 + r_3^2$ minimal. Since

$$r_1^2 + r_2^2 + r_3^2 = (a_{11}^2 + a_{21}^2 + a_{31}^2)x_1^2 + (a_{12}^2 + a_{22}^2 + a_{32}^2)x_2^2 + 2(a_{11}a_{12} + a_{21}a_{22} + a_{31}a_{32})x_1x_2$$
$$- 2(a_{11}b_1 + a_{21}b_2 + a_{31}b_3)x_1 - 2(a_{12}b_1 + a_{22}b_2 + a_{32}b_3)x_2 + (b_1^2 + b_2^2 + b_3^2)$$

the result of setting derivatives relative to x_1 and x_2 equal to zero is the pair of normal equations

$$(a_1, a_1)x_1 + (a_1, a_2)x_2 = (a_1, b)$$
$$(a_2, a_1)x_1 + (a_2, a_2)x_2 = (a_2, b)$$

in which the parentheses denote

$$(a_1, a_1) = a_{11}^2 + a_{21}^2 + a_{31}^2 \qquad (a_1, a_2) = a_{11}a_{12} + a_{21}a_{22} + a_{31}a_{32}$$

and so on. These are the *scalar products* of the various columns of coefficients in the original system, so that the normal equations may be written directly. For the general problem of m equations in n unknowns $(m > n)$,

$$a_{11}x_1 + \cdots + a_{1n}x_n = b_1$$
$$a_{21}x_1 + \cdots + a_{2n}x_n = b_2$$
$$\cdots\cdots\cdots\cdots\cdots\cdots$$
$$a_{m1}x_1 + \cdots + a_{mn}x_n = b_m$$

an almost identical argument leads to the normal equations

$$(a_1, a_1)x_1 + (a_1, a_2)x_2 + \cdots + (a_1, a_n)x_n = (a_1, b)$$
$$(a_2, a_1)x_1 + (a_2, a_2)x_2 + \cdots + (a_2, a_n)x_n = (a_2, b)$$
$$\cdots\cdots\cdots\cdots\cdots\cdots\cdots\cdots\cdots\cdots$$
$$(a_n, a_1)x_1 + (a_n, a_2)x_2 + \cdots + (a_n, a_n)x_n = (a_n, b)$$

This is a symmetric, positive definite system of equations.

It is also worth noticing that the present problem again fits the model of our general least-squares approach in Problems 21.7 and 21.8. The results just obtained follow at once as a special case, with the vector space E consisting of m-dimensional vectors such as, for instance, the column vectors of the matrix A which we denote by $a_1, a_2, \ldots, a_n$ and the column of numbers b_i which we denote by b. The subspace S is the range of the matrix A, that is, the set of vectors Ax. We are looking for a vector p in S

which minimizes

$$\|p-b\|^2 = \|Ax-b\|^2 = \sum r_i^2$$

and this vector is the orthogonal projection of b onto S, determined by $(p-b, u_k)=0$, where the u_k are some basis for S. Choosing for this basis $u_k = a_k$, $k=1,\ldots,n$, we have the usual representation $p = x_1a_1 + \cdots + x_na_n$ (the notation being somewhat altered from that of our general model) and substitution leads to the normal equations.

28.2. Find the least-squares solution of this system:

$$x_1 - x_2 = 2$$
$$x_1 + x_2 = 4$$
$$2x_1 + x_2 = 8$$

Forming the required scalar products, we have

$$6x_1 + 2x_2 = 22 \qquad 2x_1 + 3x_2 = 10$$

for the normal equations. This makes $x_1 = \frac{23}{7}$ and $x_2 = \frac{8}{7}$. The residuals corresponding to this x_1 and x_2 are $r_1 = \frac{1}{7}$, $r_2 = \frac{3}{7}$, and $r_3 = -\frac{2}{7}$, and the sum of their squares is $\frac{2}{7}$. The root-mean-square error is therefore $\rho = \sqrt{\frac{2}{21}}$. This is smaller than for any other choice of x_1 and x_2.

28.3. Suppose three more equations are added to the already overdetermined system of Problem 28.2:

$$x_1 + 2x_2 = 4$$
$$2x_1 - x_2 = 5$$
$$x_1 - 2x_2 = 2$$

Find the least-squares solution of the set of six equations.

Again forming scalar products we obtain $12x_1 = 38$, $12x_2 = 9$ for the normal equations, making $x_1 = \frac{19}{6}$, $x_2 = \frac{3}{4}$. The six residuals are 5, -1, -11, 8, 7, and -4, all divided by 12. The RMS error is $\rho = \sqrt{\frac{23}{72}}$.

28.4. In the case of a large system, how may the set of normal equations be solved?

Since the set of normal equations is symmetric and positive definite, several methods perform very well. The Gauss elimination method may be applied, and if its pivots are chosen by descending the main diagonal then the problem remains symmetric to the end. Almost half the computation can therefore be saved.

CHEBYSHEV SOLUTION

28.5. Show how the Chebyshev solution of an overdetermined system of linear equations may be found by the method of linear programming.

Once again we treat the small system of Problem 28.1, the details for larger systems being almost identical. Let r be the maximum of the absolute values of the residuals, so that $|r_1| \leqq r$, $|r_2| \leqq r$, $|r_3| \leqq r$. This means that $r_1 \leqq r$ and $-r_1 \leqq r$, with similar requirements on r_2 and r_3. Recalling the definitions of the residuals we now have six inequalities:

$$a_{11}x_1 + a_{12}x_2 - b_1 \leqq r \qquad -a_{11}x_1 - a_{12}x_2 + b_1 \leqq r$$
$$a_{21}x_1 + a_{22}x_2 - b_2 \leqq r \qquad -a_{21}x_1 - a_{22}x_2 + b_2 \leqq r$$
$$a_{31}x_1 + a_{32}x_2 - b_3 \leqq r \qquad -a_{31}x_1 - a_{32}x_2 + b_3 \leqq r$$

If we also suppose that x_1 and x_2 must be nonnegative, and recall that the Chebyshev solution is defined to be that choice of x_1, x_2 which makes r minimal, then it is evident that we have a linear programming problem. It is convenient to modify it slightly. Dividing through by r and letting $x_1/r = y_1$, $x_2/r = y_2$, $1/r = y_3$, the constraints become

$$\begin{aligned} a_{11}y_1 + a_{12}y_2 - b_1y_3 &\leqq 1 \qquad & -a_{11}y_1 - a_{12}y_2 + b_1y_3 &\leqq 1 \\ a_{21}y_1 + a_{22}y_2 - b_2y_3 &\leqq 1 \qquad & -a_{21}y_1 - a_{22}y_2 + b_2y_3 &\leqq 1 \\ a_{31}y_1 + a_{32}y_2 - b_3y_3 &\leqq 1 \qquad & -a_{31}y_1 - a_{32}y_2 + b_3y_3 &\leqq 1 \end{aligned}$$

and we must maximize y_3 or, what is the same thing, make $F = -y_3 = \text{minimum}$. This linear program can be formed directly from the original overdetermined system. The generalization for larger systems is almost obvious. The condition that the x_j be positive is often met in practice, these numbers representing lengths or other physical measurements. If it is not met, then a translation $z_j = x_j + c$ may be made, or a modification of the linear programming algorithm may be used.

28.6. Apply the linear programming method to find the Chebyshev solution of the system of Problem 28.2.

Adding one slack variable to each constraint, we have

$$\begin{aligned} y_1 - y_2 - 2y_3 + y_4 &= 1 \\ y_1 + y_2 - 4y_3 \qquad + y_5 &= 1 \\ 2y_1 + y_2 - 8y_3 \qquad\qquad + y_6 &= 1 \\ -y_1 + y_2 + 2y_3 \qquad\qquad\qquad + y_7 &= 1 \\ -y_1 - y_2 + 4y_3 \qquad\qquad\qquad\qquad + y_8 &= 1 \\ -2y_1 - y_2 + 8y_3 \qquad\qquad\qquad\qquad\qquad + y_9 &= 1 \end{aligned}$$

with $F = -y_3$ to be minimized and all y_j to be nonnegative. The starting display and three exchanges following the simplex algorithm are shown in Fig. 28-1. The six columns corresponding to the slack

Basis	b	v_1	v_2	v_3
v_4	1	1	-1	-2
v_5	1	1	1	-4
v_6	1	2	1	-8
v_7	1	-1	1	2
v_8	1	-1	-1	4
v_9	1	-2	-1	(8)
	0	0	0	1

Basis	b	v_1	v_2	v_3
v_4	$\frac{5}{4}$	($\frac{1}{2}$)	$-\frac{5}{4}$	0
v_5	$\frac{3}{2}$	0	$\frac{1}{2}$	0
v_6	2	0	0	0
v_7	$\frac{3}{4}$	$-\frac{1}{2}$	$\frac{5}{4}$	0
v_8	$\frac{1}{2}$	0	$-\frac{1}{2}$	0
v_3	$\frac{1}{8}$	$-\frac{1}{4}$	$-\frac{1}{8}$	1
	$-\frac{1}{8}$	$\frac{1}{4}$	$\frac{1}{8}$	0

Basis	b	v_1	v_2	v_3
v_1	$\frac{5}{2}$	1	$-\frac{5}{2}$	0
v_5	$\frac{3}{2}$	0	($\frac{1}{2}$)	0
v_6	2	0	0	0
v_7	2	0	0	0
v_8	$\frac{1}{2}$	0	$-\frac{1}{2}$	0
v_3	$\frac{3}{4}$	0	$-\frac{3}{4}$	1
	$-\frac{3}{4}$	0	$\frac{3}{4}$	0

Basis	b	v_1	v_2	v_3
v_1	10	1	0	0
v_2	3	0	1	0
v_6	2	0	0	0
v_7	2	0	0	0
v_8	2	0	0	0
v_3	3	0	0	1
	-3	0	0	0

Fig. 28-1

variables are omitted since they actually contain no vital information. From the final display we find $y_1 = 10$ and $y_2 = y_3 = 3$. This makes $r = 1/y_3 = \frac{1}{3}$ and then $x_1 = \frac{10}{3}$, $x_2 = 1$. The three residuals are $\frac{1}{3}, \frac{1}{3}, -\frac{1}{3}$ so that the familiar Chebyshev feature of equal error sizes is again present.

28.7. Apply the linear programming method to find the Chebyshev solution of the overdetermined system of Problem 28.3.

The six additional constraints bring six more slack variables, $y_{10}, \ldots, y_{15}$. The details are very much as in Problem 28.6. Once again the columns for slack variables are omitted from Fig. 28-2, which summarizes three exchanges of the simplex algorithm. After the last exchange we find $y_1 = \frac{13}{3}$, $y_2 = 1$, $y_3 = \frac{4}{3}$. So $r = \frac{3}{4}$ and $x_1 = \frac{13}{4}$, $x_2 = \frac{3}{4}$. The six residuals are 2, 0, −3, 3, 3, and −1, all divided by 4. Once again three residuals equal the min–max residual r, the others now being smaller. In the general problem $n + 1$ equal residuals, the others being smaller, identify the Chebyshev solution, n being the number of unknowns.

Basis	b	v_1	v_2	v_3
v_4	1	1	−1	−2
v_5	1	1	1	−4
v_6	1	2	1	−8
v_7	1	−1	1	2
v_8	1	−1	−1	4
v_9	1	−2	−1	**8** (circled)
v_{10}	1	1	2	−4
v_{11}	1	2	−1	−5
v_{12}	1	1	−2	−2
v_{13}	1	−1	−2	4
v_{14}	1	−2	1	5
v_{15}	1	−1	2	2
	0	0	0	1

Basis	b	v_1	v_2	v_3
v_4	$\frac{5}{4}$	$\frac{1}{2}$	$-\frac{5}{4}$	0
v_5	$\frac{3}{2}$	0	$\frac{1}{2}$	0
v_6	2	0	0	0
v_7	$\frac{3}{4}$	$-\frac{1}{2}$	$\frac{5}{4}$	0
v_8	$\frac{1}{2}$	0	$-\frac{1}{2}$	0
v_3	$\frac{1}{8}$	$-\frac{1}{4}$	$-\frac{1}{8}$	1
v_{10}	$\frac{3}{2}$	0	$\frac{3}{2}$	0
v_{11}	$\frac{13}{8}$	**$\frac{3}{4}$** (circled)	$-\frac{13}{8}$	0
v_{12}	$\frac{5}{4}$	$\frac{1}{2}$	$-\frac{9}{4}$	0
v_{13}	$\frac{1}{2}$	0	$-\frac{3}{2}$	0
v_{14}	$\frac{3}{8}$	$-\frac{3}{4}$	$\frac{13}{8}$	0
v_{15}	$\frac{3}{4}$	$-\frac{1}{2}$	$\frac{9}{4}$	0
	$-\frac{1}{8}$	$\frac{1}{4}$	$\frac{1}{8}$	0

Basis	b	v_1	v_2	v_3
v_4	$\frac{1}{6}$	0	$-\frac{1}{6}$	0
v_5	$\frac{3}{2}$	0	$\frac{1}{2}$	0
v_6	2	0	0	0
v_7	$\frac{11}{6}$	0	$\frac{1}{6}$	0
v_8	$\frac{1}{2}$	0	$-\frac{1}{2}$	0
v_3	$\frac{2}{3}$	0	$-\frac{2}{3}$	1
v_{10}	$\frac{3}{2}$	0	**$\frac{3}{2}$** (circled)	0
v_1	$\frac{13}{6}$	1	$-\frac{13}{6}$	0
v_{12}	$\frac{1}{6}$	0	$-\frac{7}{6}$	0
v_{13}	$\frac{1}{2}$	0	$-\frac{3}{2}$	0
v_{14}	2	0	0	0
v_{15}	$\frac{11}{6}$	0	$\frac{7}{6}$	0
	$-\frac{2}{3}$	0	$\frac{2}{3}$	0

Basis	b	v_1	v_2	v_3
v_4	$\frac{1}{3}$	0	0	0
v_5	1	0	0	0
v_6	2	0	0	0
v_7	$\frac{5}{3}$	0	0	0
v_8	1	0	0	0
v_3	$\frac{4}{3}$	0	0	1
v_2	1	0	1	0
v_1	$\frac{13}{3}$	1	0	0
v_{12}	$\frac{4}{3}$	0	0	0
v_{13}	2	0	0	0
v_{14}	2	0	0	0
v_{15}	$\frac{2}{3}$	0	0	0
	$-\frac{4}{3}$	0	0	0

Fig. 28-2

28.8. Compare the residuals of least-squares and Chebyshev solutions.

For an arbitrary set of numbers $x_1, \ldots, x_n$ let $|r|_{\max}$ be the largest residual in absolute value. Then $r_1^2 + \cdots + r_m^2 \leqq m\,|r|_{\max}^2$ so that the root-mean-square error surely does not exceed $|r|_{\max}$. But the least-squares solution has the smallest RMS error of all, so that, denoting this error by ρ, $\rho \leqq |r|_{max}$. In particular this is true when the x_j are the Chebyshev solution, in which case $|r|_{max}$ is what we have been calling r. But the Chebyshev solution also has the property that its maximum error is smallest, so if $|\rho|_{max}$ denotes the absolutely largest residual of the least-squares solution, $|r|_{max} \leqq |\rho|_{max}$. Putting the two inequalities together, $\rho \leqq r \leqq |\rho|_{max}$ and we have the Chebyshev error bounded on both sides. Since the least-squares solution is often easier to find, this last result may be used to decide if it is worth continuing on to obtain the further reduction of maximum residual which the Chebyshev solution brings.

28.9. Apply the previous problem to the systems of Problem 28.2.

We have already found $\rho = \sqrt{\frac{2}{21}}$, $r = \frac{1}{3}$, and $|\rho|_{max} = \frac{3}{7}$ which do steadily increase as Problem 28.8 suggests. The fact that one of the least-squares residuals is three times as large as another already recommends the search for a Chebyshev solution.

28.10. Apply Problem 28.8 to the system of Problem 28.3.

We have found $\rho = \sqrt{\frac{23}{72}}$, $r = \frac{3}{4}$, and $|\rho|_{max} = \frac{11}{12}$. The spread does support a search for the Chebyshev solution.

Supplementary Problems

28.11. Find the least-squares solution of this system:

$$x_1 - x_2 = -1 \qquad 2x_1 - x_2 = 2$$
$$x_1 + x_2 = 8 \qquad 2x_1 + x_2 = 14$$

Compute the RMS error of this solution.

28.12. Compare $|\rho|_{max}$ with ρ for the solution found in Problem 28.11.

28.13. Find the Chebyshev solution of the system in Problem 28.11 and compare its r value with ρ and $|\rho|_{max}$.

28.14. Find both the least-squares and Chebyshev solutions for this system:

$$x_1 + x_2 - x_3 = 5 \qquad x_1 + 2x_2 - 2x_3 = 1$$
$$2x_1 - 3x_2 + x_3 = -4 \qquad 4x_1 - x_2 - x_3 = 6$$

28.15. Suppose it is known that $-1 \leqq x_j$. Find the Chebyshev solution of the following system by first letting $z_j = x_j + 1$ which guarantees $0 \leqq z_j$. Also find the least-squares solution.

$$2x_1 - 2x_2 + x_3 + 2x_4 = 1 \qquad -2x_1 - 2x_2 + 3x_3 + 3x_4 = 4$$
$$x_1 + x_2 + 2x_3 + 4x_4 = 1 \qquad -x_1 - 3x_2 - 3x_3 + x_4 = 3$$
$$x_1 - 3x_2 + x_3 + 2x_4 = 2 \qquad 2x_1 + 4x_2 + x_3 + 5x_4 = 0$$

28.16. Find the least-squares solution of this system:

$$x_1 = 0 \qquad x_1 + x_2 = -1$$
$$x_2 = 0 \qquad .1x_1 + .1x_2 = .1$$

What is the RMS error?

28.17. Find the Chebyshev solution of the system in Problem 28.16.

28.18. Four altitudes x_1, x_2, x_3, x_4 are measured, together with the six differences in altitude, as follows. Find the least-squares values.

$$x_1 = 3.47 \qquad x_2 = 2.01 \qquad x_3 = 1.58 \qquad x_4 = .43$$

$$x_1 - x_2 = 1.42 \qquad x_1 - x_3 = 1.92 \qquad x_1 - x_4 = 3.06 \qquad x_2 - x_3 = .44$$

$$x_2 - x_4 = 1.53 \qquad x_3 - x_4 = 1.20$$

28.19. A quantity x is measured N times, the results being $a_1, a_2, \ldots, a_N$. Solve the overdetermined system

$$x = a_i \qquad i = 1, \ldots, N$$

by the least-squares method. What value of x appears?

28.20. Two quantities x and y are measured, together with their difference $x - y$ and sum $x + y$.

$$x = A \qquad y = B \qquad x - y = C \qquad x + y = D$$

Solve the overdetermined system by least-squares.

28.21. The three angles of a triangle are measured to be A_1, A_2, A_3. If x_1, x_2, x_3 denote the correct values, we are led to the overdetermined system

$$x_1 = A_1 \qquad x_2 = A_2 \qquad \pi - x_1 - x_2 = A_3$$

Solve by the method of least-squares.

28.22. The two legs of a right triangle are measured to be A and B, and the hypotenuse to be C. Let L_1, L_2, and H denote the exact values, and let $x_1 = L_1^2$, $x_2 = L_2^2$. Consider the overdetermined system

$$x_1 = A^2 \qquad x_2 = B^2 \qquad x_1 + x_2 = C^2$$

and obtain the least-squares estimates of x_1 and x_2. From these estimate L_1, L_2, and H.

28.23. Verify that the normal equations for the least-squares solution of $Ax = b$ are equivalent to $A^T A = A^T b$.

Chapter 29

Boundary Value Problems

NATURE OF THE PROBLEM

This is a subject that runs wide and deep. Volumes could be filled with its variations and algorithms. This chapter can offer only a sampling of the many ideas that have been brought to bear on it. This means that the coverage is, of necessity, superficial, but the alternative of omission seemed totally unacceptable.

A boundary value problem requires the solution of a differential equation, or system of equations, in a region R, subject to additional conditions on the boundary of R. Applications generate a broad variety of such problems. The classical two-point boundary value problem of ordinary differential equations involves a second-order equation, an initial condition, and a terminal condition.

$$y'' = f(x, y, y') \qquad y(a) = A \qquad y(b) = B$$

Here the region R is the interval (a, b) and the boundary consists of the two endpoints. A typical problem of partial differential equations is the Dirichlet problem, which asks that the Laplace equation

$$U_{xx} + U_{yy} = 0$$

be satisfied inside a region R of the xy plane and that $U(x, y)$ take specified values on the boundary of R. These examples suggest two important classes of boundary value problems.

SOLUTION METHODS

1. The **superposition principle** is useful for linear problems. For example, to solve

$$y'' = q(x)y \qquad y(a) = A \qquad y(b) = B$$

one could use the methods of Chapter 19 to solve the two initial value problems

$$y_1'' = q(x)y_1 \qquad y_1(a) = 1 \qquad y_1(b) = 0$$
$$y_2'' = q(x)y_2 \qquad y_2(a) = 0 \qquad y_2(b) = 1$$

after which

$$y(x) = Ay_1(x) + By_2(x)$$

2. **Replacement by a matrix problem** is also an option when the problem is linear. For example, replacing $y''(x_k)$ by a second difference converts the equation $y'' = q(x)y$ into the difference equation

$$y_{k-1} - (2 + h^2 q_k)y_k + y_{k+1} = 0$$

which is required to hold for $k = 1, \ldots, n$ corresponding to the arguments $x_1, \ldots, x_n$. With $y_0 = A$ and $y_{n+1} = B$, we then have a linear system of order n, producing approximate y values at the listed arguments.

Similarly, the Laplace equation $U_{xx} + U_{yy} = 0$ converts to the difference equation

$$U(x, y) = \frac{1}{4}[U(x-h, y) + U(x+h, y) + U(x, y-h) + U(x, y+h)]$$

which makes each value the average of its four neighbors in the square lattice of points $x_m = x_0 + mh$, $y_n = y_0 + nh$. Writing this equation for each interior lattice point produces a linear system of order N, where N is the number of such points. The idea can be adapted to other equations, to regions with curved boundaries, and to more dimensions. Convergence to the correct solution can be proved under fairly broad circumstances.

The classic diffusion problem

$$T_t = T_{xx} \qquad T(0, t) = T(1, t) = 0 \qquad T(x, 0) = f(x)$$

also responds to the finite difference treatment. The equation is to be satisfied inside the semi-infinite strip $0 \leqq x \leqq 1$, $0 \leqq t$. On the boundaries of the strip, T is prescribed. There is a well-known solution by Fourier series, but finite differences are useful for various modifications. Replacing derivatives by simple differences, the above equation becomes

$$T_{m,n+1} = \lambda T_{m-1,n} + (1 - 2\lambda)T_{m,n} + \lambda T_{m+1,n}$$

with $x_m = mh$, $t_n = nk$, and $\lambda = k/h^2$. A rectangular lattice of points thus replaces the strip. In the form given, the difference equation allows each T value to be computed directly from values at the previous time step, with the given initial values $f(x_m)$ triggering the process. For proper choices of h and k, tending to zero, the method converges to the true solution. However, for small k the computation is strenuous and numerous variations have been proposed for reducing the size of the job.

3. The **garden hose method** offers an intuitive approach to the classic two-point boundary value problem. We first solve the initial value problem

$$y'' = f(x, y, y') \qquad y(a) = A \qquad y'(a) = M$$

for some choice of M. The terminal value obtained will depend upon the choice of M. Call it $F(M)$. Then what we want is that $F(M) = B$. This is a problem similar to the root-finding problems of Chapter 25 and can be solved by similar methods. Successive approximations to M are found, each one bringing a new initial value problem. As with root finding, there are several ways for choosing the corrections to M, including a Newton-type method.

$$M_2 = M_1 - \frac{F(M_1) - B}{F'(M_1)}$$

4. The **calculus of variations** establishes the equivalence of certain boundary value problems with problems of optimization. To find the function $y(x)$ which has $y(a) = A$ and $y(b) = B$ and also makes

$$\int_a^b F(x, y, y')\, dx$$

a minimum (or maximum), one may solve the Euler equation

$$F_y = \frac{d}{dx} F_{y'}$$

subject to the same boundary conditions. There are also direct methods, such as the Ritz method, for minimizing the integral, which may, therefore, be considered as methods for solving the Euler equation with its boundary conditions.

For the Laplace equation a corresponding minimization problem is

$$\iint (U_x^2 + U_y^2)\, dx\, dy = \text{minimum}$$

with the double integral being taken over the region R of the boundary value problem.

For the Poisson equation $U_{xx} + U_{yy} = K$, the appropriate optimization problem is

$$\iint \left[\frac{1}{2}(U_x^2 + U_y^2) + KU\right] dx\, dy = \text{minimum}$$

5. The **finite element method** is a powerful procedure for direct solution of optimization problems. The region R is subdivided into basic pieces (triangles, squares, etc. for a

two-dimensional R) and a solution element is associated with each piece. For instance, over a set of basic triangles one might choose a set of plane triangular elements, joined to form a continuous surface. The vertical coordinates of the corners of these elements become the independent variables of the optimization. Partial derivatives relative to these variables are developed and equated to zero. The resulting system of equations must then be solved.

6. **Infinite series** provide solutions for many classic problems. They are a development of the superposition principle. Fourier series and their various generalizations are prominent.

Solved Problems

LINEAR ORDINARY DIFFERENTIAL EQUATIONS

29.1. Find a solution of the second-order equation

$$L(y)=y''(x)-p(x)y'(x)-q(x)y(x)=r(x)$$

satisfying the two boundary conditions

$$c_{11}y(a)+c_{12}y(b)+c_{13}y'(a)+c_{14}y'(b)=A$$
$$c_{21}y(a)+c_{22}y(b)+c_{23}y'(a)+c_{24}y'(b)=B$$

With linear equations, we may rely upon the superposition principle which is used in solving elementary examples by analytic methods. Assuming that elementary solutions cannot be found for the above equation, the numerical algorithms of an earlier chapter (Runge–Kutta, Adams, etc.) may be used to compute approximate solutions of these three *initial* value problems for $a \leqq x \leqq b$.

$$\begin{array}{lll} L(y_1)=0 & L(y_2)=0 & L(Y)=r(x) \\ y_1(a)=1 & y_2(a)=0 & Y(a)=0 \\ y_1'(a)=0 & y_2'(a)=1 & Y'(a)=0 \end{array}$$

The required solution is then available by superposition,

$$y(x)=C_1y_1(x)+C_2y_2(x)+Y(x)$$

where to satisfy the boundary conditions we determine C_1 and C_2 from the equations

$$[c_{11}+c_{12}y_1(b)+c_{14}y_1'(b)]C_1+[c_{13}+c_{12}y_2(b)+c_{14}y_2'(b)]C_2=A-c_{12}Y(b)-c_{14}Y'(b)$$
$$[c_{21}+c_{22}y_1(b)+c_{24}y_1'(b)]C_1+[c_{23}+c_{22}y_2(b)+c_{24}y_2'(b)]C_2=B-c_{22}Y(b)-c_{24}Y'(b)$$

In this way the linear boundary value problem is solved by our algorithms for *initial* value problems. The method is easily extended to higher-order equations or to linear systems. We assume that the given problem has a unique solution and that the functions y_1, y_2, etc., can be found with reasonable accuracy. The equations determining C_1, C_2, etc., will then also have a unique solution.

29.2. Show how a linear boundary value problem may be solved approximately by reducing it to a linear algebraic system.

Choose equally spaced arguments $x_j=a+jh$ with $x_0=a$ and $x_{N+1}=b$. We now seek to determine the corresponding values $y_j=y(x_j)$. Replacing $y''(x_j)$ by the approximation

$$y''(x_j)\simeq\frac{y_{j+1}-2y_j+y_{j-1}}{h^2}$$

and $y'(x_j)$ by

$$y'(x_j)\simeq\frac{y_{j+1}-y_{j-1}}{2h}$$

the differential equation $L(y) = r(x)$ of Problem 29.1 becomes, after slight rearrangement,

$$\left(1 - \frac{1}{2}hp_j\right)y_{j-1} + (-2 + h^2 q_j)y_j + \left(1 + \frac{1}{2}hp_j\right)y_{j+1} = h^2 r_j$$

If we require this to hold at the interior points $j = 1, \ldots, N$, then we have N linear equations in the N unknowns $y_1, \ldots, y_N$, assuming the two boundary values to be specified as $y_0 = y(a) = A$, $y_{N+1} = y(b) = B$. In this case the linear system takes the following form:

$$\begin{aligned}
&\beta_1 y_1 + \gamma_1 y_2 &&= h^2 r_1 - \alpha_1 A \\
&\alpha_2 y_1 + \beta_2 y_2 + \gamma_2 y_3 &&= h^2 r_2 \\
&\qquad \alpha_3 y_2 + \beta_3 y_3 + \gamma_3 y_4 &&= h^2 r_3 \\
&\qquad \cdots\cdots\cdots\cdots\cdots\cdots\cdots && \\
&\qquad\qquad \alpha_N y_{N-1} + \beta_N y_n &&= h^2 r_N - \gamma_N b
\end{aligned}$$

where

$$\alpha_j = 1 - \frac{1}{2}hp_j \qquad \beta_j = -2 + h^2 q_j \qquad \gamma_j = 1 + \frac{1}{2}hp_j$$

The *band matrix* of this system is typical of linear systems obtained by discretizing differential boundary value problems. Only a few diagonals are nonzero. Such matrices are easier to handle than others which are not so sparse. If Gaussian elimination is used, with the pivots descending the main diagonal, the band nature will not be disturbed. This fact can be used to abbreviate the computation. The iterative Gauss–Seidel algorithm is also effective. If the more general boundary conditions of Problem 29.1 occur these may also be discretized, perhaps using

$$y'(a) \simeq \frac{y_1 - y_0}{h} \qquad y'(b) \simeq \frac{y_{N+1} - y_N}{h}$$

This brings a system of $N + 2$ equations in the unknowns $y_0, \ldots, y_{N+1}$.

In this and the previous problem we have alternative approaches to the same goal. In both cases the output is a finite set of numbers y_j. If either method is reapplied with smaller h, then hopefully the larger output will represent the true solution $y(x)$ more accurately. This is the question of *convergence*.

29.3. Show that for the special case

$$y'' + y = 0 \qquad y(0) = 0 \qquad y(1) = 1$$

the method of Problem 29.2 is convergent.

The exact solution function is $y(x) = (\sin x)(\sin 1)$. The approximating difference equation is

$$y_{j-1} + (-2 + h^2)y_j + y_{j+1} = 0$$

and this has the exact solution

$$y_j = \frac{\sin(\alpha x_j/h)}{\sin(\alpha/h)}$$

for the same boundary conditions $y_0 = 0$, $y_{N+1} = 1$. Here $x_j = jh$ and $\cos\alpha = 1 - \frac{1}{2}h^2$. These facts may be verified directly or deduced by the methods of our section on difference equations. Since $\lim(\alpha/h)$ is 1 for h tending to zero, we now see that $\lim y_j = y(x_j)$, that is, solutions of the difference problem for decreasing h converge to the solution of the differential problem. In this example both problems may be solved analytically and their solutions compared. The proof of convergence for more general problems must proceed by other methods.

29.4. Illustrate the reduction of a linear differential eigenvalue problem to an approximating algebraic system.

Consider the problem

$$y'' + \lambda y = 0 \qquad y(0) = y(1) = 0$$

This has the exact solutions $y(x) = C \sin n\pi x$, for $n = 1, 2, \ldots$. The corresponding eigenvalues are $\lambda_n = n^2\pi^2$. Simply to illustrate a procedure applicable to other problems for which exact solutions are not so easily found, we replace this differential equation by the difference equation

$$y_{j-1} + (-2 + \lambda h^2)y_j + y_{j+1} = 0$$

Requiring this to hold at the interior points $j = 1, \ldots, N$, we have an algebraic eigenvalue problem $Ay = \lambda h^2 y$ with the band matrix

$$A = \begin{bmatrix} -2 & 1 & & & \\ 1 & -2 & 1 & & \\ & 1 & -2 & & \\ \cdots & \cdots & \cdots & \cdots & \cdots \\ \cdots & \cdots & \cdots & \cdots & 1 \\ & & & 1 & -2 \end{bmatrix}$$

all other elements being zero, and $y^T = (y_1, \ldots, y_N)$. The exact solution of this problem may be found to be

$$y_j = C \sin n\pi x_j \qquad \text{with} \qquad \lambda_n = \frac{4}{h^2} \sin^2 \frac{n\pi h}{2}$$

Plainly, as h tends to zero these results converge to those of the target differential problem.

NONLINEAR ORDINARY DIFFERENTIAL EQUATIONS

29.5. What is the *garden-hose method*?

Given the equation $y'' = f(x, y, y')$, we are to find a solution which satisfies the boundary conditions $y(a) = A$, $y(b) = B$.

One simple procedure is to compute solutions of the initial value problem

$$y'' = f(x, y, y') \qquad y(a) = A \qquad y'(a) = M$$

for various values of M until two solutions, one with $y(b) < B$ and the other with $y(b) > B$, have been found. If these solutions correspond to initial slopes of M_1 and M_2, then interpolation will suggest a new M value between these and a better approximation may then be computed (see Fig. 29-1). Continuing this process leads to successively better approximations and is essentially the *regula falsi* algorithm used for nonlinear algebraic problems. Here our computed terminal value is a function of M, say $F(M)$, and we do have to solve the equation $F(M) = B$. However, for each choice of M the calculation of $F(M)$ is no longer the evaluation of an algebraic expression but involves the solution of an initial value problem of the differential equation.

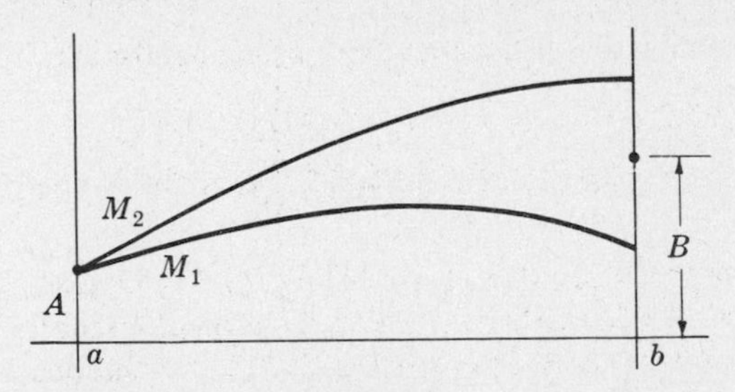

Fig. 29-1

29.6. How may the garden-hose method be refined?

Instead of using the equivalent of *regula falsi*, we may adapt Newton's method to the present problem, presumably obtaining improved convergence to the correct M value. To do this we need to know $F'(M)$. Let $y(x, M)$ denote the solution of

$$y'' = f(x, y, y') \qquad y(a) = A \qquad y'(a) = M$$

and for brevity let $z(x, M)$ be its partial derivative relative to M. Differentiating relative to M brings

$$z'' = f_y(x, y, y')z + f_{y'}(x, y, y')z' \tag{1}$$

if we freely reverse the orders of the various derivatives. Also differentiating the initial conditions, we have

$$z(a, M) = 0 \qquad z'(a, M) = 1$$

Let M_1 be a first approximation to M and solve the original problem for the approximate solution $y(x, M_1)$. This may then be substituted for y in equation (1) and the function $z(x, M_1)$ computed. Then $F'(M) = z(b, M_1)$. With this quantity available the Newton method for solving $F(M) - B = 0$ now offers us the next approximation to M:

$$M_2 = M_1 - \frac{F(M_1) - B}{F'(M_1)}$$

With this M_2 a new approximation $y(x, M_2)$ may be computed and the process repeated. The method may be extended to higher-order equations or to systems, the central idea being the derivation of an equation similar to (1), which is called the *variational equation*.

OPTIMIZATION

29.7. Reduce the problem of maximizing or minimizing $\int_a^b F(x, y, y')\,dx$ to a boundary value problem.

This is the classical problem of the calculus of variations. If the solution function $y(x)$ exists and has adequate smoothness, then it is required to satisfy the Euler differential equation $F_y = (d/dx)F_{y'}$. If boundary conditions such as $y(a) = A$, $y(b) = B$ are specified in the original optimization problem, then we already have a second-order boundary value problem. If either of these conditions is omitted, then the variational argument shows that $F_{y'} = 0$ must hold at that end of the interval. This is called the *natural boundary condition*.

29.8. Minimize $\int_0^1 (y^2 + y'^2)\,dx$ subject to $y(0) = 1$.

The Euler equation is $2y = 2y''$ and the natural boundary condition is $y'(1) = 0$. The solution is now easily found to be $y = \cosh x - \tanh 1 \sinh x$ and it makes the integral equal to tanh 1, which is about .76. In general the Euler equation will be nonlinear and the garden-hose method may be used to find $y(x)$.

29.9. Illustrate the Ritz method of solving a boundary value problem.

The idea of the Ritz method is to solve an equivalent minimization problem instead. Consider

$$y'' = -x^2 \qquad y(0) = y(1) = 0$$

sometimes called a Poisson problem in one variable, but in fact requiring only two integrations to discover the solution

$$y(x) = \frac{x(1 - x^3)}{12}$$

Methods are available for finding an equivalent minimization problem for a given boundary problem, but here one is well known.

$$J(y) = \int_0^1 \left[\frac{1}{2}(y')^2 - x^2 y\right] dx = \text{minimum}$$

The Euler equation for this integral proves to be our original differential equation.

To approximate the solution by the Ritz method, we need a family of functions satisfying the boundary conditions. Suppose we choose

$$\phi(x) = cx(1 - x)$$

which is probably the simplest such family for this problem. Substituting ϕ for y in the integral, an easy calculation yields

$$J(\phi)=\frac{c^2}{6}-\frac{c}{20}=f(c)$$

which we minimize by setting $f'(c)=0$. The resulting $c=\frac{3}{20}$ gives us the approximation

$$\phi(x)=\frac{3}{20}x(1-x)$$

which is compared with the true solution in Fig. 29-2. More accurate approximations are available through the use of a broader family of approximating functions, perhaps

$$\phi(x)=x(1-x)(c_0+c_1x+c_2x^2+\cdots+c_nx^n)$$

leading to a linear system for determining the coefficients c_i. The central idea of the Ritz method is the search for the optimum function among members of a restricted family $\phi(x)$, rather than among all $y(x)$ for which the given integral exists.

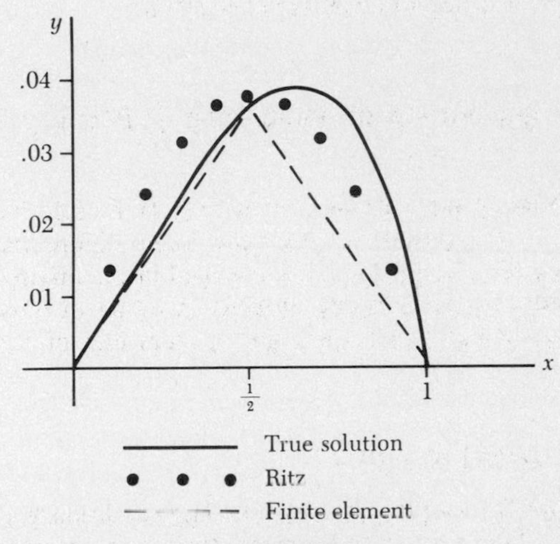

Fig. 29-2

29.10. Use the same boundary value problem solved in Problem 29.9 to illustrate a finite element solution method.

The basic idea is the same. It is the nature of the approximating family that identifies a finite element method. Suppose we divide our interval (0, 1) into halves and use two line segments

$$\phi_1(x)=2Ax \qquad \phi_2(x)=2A(1-x)$$

meeting at point $(\frac{1}{2}, A)$ to approximate $y(x)$. In fact, we have a family of such approximations, with parameter A to be selected. The two line segments are called finite elements, and the approximating function is formed by piecing them together. As before we substitute into the integral, and we easily compute

$$J(\phi)=\int_0^{1/2}\phi_1\,dx+\int_{1/2}^{1}\phi_2\,dx=2A^2-\frac{7}{48}A=f(A)$$

which we minimize by setting $f'(A)=0$. This makes $A=\frac{7}{192}$. A quick calculation shows that this is actually the correct value of the solution at $x=\frac{1}{2}$. (See Fig. 29-2.) It has been shown that if line segments are used as finite elements (in a one-dimensional problem, of course) correct values are systematically produced at the joins.

29.11. Extend the procedure of the preceding problem to include more finite elements.

Divide the interval (0, 1) into n parts, with endpoints at $0=x_0, x_1, x_2, \ldots, x_n=1$. Let $y_1, \ldots, y_{n-1}$ be corresponding and arbitrary ordinates, with $y_0=y_n=0$. Define linear finite elements $\phi_1, \ldots, \phi_n$ in the obvious way. (See Fig. 29-3.) Then

$$\phi_i(x)=y_{i-1}\frac{x_i-x}{x_i-x_{i-1}}+y_i\frac{x-x_{i-1}}{x_i-x_{i-1}}$$

$$=y_{i-1}\frac{x_i-x}{h}+y_i\frac{x-x_{i-1}}{h}$$

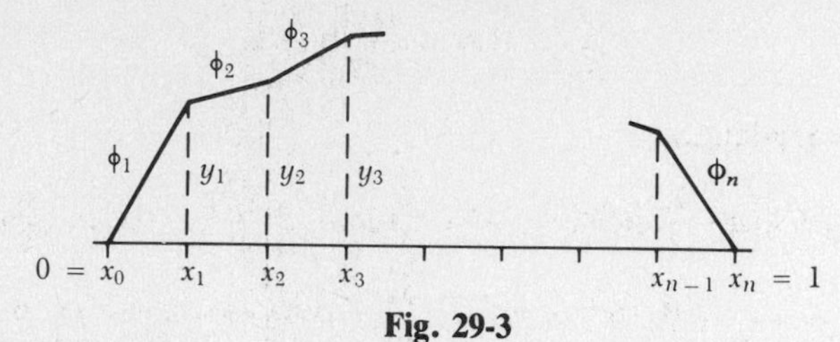

Fig. 29-3

the second equality holding if the x_i are equally spaced. We also have

$$\phi_i'(x)=\frac{y_i-y_{i-1}}{x_i-x_{i-1}}=\frac{y_i-y_{i-1}}{h}$$

Now consider the integral

$$J(\phi)=\sum_{i=1}^{n}\int_{x_{i-1}}^{x_i}\left[\frac{1}{2}(\phi_i')^2-x^2\phi_i\right]dx=\sum_{i=1}^{n}J_i$$

$$=f(y_1,\ldots,y_{n-1})$$

To minimize this, we could obtain f explicitly in terms of the y_i and then compute the partial derivatives, setting them to zero and solving the resulting system of equations. This is what was just done in the simpler case. Here suppose we take derivatives first, integrate second, and then form the ultimate system. The dependence of f upon a particular ordinate y_k is through only two of the component terms J_k and J_{k+1}. Accordingly, for $k=1,\ldots,n-1$,

$$\frac{\partial f}{\partial y_k}=\int_{x_{k-1}}^{x_k}\left[\frac{y_k-y_{k-1}}{h}\left(\frac{1}{h}\right)-x^2\frac{x-x_{k-1}}{h}\right]dx+\int_{x_k}^{x_{k+1}}\left[\frac{y_{k+1}-y_k}{h}\left(\frac{-1}{h}\right)-x^2\frac{x_{k+1}-x}{h}\right]dx$$

and the integrals being elementary we soon have the system

$$-y_{k-1}+2y_k-y_{k+1}=\frac{1}{12}x_{k-1}^4+\frac{1}{2}x_k^4+\frac{1}{12}x_{k+1}^4-\frac{1}{3}x_{k-1}x_k^3-\frac{1}{3}x_{k+1}x_k^3$$

for $k=1,\ldots,n-1$.

With $n=2$ and $k=1$, this quickly reproduces the $y_1=\frac{7}{192}$ found before. With $n=3$, the system becomes

$$2y_1-\ y_2=\frac{7}{486}$$

$$-y_1+2y_2=\frac{25}{486}$$

from which come $y_1=\frac{13}{486}$ and $y_2=\frac{19}{486}$, both of which agree with the true solution values for these positions.

THE DIFFUSION EQUATION

29.12. Replace the diffusion problem involving the equation

$$\frac{\partial T}{\partial t}=a\left(\frac{\partial^2 T}{\partial x^2}\right)+b\left(\frac{\partial T}{\partial x}\right)+cT$$

and the conditions $T(0, t) = f(t)$, $T(l, t) = g(t)$, $T(x, 0) = F(x)$ by a finite difference approximation.

Let $x_m = mh$ and $t_n = nk$, where $x_{M+1} = l$. Denoting the value $T(x, t)$ by the alternate symbol $T_{m,n}$, the approximations

$$\frac{\partial T}{\partial t} \simeq \frac{T_{m,n+1} - T_{m,n}}{k} \qquad \frac{\partial T}{\partial x} \simeq \frac{T_{m+1,n} - T_{m-1,n}}{2h}$$

$$\frac{\partial^2 T}{\partial x^2} \simeq \frac{T_{m+1,n} - 2T_{m,n} + T_{m-1,n}}{h^2}$$

convert the diffusion equation to

$$T_{m,n+1} = \lambda\left(a - \frac{1}{2}bh\right)T_{m-1,n} + [1 - \lambda(2a + ch^2)]T_{m,n} + \lambda\left(a + \frac{1}{2}bh\right)T_{m+1,n}$$

where $\lambda = k/h^2$, $m = 1, 2, \ldots, M$ and $n = 1, 2, \ldots$. Using the same initial and boundary conditions above, in the form $T_{0,n} = f(t_n)$, $T_{M+1,n} = g(t_n)$ and $T_{m,0} = F(x_m)$, this difference equation provides an approximation to each interior $T_{m,n+1}$ value in terms of its three nearest neighbors at the previous time step. The computation therefore begins at the (given) values for $t = 0$ and proceeds first to $t = k$, then to $t = 2k$, and so on. (See the next problem for an illustration.)

29.13. Apply the procedure of the preceding problem to the case $a = 1$, $b = c = 0$, $f(t) = g(t) = 0$, $F(x) = 1$, and $l = 1$.

Suppose we choose $h = \frac{1}{4}$ and $k = \frac{1}{32}$. Then $\lambda = \frac{1}{2}$ and the difference equation simplifies to

$$T_{m,n+1} = \frac{1}{2}(T_{m-1,n} + T_{m+1,n})$$

A few lines of computation are summarized in Table 29.1(a). The bottom line and the side columns are simply the initial and boundary conditions. The interior values are computed from the difference equation line by line, beginning with the looped ① which comes from averaging its two lower neighbors, also looped. A slow trend toward the ultimate "steady state" in which all T values are zero may be noticed, but not too much accuracy need be expected of so brief a calculation.

For a second try we choose $h = \frac{1}{8}$, $k = \frac{1}{128}$, keeping $\lambda = \frac{1}{2}$. The results appear in Table 29.1(b). The top line of this table corresponds to the second line in Table 29.1(a) and is in fact a better approximation to $T(x, \frac{1}{32})$. This amounts to a primitive suggestion that the process is starting to *converge* to the correct $T(x, t)$ values.

In Table 29.1(c) we have the results if $h = \frac{1}{4}$, $k = \frac{1}{16}$ are chosen, making $\lambda = 1$. The difference equation for this choice is

$$T_{m,n+1} = T_{m-1,n} - T_{m,n} + T_{m+1,n}$$

Table 29.1

0	$\frac{1}{4}$	$\frac{1}{2}$	$\frac{1}{4}$	0
0	$\frac{1}{2}$	$\frac{1}{2}$	$\frac{1}{2}$	0
0	$\frac{1}{2}$	1	$\frac{1}{2}$	0
0	①	1	1	0
①	1	①	1	1

(a)

0	$\frac{3}{8}$	$\frac{3}{4}$	$\frac{7}{8}$	1	$\frac{7}{8}$	$\frac{3}{4}$	$\frac{3}{8}$	0
0	$\frac{1}{2}$	$\frac{3}{4}$	1	1	1	$\frac{3}{4}$	$\frac{1}{2}$	0
0	$\frac{1}{2}$	1	1	1	1	1	$\frac{1}{2}$	0
0	1	1	1	1	1	1	1	0
1	1	1	1	1	1	1	1	1

(b)

0	−12	17	−12	0
0	5	−7	5	0
0	−2	3	−2	0
0	1	−1	1	0
0	0	1	0	0
0	1	1	1	0
1	1	1	1	1

(c)

The start of an explosive oscillation can be seen. This does not at all conform to the correct solution, which is known to decay exponentially. Later we shall see that unless $\lambda \leqq \frac{1}{2}$ such an explosive and unrealistic oscillation may occur. This is a form of numerical *instability*.

29.14. What is the *truncation error* of this method?

As earlier we apply Taylor's theorem to the difference equation, and find that our approximation has introduced error terms depending on h and k. These terms are the truncation error

$$\frac{1}{2}kT_{tt} - \frac{1}{12}ah^2T_{xxxx} + \frac{1}{6}bh^2T_{xxx} + 0(h^4)$$

subscripts denoting partial derivatives. In the important special case $a=1$, $b=0$, we have $T_{tt} = T_{xxxx}$ so that the choice $k = h^2/6$ (or $\lambda = \frac{1}{6}$) seems especially desirable from this point of view, the truncation error then being $0(h^4)$.

29.15. Show that the method of Problem 29.12 is convergent in the particular case

$$\frac{\partial T}{\partial t} = \frac{\partial^2 T}{\partial x^2} \qquad T(0, t) = T(\pi, t) = 0 \qquad T(x, 0) = \sin px$$

where p is a positive integer.

The exact solution may be verified to be $T(x, t) = e^{-p^2t} \sin px$. The corresponding difference equation is

$$T_{m,n+1} - T_{m,n} = \lambda(T_{m+1,n} - 2T_{m,n} + T_{m-1,n})$$

and the remaining conditions may be written

$$T_{m,0} = \sin\frac{mp\pi}{M+1} \qquad T_{0,n} = T_{M+1,n} = 0$$

This finite difference problem can be solved by "separation of the variables." Let $T_{m,n} = u_m v_n$ to obtain

$$\frac{v_{n+1} - v_n}{v_n} = \lambda\left(\frac{u_{m+1} - 2u_m + u_{m-1}}{u_m}\right) = -\lambda C$$

which defines C. But comparing C with the extreme left member we find it independent of m, and comparing it with the middle member we find it also independent of n. It is therefore a constant and we obtain separate equations for u_m and v_n in the form

$$v_{n+1} = (1 - \lambda C)v_n \qquad u_{m+1} - (2 - C)u_m + u_{m-1} = 0$$

These are easily solved by our difference equation methods. The second has no solution with $u_0 = u_{M+1} = 0$ (except u_m identically zero) unless $0 < C < 4$, in which case

$$u_m = A \cos \alpha m + B \sin \alpha m$$

where A and B are constants, and $\cos \alpha = 1 - \frac{1}{2}C$. To satisfy the boundary conditions, we must now have $A = 0$ and $\alpha(M+1) = j\pi$, j being an integer. Thus

$$u_m = B \sin\frac{mj\pi}{M+1}$$

Turning toward v_n, we first find that $C = 2(1 - \cos \alpha) = 4\sin^2\{j\pi/[2(M+1)]\}$ after which

$$v_n = \left[1 - 4\lambda \sin^2\frac{j\pi}{2(M+1)}\right]^n v_0$$

It is now easy to see that choosing $B = v_0 = 1$ and $j = p$ we obtain a function

$$T_{m,n} = u_m v_n = \left[1 - 4\lambda \sin^2\frac{p\pi}{2(M+1)}\right]^n \sin\frac{mp\pi}{M+1}$$

which has all the required features. For comparison with the differential solution we return to the symbols $x_m = mh$, $t_n = nk$.

$$T_{m,n} = \left(1 - 4\lambda \sin^2 \frac{ph}{2}\right)^{t_n/\lambda h^2} \sin px_m$$

As h now tends to zero, assuming $\lambda = k/h^2$ is kept fixed, the coefficient of $\sin px_m$ has limit $e^{-p^2 t_n}$ so that convergence is proved. Here we must arrange that the point (x_m, t_n) also remain fixed, which involves increasing m and n as h and k diminish, in order that the $T_{m,n}$ values be successive approximations to the same $T(x, t)$.

29.16. Use the previous problem to show that for the special case considered an explosive oscillation may occur unless $\lambda \leqq \frac{1}{2}$.

The question now is not what happens as h tends to zero, but what happens for fixed h as the computation is continued to larger n arguments. Examining the coefficient of $\sin px_m$ we see that the quantity in brackets may be less than -1 for some values of λ, p, and h. This would lead to an explosive oscillation with increasing t_n. The explosion may be avoided by requiring that λ be no greater than $\frac{1}{2}$. Since this makes $k \leqq h^2/2$ the computation will proceed very slowly, and if results for large t arguments are wanted it may be useful to use a different approach. (See the next problem.)

29.17. Solve Problem 29.12 by means of a Fourier series.

This is the classical procedure when a is constant and $b = c = 0$. We first look for solutions of the diffusion equation having the product form $U(x)V(t)$. Substitution brings $V'/V = U''/U = -\alpha^2$ where α is constant. (The negative sign will help us satisfy the boundary conditions.) This makes

$$V = Ae^{-\alpha^2 t} \qquad U = B \cos \alpha x + C \sin \alpha x$$

To make $T(0, t) = 0$, we choose $B = 0$. To make $T(1, t) = 0$, we choose $\alpha = n\pi$ where n is a positive integer. Putting $C = 1$ arbitrarily and changing the symbol A to A_n, we have the functions

$$A_n e^{-n^2\pi^2 t} \sin n\pi x \qquad n = 1, 2, 3, \ldots$$

each of which meets all our requirements except for the initial condition. The series

$$T(x, t) = \sum_{n=1}^{\infty} A_n e^{-n^2\pi^2 t} \sin n\pi x$$

if it converges properly will also meet these requirements, and the initial condition may also be satisfied by suitable choice of the A_n. For $F(x) = 1$ we need

$$T(x, 0) = F(x) = \sum_{n=1}^{\infty} A_n \sin n\pi x$$

and this is achieved by using the Fourier coefficients for $F(x)$,

$$A_n = 2 \int_0^1 F(x) \sin n\pi x \, dx$$

The partial sums of our series now serve as approximate solutions of the diffusion problem. The exact solution used in Problem 29.15 may be viewed as a one-term Fourier series.

THE LAPLACE EQUATION

29.18. Replace the Laplace equation

$$\frac{\partial^2 T}{\partial x^2} + \frac{\partial^2 T}{\partial y^2} = 0 \qquad 0 \leqq x \leqq l \qquad 0 \leqq y \leqq l$$

by a finite difference approximation. If the boundary values of $T(x, y)$ are assigned on all four sides of the square, show how a linear algebraic system is encountered.

The natural approximations are

$$\frac{\partial^2 T}{\partial x^2} \simeq \frac{T(x-h, y) - 2T(x, y) + T(x+h, y)}{h^2}$$

$$\frac{\partial^2 T}{\partial y^2} \simeq \frac{T(x, y-h) - 2T(x, y) + T(x, y+h)}{h^2}$$

and they lead at once to the difference equation

$$T(x, y) = \frac{1}{4}[T(x-h, y) + T(x+h, y) + T(x, y-h) + T(x, y+h)]$$

which requires each T value to be the average of its four nearest neighbors. Here we focus our attention on a square lattice of points with horizontal and vertical separation h. Our difference equation can be abbreviated to

$$T_Z = \frac{1}{4}(T_A + T_B + T_C + T_D)$$

with points labeled as in Fig. 29-4. Writing such an equation for each interior point Z (where T is unknown), we have a linear system in which each equation involves five unknowns, except when a known boundary value reduces this number.

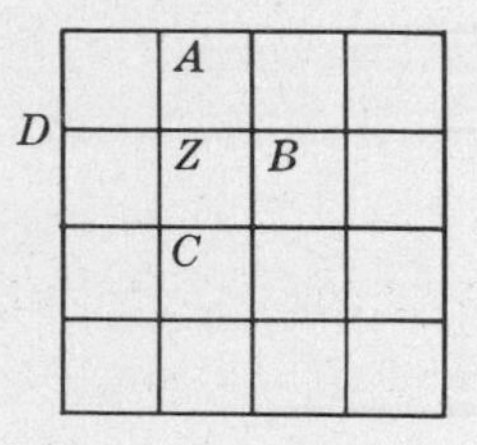

Fig. 29-4

29.19. Apply the method of the previous problem when $T(x, 0) = 1$, the other boundary values being 0.

For simplicity we choose h so that there are only nine interior points, as in Fig. 29-4. Numbering these points from left to right, top row first, our nine equations are these:

$$T_1 = \frac{1}{4}(0 + T_2 + T_4 + 0) \qquad T_6 = \frac{1}{4}(T_3 + 0 + T_9 + T_5)$$

$$T_2 = \frac{1}{4}(0 + T_3 + T_5 + T_1) \qquad T_7 = \frac{1}{4}(T_4 + T_8 + 1 + 0)$$

$$T_3 = \frac{1}{4}(0 + 0 + T_6 + T_2) \qquad T_8 = \frac{1}{4}(T_5 + T_9 + 1 + T_7)$$

$$T_4 = \frac{1}{4}(T_1 + T_5 + T_7 + 0) \qquad T_9 = \frac{1}{4}(T_6 + 0 + 1 + T_8)$$

$$T_5 = \frac{1}{4}(T_2 + T_6 + T_8 + T_4)$$

The system could be rearranged for Gaussian elimination, but as it stands the Gauss–Seidel iteration seems natural. Starting from the very poor initial approximation of zero for each interior T_i the successive results given in Table 29.2 are obtained. Ten iterations bring three-place accuracy for this linear system. (For a discussion of convergence of the Gauss–Seidel iteration see Problem 26.34.)

Table 29.2

Iteration	T_1	T_2	T_3	T_4	T_5	T_6	T_7	T_8	T_9
0	0	0	0	0	0	0	0	0	0
1	0	0	0	0	0	0	.250	.312	.328
2	0	0	0	.062	.078	.082	.328	.394	.328
3	.016	.024	.027	.106	.152	.127	.375	.464	.398
4	.032	.053	.045	.140	.196	.160	.401	.499	.415
5	.048	.072	.058	.161	.223	.174	.415	.513	.422
6	.058	.085	.065	.174	.236	.181	.422	.520	.425
7	.065	.092	.068	.181	.244	.184	.425	.524	.427
8	.068	.095	.070	.184	.247	.186	.427	.525	.428
9	.070	.097	.071	.186	.249	.187	.428	.526	.428
10	.071	.098	.071	.187	.250	.187	.428	.526	.428

A CONVERGENCE PROOF

29.20. Prove that the linear system encountered in Problem 29.18 will always have a unique solution.

The point is that, since we base our approximation on this system, it is important that it be nonsingular. Denoting the unknown interior values $T_1, \ldots, T_N$, we may rewrite the system in the form

$$\sum_{k=1}^{N} a_{ik} T_k = b_i \tag{1}$$

where the b_i depend upon the boundary values. If all boundary values were zero, then all b_i would be zero also:

$$\sum_{k=1}^{N} a_{ik} T_k = 0 \tag{2}$$

By the fundamental theorem of linear algebra the system (*1*) will have a unique solution provided that (*2*) has only the zero solution. Accordingly, we suppose all boundary values are zero. If the maximum T_k value occurred at an interior point Z, then because of $T_Z = \frac{1}{4}(T_A + T_B + T_C + T_D)$ it would also have to occur at A, B, C, and D, the neighbors of Z. Similarly this maximum would occur at the neighboring points of A, B, C, and D themselves. By continuing this argument we find that the maximum T_k value must also occur at a boundary point and so must be zero. An identical argument proves that the minimum T_k value must occur on the boundary and so must be zero. Thus all T_k in system (*2*) are zero and the fundamental theorem applies. Notice that our proof includes a bonus theorem. The maximum and minimum T_k values for both (*1*) and (*2*) occur at boundary points.

29.21. Prove that the solution of system (*1*) of Problem 29.20 converges to the corresponding solution of Laplace's equation as h tends to zero.

Denote the solution of (*1*) by $T(x, y, h)$ and that of Laplace's equation by $T(x, y)$, boundary values of both being identical. We are to prove that at each point (x, y) as h tends to zero

$$\lim T(x, y, h) = T(x, y)$$

For convenience we introduce the symbol

$$L[F] = F(x+h, y) + F(x-h, y) + F(x, y+h) + F(x, y-h) - 4F(x, y)$$

By applying Taylor's theorem on the right we easily discover that for $F = T(x, y)$, $|L[T(x, y)]| \leqq Mh^4/6$ where M is an upper bound of $|T_{xxxx}|$ and $|T_{yyyy}|$. Moreover, $L[T(x, y, h)] = 0$ by the definition of $T(x, y, h)$. Now suppose the origin of x, y coordinates to be at the lower left corner of our square. This can always be arranged by a coordinate shift, which does not alter the Laplace equation. Introduce the function

$$S(x, y, h) = T(x, y, h) - T(x, y) - \frac{\Delta}{2D^2}(D^2 - x^2 - y^2) - \frac{\Delta}{2}$$

where Δ is an arbitrary positive number and D is the diagonal length of the square. A direct computation now shows

$$L[S(x, y, h)] = \frac{2h^2\Delta}{D^2} + 0\left(\frac{Mh^4}{6}\right)$$

so that for h sufficiently small, $L[S] > 0$. This implies that S cannot take its maximum value at an interior point of the square. Thus the maximum occurs on the boundary. But on the boundary $T(x, y, h) = T(x, y)$ and we see that S is surely negative. This makes S everywhere negative and we easily deduce that $T(x, y, h) - T(x, y) < \Delta$. A similar argument using the function

$$R(x, y, h) = T(x, y) - T(x, y, h) - \frac{\Delta}{2D^2}(D^2 - x^2 - y^2) - \frac{\Delta}{2}$$

proves that $T(x, y) - T(x, y, h) < \Delta$. The two results together imply $|T(x, y, h) - T(x, y)| < \Delta$ for arbitrarily small Δ, when h is sufficiently small. This is what convergence means.

29.22. Prove that the Gauss–Seidel method, as applied in Problem 29.19, converges to the exact solution $T(x, y, h)$ of system (*1*), Problem 29.20.

This is, of course, an altogether separate matter from the convergence result just obtained. Here we are concerned with the actual computation of $T(x, y, h)$ and have selected a method of successive approximations. Suppose we number the interior points of our square lattice from 1 to N as follows. First we take the points in the top row from left to right, then those in the next row from left to right, and so on. Assign arbitrary initial approximations T_i^0 at all interior points, $i = 1, \ldots, N$. Let the succeeding approximations be called T_i^n. We are to prove

$$\lim T_i^n = T_i = T(x, y, h)$$

as n tends to infinity. Let $S_i^n = T_i^n - T_i$. Now it is our aim to prove $\lim S_i^n = 0$. The proof is based on the fact that each S_i is the average of its four neighbors, which is true since both T_i^n and T_i have this property. (At boundary points we put S equal to zero.) Let M be the maximum $|S_i^0|$. Then, since the first point is adjacent to at least one boundary point,

$$|S_1'| \leqq \frac{1}{4}(M + M + M + 0) = \frac{3}{4}M$$

And since each succeeding point is adjacent to at least one earlier point,

$$|S_{i+1}'| \leqq \frac{1}{4}(M + M + M + |S_i'|)$$

Assuming for induction purposes that $|S_i'| \leqq [1 - (\frac{1}{4})^i]M$ we have at once

$$|S_{i+1}'| \leqq \frac{3}{4}M + \frac{1}{4}\left[1 - \left(\frac{1}{4}\right)^i\right]M = \left[1 - \left(\frac{1}{4}\right)^{i+1}\right]M$$

The induction is already complete and we have $|S_N'| \leqq [1 - (\frac{1}{4})^N]M = \alpha M$ which further implies

$$|S_i'| \leqq \alpha M \qquad i = 1, \ldots, N$$

Repetitions of this process then show that $|S_i^n| \leqq \alpha^n M$, and since $\alpha < 1$ we have $\lim S_i^n = 0$ as required. Though this proves convergence for arbitrary initial T_i^0, surely good approximations T_i^n will be obtained more rapidly if accurate starting values can be found.

29.23. Develop the basic formulas for a finite element method using triangular elements and the

Poisson equation

$$U_{xx} + U_{yy} = K \qquad (K \text{ a constant})$$

The region over which this equation is to hold must first be divided up into triangular pieces, making approximations where necessary. Let (x_i, y_i), (x_j, y_j), (x_k, y_k) be the vertices of one such triangle. The solution surface above this triangle is to be approximated by a plane element $\phi^{(e)}(x, y)$, the superscript referring to the element in question. If z_i, z_j, z_k are the distances up to this plane at the triangle corners, or nodes, then

$$\phi^{(e)} = L_i^{(e)} z_i + L_j^{(e)} z_j + L_k^{(e)} z_k$$

where $L_i^{(e)}$ is equal to 1 at node i and 0 at the other two nodes, with corresponding properties for $L_j^{(e)}$ and $L_k^{(e)}$. Let Δ_e be the area of the base triangle, formed by the three nodes. Then

$$2\Delta_e = \begin{vmatrix} 1 & x_i & y_i \\ 1 & x_j & y_j \\ 1 & x_k & y_k \end{vmatrix}$$

which leads quickly to the following representations:

$$L_i^{(e)} = \frac{1}{2\Delta_e} \begin{vmatrix} 1 & x & y \\ 1 & x_j & y_j \\ 1 & x_k & y_k \end{vmatrix} \qquad L_j^{(e)} = \frac{1}{2\Delta_e} \begin{vmatrix} 1 & x & y \\ 1 & x_k & y_k \\ 1 & x_i & y_i \end{vmatrix} \qquad L_k^{(e)} = \frac{1}{2\Delta_e} \begin{vmatrix} 1 & x & y \\ 1 & x_i & y_i \\ 1 & x_j & y_j \end{vmatrix}$$

If we also write

$$L_i^{(e)} = \frac{1}{2\Delta_e}(a_i + b_i x + c_i y)$$

then from the determinants

$$a_i = x_j y_k - x_k y_j \qquad b_i = y_j - y_k \qquad c_i = x_k - x_j$$

with these parallel formulas coming from $L_j^{(e)}$ and $L_k^{(e)}$.

$$a_j = x_k y_i - x_i y_k \qquad b_j = y_k - y_i \qquad c_j = x_i - x_k$$

$$a_k = x_i y_j - x_j y_i \qquad b_k = y_i - y_j \qquad c_k = x_j - x_i$$

All these a, b, c coefficients should have the superscript (e) but for simplicity it has been suppressed.

It is now time to consider the minimization problem equivalent to the Poisson equation. It is

$$J(U) = \iint \left[\frac{1}{2}(U_x^2 + U_y^2) + KU\right] dx\, dy = \text{minimum}$$

with the double integral to be evaluated over the given region R of the boundary value problem. We are approximating U by a function ϕ, a composite of plane triangular elements each defined over a triangular portion of R. So we consider the substitute problem of minimizing

$$J(\phi) = \sum J_e(\phi^{(e)})$$

with each term of the sum evaluated over its own base triangle. We want to set the appropriate derivatives of $J(\phi)$ to zero and to this end require the derivatives of the J_e components. Note that

$$\phi_x^{(e)} = \frac{1}{2\Delta_e}(b_i z_i + b_j z_j + b_k z_k)$$

$$\phi_y^{(e)} = \frac{1}{2\Delta_e}(c_i z_i + c_j z_j + c_k z_k)$$

so that, suppressing the superscript,

$$J_e = \iint \left[\frac{1}{2}(\phi_x^2 + \phi_y^2) + K\phi\right] dx\, dy = f(z_i, z_j, z_k)$$

The differentiations are straightforward. For example,

$$\frac{\partial f}{z_i} = \iint \left\{ \phi_x \frac{b_i}{2\Delta_e} + \phi_y \frac{c_i}{2\Delta_e} + KL_i \right) dx\,dy = \frac{1}{\Delta_e} \left(\frac{b_i^2 + c_i^2}{4} z_i + \frac{b_i b_j + c_i c_j}{4} z_j + \frac{b_i b_k + c_i c_k}{4} z_k \right) + \frac{4}{3} \Delta_e$$

with very similar results for $\partial f / z_j$ and $\partial f / z_k$. The three can be grouped neatly in matrix form:

$$\begin{bmatrix} \partial f / z_i \\ \partial f / z_j \\ \partial f / z_k \end{bmatrix} = \frac{1}{4\Delta_e} \begin{bmatrix} b_i^2 + c_i^2 & b_i b_j + c_i c_j & b_i b_k + c_i c_k \\ b_i b_j + c_i c_j & b_j^2 + c_j^2 & b_j b_k + c_j c_k \\ b_i b_k + c_i c_k & b_j b_k + c_j c_k & b_k^2 + c_k^2 \end{bmatrix} \begin{bmatrix} z_i \\ z_j \\ z_k \end{bmatrix} + \frac{4}{3} \Delta_e \begin{bmatrix} 1 \\ 1 \\ 1 \end{bmatrix}$$

The fact that K has been assumed constant makes the integrations needed to achieve this result easy enough. Note also that the integral of each L function is $\frac{1}{3}$, by elementary calculus.

The above matrix equation contains the ingredients needed to assemble the partial derivatives of $J(\phi)$. It remains, in a particular application, to do the assembling properly. Specifically, for each element $\phi^{(e)}$ the active nodes i, j, k must be noted and contributions recorded for derivatives relative to the corresponding variables among the $z_1, z_2, z_3, \ldots$.

29.24. Apply the finite element method of the preceding problem given that the region R is the unit square of Fig. 29-5, with the boundary values indicated. The exact solution is easily seen to be $U(x, y) = x^2 + y^2$, since this satisfies $U_{xx} + U_{yy} = 4$.

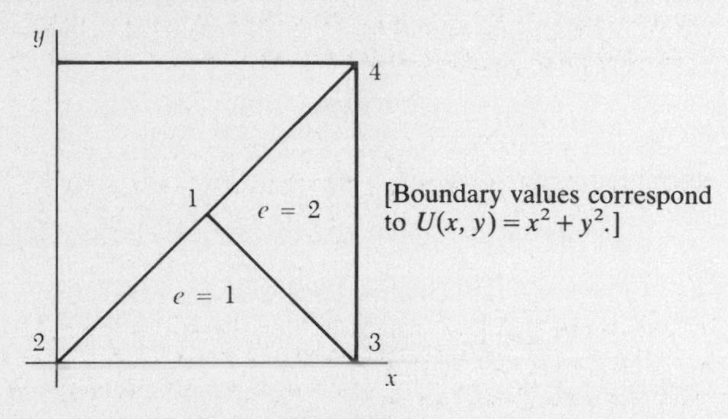

Fig. 29-5

By symmetry only the lower right half of the square needs to be considered, and this has been split into two triangles. The nodes are numbered 1 to 4, and the two triangles are identified by the node numbers involved.

Node	x	y
1	$\frac{1}{2}$	$\frac{1}{2}$
2	0	0
3	1	0
4	1	1

Elements (by node numbers)
1 2 3 $(e = 1)$
1 3 4 $(e = 2)$
$\Delta_1 = \Delta_2 = \frac{1}{4}$

From this basic input information we first compute the a, b, c coefficients. Each column below corresponds to a node (i, j, k).

	$e=1$			$e=2$		
a	0	$\frac{1}{2}$	0	1	0	$-\frac{1}{2}$
b	0	$-\frac{1}{2}$	$\frac{1}{2}$	-1	$\frac{1}{2}$	$\frac{1}{2}$
c	1	$-\frac{1}{2}$	$-\frac{1}{2}$	0	$-\frac{1}{2}$	$\frac{1}{2}$

It is useful to verify that columns do provide the desired $L_i^{(e)}$ functions. For instance, the first column gives

$$L_1^{(1)} = 2[0 + (0)x + (1)y]$$

where the leading 2 is the $1/2\Delta_e$. At node 1 this does produce the value 1, while at nodes 2 and 3 it manages 0. The other columns verify in similar fashion.

For clarity the process of assembling the partial derivatives of $J(\phi) = f(z_1, z_2, z_3, z_4)$ will now be presented in more detail than is probably needed. The matrix equation of the preceding problem contains the contributions to these derivatives from each of our two elements. From element 1 comes

	z_1	z_2	z_3	
$\partial f/z_1$	1	$-\frac{1}{2}$	$-\frac{1}{2}$	$\frac{1}{3}$
$\partial f/z_2$	$-\frac{1}{2}$	$\frac{1}{2}$	0	$\frac{1}{3}$
$\partial f/z_3$	$-\frac{1}{2}$	0	$\frac{1}{2}$	$\frac{1}{3}$

the last column containing constants. Element 2 provides these pieces.

	z_1	z_3	z_4	
$\partial f/z_1$	1	$-\frac{1}{2}$	$-\frac{1}{2}$	$\frac{1}{3}$
$\partial f/z_3$	$-\frac{1}{2}$	$\frac{1}{2}$	0	$\frac{1}{3}$
$\partial f/z_4$	$-\frac{1}{2}$	0	$\frac{1}{2}$	$\frac{1}{3}$

Assembling the two matrices we have this finished product:

	z_1	z_2	z_3	z_4	
$\partial f/z_1$	2	$-\frac{1}{2}$	-1	$-\frac{1}{2}$	$\frac{2}{3}$
$\partial f/z_2$	$-\frac{1}{2}$	$\frac{1}{2}$	0	0	$\frac{1}{3}$
$\partial f/z_3$	-1	0	1	0	$\frac{2}{3}$
$\partial f/z_4$	$-\frac{1}{2}$	0	0	$\frac{1}{2}$	$\frac{1}{3}$

Having thus illustrated the process of assembling elements, it must now be confessed that for the present case only the top row is really needed. The values of z_2, z_3, z_4 are boundary values and given as 0, 1, 2. They are not independent variables, and the function f depends only upon z_1. Setting this one derivative to zero and inserting the boundary values, we have

$$2z_1 - \frac{1}{2}(0) - (1) - \frac{1}{2}(2) + \frac{2}{3} = 0$$

making $z_1 = \frac{2}{3}$. The correct value is, of course, $\frac{1}{2}$.

29.25. Rework the preceding problem using the finer network of triangles shown in Fig. 29-6.

We have these input ingredients: first, the nodes 1 to 4 where the coordinates (x, y) are $(\frac{1}{2}, \frac{1}{2})$, $(\frac{1}{4}, \frac{1}{4})$, $(\frac{3}{4}, \frac{1}{4})$, and $(\frac{3}{4}, \frac{3}{4})$ with corresponding z coordinates to be determined; second, the nodes 5 to 9 at which boundary values are assigned making (x, y, z) coordinates $(1, 1, 2)$, $(1, \frac{1}{2}, \frac{5}{4})$, $(1, 0, 1)$, $(\frac{1}{2}, 0, \frac{1}{4})$, and $(0, 0, 0)$; and third, the eight basic triangles designated by node numbers:

2 9 8 2 8 1 1 8 3 3 8 7 3 7 6 1 3 6 1 6 4 4 6 5

A computer program to run the finite element algorithm as described would need this input information.

Suppose we begin a manual run, carrying it through only one of the eight elements, the first. The a,

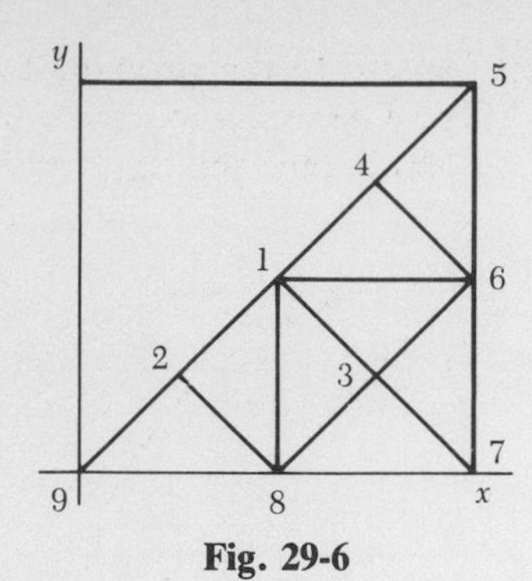

Fig. 29-6

b, c coefficients prove to be as follows:

a	0	$\frac{1}{8}$	0
b	0	$-\frac{1}{4}$	$\frac{1}{4}$
c	$\frac{1}{2}$	$-\frac{1}{4}$	$-\frac{1}{4}$

This may be checked as in the preceding problem, the columns representing the three nodes in the given order. The area of each basic triangle is $\frac{1}{16}$. Since partial derivatives will be needed only relative to z_1 to z_4, we can shorten our manual effort by finding only the terms contributing to these. For this element, we have

$$b_i^2 + c_i^2 = 0 + \frac{1}{4} = \frac{1}{4}$$

$$b_i b_j + c_i c_j = 0 - \frac{1}{8} = -\frac{1}{8} \qquad b_i b_k + c_i c_k = 0 - \frac{1}{8} = -\frac{1}{8}$$

which, after multiplication by $1/4\Delta_e = 4$, we enter into columns 2, 8, and 9 of the partial derivative matrix. The constant $4\Delta_e/3 = \frac{1}{12}$ is also recorded, all entries in row 2 which pertains to $\partial f/z_2$.

	z_1	z_2	z_3	z_4	z_5	z_6	z_7	z_8	z_9	
$\partial f/z_1$	$\frac{1}{2}$	$-\frac{1}{2}$						0		
$\partial f/z_2$		1						$-\frac{1}{2}$	$-\frac{1}{2}$	$\frac{1}{12}$
$\partial f/z_3$										
$\partial f/z_4$										

It remains to find the similar contributions of the other seven elements and to assemble them into the above matrix. It is useful to verify that the second element introduces the terms shown in row 1 and to find its further contributions to row 2. The rest of the assembly process will be left to the computer as will the substitution of boundary values and solution of the resulting fourth-order linear system. The following output was obtained:

Node	Computed	True
1	.500000	$\frac{1}{2}$
2	.166667	$\frac{1}{8}$
3	.666667	$\frac{5}{8}$
4	1.166667	$\frac{9}{8}$

The bull's-eye at node 1 is interesting.

29.26. Apply the same finite element method to the problem of a quarter circle, using just a single element as shown in Fig. 29-7. The Poisson equation is again to be used, as are the boundary values $x^2 + y^2 = 1$. The true solution is thus the same $x^2 + y^2$.

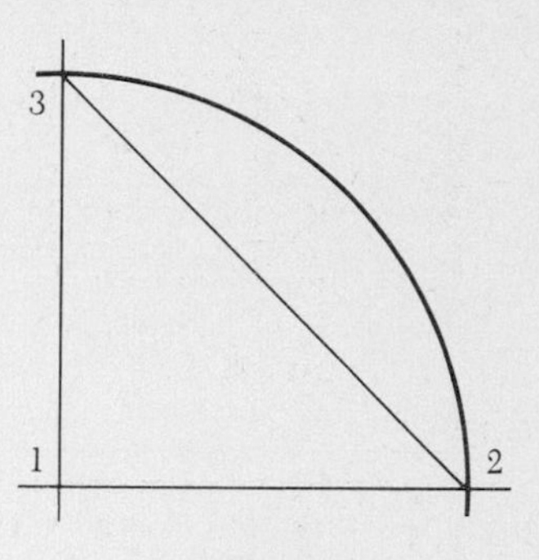

Fig. 29-7

The problem illustrates the approximation of a curved boundary by a straight-line segment. In general, many such segments would be used. The three nodes have these coordinates:

Node	x	y	z
1	0	0	—
2	1	0	1
3	0	1	1

The value of z_1 is the independent variable of the optimization. The a, b, c coefficients are

	Node 1	Node 2	Node 3
a	1	0	0
b	−1	1	0
c	−1	0	1

and lead to

$$\frac{\partial f}{z_1} = z_1 - \frac{1}{2}z_2 - \frac{1}{2}z_3 + \frac{2}{3} = 0$$

from which $z_1 = \frac{1}{3}$ follows at once. The true value is, of course, zero. By symmetry the same result would be found for the full circle by using four such triangles.

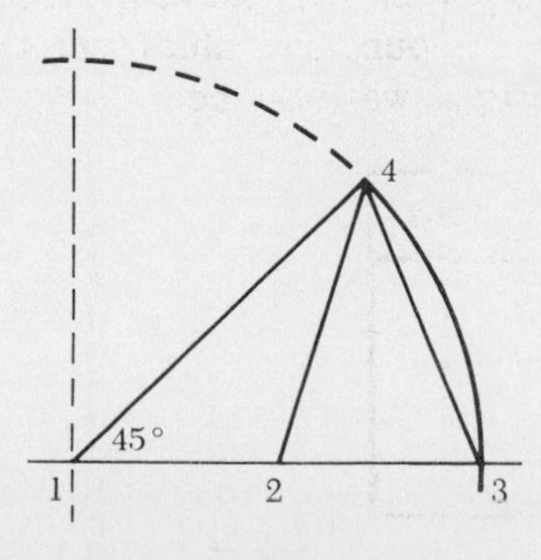

Fig. 29-8

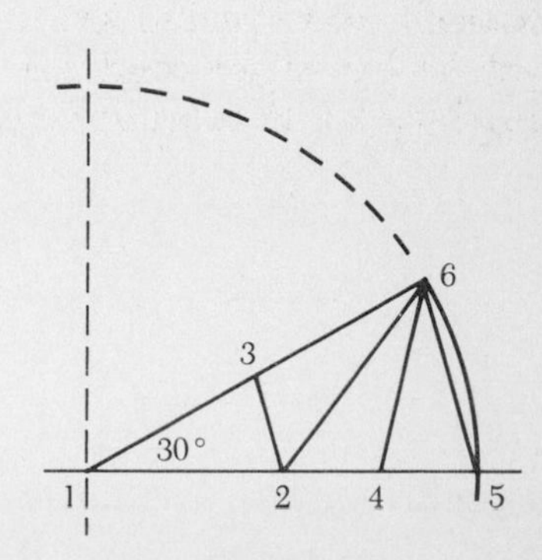

Fig. 29-9

29.27. Illustrate the concept of convergence, as it applies to finite element methods, by comparing the crude approximation just found with results from two-triangle and four-triangle efforts based upon the arrangements shown in Figs. 29-8 and 29-9.

Needless to say, all these efforts are relatively crude, but it is interesting to observe the results.

Node	(0, 0)	($\frac{1}{2}$, 0)
Fig. 29-7	.33	—
Fig. 29-8	−.08	.35
Fig. 29-9	−.03	.26
True	0	.25

Things have begun to move in a good direction. Finite element methods have been shown to be convergent provided the process of element refinement is carried out in a reasonable way.

THE WAVE EQUATION

29.28. Apply finite difference methods to the equation

$$\frac{\partial^2 U}{\partial t^2} - \frac{\partial^2 U}{\partial x^2} = F[t, x, U, U_t, U_x] \qquad -\infty < x < \infty,\ 0 \leqq t$$

with initial conditions $U(x, 0) = f(x)$, $U_t(x, 0) = g(x)$.

Introduce a rectangular lattice of points $x_m = mh$, $t_n = nk$. At $t = n = 0$ the U values are given by the initial conditions. Using

$$\frac{\partial U}{\partial t} \simeq \frac{U(x, t+k) - U(x, t)}{k}$$

at $t = 0$ we have $U(x, k) \simeq f(x) + kg(x)$. To proceed to higher t levels we need the differential equation, perhaps approximated by

$$\frac{U(x, t+k) - 2U(x, t) + U(x, t-k)}{k^2} - \frac{U(x+h, t) - 2U(x, t) + U(x-h, t)}{h^2}$$
$$= F\left[t, x, U, \frac{U(x, t) - U(x, t-k)}{k}, \frac{U(x+h, t) - U(x-h, t)}{2h}\right]$$

which may be solved for $U(x, t+k)$. Applied successively with $t = k, k+1, \ldots,$ this generates U values to any t level and for all x_m.

29.29. Illustrate the above method in the simple case $F = 0$, $f(x) = x^2$, $g(x) = 1$.

The basic difference equation may be written (see Fig. 29-10)

$$U_A = 2(1 - \lambda^2)U_C + \lambda^2(U_B + U_D) - U_E$$

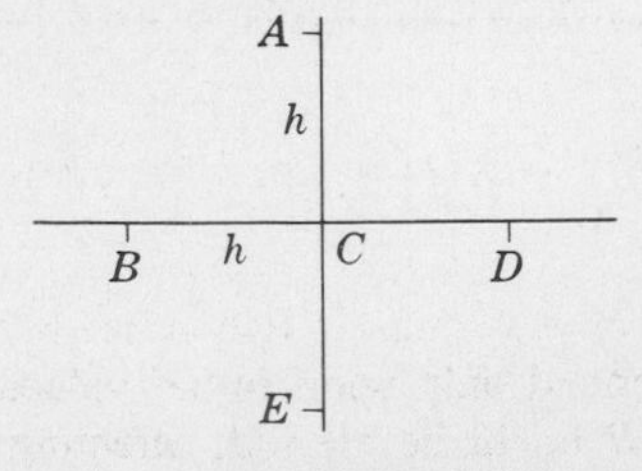

Fig. 29-10

where $\lambda = k/h$. For $\lambda = 1$ this is especially simple, and results of computation with $h = k = .2$ are given in Table 29.3. Note that the initial values for $x = 0$ to 1 determine the U values in a roughly triangular region. This is also true of the differential equation, the value $U(x, t)$ being determined by initial values between $(x - t, 0)$ and $(x + t, 0)$. (See Problem 29.30.)

Table 29.3

t/x	0	.2	.4	.6	.8	1.0
.6			1.00	1.20		
.4		.52	.64	.84	1.12	
.2	.20	.24	.36	.56	.84	1.20
0	.00	.04	.16	.36	.64	1.00

29.30. Show that the exact solution value $U(x, t)$ of $U_{tt} = U_{xx}$, $U(x, 0) = f(x)$, $U_t(x, 0) = g(x)$ depends upon initial values between $(x - t, 0)$ and $(x + t, 0)$.

For this old familiar problem, which is serving us here as a test case, the exact solution is easily verified to be

$$U(x, t) = \frac{f(x+t) + f(x-t)}{2} + \frac{1}{2}\int_{x-t}^{x+t} g(\xi)\, d\xi$$

and the required result follows at once. A similar result holds for more general problems.

29.31. Illustrate the idea of *convergence* for the present example.

Keeping $\lambda = 1$, we reduce h and k in steps. To begin, a few results for $h = k = .1$ appear in Table 29.4. One looped entry is a second approximation to $U(.2, .2)$ so that .26 is presumably more accurate than .24. Using $h = k = .05$ would lead to the value .27 for this position. Since the exact solution of the differential problem may be verified to be

$$U(x, t) = x^2 + t^2 + t$$

we see that $U(.2, .2) = .28$ and that for diminishing h and k our computations seem to be headed toward this exact value. This illustrates, but by no means proves, convergence. Similarly, another looped entry is a second approximation to $U(.4, .4)$ and is better than our earlier .64 because the correct value is .72.

Table 29.4

t/x	0	.1	.2	.3	.4	.5	.6	.7
.4				.61	(.68)			
.3			.40	.45	.52	.61		
.2		.23	(.26)	.31	.38	.47	.58	
.1	.10	.11	.14	.19	.26	.35	.46	.59
0	.00	.01	.04	.09	.16	.25	.36	.49

29.32. Why is a choice of $\lambda = k/h > 1$ not recommended, even though this proceeds more rapidly in the t direction?

The exact value of $U(x, t)$ depends upon initial values between $(x - t, 0)$ and $(x + t, 0)$. If $\lambda > 1$ the computed value at (x, t) will depend only upon initial values in subset AB of this interval. (See Fig. 29-11.) Initial values outside AB could be altered, affecting the true solution, but not affecting our computed value at (x, t). This is unrealistic.

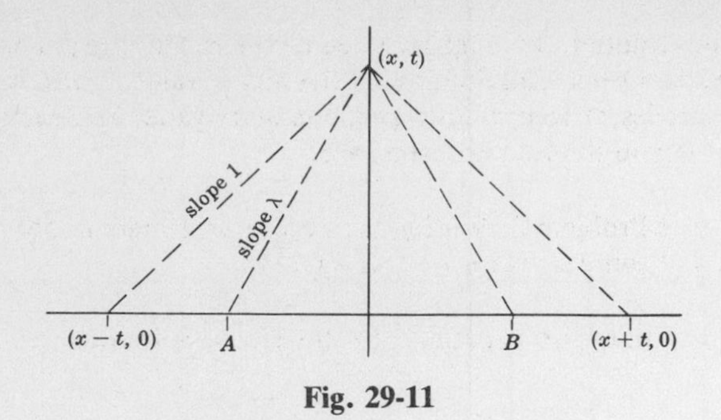

Fig. 29-11

Supplementary Problems

29.33. Solve the equation $y'' + y' + xy = 0$ with $y(0) = 1$ and $y(1) = 0$ by the method of Problem 29.1.

29.34. Solve the previous problem by the method of Problem 29.2. Which approach do you find more convenient?

29.35. Solve $y'' + \sqrt{x}\, y' + y = e^x$ with $y(0) = 0$ and $y(1) = 0$.

29.36. Apply the method of Problem 29.4 to $y'' + \lambda y = 0$ with $y(0) = y$ and $y'(1) = 0$. Prove convergence to the exact solution $y = \sin (2n + 1)(\pi x/2)$, $\lambda_n = [(2n + 1)(\pi/2)]^2$.

29.37. Apply the method of Problem 29.4 to obtain the largest eigenvalue of $y'' + \lambda xy = 0$ with $y(0) = y(1) = 0$.

29.38. Apply the method of Problem 29.5 to $y'' = y^2 + (y')^2$, $y(0) = 0$, $y(1) = 1$.

29.39. An object climbs from ground level to height 100 feet in 1 second. Assuming an atmospheric drag which makes the equation of motion $y'' = -32 - .1\sqrt{y'}$, what was the initial velocity?

29.40. An object climbs from (0, 0) to (2000, 1000) in 1 second, distances being in feet. If the equations of motion are

$$x''(t) = -.1\sqrt{v} \cos \alpha \qquad y''(t) = -32 - .1\sqrt{v} \sin \alpha$$

where $v^2 = (x')^2 + (y')^2$ and $\alpha = \arctan (y'/x')$, find the initial velocity.

29.41. Find the function $y(x)$ which minimizes $\int_0^1 [xy^2 + (y')^2]\, dx$ and satisfies $y(0) = 0$, $y(1) = 1$. Use the method of Problem 29.7.

29.42. Apply the method of Problem 29.12 to the case $a = c = 1$, $b = 0$, $l = 1$, $f(t) = g(t) = 0$, $F(x) = x(1 - x)$. Diminish h, obtaining successive approximations until you feel you have results correct to two decimal places. Use $\lambda = \frac{1}{2}$.

29.43. Repeat the previous problem with $\lambda = \frac{1}{6}$. Are satisfactory results obtained more economically or not? Try $\lambda = 1$.

29.44. Show that replacement of derivatives by simple finite differences converts the two-dimensional diffusion equation $T_t = T_{xx} + T_{yy}$ into

$$T_{l,m,n+1} = (1 - 4\lambda)T_{l,m,n} + \lambda(T_{l+1,m,n} + T_{l-1,m,n} + T_{l,m+1,n} + T_{l,m-1,n})$$

and obtain a similar approximation to the three-dimensional diffusion equation $T_t = T_{xx} + T_{yy} + T_{zz}$.

29.45. Find an approximate solution to Laplace's equation in the region $0 \leqq x$, $0 \leqq y$, $y \leqq 1 - x^2$ with $T(0, y) = 1 - y$, $T(x, 0) = 1 - x$ and the other boundary values zero. Use the simplest method for handling curved boundaries, merely transferring boundary values to nearby lattice points. Try $h = \frac{1}{4}$ and $h = \frac{1}{8}$. How accurate do you think your results are?

29.46. Repeat the procedure of Problem 29.9 using the Ritz approximation $\phi(x) = x(1 - x)(c_0 + c_1 x)$. Plot the corresponding curve and compare with the true solution.

29.47. Write out the linear system of Problem 29.11 for the case $n = 4$. Solve it and verify that exact values are found.

29.48. Verify the partial derivatives of f relative to z_i, z_j, z_k as given in Problem 29.23.

29.49. Complete the verifications of the a, b, c coefficients, as suggested in Problem 29.24.

29.50. Verify the contributions of the second finite element, as suggested in Problem 29.25.

29.51. Verify the results given in Problem 29.27 for the two-triangle and four-triangle configurations.

29.52. Apply the finite element method to the Laplace equation (set $K = 0$ instead of 4) on the triangle with vertices $(0, 0)$, $(1, 1)$, $(-1, 1)$ with boundary values given by $y^2 - x^2$. Note that this makes $U(x, y) = y^2 - x^2$ the true solution. From the symmetry it will be enough to work with the right half of the triangle. Use two interior nodes, at $(0, \frac{1}{3})$ and $(0, \frac{2}{3})$, joining these to $(1, 1)$ to form three basic triangles. The true values of U at the two interior nodes are, of course, $\frac{1}{9}$ and $\frac{4}{9}$. What values do these three elements produce?

29.53. Suggest a simple finite difference approximation to $T_{xx} + T_{yy} + T_{zz} = 0$.

29.54. The boundary value problem $y'' = n(n-1)y/(x-1)^2$, $y(0) = 1$, $y(1) = 0$ has an elementary solution. Ignore this fact and solve by the garden-hose method, using $n = 2$.

29.55. Try the previous problem with $n = 20$. What is the troublesome feature?

29.56. The boundary value problem $y'' - n^2 y = -n^2/(1 - e^{-n})$, $y(0) = 0$, $y(1) = 1$ has an elementary solution. Ignore this fact and solve by one of our approximation methods, using $n = 1$.

29.57. Try the previous problem with $n = 100$. What is the troublesome feature?

29.58. The boundary value problem

$$U_{tt} + U_{xxxx} = 0 \qquad 0 < x \qquad 0 < t \qquad U(x, 0) = U_t(x, 0) = U_{xx}(0, t) = 0 \qquad U(0, t) = 1$$

represents the vibration of a beam, initially at rest on the x axis, and given a displacement at $x = 0$. This problem can be solved using Laplace transforms, the result appearing as a Fresnel integral which must then be computed by numerical integration. Proceed, however, by one of our finite difference methods.

Chapter 30

Monte Carlo Methods

RANDOM NUMBERS

For our purposes, random numbers are not numbers generated by a random process such as the flip of a coin or the spin of a wheel. Instead they are numbers generated by a completely deterministic arithmetical process, the resulting set of numbers having various statistical properties which together are called randomness. A typical mechanism is

$$x_{n+1} = rx_n(\text{mod } N)$$

with r and N specified, and x_0 the "seed" of the sequence of "random" numbers x_n. Such modular multiplicative methods are commonly used as random number generators. With decimal computers

$$x_{n+1} = 7^9 x_n(\text{mod } 10^s) \qquad x_0 = 1$$

has been used, and with binary computers

$$x_{n+1} = (8t - 3)x_n(\text{mod } 2^s) \qquad x_0 = 1$$

with t some large number. Some generators include an additive element in this way:

$$x_{n+1} = (rx_n + s)(\text{mod } N)$$

A simple example suitable for practice problems is

$$x_{n+1} = (25{,}173x_n + 13{,}849)(\text{mod } 65{,}536)$$

which produces a well-scrambled arrangement of the integers from 0 to 65,535.

To be considered random, the sequence of x_n numbers must pass a set of statistical tests. They must be evenly distributed over the interval $(0, N)$, must have the expected number of upward and downward double runs (13, 69, 97, for example), triple runs (09, 17, 21, 73), and so on. Sometimes a successful sequence is said to consist of *pseudorandom* numbers, presumably to reserve the word random for the output of truly random devices (roulette wheels?). In this chapter *randomness* will refer to the qualities of the output, not to the nature of the generator. This will cover the apparent contradiction in terms, which has a thoroughly deterministic mechanism producing random output.

Many programming languages (Fortran, for instance) have a built-in random number generator subject to call. Very likely it is constructed to a modular multiplicative design.

APPLICATIONS

Monte Carlo methods solve certain types of problems through the use of random numbers. Although in theory the methods ultimately converge to the exact results, in practice only modest accuracy is attainable. This is due to the extremely slow rates of convergence. Sometimes Monte Carlo methods are used to obtain good starting approximations for speedier refinement algorithms. Two types of applications are offered.

1. **Simulation** refers to methods of providing arithmetical imitations of "real" phenomena. In a broad sense this describes the general idea of applied mathematics. A differential equation may, for example, simulate the flight of a missile. Here, however, the term *simulation* refers to the imitation of random processes by Monte Carlo methods. The classic example is the simulation of a neutron's motion into a reactor wall, its zigzag path being imitated by an arithmetical *random walk*. (See Problems 30.2 and 30.4.)
2. **Sampling** refers to methods of deducing properties of a large set of elements by studying only a small, random subset. Thus the average value of $f(x)$ over an interval may be

estimated from its average over a finite, random subset of points in the interval. Since the average of $f(x)$ is actually an integral, this amounts to a Monte Carlo method for approximate integration. As a second example, the location of the center of gravity of a set of N random points on the unit circle may be studied by using a few hundred or a few thousand such sets as a sample. (See Problem 30.5.)

Solved Problems

30.1. What are *random numbers* and how may they be produced?

For a simple but informative first example begin with the number 01. Multiply by 13 to obtain 13. Again multiply by 13, but discard the hundred, to obtain 69. Now continue in this way, multiplying continually by 13 modulo 100, to produce the following sequence of two-digit numbers.

$$01,\ 13,\ 69,\ 97,\ 61,\ 93,\ 09,\ 17,\ 21,\ 73,\ 49,\ 37,\ 81,\ 53,\ 89,\ 57,\ 41,\ 33,\ 29,\ 77$$

After 77 the sequence begins again at 01.

There is nothing random about the way these numbers have been generated, and yet they are typical of what are known as random numbers. If we plot them on a scale from 00 to 99 they show a rather uniform distribution, no obvious preference for any part of the scale. Taking them consecutively from 01 and back again, we find ten increases and ten decreases. Taking them in triples, we find double increases (such as $01, 13, 69$) together with double decreases occurring about half the time, as probability theory suggests they should. The term *random numbers* is applied to sequences which pass a reasonable number of such probability tests of randomness. Our sequence is, of course, too short to stand up to tests of any sophistication. If we count triple increases (runs such as 01, 13, 69, 97) together with triple decreases, we find them more numerous than they should be. So we must not expect too much. As primitive as it is, the sequence is better than what we would get by using 5 as multiplier $(01, 05, 25, 25, 25, \ldots$ which are in no sense random numbers). A small multiplier such as 3 leads to $01, 03, 09, 27, 81, \ldots$ and this long upward run is hardly a good omen. It appears that a well-chosen large multiplier may be best.

30.2. Use the random numbers of the preceding problem in a *simulation* of the movement of neutrons through the lead wall of an atomic reactor.

For simplicity we assume that each neutron entering the wall travels a distance D before colliding with an atom of lead, that the neutron then rebounds in a random direction and travels distance D once again to its next collision, and so on. Also suppose the thickness of the wall is $3D$, though this is far too flimsy for adequate shielding. Finally suppose that ten collisions are all a neutron can stand. What proportion of entering neutrons will be able to escape through this lead wall? If our random numbers are interpreted as directions (Fig. 30-1) then they may serve to predict the random directions of rebound. Starting with 01, for example, the path shown by the broken line in Fig. 30-2 would be followed. This neutron gets through, after four collisions. A second neutron follows the solid path in Fig. 30-2, and after ten collisions stops inside the wall. It is now plain that we do not have enough random numbers for a realistic effort, but see Problem 30.3.

30.3. How may a more extensive supply of random numbers be produced?

There are quite a few methods now available, but most of the best use the modular multiplication idea of Problem 30.1. For example, the recursion

$$x_{n+1} = 7^9 x_n \pmod{10^s} \qquad x_0 = 1$$

generates a sequence of length $5 \cdot 10^{s-3}$ having quite satisfactory statistical behavior. It is suitable for

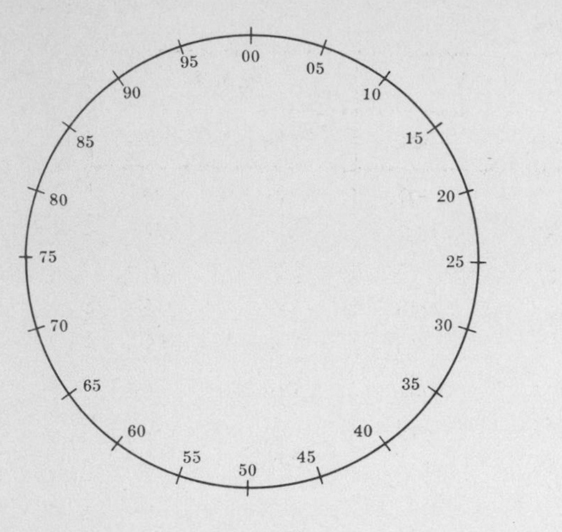

Fig. 30-1

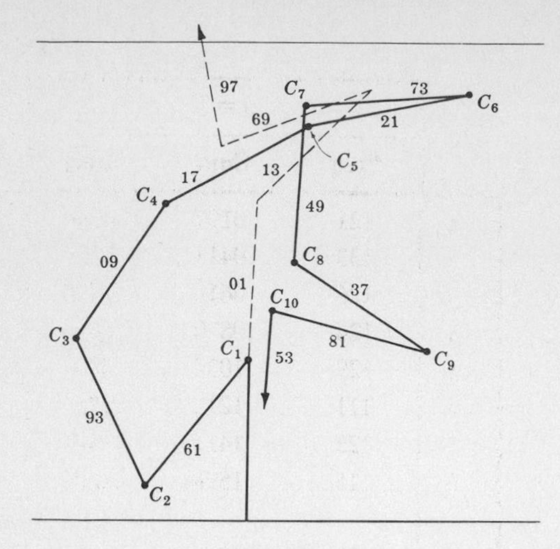

Fig. 30-2

decimal machines. The recursion

$$x_{n+1} = (8t - 3)x_n (\text{mod } 2^s) \qquad x_0 = 1$$

generates a permutation of the sequence $1, 5, 9, \ldots, 2^s - 3$, again with adequate statistical behavior. It is suitable for binary machines. The number t is arbitrary but should be chosen large to avoid long upward runs. In both these methods s represents the standard word length of the computer involved, perhaps $s = 10$ in a decimal machine and $s = 32$ in a binary machine.

30.4. Continue Problem 30.2 using a good supply of random numbers.

Using the first sequence of Problem 30.3 on a ten-digit machine ($s = 10$), the results given below were obtained. These results are typical of Monte Carlo methods, convergence toward a precision answer being very slow. It appears that about 28 percent of the neutrons will get through, so that a much thicker wall is definitely in order.

Number of trials	5,000	10,000	15,000	20,000
Percent penetration	28.6	28.2	28.3	28.4

30.5. Suppose N points are selected at random on the rim of the unit circle. Where may we expect their center of gravity to fall?

By symmetry the angular coordinate of the center of gravity should be uniformly distributed, that is, one angular position is as likely as another. The radial coordinate is more interesting and we approach it by a *sampling* technique. Each random number of the Problem 30.3 sequences may be preceded by a decimal (or binary) point and multiplied by 2π. The result is a random angle θ_i between 0 and 2π, which we use to specify one random point on the unit circle. Taking N such random points together, their center of gravity will be at

$$X = \frac{1}{N}\sum_{i=1}^{N} \cos\theta_i \qquad Y = \frac{1}{N}\sum_{i=1}^{N} \sin\theta_i$$

and the radial coordinate will be $r = \sqrt{X^2 + Y^2}$. Dividing the range $0 \leqq r \leqq 1$ into subintervals of length $\frac{1}{32}$, we next discover into which subinterval this particular r value falls. A new sample of N random points is then taken and the process repeated. In this way we obtain a discrete approximation to the distribution of the radial coordinate. Results of over 6000 samples for the cases $N = 2$, 3, and 4 are given in Table 30.1. The columns headed Freq give the actual frequency with which the center of gravity

Table 30.1

	$n=2$			$n=3$		$n=4$	
	Freq	**Cum**	**Exact**	**Freq**	**Cum**	**Freq**	**Cum**
1	121	.0197	.0199	7	.001	36	.005
2	133	.0413	.0398	37	.007	87	.018
3	126	.0618	.0598	58	.017	128	.038
4	124	.0820	.0798	67	.028	169	.063
5	129	.1030	.0999	95	.043	209	.094
6	111	.1211	.1201	113	.061	192	.123
7	123	.1411	.1404	141	.084	266	.163
8	115	.1598	.1609	172	.112	289	.207
9	129	.1808	.1816	224	.149	238	.242
10	142	.2039	.2023	336	.203	316	.290
11	123	.2240	.2234	466	.279	335	.340
12	138	.2464	.2447	344	.335	360	.394
13	126	.2669	.2663	291	.383	357	.448
14	157	.2925	.2883	285	.429	365	.503
15	126	.3130	.3106	269	.473	365	.558
16	125	.3333	.3333	255	.514	405	.618
17	150	.3577	.3565	223	.551	353	.672
18	158	.3835	.3803	189	.581	255	.710
19	135	.4054	.4047	208	.615	275	.751
20	148	.4295	.4298	185	.645	262	.790
21	157	.4551	.4558	215	.680	182	.818
22	158	.4808	.4826	197	.712	159	.842
23	173	.5090	.5106	183	.742	163	.866
24	190	.5399	.5399	201	.775	168	.892
25	191	.5710	.5708	188	.805	167	.917
26	211	.6053	.6038	183	.835	131	.936
27	197	.6374	.6393	163	.862	102	.952
28	247	.6776	.6783	176	.890	87	.965
29	262	.7202	.7221	170	.918	87	.978
30	308	.7703	.7737	162	.944	76	.989
31	424	.8394	.8407	163	.971	45	.996
32	987	1.0000	1.0000	178	1.000	27	1.000

appeared in each subinterval, from the center outward. Columns headed Cum give the cumulative proportions. For the case $N=2$ this cumulative result also happens to be exactly $(2/\pi)\arcsin(r/2)$ which serves as an accuracy check. Note that we seem to have about three-place accuracy.

30.6. Solve the boundary value problem

$$T_{xx}+T_{yy}=0 \qquad T(0,y)=T(1,y)=T(x,1)=0 \qquad T(x,0)=1$$

by a sampling method which uses *random walks*.

This is an example of a problem, with no obvious statistical flavor, which can be converted to a form suitable for Monte Carlo methods. The familiar finite difference approximations lead to a discrete set of points (say the nine in Fig. 30-3), and at each of these points an equation such as

$$T_5 = \frac{1}{4}(T_2 + T_4 + T_6 + T_8)$$

makes each T value the average of its four neighbors. This same set of nine equations was encountered in Problem 26.29, each unknown standing for the probability that a lost dog will eventually emerge on the south side of our diagram, reinterpreted as a maze of corridors! Though a sampling approach is hardly the most economical here, it is interesting to see what it manages. Starting a fictitious dog at position 1, for example, we generate a random number. Depending on which of the four subintervals $(0, \frac{1}{4})$, $(\frac{1}{4}, \frac{1}{2})$, $(\frac{1}{2}, \frac{3}{4})$, or $(\frac{3}{4}, 1)$ this random number occupies, our dog moves north, east, south, or west to the next intersection. We check to see if this brings him outside the maze. If it does not, another random number is generated and a second move follows. When the dog finally emerges somewhere, we record whether it was at the south side or not. Then we start a new fictitious dog at position 1 and repeat the action. The result of 10,000 such computer samples was 695 successful appearances at a south exit. This makes the probability of success .0695 and should be compared with the result .071 found by the Gauss–Seidel iteration. The latter is more accurate, but the possibility of solving differential boundary value problems by sampling methods may be useful in more complicated circumstances.

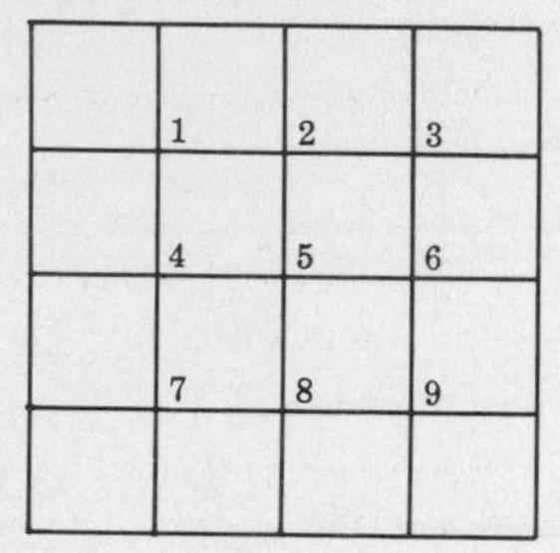

Fig. 30-3

30.7. Illustrate approximate integration by Monte Carlo methods.

Perhaps the simplest procedure is the approximation of the integral by an average,

$$\int_a^b f(x)\,dx = \frac{1}{N}\sum_{i=1}^{N} f(x_i)$$

where the x_i are selected at random in (a, b). For example, if we use just the first five random numbers of Problem 30.1, all preceded by a decimal point, then we have

$$\int_0^1 x\,dx \simeq \frac{1}{5}(2.41) \simeq .48$$

where the correct result is $\frac{1}{2}$, and we also find $\int_0^1 x^2\,dx \simeq .36$ where the correct result is $\frac{1}{3}$. For the same integrals, with $N = 100$ and using the longer sequences of Problem 30.3, the results .523 and .316 are obtained, the errors being about 5 percent. This is not great accuracy, but in the case of integration in several dimensions the same accuracy holds and Monte Carlo methods compete well with other integration algorithms.

Supplementary Problems

30.8. Generate a sequence of 20 random numbers using $x_{n+1} = rx_n \pmod{100}$, selecting your own multiplier r. Use these numbers to simulate three or four neutron paths as in Problem 30.2.

30.9. Using a sequence of the sort in Problem 30.3, simulate 1000 neutron paths as in Problem 30.4. Repeat for lead walls of thickness $5D$, $10D$, and $20D$. How does the shielding efficiency seem to grow?

30.10. Simulate 1000 random walks in a plane, each walk being 25 steps long, steps having equal lengths. Let each walk start at $(0, 0)$ and each step be in a random direction. Compute the average distance from $(0, 0)$ after 4, 9, 16, and 25 steps.

30.11. Approximate this integral using random numbers: $\int_0^\pi \sin x\, dx$.

30.12. Approximate this integral using random numbers:

$$\int_0^1 \int_0^1 \int_0^1 \int_0^1 \int_0^1 \int_0^1 \frac{dA\, dB\, dC\, dD\, dE\, dF}{1 + A + B + C + D + E + F}$$

30.13. Golfers A and B have the following records:

Score	80	81	82	83	84	85	86	87	88	89
A	5	5	60	20	10					
B				5	5	10	40	20	10	10

The numbers in the A and B rows indicate how many times each man has shot the given score. Assuming they continue this quality of play and that A allows B four strokes per round (meaning that B can subtract four strokes from his scores), simulate 1000 matches between these men. How often does A defeat B? How often do they tie?

30.14. A, B, and C each has an ordinary pack of cards. They shuffle the packs and each exposes one card, at random. The three cards showing may include 1, 2, or 3 different suits. The winner is decided as follows:

Number of suits showing	1	2	3
Winner is	A	B	C

The exposed cards are replaced and this completes one play. If many such plays are made, how often should each man win? The answer can be found by elementary probability, but simulate the actual play by generating three random numbers at a time, determining suits according to this scheme:

x falls inside interval	$(0, \frac{1}{4})$	$(\frac{1}{4}, \frac{1}{2})$	$(\frac{1}{2}, \frac{3}{4})$	$(\frac{3}{4}, 1)$
Suit is	S	H	D	C

30.15. A baseball batter with average .300 comes to bat four times in a game. What are his chances of getting 0, 1, 2, 3, and 4 hits, respectively? The answer can be found by elementary probability, but proceed by simulation.

30.16. In the "first man back to zero" game two players take turns moving the same marker back and forth across the board.

10	9	8	7	6	5	4	3	2	1	0	1	2	3	4	5	6	7	8	9	10

The marker is started at 0. Player A starts and always moves to the right and B to the left, the number of squares moved being determined by the throw of one die. The first man to stop on zero exactly is the winner. If the marker goes off either end of the board the game is a tie, the marker is returned to 0 and a new game is started by player A. What are the chances of A winning? The answer is not so easy to find by probability theory. Proceed by simulation.

30.17. The integers 1 to N are arranged in a random order. What are the chances that no integer is in its natural place? This is the famous "probleme des rencontres" and is solved by probability theory. But choose some value of N and proceed by simulation.

30.18. Generate three random numbers. Arrange them in increasing order $x_1 < x_2 < x_3$. Repeat many times and compute the average x_1, average x_2, and average x_3.

30.19. Suppose that random numbers y with nonuniform distribution are required, the density to be $f(y)$. Such numbers can be generated from a uniform distribution of random numbers x by equating the cumulative distributions, that is,

$$\int_0^x 1 \cdot dx = \int_0^y f(y)\, dy$$

For the special case $f(y) = e^{-y}$, show how y may be computed from x.

30.20. For the normal distribution $f(y) = e^{-y^2}/\sqrt{2\pi}$ the procedure of the preceding problem is troublesome. A popular alternative is to generate 12 random numbers x, from a uniform distribution over $(0, 1)$, to sum these and, since a mean value of zero is often preferred for the normal distribution, to subtract 6. This process depends upon the fact that the sum of several uniformly distributed random numbers is close to normally distributed. Use it to generate 100 or 1000 numbers

$$y = \left(\sum_{i=1}^{12} x_i\right) - 6$$

Then check the distribution of the y numbers generated. What fraction of them are in the intervals $(0, 1)$, $(1, 2)$, $(2, 3)$, and $(3, 4)$? The corresponding negative intervals should have similar shares.

Answers to Supplementary Problems

CHAPTER 1

1.39. $1 + .018$, only two terms being needed.

1.40. $-.009$

1.41. $N = 100$, $N = 10{,}000$

1.42. .114904, .019565, .002486, .000323, .000744, .008605

1.43. .008605

1.44. Computed $J_8 = .119726$.

1.48. .1494 approx.

1.49. Above $\frac{15}{8}$ there is overflow; below $\frac{1}{4}$, underflow.

1.56. Pi in binary, approx.

1.57. L_1 for taxicabs, L_∞ for the king.

CHAPTER 2

2.11. $(x-1)(x^2+1)$

2.12. 3, -3, 3, -3, 3

2.13. $p(x) = 2x - x^2$

2.15. Est. max. error = .242; actual error = .043.

2.16. $y' = 1.11$, $p' = 1$

2.17. $y'' = -1.75$, $p'' = -2$

2.18. $4/\pi$, $\frac{4}{3}$

2.19. $y = x + 7x(x-1) + 6x(x-1)(x-2)$

2.20. $\pi(x) = x(x-1)(x-2)(x-3)$

2.21. 1

CHAPTER 3

3.13. Fourth differences are all 24.

3.14. $\Delta^5 y_0 = \Delta^4 y_1 - \Delta^4 y_0$ and now use our result for fourth differences.

3.15. $\dfrac{u_{k+1}}{v_{k+1}} - \dfrac{u_k}{v_k} = \dfrac{v_k u_{k+1} - u_k v_{k+1}}{v_{k+1} v_k}$, etc.

3.16. Fifth differences are 5, 0, -5.

3.17. Change y_2 to 0.

3.22. 1, 3, 7, 14, 25, 41

3.23. $\Delta y_k = 0, 1, 5, 18, 36, 60$; $y_k = 0, 0, 1, 6, 24, 60, 120$

3.24. $\Delta^2 y_k = 24, 30, 36$; $\Delta y_k = 60, 90, 126$; $y_k = 120, 210, 336$

3.25. Change 113 to 131.

3.26. $\Delta^2 y_1 = y_3 - 2y_2 + y_1$; $\Delta^2 y_2 = y_4 - 2y_3 + y_2$

3.27. 3^k

3.28. 4^k, $(-2)^k$

3.29. $\frac{1}{6}[4^k - (-2)^k]$

3.30. Apply the identity for the sine of a difference.

3.31. Apply the identity for the cosine of a difference.

CHAPTER 4

4.23. $120, 720, 0, -\frac{2}{9}, \frac{10}{27}, -\frac{80}{81}$

4.24. $\frac{1}{7}, \frac{1}{56}, \frac{1}{504}, \frac{3}{4}, \frac{9}{28}, \frac{27}{280}$

4.25. $20, 1, 0, -\frac{1}{9}, \frac{5}{81}, -\frac{10}{243}$

4.26. Fourth differences are all 24.

4.27. $4k^{(3)}, 12k^{(2)}, 24k, 24$

4.28. $5k^{(4)}, 20k^{(3)}, 60k^{(2)}, 120k, 120$

4.29. $2k^3 - 7k^2 + 9k - 7$

4.30. $k^6 - 15k^5 + 85k^4 - 224k^3 + 271k^2 - 118k + 1$

4.31. $\frac{2}{3}k^{(4)} + 4k^{(3)} + 2k^{(2)} - 2k^{(1)} + 1$

4.32. $3k^{(5)} - 25k^{(3)} + 75k^{(2)} + 53k^{(1)}$

4.33. $\Delta y_k = 53 + 135k + 90k^2 - 90k^3 + 15k^4$

4.34. $\Delta^2 y_k = 150 - 30k - 180k^2 + 60k^3$

4.35. 31, 129, 351

4.36. 10, 45, 126

4.37. 2

4.38. 4

4.39. $k^{(3)}/3$

4.40. $k^{(4)}/4$

4.41. $\frac{1}{3}k^{(3)} + \frac{1}{2}k^{(2)}$

4.42. $\frac{1}{2}k^{(2)} + k^{(3)} + \frac{1}{4}k^{(4)}$

4.43. $-1/(k+1)$

CHAPTER 5

5.9. $\frac{1}{2}[(n+1)^{(2)} - 1^{(2)}]$

5.10. $n^2(n+1)^2/4$

5.11. Use the fact that $A^i = \Delta[A^i/(A-1)]$.

5.12. Use the fact that $\binom{i}{k} = i^{(k)}/k! = \Delta[i^{(k+1)}/(k+1)!]$.

5.13. $\frac{1}{4}$

5.14. $\frac{3}{4}$

5.15. $(R^3 + 4R^2 + R)/(1-R)^4$

5.16. 26

5.17. $-\frac{1}{3}$

5.18. $\log(n+1)$

5.19. $\sum_{j=1}^{n} \{s_j^{(n)}[(N+1)^{(j+1)}]/(j+1)\}$

5.20. $\frac{1}{n}\left(1 + \frac{1}{2} + \frac{1}{3} + \cdots + \frac{1}{n}\right)$

5.21. Denote the sum by $S_n(R)$. Then $S_{n+1}(R) = RS'_n(R)$ which may be used to compute each sum in its turn.

5.22. $y_k = 1 + \frac{1}{2} + \frac{1}{3} + \cdots + \frac{1}{k-1}$

5.23. $y_k = \log 2 + \log 3 + \cdots + \log(k-1)$

CHAPTER 6

6.8. $[(x-2)(x-4)/64][8 - 4(x-6) + (x-6)(x-8)]$

6.9. $1 + x + \frac{1}{2}x(x-1)$

6.10. $6 + 18(x-3) + 9(x-3)(x-4) + (x-3)(x-4)(x-5)$

6.11. Degree four suffices, $x(x-1)[\frac{1}{2} - \frac{1}{3}(x-2) + \frac{1}{12}(x-2)(x-3)]$.

6.12. $1 + x + \frac{1}{2}x(x-1) + \frac{1}{6}x(x-1)(x-2)$

6.14. $7x^2 - 6x$

6.15. $\frac{1}{3}x^3 - 2x^2 + \frac{8}{3}x$; collocation at $x = 4$, but not at $x = 5$.

6.16. No, degree three.

6.17. No, degree one.

6.18. $(7x^2 - x^4)/6$; greater in $(-2, -1)$ and $(1, 2)$.

6.19. $(7x - x^2)/6$; arguments are not equally spaced.

6.20. $y_k = \frac{1}{6}k(k-1)/(k-2)$

CHAPTER 7

7.33. $1 + 2k + 2k(k+1) + \frac{4}{3}k(k+1)(k+2) + \frac{2}{3}k(k+1)(k+2)(k+3)$

7.34. $120 + 60k + 12k(k+1) + k(k+1)(k+2)$

7.36. $2x - 3x^2 + x^3$

7.37. $1 - k - k(k-1) + \frac{1}{2}(k+1)k(k-1) + \frac{1}{4}(k+1)k(k-1)(k-2)$

7.38. $1 + k - (k+1)k - \frac{1}{2}(k+1)k(k-1) + \frac{1}{4}(k+2)(k+1)k(k-1)$

7.39. $24 + 36k + 9k(k-1) + (k+1)k(k-1)$

7.40. $1 - \frac{1}{2}k(k-1) + \frac{1}{12}(k+1)k(k-1)(k-2)$

7.42. $1 - k^2 + \frac{1}{4}(k+1)k^2(k-1)$

7.43. With $k = 0$ at $x = 1$, $y = 2 + \frac{3}{2}k + \frac{1}{2}k^2$.

7.44. $60k - 24(k-1) + 4(k+1)k(k-1) - 3k(k-1)(k-2)$

7.45. $1 - \frac{1}{6}[(k+1)k(k-1) - k(k-1)(k-2)] + \frac{1}{60}[(k^2-4)(k^2-1)k - (k^2-1)k(k-2)(k-3)]$

7.46. $4k - 2(k-1) + \frac{1}{6}[(k^2-1)k - k(k-1)(k-2)]$

7.47. $42 + 36(k - \frac{1}{2}) + \frac{21}{2}k(k-1) + (k - \frac{1}{2})k(k-1)$

7.48. $1 - \frac{1}{2}k(k-1) + \frac{1}{12}(k+1)k(k-1)(k-2)$

CHAPTER 8

8.15. $\dfrac{(x-1)(x-4)(x-6)}{-24} - \dfrac{x(x-4)(x-6)}{15} + \dfrac{x(x-1)(x-6)}{-24} - \dfrac{x(x-1)(x-4)}{60}$; $y(2) = -1$, $y(3) = 0$, $y(5) = 1$

8.16. $-\dfrac{4x(x-2)(x-4)(x-5)}{3} + 4x(x-1)(x-4)(x-5) - 11\dfrac{x(x-1)(x-2)(x-5)}{3}$; $y(3) = 84$

8.18. $a_0 = \frac{5}{2}$, $a_1 = -15$, $a_2 = \frac{31}{2}$

8.19. $\dfrac{\frac{2}{35}}{x+1} + \dfrac{\frac{4}{15}}{x-1} - \dfrac{\frac{41}{30}}{x-4} + \dfrac{\frac{73}{70}}{x-6}$

8.22. $\dfrac{(x-x_1)^2}{(x_0-x_1)^2}\left[\left(1 - \dfrac{2(x-x_0)}{x_0-x_1}\right)y_0 + (x-x_0)y_0'\right] + \dfrac{(x-x_0)^2}{(x_1-x_0)^2}\left[\left(1 - \dfrac{2(x-x_1)}{x_1-x_0}\right)y_1 + (x-x_1)y_1'\right]$

8.23. First order, -2, $\frac{2}{3}$, -1; second order, $\frac{2}{3}$, $-\frac{1}{3}$; third order, $-\frac{1}{6}$.

8.24. $1 - 2x + \frac{2}{3}x(x-1) - \frac{1}{6}x(x-1)(x-4)$

8.25. First order, $\frac{2}{3}$, 0, $-\frac{1}{3}$; second order, $-\frac{1}{3}$, $\frac{1}{3}$; third order, $-\frac{1}{6}$.

8.26. -1

8.27. $16x + 8x(x-1) - 3x(x-1)(x-2) - x(x-1)(x-2)(x-4)$; $y(3) = 84$

CHAPTER 9

9.22. $C_0 = C_4 = 0$, $C_1 = C_3 = \frac{18}{7}$, $C_2 = -\frac{30}{7}$

9.23. $S_2(x) = (2-x)^3/6 - 7(x-1)^3/12 - (2-x)/6 + 19(x-1)/12$;
$S_3(x) = -7(3-x)^3/12 + (x-2)^3/6 + 19(3-x)/12 - (x-2)/6$

9.24. The d_i are all zero.

9.25. The d_i are six times the second divided difference of y, which is a constant. All equations except the end conditions reduce to $3C = d_i$.

CHAPTER 10

10.8. $2x^2 - x^3$

10.9. $x^4 - 4x^3 + 4x^2$

10.10. $3x^5 - 8x^4 + 6x^3$

10.11. $p_1(x) = \frac{1}{4}x^2,\ p_2(x) = 2 - \frac{1}{4}(4-x)^2$

10.12. $p_1(x) = x^3(4-x)/16,\ p_2(x) = 2 - (4-x)^3x/16$

10.15. $x^4 - 2x^2 + 1$

10.16. $2x^4 - x + 1$

10.17. $x^3 - x^2 + 1$

CHAPTER 11

11.20. $\sin x = x - x^3/3! + x^5/5! - x^7/7! + \cdots$ to odd degree n;
$\cos x = 1 - x^2/2! + x^4/4! - x^6/6! + \cdots$ to even degree n.

11.21. $\pm \sin \xi \cdot x^{n+1}/(n+1)!$ for both functions.

11.22. $n = 7$

11.23. $n = 8,\ n = 12$

11.24. $\sum_{i=1}^{\infty} D^i/i!$

11.27. $\delta + \frac{1}{2}\delta^2 + \frac{1}{8}\delta^3 - \frac{1}{128}\delta^5 + \frac{1}{1024}\delta^7 + \cdots$

CHAPTER 12

12.31. 1.0060, 1.0085, no

12.32. 1.0291

12.33. 1.01489

12.34. 1.12250

12.35. 1.05830

12.36. .12451559

12.37. .1295

12.38. 1.4975

12.39. 1.4975

12.40. .1714, .1295, .0941

12.41. .02

12.42. .006

12.43. .25, .12

12.45. About 1

12.48. About $h = .15$ for $x > 1$.

12.49. $\frac{5}{4}$

12.51. 15.150

12.52. .841552021

12.54. 1.16190, 1.18327, 1.20419, the last being 3 units off.

12.55. 1.20419, 1.22390, both being somewhat in error.

12.56. Error $= x^4 - 7x^2 + 6x$; $\xi = 0$ explains the zero error.

12.57. Fortunate value of ξ.

12.58. 0

12.59. 24

12.60. 0 and 1

CHAPTER 13

13.22. $hp' = \delta y_{1/2} + (k - \frac{1}{2})\mu\delta^2 y_{1/2} + \dfrac{6k^2 - 6k + 1}{12}\delta^3 y_{1/2} + \dfrac{4k^3 - 6k^2 - 2k + 2}{24}\mu\delta^4 y_{1/2} + \dfrac{5k^4 - 10k^3 + 5k - 1}{120}\delta^5 y_{1/2}$

$h^2p^{(2)} = \mu\delta^2 y_{1/2} + (k - \frac{1}{2})\delta^3 y_{1/2} + \dfrac{12k^2 - 12k - 2}{24}\mu\delta^4 y_{1/2} + \dfrac{4k^3 - 6k^2 + 1}{24}\delta^5 y_{1/2}$

$h^3(p)^{(3)} = \delta^3 y_{1/2} + (k - \frac{1}{2})\mu\delta^4 y_{1/2} + \frac{1}{2}(k^2 - k)\delta^5 y_{1/2}$

$h^4p^{(4)} = \mu\delta^4 y_{1/2} + (k - \frac{1}{2})\delta^5 y_{1/2}$ $\qquad h^5p^{(5)} = \delta^5 y_{1/2}$

13.23. .4714, −.208, .32

13.24. Predicted error approx. 10^{-9}; actual error .000038.

13.25. Max. r.o. error is about $2.5E/h$; for Table 13.1 this becomes .00025.

13.28. Exact result is $x = \pi/2$, $y = 1$.

13.29. 1.57

13.31. $h^5 = 3E/8A$; $h \simeq .11$

CHAPTER 14

14.41. $h \simeq \sqrt{3}/100$

14.42. $A_2 = .69564$, $A_1 = .69377$, $(4A_1 - A_2)/3 = .69315$

14.43. .69315

14.44. .6931, no corrections needed.

14.45. $h = .14$

14.46. $\sqrt{3}/10^4$ trapezoidal, .014 Simpson.

14.52. Exact value is $\pi/4 = .7853982$.

14.53. Correct value is 1.4675.

14.58. .36422193

14.60. 9.68848

14.62. $a_{-1} = a_1 = \frac{7}{15}$, $a_0 = \frac{16}{15}$, $b_0 = 0$, $b_{-1} = -b_1 = \frac{1}{15}$

14.67. .807511

CHAPTER 15

15.56. 1.0000081

15.57. 1.5

15.61. $L_0 = 1$, $L_1 = 1 - x$, $L_2 = 2 - 4x + x^2$, $L_3 = 6 - 18x + 9x^2 - x^3$,
$L_4 = 24 - 96x + 72x^2 - 16x^3 + x^4$, $L_5 = 120 - 600x + 600x^2 - 200x^3 + 25x^4 - x^5$

15.68. Exact value is .5.

15.69. Correct value to five places is .59634.

15.71. $H_0 = 1$, $H_1 = 2x$, $H_2 = 4x^2 - 2$, $H_3 = 8x^3 - 12x$, $H_4 = 16x^4 - 48x^2 + 12$, $H_5 = 32x^5 - 160x^3 + 120x$

15.73. $(\sqrt{\pi}/6)[y(-\sqrt{\frac{3}{2}}) + y(\sqrt{\frac{3}{2}}) + 4y(0)]$; $3\sqrt{\pi}/4$

15.77. 2.128

15.78. .587

15.80 2.404

15.81. 3.82

CHAPTER 16

16.13. .5 and $-.23$, compared with the exact values .5 and $-.25$

16.15. 1.935

16.18. $-.797$

CHAPTER 17

17.50. $n(n-1)(n-2)/3$

17.51. $(n+1)^2n^2(2n^2+2n-1)/12$

17.52. $\dfrac{3}{4} - \dfrac{2n+3}{2(n+1)(n+2)}$

17.55. $\dfrac{11}{18} - \dfrac{1}{3}\left(\dfrac{1}{n+1} + \dfrac{1}{n+2} + \dfrac{1}{n+3}\right)$

17.57. .6049

17.61. About $x = .7$.

17.62. At most eight.

17.63. About $x = .7$.

17.64. $\dfrac{x^{2n+1}}{(2n+1)!} \cdot \dfrac{(2n+2)^2}{(2n+2)^2 - x^2}$; about $x = 10$.

17.65. .798

17.66. .687

17.67. .577

17.68. 1.1285

17.73. $Q_i = x^i$

17.78. After four terms; this method yields $C \simeq .5769$.

17.86. After seven terms.

CHAPTER 18

18.31. $y_k = \left[A + \dfrac{1}{(1-r)^2}\right]r^k + \dfrac{k}{1-r} - \dfrac{1}{(1-r)^2}$, except when $r = 1$.

18.32. 1, 3, 1, 3, etc.; $2 - (-1)^k$; $(y_0 - 2)(-1)^k + 2$

18.35. Let $y_k = (k-1)!\,A(k)$ to obtain $y_k = (k-1)!\,(2^k - 1)$ for $k > 0$.

18.36. $\frac{127}{64}$

18.37. $\left(\left(\left(\left(\dfrac{x^2}{9\cdot 8} - 1\right)\dfrac{x^2}{7\cdot 6} + 1\right)\dfrac{x^2}{5\cdot 4} - 1\right)\dfrac{x^2}{3\cdot 2} + 1\right)x$

18.40. $1/(k-1)!$

18.41. $\psi^{(3)}(0) = 3!\,\pi^4/90$, $\psi^{(3)}(n) = 3!\left[\dfrac{\pi^4}{90} - \sum_{k=1}^{n}\dfrac{1}{k^4}\right]$

18.42. $\frac{3}{4}$

18.43. $\pi^2/12 - \frac{11}{16}$

18.44. $\psi(\frac{1}{2}) = .0365$, $\psi(\frac{3}{2}) = .7032$, $\psi(-\frac{1}{2}) = 1.9635$

18.45. It takes arbitrarily large negative values.

18.46. $\frac{2}{3}\psi(0) - \frac{1}{3}\psi(\sqrt{\frac{3}{5}}) - \frac{1}{3}\psi(-\sqrt{\frac{3}{5}})$

18.47. $\frac{1}{3}\psi(0) - \frac{1}{6}\psi(\sqrt{\frac{3}{4}}) - \frac{1}{6}\psi(-\sqrt{\frac{3}{4}})$

18.50. $5(-1)^k - 3(-2)^k$

18.52. $A + B(-1)^k$

18.53. $A4^k + B3^k + (a\cos k + b\sin k)/(a^2 + b^2)$,

where $a = \cos 2 - 7\cos 1 + 12$, $b = \sin 2 - 7\sin 1$

$A = (3a - a\cos 1 - b\sin 1)/(a^2 + b^2)$

$B = (-4a + a\cos 1 + b\sin 1)/(a^2 + b^2)$

18.54. $[-4(-\frac{1}{2})^k + 2k(-\frac{1}{2})^k + 3k^2 - 8k + 4]/27$

18.56. $\frac{2}{3}[2^k - (\frac{1}{2})^k]$

18.57. $[5^k(-\cos k\theta - \frac{5}{4}\sin k\theta) + 2^k]/41$, $\cos\theta = -\frac{3}{5}$, $\sin\theta = \frac{4}{5}$

18.59. $a < 0$

18.60. $\frac{1}{8}(3^k) - \frac{1}{16}(-1)^k - \frac{3}{8}k^2 - \frac{1}{16}$

18.61. Oscillatory, linear, exponential.

18.65. $\frac{1}{2}[1 - (-1)^k]$

CHAPTER 19

19.76. Exact value is 1.

19.77. 1.4060059

19.78. Exact solution is $x^3y^4 + 2y = 3x$.

19.79. Exact solution is $x^2y + xe^y = 1$.

19.80. Exact solution is $\log(x^2 + y^2) = \arctan y/x$.

19.81. 4 days, 18 hours, 10 minutes

19.82. 4

19.83. Exact value is $\frac{1}{8}\arctan\frac{1}{4}$.

19.84. Exact solution is $x = -\sqrt{1-y^2} + \log(1 + \sqrt{1-y^2})/y$.

CHAPTER 20

20.16. See Problem 19.87.
20.19. $a_0 = a_1 = 1,\ k^2 a_k - (2k-1)a_{k-1} + a_{k-2} = 0$ for $k > 1$
20.20. Fourth-degree Taylor approximation to e^{-21h} is 6.2374 compared with the correct .014996.

CHAPTER 21

21.57. $y = .07h + 4.07$
21.58. 4.49, 4.63, 4.77, 4.91, 5.05, 5.19, 5.33, 5.47, 5.61, 5.75
21.59. .07.
21.60. No.
21.62. Very little.
21.63. They alternate.
21.65. $A = 84.8,\ M = -.456$
21.67. Five-point formula does better here.
21.69. Results are almost the same as from five-point formula.
21.85. $p(x) = \frac{1}{3}$
21.86. $p(x) = 3x/5$
21.87. $p(x) = 3x/5$
21.88. $p(x) = .37 + .01x - .225(3x^2 - 1)/2$
21.90. $p(x) = \frac{1}{2}$
21.91. $p(x) = 3x/4$
21.92. Drop two terms and have $1.2660T_0 - 1.1303T_1 + .2715T_2 - .0444T_3 + .0055T_4 - .0005T_5$.
21.102. $(81 + 72x)/64$; over $(-1, 1)$ this is only slightly worse than the quadratic.
21.106. $3x/4$
21.107. Min. integral parabola is $p = \frac{2}{\pi} + \frac{4}{3\pi}(2x^2 - 1)$.
21.109. .001, .125, .217, .288, .346, .385, .416, .438, .451, .459, .466
21.110. −.8, 19.4, 74.4, 143.9, 196.6, 203.9, 108.2, 143.4, 126.7, 118.4, 112.3, 97.3, 87.0, 73.3, 56.5, 41.8, 33.4, 26.5, 15.3, 6.6, 1.2
21.111. $5.045 - 4.043x + 1.009x^2$

CHAPTER 22

22.34. $P = 4.44e^{.45x}$
22.37. $p = \frac{5 - 3\sqrt{3}}{16} + \frac{3}{\pi}\left(\sqrt{3} - \frac{1}{2}\right)x + \frac{9}{\pi^2} \cdot \frac{1 - \sqrt{3}}{2} x^2$
22.38. $p = (1 - 18x + 48x^2)/32;\ h = \frac{1}{32}$
22.41. $(10T_0 + 15T_2 + 6T_4)/32;\ \frac{1}{32}$
22.42. $T_0 + T_1 + T_2$; 1
22.43. $\frac{1763}{2304} T_0 - \frac{353}{1536} T_2 + \frac{19}{3840} T_4$; 1/23,040
22.44. $p = 2x/\pi - 1.10525$
22.45. Method fails, x_2 becoming the point of discontinuity.
22.46. $p = -2x/\pi + 1.105$
22.50. $1.6476 + .4252x + .0529x^2$; .0087
22.51. Degree four.
22.52. Not more than .000005.
22.53. Degree four.
22.54. Degree two.

CHAPTER 23

23.18. $3/x$; no, the method produces $4-x$.

23.19. $90/(90+97x-7x^2)$; no, the method produces $(20+7x)/(20+34x)$.

23.20. $(x^2-1)/(x^2+1)$

23.21. $x^2/(1+x)$

23.22. $(x+1)/(x+2)$

23.24. $1/(2-x^2)$

23.25. $-\frac{1}{2}$

23.28. $4(1-x+x^2)/(1+x)$

23.29. $12(x+1)/(4-x^2)$

23.30. $(x^2+x+2)/(x^2+x+1)$

23.31. $1/(\sin 1°30') \simeq 38.201547$

23.32. $(1680-2478x+897x^2-99x^3)/(140+24x-17x^2)$

23.33. $(24+18x+6x^2+x^3)/(24-6x)$

23.34. $(24+6x)/(24-18x+6x^2-x^3)$

CHAPTER 24

24.40. $a_0=1.6,\ a_1=-.8472,\ a_2=.0472,\ b_1=.6155,\ b_2=-.1454$

24.42. $a_0=2,\ a_1=-1,\ a_2=a_3=0,\ b_1=\sqrt{3}/3,\ b_2=0$

24.43. $.8;\ .8-.8472\cos(2\pi x/5)+.6155\sin(2\pi x/5)$

24.45. $T_0(x)=1;\ T_1(x)=1-\cos(\pi x/3)+(\sqrt{3}/3)\sin(\pi x/3)=y(x)$

24.46. $[(\sqrt{2}+2)/2]\sin(\pi x/4)+[(\sqrt{2}-2)/2]\sin(3\pi x/4)$

24.47. $1-\cos(\pi x/2)$

24.49. $\pi^2/12$ and $\pi^2/6$

24.50. $\pi^2/8$

24.52. $\pi^3/32$

24.56. $1-\omega^2,\ 0,\ 1-\omega$

24.57. $V^T=(3,-2,0,-1,0,-2)$

24.58. $V^T=(5,1,5,1,-3,1,-3,1)$

CHAPTER 25

25.51. About 1.839.

25.52. Two; three; .567143

25.53. 1.83929

25.54. 1.732051

25.55. 1.245731

25.60. 1.618034

25.69. $x=.772,\ y=.420$

25.72. 3 and -2.

25.74. $x^2+1.9413x+1.9538$

25.75. 4.3275

25.76. 1.123106 and 1.121320

25.77. 1.79632

25.78. .44880

25.79. 1.895494267

25.80. $-.9706 \pm 1.0058i$

25.81. $x=7.4977,\ y=2.7687$

25.82. $x=1.8836,\ y=2.7159$

25.83. .94775

25.84. $x=2.55245$

25.85. 1.4458

25.86. $x=1.086,\ y=1.944$

25.87. 1.85558452522

25.88. .58853274

25.89. $(x^2+2.90295x-4.91774)(x^2+2.09705x+1.83011)$

25.90. 1.497300

25.91. 7.87298, -1.5, .12702

25.92. 1.403602

25.93. 1.7684 and 2.2410

CHAPTER 26

26.86. Exact solution is .8, .6, .4, .2.

26.88. Exact solution is given in Problem 26.55.

26.91. Exact solution is 5, −10, 10, −5, 1.

26.92. Exact inverse is $\begin{bmatrix} 5 & -10 & 10 & -5 & 1 \\ -10 & 30 & -35 & 19 & -4 \\ 10 & -35 & 46 & -27 & 6 \\ -5 & 19 & -27 & 17 & -4 \\ 1 & -4 & 6 & -4 & 1 \end{bmatrix}$

26.96. Exact inverse is $\begin{bmatrix} 25 & -41 & 10 & -6 \\ -41 & 68 & -17 & 10 \\ 10 & -17 & 5 & -3 \\ -6 & 10 & -3 & 2 \end{bmatrix}$

26.101. $2160\lambda^3 - 3312\lambda^2 + 381\lambda - 1 = 0$

26.109. $(0, -i, i)$

26.110. $\begin{bmatrix} 0 & i & 1 \\ -i & -1 & i \\ 1 & -i & 0 \end{bmatrix}$

26.119. 2.18518, −.56031, 2.00532, −.36819

26.120. 1.62772, 3, 7.37228

26.121. 8.3874, $C(.8077, .7720, 1)$; 4.4867, $C(.2170, 1, -.9473)$; 2.1260, $C(1, -.5673, -.3698)$; C being any constant.

26.122. $\begin{bmatrix} 5 & -10 & 10 & -5 & 1 \\ -10 & 30 & -35 & 19 & -4 \\ 10 & -35 & 46 & -27 & 6 \\ -5 & 19 & -27 & 17 & -4 \\ 1 & -4 & 6 & -4 & 1 \end{bmatrix}$

26.123. $\dfrac{15}{64}\begin{bmatrix} 15 & -70 & 63 \\ -70 & 588 & -630 \\ 63 & -630 & 735 \end{bmatrix}$

26.124. $\dfrac{1}{6}\begin{bmatrix} 6-8i & -2+4i \\ -3+10i & 1-5i \end{bmatrix}$

26.125. 98.522

26.126. 12.054; $[1, .5522i, .0995(3+2i)]$

26.127. 19.29, −7.08

26.129. .625, 1.261, 1.977, 4.136

26.130. $.227 =$ smallest λ

26.131. No

CHAPTER 27

27.18. (0, 0), (0, 1), $(\frac{2}{3}, \frac{5}{3})$, (2, 1), (3, 0); min. of $-\frac{8}{3}$ at $(\frac{2}{3}, \frac{5}{3})$; max. of 3 at (3, 0).

27.19. See Problem 27.18.

27.20. $-4y_1 - y_2 - 3y_3 =$ max.; y_1, y_2, y_3 nonnegative; $-y_1 + y_2 - y_3 \leqq 1$, $-2y_1 - y_2 - y_3 \leqq -2$.

27.21. See Problem 27.18.

27.22. $4y_1 + y_2 + 3y_3 = \text{min.}$; y_1, y_2, y_3 nonnegative; $y_1 - y_2 + y_3 \geqq 1$, $2y_1 + y_2 + y_3 \geqq -2$; solution at $(0, 0, 1)$.

27.23. See Problems 27.18 and 27.20.

27.24. $x_1 = \frac{3}{5}$, $x_2 = \frac{6}{5}$

27.25. Extreme solution points are $(0, 1)$ and $(\frac{2}{3}, \frac{5}{3})$.

27.27. Payoff is 2.5; $R(\frac{1}{2}, \frac{1}{2})$, $C(\frac{1}{4}, \frac{3}{4})$.

27.30. $\frac{37}{16} + \frac{17}{12}x + \frac{15}{8}x^2 + \frac{1}{12}x^3$; 1.3125; −2, −1, 0, 1, 2

27.31. $1.04508 - 2.47210x + 1.52784x^2$; .04508; 0, .08, .31, .73, 1

27.32. Same result; five positions of maximum error.

27.33. Max. = 4.4 for $x = (4.4, 0, 0, .6)$.

27.34. Min. $(5y_1 + 2y_2) = 4.4$.

27.35.

A	0	3	6	9	12
Max.	0	2	2	10	10

27.36. $\frac{3}{8}, \frac{5}{8}$

27.37. $R(\frac{1}{3}, \frac{2}{3})$, $C(\frac{2}{3}, \frac{1}{3})$

CHAPTER 28

28.11. $x_1 = 3.90$, $x_2 = 5.25$, error = .814

28.12. $p = .814$, $|\rho|_{max} = 1.15$

28.16. $x_1 = -.3278 = x_2$, error = .3004

28.17. $x_1 = -\frac{1}{3} = x_2$

28.18. 3.472, 2.010; 1.582; .426

28.19. The average $(\sum a_i)/N$.

28.20. $x = (A + C + D)/3$, $y = (B - C + D)/3$

28.21. $x_i = A_i + \frac{1}{3}(\pi - A_1 - A_2 - A_3)$

28.22. $L_1^2 = A^2 - D$, $L_2^2 = B^2 - D$, $H^2 = C^2 + D$ where $D = \frac{1}{3}(A^2 + B^2 - C^2)$

CHAPTER 29

29.46. $c_0 = \frac{1}{15}$, $c_1 = \frac{1}{6}$

29.52. .2, .5

29.53. $T(x, y, z) = \frac{1}{6}[T(x + h, y, z) + T(x - h, y, z) + T(x, y + h, z) + \text{etc.}]$

29.54. $y = (x - 1)^n$

29.55. A near-singularity at $x = 0$.

29.56. $y = (1 - e^{-nx})/(1 - e^{-n})$

29.57. A near-singularity at $x = 0$.

29.58. Exact solution is $1 - \sqrt{2/\pi} \int_0^{x/\sqrt{t}} [\cos(u^2) + \sin(u^2)]\, du$.

CHAPTER 30

30.10. Theoretical values are 2, 3, 4, and 5 step lengths.

30.11. Exact value is 2.

30.14. Theoretical values are $\frac{1}{16}, \frac{9}{16}, \frac{6}{16}$.

30.15. Theoretical values are .2401, .4116, .2646, .0756, .0081.

30.17. For $N \to \infty$ the theoretical value is $1/e$.

30.18. Theoretical values are $\frac{1}{4}, \frac{1}{2}, \frac{3}{4}$.

30.19. $y = -\log(1 - x)$ or equally well $y = -\log x$.

30.20. Theoretical values are .3413, .1359, .0215, .0013.

INDEX